Lecture Notes in Artificial Intelligence   5001

Edited by R. Goebel, J. Siekmann, and W. Wahlster

Subseries of Lecture Notes in Computer Science

Ubbo Visser   Fernando Ribeiro
Takeshi Ohashi   Frank Dellaert (Eds.)

RoboCup 2007:
Robot Soccer
World Cup XI

 Springer

Series Editors

Randy Goebel, University of Alberta, Edmonton, Canada
Jörg Siekmann, University of Saarland, Saarbrücken, Germany
Wolfgang Wahlster, DFKI and University of Saarland, Saarbrücken, Germany

Volume Editors

Ubbo Visser
Universität Bremen
TZI - Center for Computing Technologies
Am Fallturm 1, 28359 Bremen, Germany
E-mail: visser@informatik.uni-bremen.de

Fernando Ribeiro
Universidade do Minho
Dep. de Electronica Industrial
Campus de Azurem, 4800 Guimaraes, Portugal
E-mail: fernando@dei.uminho.pt

Takeshi Ohashi
Kyushu Institute of Technology
Department of Artificial Intelligence
Fukuoka Prefecture 820-8502, Iizuka City, Kawazu 680-4, Japan
E-mail: ohashi@ai.kyutech.ac.jp

Frank Dellaert
Georgia Institute of Technology
College of Computing
801 Atlantic Drive, Atlanta, Georgia 30332-0280, USA
E-mail: frank@cc.gatech.edu

aitainment GmbH holds the copyright for the illustration on the cover of this book.

Library of Congress Control Number: 2008930109
CR Subject Classification (1998): I.2, C.2.4, D.2.7, H.5, I.5.4, J.4
LNCS Sublibrary: SL 7 – Artificial Intelligence

ISSN   0302-9743
ISBN-10   3-540-68846-3 Springer Berlin Heidelberg New York
ISBN-13   978-3-540-68846-4 Springer Berlin Heidelberg New York

Springer is a part of Springer Science+Business Media

springer.com

© Springer-Verlag Berlin Heidelberg 2008
Printed in Germany

Typesetting: Camera-ready by author, data conversion by Scientific Publishing Services, Chennai, India
Printed on acid-free paper   SPIN: 12277568   06/3180   5 4 3 2 1 0

Preface

The 11th RoboCup International Symposium was held during July 9–10, 2007 at the Fox Theatre in Atlanta, GA, immediately after the 2007 Soccer, Rescue and Junior Competitions. The RoboCup community has observed an increasing interest from other communities over the past few years, e.g., the robotics community. RoboCup is seen as a significant approach to the evaluation of newly-developed methods to many difficult problems in robotics. Atlanta was also the location of a RoboCup@Space demonstration, which reflected the role of AI and robotics in space exploration. Prior to the symposium, space agencies had expressed an interest in cooperating with RoboCup. A first step in this direction was a successful demonstration at RoboCup 2007, which was accompanied with an invited talk given by a leading scientist from the Japan Aerospace Exploration Agency JAXA.

The symposium represented the core meeting for the presentation and discussion of scientific contributions in diverse areas related to the main threads within RoboCupSoccer, RoboCupRescue and RoboCupJunior. Its scope encompassed, but was not restricted to, research and education activities within the fields of artificial intelligence and robotics. The RoboCup International Symposium 2007 featured 18 full papers for oral presentation and 42 posters. These were selected by the program committee from a total of 133 submissions, which meant an acceptance rate of 13,5% for full papers. Each paper was reviewed by at least three program committee members, usually two experts and one person outside the immediate area of the paper. After the initial reviewing period, papers that had not received unanimous recommendations were discussed among the reviewers, moderated by the Co-chairs, and a consensus was reached in all cases. The final decisions were made by the Co-chairs.

In addition to the paper and poster presentations, which cover the state of the art in a broad range of topics central to the RoboCup community, we had two outstanding invited speakers, Takashi Kubota from the Institute of Space and Astronautical Science, JAXA, Japan and Christine Lisetti from Florida International University, USA. Kubota's talk was about Japanese space activities in the field of automation and robotics, and Lisetti talked about building communicative interfaces for natural human-robot interaction.

March 2008

Ubbo Visser
Fernando Ribeiro
Takeshi Ohashi
Frank Dellaert

Organization

Symposium Co-chairs

Ubbo Visser   Center for Computing Technologies (TZI),
University of Bremen, Germany
Fernando Ribeiro   Universidade do Minho, Portugal
Takeshi Ohashi   Kyushu Institute of Technology, Japan
Frank Dellaert   Georgia Tech, USA

International Symposium Program Committee

Akin, Levent H., Turkey
Almeida, Luis, Portugal
Amigoni, Francesco, Italy
Baltes, Jacky, Canada
Beetz, Michael, Germany
Behnke, Sven, Germany
Birk, Andreas, Germany
Bonarini, Andrea, Italy
Bredenfeld, Ansgar, Germany
Brena, Ramon, Mexico
Browning, Brett, USA
Bruce, James, USA
Burkhard, Hans-Dieter, Germany
Caglioti, Vincenzo, Italy
Cardeira, Carlos, Portugal
Carpin, Stefano, USA
Cassinis, Riccardo, Italy
Chaib-draa, Brahim, Canada
Chen, Xiaoping, China
Coradeschi, Silvia, Sweden
Costa, Anna, Brazil
Dias, Bernadine M., USA
Eguchi, Amy, USA
Folgheraiter, Michele, Italy
Frese, Udo, Germany
Frontoni, Emanuele, Italy
Gini, Giuseppina, Italy
Goncalves, Luiz, Brazil
Gutmann, Steffen, Japan
Indiveri, Giovanni, Italy

Iocchi, Luca, Italy
Jahshan, David, Australia
Jamzad, Mansour, Iran
Karlapalem, Kamal, India
Kimura, Testuya, Japan
Kraetzschmar, Gerhard, Germany
Kuwata, Yoshitaka, Japan
Lakemeyer, Gerhard, Germany
Lee, Daniel, USA
Levy, Simon, USA
Lima, Pedro, Portugal
Matteucci, Matteo, Italy
Mayer, Gerd, Germany
Menegatti, Emanuele, Italy
Nakashima, Tomoharu, Japan
Nardi, Daniele, Italy
Noda, Itsuki, Japan
Nomura, Tairo, Japan
Ouerhani, Nabil, Switzerland
Pagello, Enrico, Italy
Paiva, Ana, Portugal
Parsons, Simon, USA
Pirri, Fiora, Italy
Polani, Daniel, UK
Prokopenko, Mikhail, Australia
Rahimi, Arash, Iran
Reis, Paulo Luis, Portugal
Restelli, Marcello, Italy
Ribeiro, Carlos, Brazil
Röfer, Thomas, Germany

Rojas, Raul, Germany
Ruiz-del-Solar, Javier, Chile
Sammut, Claude, Australia
Schurr, Nathan, USA
Schut, Martijn, The Netherlands
Shiry, Saeed, Iran
Sklar, Elizabeth, USA
Sorrenti, Domenico, Italy
Sridharan, Mohan, USA
Takahashi, Tomoichi, Japan
Takahashi, Yashutake, Japan
Tawfik, Ahmed, Canada

Vail, Douglas, USA
Velastin, Sergio A., UK
Verner, Igor, Israel
Weitzenfeld, Alfredo, Mexico
Williams, Mary-Anne, Australia
Wisse, Martijn, The Netherlands
Wotawa, Franz, Austria
Wyeth, Gordon, Australia
Wyeth, Peta, Australia
Yanco, Holly, USA
Zell, Andreas, Germany
Zhou, Changjiu, Singapore

Additional Reviewers

Bastos, Guilherme
Friske, Leticia

Mandel, Christian
Silva, Anderson

Wagner, Thomas

Table of Contents

Short Papers (Posters)

Instance-Based Action Models for Fast Action Planning

Mazda Ahmadi and Peter Stone

Department of Computer Sciences,
The University of Texas at Austin
{mazda,pstone}@cs.utexas.edu

Abstract. Two main challenges of robot action planning in real domains are uncertain action effects and dynamic environments. In this paper, an *instance-based* action model is learned empirically by robots trying actions in the environment. Modeling the action planning problem as a Markov decision process, the action model is used to build the transition function. In static environments, standard value iteration techniques are used for computing the optimal policy. In dynamic environments, an algorithm is proposed for fast replanning, which updates a subset of state-action values computed for the static environment. As a test-bed, the goal scoring task in the RoboCup 4-legged scenario is used. The algorithms are validated in the problem of planning kicks for scoring goals in the presence of opponent robots. The experimental results both in simulation and on real robots show that the instance-based action model boosts performance over using parametric models as done previously, and also incremental replanning significantly improves over original off-line planning.

1 Introduction

In many robotic applications, robots need to plan a series of actions to achieve their goals. In comparison to classical planning, planning on-board robotic agents introduces several new challenges, including (1) exceedingly noisy actions effects, often with irregular noise distributions; (2) dynamically changing environments; and (3) a need for real-time decision-making despite limited processing power. In this paper, the problem of robot action planning in dynamic environments with uncertain action effects is considered. We introduce an *instance-based* action model that captures arbitrary distributions of action effects and use it for action planning. To cope with dynamically changing environments, we introduce an efficient on-line incremental replanning method that modifies the transition model to account for the effects of other agents and then replans only for the affected states.

Learning action models has been studied in classical planning (e.g. see [1,2]), and also in probabilistic planning (e.g. see prioritized sweeping [3]). But those methods use many trials to learn the model; instead we use domain heuristics to learn the model with few experiments prior to planning.

U. Visser et al. (Eds.): RoboCup 2007, LNAI 5001, pp. 1–16, 2008.

A common shortcoming of prior methods for learning probabilistic action models is the assumption that the noise is normally distributed, which in many cases is not true. To overcome that shortcoming, we propose an instance-based approach for learning the action model. The action model is built empirically by trying actions in the environment. Each sample effect is stored and is considered individually for planning purposes.

The planning problem is modeled as a Markov Decision Process (MDP). The transition function of the MDP is built with the help of the learned action model. Using value iteration [4] with state aggregation, a plan which maximizes the received reward is generated. When the environment is static, the value iteration algorithm can be run offline. In dynamic environments, the planning must be performed online. The online algorithm must be fast enough to be within computational bounds of the robots. Though fast replanning algorithms for robotic applications have been studied for classical planning problems (e.g. see [5,6]), the probabilistic replanning algorithms that we know of (e.g. [7]), are computationally expensive.

When using an instance-based approach, the observation of each dynamic factor changes the modeled transition function of the MDP. But it only changes the values of a small subset of state-action pairs. In the replanning algorithm, using domain-dependent heuristics, the state-action pairs that are affected by the dynamic factors are discovered, and only the values of those state-action pairs are updated.

To evaluate our methods, we use a goal scoring task from the 4-legged RoboCup competitions. Chernova and Veloso [8] learn models of kicks for the 4-legged robots, however they do not discuss how they use this model. Their model consists of the speed and direction of the ball in the first second. We extend their work by introducing an instance-based model instead of their parametric model, and show the advantages of using such an instance-based model. Furthermore, we use the kick model for action planning in the presence of opponent robots.

The two main contributions of this paper are:

- An approach to dynamic replanning of action sequences based on an instance-based representation of action effects that is fully-implemented and tested on a physical robot.
- An empirical comparison of an instance-based action model and the more popular parametric action models on a physical robot.

The remainder of this paper is organized as follows. Related work is discussed in Section 2. In Section 3 the RoboCup test-bed is presented. In the next three sections, we consider the problem as an abstract MDP. In Section 4 the details of the instance-based action model, and how to learn in it, are provided; Section 5 introduces the planning algorithm for static environments; and Section 6 extends the planning problem to dynamic environments. The implementation details of the RoboCup domain are presented in Section 7. In Section 8, experimental results are provided, and Section 9 concludes.

2   Related Work

In recent years there has been significant progress in building autonomous mobile robots. For example Burgard et al. have developed an autonomous tour guide robot [9]. Also, there have been many advances in robot soccer playing agents for the RoboCup competitions. However the focus of the research in the RoboCup 4-legged community is mainly on the lower level components such as vision, localization or locomotion. The action selection has generally been reactive or the result of shallow look-ahead. The main reason is that the action effects are highly uncertain and creating an accurate action model is challenging. In this paper using the lower level components, we tackle the problem of building an accurate action model and use that to incorpate a full-blown planning approach based on an MDP representation.

Researchers have used planning methods for robotic applications. For example Farritor and Dubowsky build a climbing robot which uses planning to find a sequence of actions which achieves its goal [10]. In earlier work Frommherz and Werling propose using heuristic search for planning and show experiments in a simple assembly task [11]. Compared to the problem considered in this paper, in the mentioned papers the action effects are not nearly as uncertain. Thus a parametric action model performs sufficiently well. Also, the environment is not dynamic, so there is no need for replanning.

Instance-based methods have been used for various fields such as health care, assessment and design [12]. Also Planning tasks have been solved with instance-based methods [12]. Atkeson [13] investigates the use of an instance-based method (locally weighted regression) to learn task models for control. Gabel and Riedmiller use an instance based method for value function approximation of a reinforcement learning problem in the soccer simulation domain [14]. Atkeson and Santamaria compare model based (using instance-based method) and model-free reinforcement learning in a simple robotic task (pendulum swing up) and conclude that model based methods learn faster [15]. Note that in the mentioned robotic applications the uncertainty of the action effects is low and also the environment is static. We take the next step of evaluating instance-based action models in a dynamic robotic environment with highly uncertain action effects.

3   Problem Description

As a test-bed domain for our research we use a subtask of the RoboCup four legged league. In this work we consider single robot goal scoring possibly against multiple opponents. We assume the opponents only block the ball, and do not kick or grab it.

As baseline software, we use the UT Austin Villa code base [16], which provides robust color-based vision, fast locomotion, and reasonable localization within a 3.6m × 5.4m area[1] via a particle filtering approach. Even so, the robot

---

[1] The field is as specified in the 2005 rules of the RoboCup Four-Legged Robot League:
http://www.tzi.de/4legged

(a)   (b)

Fig. 1. (a) Representation of FALL-KICK from left to right (b) Representation of HEAD-KICK from left to right

is not, in general, perfectly localized, as a result of both noisy sensations and noisy actions effects. The robot also has limited processing power, which limits the algorithms that can be designed for it.

The baseline software provides different types of kicks. The two that are considered in this work are called FALL-KICK (see Figure 1(a)) and HEAD-KICK (see Figure 1(b)). The effects of these kicks are probabilistic based on how accurately they are executed and what exact part of the robot hits the ball.

4   Instance-Based Action Model

The first step of planning with any action is understanding its effects. We build the action model empirically by trying actions in the domain. In most real robot applications, actions have probabilistic effects. Thus the model must be able to represent uncertainty in action effects.

Previous methods (e.g. [8]) use parametric models of actions. Most popular methods for modeling the noise assume a Gaussian distribution of noise for each of the parameters of the action model. Instead we take an *instance-based* approach, where each action effect from experiments is called a sample action effect, and is stored in the model.

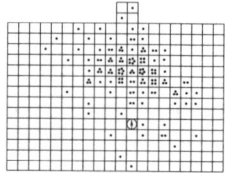

Fig. 2. Representation of the model of FALL-KICK. The position of the robot and kick direction is marked by an arrow. Each dot represents a sample action effect.

We claim and show in the experiments, that our instance-based action model is more effective than a parametric action model.

In addition to noisy action effects, robots are faced with noise from other sources (e.g. uncertain localization). Previous methods of building action models (e.g. [8]) try to minimize the effects of such *other* noises on the action model. If the action model does not represent the noise from all sources, the effects of the environment's noise must be accounted for in some other way for action planning (e.g. an expensive way of accounting for localization errors is by planning from a group of possible positions). Instead, we aim at having noise from all sources captured by the action model. In this way, if the action model is used with any planning algorithm, all the sources of noise are also considered. This requires the samples to be collected in a situation similar to the one that the robot faces in the real environment, not in some other controlled setting.

An example of an instance-based action model for the FALL-KICK in the RoboCup domain is shown in Figure 2.

5 Planning

In this section, we show how the instance-based action model can be used for action planning. We model the problem as a Markov decision process (MDP) and use a value iteration method to solve it. In this section the environment is assumed to be static. Dynamic environments are considered in Section 6.

The planning problem is modeled as an MDP $(S, A, Pr(s'|s, a), R)$, where:

- S is a discrete set of *states*. If the state space is continuous, it should be discretized. We show that discretizing the state space does not have much of a detrimental effect in the RoboCup goal shooting test-bed.
- A is the set of possible actions.
- $Pr(s'|s, a)$ is the true continuous state transition probability function. It gives the probability of getting to state s' after taking action a in state s. Because of noise in the environment and uncertainty of action effects, the transition function is stochastic
- $R(s, a) \in \mathbb{R}$ is the *reward function*.

The goal of the robot is to find a *policy* $\pi : S \mapsto A$ that maximizes the received reward. The policy determines which action is chosen by the robot from each state.

$Pr(s'|s, a)$ is not known to the robot, however the robot can use the action model to approximate it. The approximation of $Pr(s'|s, a)$ is called $\widetilde{Pr}(s'|s, a)$. The approximation is based on the action model. For computing $\widetilde{Pr}(s'|s, a)$, where s is a discrete state, a representative of s is used for computing $\widetilde{Pr}(s'|s, a)$ (center of the cell in grid discretization); when s is a continuous state, the true state (TS) is used to computed $\widetilde{Pr}(s'|TS, a)$. $R(s, a)$ is also computed with the help of the action model. The details of computing $\widetilde{Pr}(s'|s, a)$ and $R(s, a)$ for the RoboCup goal shooting test-bed are presented in Section 7.

For state s, the value $V^\pi(s)$ is defined as the expected sum of rewards from s until the end of the episode, while following policy π. $V^*(s)$ is the optimal policy if $V^*(s) \geq V^\pi(s)$ for all policies π and all $s \in S$. $Q^\pi(s, a)$ is defined as the sum of the rewards received from state s until the end of the episode, while following policy π (see Equation 1), if the first action executed is a. $Q^*(s, a)$ is the optimal such policy. The advantage of using $Q^\pi(s, a)$ over $V^\pi(s, a)$ is presented in the next section, where only a subset of Q values needs to be updated.

We use following two equations to perform standard value iteration [17]:

$$Q(s, a) = R(s, a) + \sum_{s' \in S} \widetilde{Pr}(s'|s, a) V^*_{t-1}(s') \tag{1}$$

$$V(s, a) = \max_{a \in A} Q(s, a) \qquad (2)$$

When the system is in state s, a common practice for action selection is to discretize the state space (e.g. using a grid instead of the exact position) and to choose action a such that it maximizes $Q(s, a)$. Discretization makes solving the MDP tractable. But enforcing that each position in a discrete grid-cell (state) takes the same action can lead to dramatic sub-optimality. In order to alleviate this effect, we take the middle ground: for *action selection*, instead of the grid state, the robot uses its true state estimate (e.g. true position) that maximizes the following value:

$$R(TS, a) + \sum_{s' \in S} \widetilde{Pr}(s'|TS, a)V^*(s') \qquad (3)$$

where TS is the true state estimate of the robot. Note that TS is only used for action selection, not for the learning process.

This way, the effects of discretizing the state are deferred to the next step, and it results in a better policy. In [16], we empirically show that in the RoboCup test-bed, discretizing the environment with the true state for action selection is very close in performance to using the true state in planning.

Note that with discretized state (grid position) the policy can be directly derived from the V-values. However with the true state, all possible actions must be evaluated, requiring significant, but manageable (on the Aibo) computational resources. In the experiments section the advantage of this action selection method is showed empirically in simulation.

6   Replanning

In the previous section, the environment was assumed to be static, and the value function could be computed offline. In this section, the possibility of the presence of dynamic factors (e.g. the presence of opponent robots in the RoboCup test-bed) is considered. Existence of dynamic factors changes the transition function of the MDP, and the value function for the new MDP needs to be computed online. Because of the robot's limited processing power and need for real-time decision making, performing full online value iteration for consideration of the dynamic factors is not possible. In this section a fast replanning algorithm is presented, which leverages the Q-values that are computed for the static environment.

If the dynamic factors are considered in computing $\widetilde{Pr}(s'|s, a)$ in Equation 4, the value iteration and action selection algorithms described in Section 5 can also be applied to the dynamic environment. However dynamic factors, by their nature, are not known in advance.

Similar to $Q^\pi(s, a)$, $Q^\pi(s, a|F)$ is defined as the sum of the received reward in the presence of dynamic factors F, from state s until the end of the episode, while following policy π (see Equation 3), if the first action to execute is a. $Q^*(s, a|F)$ is the optimal such policy.

In the systems that we are considering, the difference between $Q^*(s, a)$ and $Q^*(s, a|F)$ is only substantial for states where F has a direct effect on $Q(s, a)$. For the rest of the states, $Q^*(s, a|F) \approx Q^*(s, a)$. This fact, which is typical of many dynamic systems, where the dynamic factors have an effect on only a subset of the Q-values, is the basis of the proposed replanning algorithm.

Assuming $Q(s, a)$ is the current Q-function, the algorithm for updating the Q-values in the event of observing dynamic factor f is as follows:

1. Flag the (s, a) pairs that are potentially affected by f (using domain-dependent heuristics).
2. For flagged pair (s, a), there is a good chance that $Q(s, a|f)$ is very different from $Q(s, a)$. Thus, if (s, a) is flagged, $Q(s, a|f)$ is initialized to zero, and otherwise, it is initialized to $Q(s, a)$.
3. Only for flagged pairs (s, a), the $Q(s, a|f)$ are updated using Equation 1. Notice that only one round of updates is performed. After all the Q-updates, the V values are re-computed using Equation 2.

Action selection is the same as in the previous section. The two main benefits of our replanning algorithm are that it is fast and the robot can distribute the value iteration steps over several decision cycles, so the robot does not miss action opportunities.

Recall that the robot does another level of inference on the effects of actions in the action selection step, so the effects of adding f is effectively backed up two steps.

7   Implementation Details

In this section the implementation details of the instance-based action model (Section 4), planning (Section 5), and replanning (Section 6) algorithms for the goal scoring test-bed (Section 3) are presented.

7.1   Learning a Kick Model

The main action for scoring goals is kicking (assuming the robot walks to the new position of the ball after each kick action). Kicks (as shown in Figures 1(a) and 1(b)) have uncertain (i.e. probabilistic) effects on the final position of the ball, which is based on the exact point of contact between the robot and the ball. In this section we present the implementation details of learning the instance-based kick model.

Chernova and Veloso [8] use the average and standard deviation for the speed and angle of each kick to model the kick. They measure the angle and distance that the ball travels in one second after the kick. By just considering the kick in the first second, and also setting the initial position of the ball by hand, they try to minimize the noise in their kick model. They do not provide details of how they use this model. But a popular way of using average and standard deviation is modeling the parameters (angle and distance) with Gaussian distributions.

In contrast, for creating our kick model, the robot approaches the ball, grabs it, and kicks it to the center of the field. Right before kicking, it records its position (kick position), which includes possible localization errors. Afterwards, the robot follows the ball, and when the ball stops moving, a human observer sends a signal to the robot and the robot evaluates the ball position and stores it (final ball position) [2]. The gathered sample kick effects (kick position, final ball position) are processed by an offline program, and for each kick sample, the difference between the final ball position and the kick position is computed. These changes in x and y coordinates get discretized and are stored in a grid.

The learned action model is a three dimensional array KT, where for kick type k, $KT[k][x][y]$ represents the number of kicks that changed the position of the ball for x grid cells in the x-axis and y grid cells in the y-axis. Figure 2 shows $KT[\text{FALL-KICK}]$ where the position of the robot (kick position) and kick direction is shown with the arrow, and each black dot represents one sample action effect resulting in the ball ending up in that grid cell. The main rectangular grid represents the size of the legged soccer field.

Two fundamental differences between our model and Chernova and Veloso's [8] as well as other usual action models (e.g. [1,2]) are that ours is (1) instance-based, and (2) unlike usual action models, where the designers make an effort to reduce the noise (e.g. tracking for one second, and putting the ball in front of the robot in [8]), we aim at designing an action model which captures the full environmental noise.

7.2 Planning

We begin by dividing the robot's environment into the disjoint grid G. Dotted lines in Figure 3 show the actual grid used in the experiments. KT is the set of different kick types available to the robot, and D is a discrete set of possible kick directions.

In Section 8, we empirically show that discretizing the field does not have much of a determental effect on the effectiveness of the algorithm.

The MDP $(S, A, Pr(s'|s, a), R)$ for the test-bed problem is defined as:

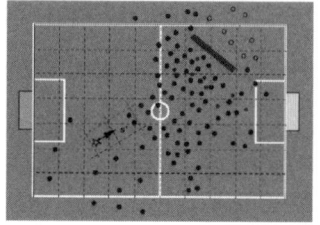

Fig. 3. Soccer field which is divided into a grid. A sample kick with the possible effects is also shown in presence of an opponent robot.

$S = G$ is a set of *states*, representing the grid cell that the ball is in. The center of a cell is assumed as the position of the grid cell. In the rest of the paper, the state is also used to point at the grid cell where the ball is located in that state. $A = KT \times D$ is the set of possible actions, which is kicking with a specific kick type (from KT) at a direction (from D). $Pr(s'|s, a)$ is the state transition probabilities. $R(s, a)$ is the *reward function* and is the probability of scoring a goal (with just one kick) from state s using action a.

---

[2] In principle it is possible for the robot to recognize when the ball has stopped moving, but our robots do not have that capability.

The kick model (KT) is used to approximate $\widetilde{Pr}(s'|s,a)$. KT is projected to the starting point of the kick with the kick direction to get a distribution over kick effects (i.e. the possible cells that the ball can land in). This new distribution is $\widetilde{Pr}(s'|s,a)$. More precisely $\widetilde{Pr}(s'|s,a)$ is computed using the following equation:

$$\widetilde{Pr}(s'|s,a) = \frac{KT[k][s'_{xd} - s_{xd}][s'_{yd} - s_{yd}]}{N[k]} \tag{4}$$

where s_{xd} (s_{yd}) is the x-index[3] (y-index) of the grid cell containing the ball in s, after the transformation that transforms the kick direction to the x-axis. For example if the a is the kick action shown in Figure 3, and s' is the grid cell shown on the field (Figure 3), then $s'_{xd} - s_{xd} = 0, s'_{yd} - s_{yd} = +1$. Thus $\widetilde{Pr}(s'|s,a) = \frac{KT[k][0][1]}{N[k]} = \frac{2}{118}$ For each state s, the center of the cell is used for computing s_{xd} and s_{yd}. k is the type of action a, and $N[k]$ is the number of kick samples of type k.

$R(s,a)$ is also computed with the help of the action model, that is, $R(s,a)$ is equal to the percentage of the sample kick effects from the kick model that result in a goal from state s by taking action a. A line is computed from the robot position to the final kick effect's position, if that lines intersects the goal line, it is considered a goal.

The value iteration and action selection algorithms described in Section 5 are used for computing Q-values and action selection. Note that for static systems the value iterations are performed offline. Each round of value iteration on the Aibo robot's processor roughly takes 2 seconds, and each decision cycle is around 33 milliseconds. This shows why performing multiple rounds of value iteration is not feasible for dynamic systems, and therefore there is a need for a fast replanning method.

7.3   Replanning

Opponent robots in the goal scoring test-bed are considered as dynamic factors. We assume that opponent robots only block the ball and do not grab or kick it. The robot models the blocking as ball reflection from the opponent robots. That is, if the kick trajectory of a kick sample in the kick model would hit any of the opponent robots, the reflection from the opponent robot is assumed by the robot as the final position of the ball. Sample kick effects of a FALL-KICK (see Figure 1(a)) in presence of an opponent robot are shown in Figure 3. The unfilled circles are the possible kick effects when the opponent does not exist.

The replanning algorithm presented in Section 6 is used to update the Q values. In the first step of the algorithm, a pair (s,a) in the presence of the opponent robot f is flagged for recomputation of its value if f is reachable by an average kick a from s (i.e. instead of the kick model of a, it uses the average distance and angle of kick a).[4]

---

[3] x-index of the grid-cells in the ith row of the grid is i.

[4] More elaborate techniques that consider all kick samples proved to need heavy computation, which is not available on the robots.

One special case to consider is when the opponent robot o intersects with a grid cell g, and based on a sample action effect, the ball also ends up in grid cell g. The value of a point in cell g is highly dependent on which side of opponent o the point is located. If the final ball point is on the same side of o as the center of cell g, $V^*(g)$ is used, and if not, average V values of the cells adjacent to g and on the same side of o as the ball are used.

One round of full updates with the provided incremental planning roughly takes 1 second, and we distribute that process over 25 decision cycles. Given the fact that most of the 33 milliseconds of the decision cycle is already taken by the vision processing, the robot on average loses every other cycle while performing the incremental update (effectively the decision cycle while performing the update is closer to 66 milliseconds).

8   Experimental Results

The algorithms are evaluated both on a custom-built AIBO simulator [16] and also on the real AIBO robots. The simulator, though abstract with respect to locomotion, provides a reasonable representation of the AIBO's visual and localization capabilities, and allows for a more thorough experimentation with significant results. The kick model used in the simulator is the same as the one used on the robot. Thus, the simulation results are based on the assumption of a correct kick model. There are methods of active localization (e.g. [18]) to reduce the localization errors, which are not considered here and can be used orthogonally with this work. Thus, robots should deal with large localization errors (in the simulation, this is reflected in the learned kick model), and that results in lower scoring percentages.

Five different algorithms are compared in this section. The considered algorithms include:

- ATGOAL: In this algorithm, the robot always shoots directly at the goal using FALL-KICK which is the more accurate kick. It is used as a baseline algorithm.
- PLAN: This is the planning algorithm (Section 5) for the clear field, where no opponent robot is present. In clear fields, this is the algorithm of choice, but it is also used to compare to REPLAN in the presence of opponent robots.
- REPLAN: This is the planning algorithm presented in Section 6, where the robot observes the position of the opponent robots online.
- FULLPLAN: This algorithm is used for comparison with REPLAN. In FULL-PLAN, it is assumed that the position of the opponent robots is known as a priori, and an offline algorithm performs the full value iteration, and passes the Q-values to the robot.
- PARAMPLAN (PARAMFULLPLAN): This is similar to the PLAN (FULLPLAN for the case of PARAMFULLPLAN) algorithm, but instead of the full instance-based kick model, a parametric kick model is used. Average and standard deviation (similar to [8]) for distance and angle of each kick sample is computed. Two different Gaussians are assumed, one for the angle, and the other

for the distance of the kick samples. Each time the robot considers a kick, it draws n random samples from each of the Gaussians, where n is equal to the number of different kick samples that it would have considered for the instance-based kick model. For each of the n pairs of (angle, distance), it evaluates the kick. The final evaluation is based on the average of the n evaluations. This experiment is used to show the power of the instance-based kick model compared to the parametric kick model with normalized noise.

Comparing REPLAN with ATGOAL shows the general effectiveness of our approach. The advantage of REPLAN compared to PLAN shows the benefit of the replanning algorithm. Experiments also show that performance of REPLAN is close to FULLPLAN which highlights the effectiveness of the proposed fast replanning algorithm. Comparing REPLAN with PARAMPLAN demonstrates the benefit of using instance-based action models.

The grid used in the experiments is 7×10 and is shown in Figure 3. The number of rounds of value iteration is set at 20.

8.1   Simulation Results

In the first experiment, the environment is assumed to be static. In later experiments, the algorithms are evaluated in dynamic environments. At the end of this section, the effects of considering the true position for action selection (see Section 5) are investigated. While doing so, we also argue that, the effects of assuming a grid for representing the position are minor.

Each episode starts with the ball positioned in a starting point, and is finished when a goal is scored or the ball goes out of bounds. Each trial consists of 100 episodes. The percentage of the episodes which resulted in goals and the average number of kicks per episode are computed for each trial. The reported data is averaged over 28 trials.

Clear Field Experiment. We start the experiments with no opponents (Figure 4). Two starting points for the ball are considered: center point (Figure 4(a)) and the upper-left point (Figure 4(b)).

Recall that the ATGOAL algorithm only uses FALL-KICK. For a fair comparison between ATGOAL and PLAN, the result for the PLAN*(FALL-KICK)* algorithm, which is similar to PLAN, but only uses FALL-KICK, is also presented. As the results in Table 1 suggest, using the HEAD-KICK does not make much of a difference for the PLAN algorithm. For that reason, in the next experiments PLAN(FALL-KICK) is no longer considered. However, one of the benefits of the algorithms presented in this paper is that they enable the addition of newly designed kicks. All that is needed is their instance-based models.

The scoring percentage and average number of kicks per episode for ATGOAL, PLAN and PARAMPLAN are presented in Table 1. As shown in the table, when the starting ball point is at the center of the field, planning significantly increases performance over the ATGOAL algorithm by 30%, and increases the performance by 76% when the starting ball position is the upper-left point. For the planning algorithm, the average number of kicks is also higher, which is a compromise for achieving a better scoring percentage.

The effectiveness of the instance-based kick model is shown by the significant advantage of the PLAN algorithm compared to PARAMPLAN in Table 1. Using the instance-based kick model for the center and upper-left starting points increases the performance by 43% and 42% compared to PARAMPLAN, respectively.

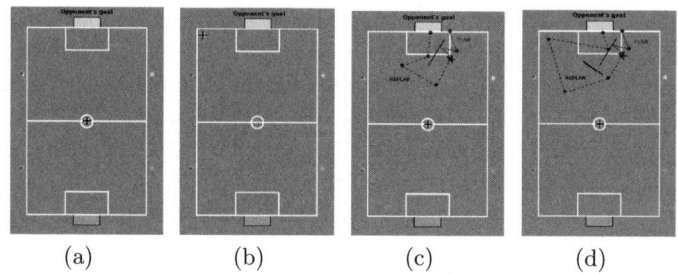

(a)   (b)   (c)   (d)

Fig. 4. Clear field. (a) The center of the field is the starting point of the ball. (b) The upper-left point is the starting point for the ball. (c) The field with one stationary opponent robot. Sample sequences for REPLAN and PLAN are shown for the ⋆ starting point. (d) The field with two stationary opponent robots.

Table 1. Comparing different algorithms for the two starting ball points in the clear field experiment (see Figure 4)

Starting Point: Center (Fig. 4(a))

Algorithm	ATGOAL	PLAN	PLAN(FALL-KCK)	PARAMPLAN
Scoring%	46.2 ± 4.8	**60.5 ± 4.8**	**58.7 ± 5.0**	42.1 ± 5.7
Kicks/episode	3.4 ± 0.1	9.7 ± 0.9	9.6 ± 0.9	10.1 ± 0.8

Starting Point: Upper-Left (Fig. 4(b))

Algorithm	ATGOAL	PLAN	PLAN(FALL-KCK)	PARAMPLAN
Scoring%	29.1 ± 5.0	**51.3 ± 5.4**	**53.1 ± 5.3**	39.0 ± 5.2
Kicks/episode	1.7 ± 0.1	6.3 ± 0.6	6.1 ± 0.8	7.1 ± 0.7

One Opponent Experiment. In this experiment, a stationary opponent robot is placed in the field. The field with the opponent robot is shown in Figure 4(c). The ball's starting point is at the center of the field. The algorithms ATGOAL, REPLAN, FULLPLAN, PLAN, and PARAMFULLPLAN are compared. Success percentage and average number of kicks is presented for the above-mentioned algorithms in Table 2.

The REPLAN algorithm significantly improves scoring percentage compared to ATGOAL, PLAN and PARAMFULLPLAN by 104%, 13%, and 45% respectively. REPLAN is also very close in performance to FULLPLAN (non-significant difference of 1.5%), where the transition function is assumed to be known a priori and the Q-values are computed offline without computational limitations. The

Table 2. Scoring percentage and average number of kicks for different algorithms, in the one opponent robot scenario (see Figure 4(c))

Algorithm	AtGoal	RePlan	FullPlan	Plan	ParamFullPlan
Scoring %	32.00 ± 5.83	**65.92 ± 4.20**	**67.10 ± 3.92**	57.42 ± 3.76	45.17 ± 4.59
# Kicks/episode	1.77 ± 0.11	11.74 ± 1.00	8.22 ± 0.67	8.42 ± 0.63	9.59 ± 0.69

average number of kicks per episode is the most for the RePLAN algorithm (see Figure 4(c) for a sample kick sequence for RePLAN), but in this domain, scoring efficiency is of greater importance.

Two Opponents Experiment. In this experiment, an additional robot is positioned on the field. The field is shown in Figure 4(d). The scoring rate and average number of kicks for different algorithms is presented in Table 3. The trend in the result is consistent with the observation in the one opponent robot scenario in the previous experiment (Section 8.1).

Table 3. Comparing different algorithms for playing against the two opponents case (Fig. 4(a))

Algorithm	AtGoal	RePlan	FullPlan	Plan	ParamFullPlan
Scoring %	21.85 ± 4.64	**54.25 ± 5.58**	**54.64 ± 5.59**	46.46 ± 5.10	38.07 ± 4.39
# Kicks/episode	5.09 ± 0.32	19.45 ± 1.97	11.43 ± 0.90	13.04 ± 1.14	11.84 ± 0.96

Moving Opponents Experiment. In this experiment two *moving* opponent robots are present on the field. In each episode, the opponent robots start from the position of the opponent robots in the previous experiment (Figure 4(d)), and after each kick, they move $150cm$[5] randomly in one of the four main directions (i.e. left, right, up or down). In an effort to reduce the noise in the result, the seed of the pseudo-random generator, which determines what direction opponent robots move is fixed for all trials (not episodes).

The scoring percentage and average number of kicks for AtGoal, RePlan, and Plan algorithms is provided in Table 4. The performance of RePLAN is 178% better than AtGoal and 6% better than the Plan algorithm. Since the robot movement is random, the position of the opponent robots can not be known as a priori, and no offline algorithm like FullPlan can be developed.

8.2   Real Robots

In this section, experiments on a real robot are reported. Robot experiments are time consuming and it is not possible to do as many trials as in simulation. First, experiments in the clear field case (Figure 4(b)) and then against two opponent robots (Figure 4(d)) are described.

---

[5] Recall that the size of the field is $5400cm \times 3600cm$.

Table 4. Comparing AtGoal, Plan and RePlan algorithms in the presence of two moving opponent robots

Algorithm	AtGoal	RePlan	Plan
Scoring %	20.60 ± 5.14	**57.32 ± 4.82**	54.14 ± 3.98
# Kicks/episode	4.52 ± 0.37	9.51 ± 0.65	9.51 ± 0.67

Real Robot on a Clear Field. The configuration is the same as the one shown in Figure 4(b). Each trial consists of 50 episodes, and the result for one trial is reported in Table 5. The trend is similar to the simulation experiment of the same configuration, and Plan increases performance over AtGoal and ParamPlan by 80% and 20% respectively.

Table 5. Comparing different algorithms for upper-left starting ball points for the clear field experiment (see Figure 4(b)) with a real robot

Algorithm	AtGoal	Plan	ParamPlan
Scoring %	20	36	30

Real Robot Against Two Opponent Robots. In this experiment the configuration of the field in Figure 4(d) is used for a real robot. Each trial consists of 25 episodes, and the results are reported in Table 6.[6] The results show the same trend as the simulation on this field: RePlan is better than Plan and AtGoal.

Table 6. Comparing different algorithms for real robot in the experiment against two opponent robots (see Figure 4(d))

Algorithm	AtGoal	Plan	RePlan
Scoring %	4	16	24

9   Conclusion

This paper considers the action planning problem in noisy environments, modeling it as an MDP. An instance-based action model is introduced to model noisy action effects. The action model is then used to build the transition function of a MDP. Furthermore a fast algorithm for action planning in dynamic environments, where dynamic factors have effect on only a subset of state-action pairs is introduced. To evaluate these approaches, goal scoring in the RoboCup 4-legged league is used as a test-bed. The experiments show the advantage of

---

[6] Since in base code used, the robots do not walk around opponent robots, in this experiment whenever the robot attempts to walk through the opponent, the human temporarily removes the opponent robot. Were the robot equipped with obstacle avoidance, it would score in roughly the same trials — it would just take a bit longer.

using an instance-based approach compared to parametric action models. They also show that the fast replanning algorithm outperforms off-line planning and approaches the best possible performance assuming the full transition model is known a priori.

In future work, the performance of the incremental replanning algorithm may be improved by learning the effects of dynamic factors on the transition model. Also, extending this approach to team behaviors, where the value of each state also depends on the positions of teammates, is an interesting direction for future consideration.

Acknowledgment

The authors would like to thank the members of the UT Austin Villa team for their efforts in developing the software used as a basis for the work reported in this paper. Special thanks to Gregory Kuhlmann for developing the simulator. This research was supported in part by NSF CAREER award IIS-0237699 and ONR YIP award N00014-04-1-0545.

References

1. Yang, Q., Wu, K., Jiang, Y.: Learning action models from plan examples with incomplete knowledge. In: Proceedings of the Fifteenth International Conference on Automated Planning and Scheduling (June 2005)
2. Wang, X.: Learning planning operators by observation and practice. In: Artificial Intelligence Planning Systems, pp. 335–340 (1994)
3. Moore, A.W., Atkeson, C.G.: Prioritized sweeping: Reinforcement learning with less data and less time. Machine Learning 13, 103–130 (1993)
4. Bellman, R.E.: Dynamic Programming. Princeton University Press, Princeton (1957)
5. Stentz, A.T.: The focussed d* algorithm for real-time replanning. In: Proceedings of the International Joint Conference on Artificial Intelligence (August 1995)
6. Koenig, S., Likhachev, M.: Improved fast replanning for robot navigation in unknown terrain. In: Proceedings of the 2002 IEEE International Conference on Robotics and Automation (May 2002)
7. Wilkins, D.E., Myers, K.L., Lowrance, J.D., Wesley, L.P.: Planning and reacting in uncertain and dynamic environments. Journal of Experimental and Theoretical AI 7(1), 197–227 (1995)
8. Chernova, S., Veloso, M.: Learning and using models of kicking motions for legged robots. In: Proceedings of International Conference on Robotics and Automation (ICRA 2004) (May 2004)
9. Burgard, W., Cremers, A., Fox, D., Hähnel, D., Lakemeyer, G., Schulz, D., Steiner, W., Thrun, S.: Experiences with an interactive museum tour-guide robot. Artificial Intelligence 114(1-2), 3–55 (1999)
10. Bevly, D., Farritor, S., Dubowsky, S.: Action module planning and its application to an experimental climbing robot. In: International Conference on Robotics and Automation (2000)

11. Frommherz, B., Werling, G.: Generating robot action plans by means of an heuristic search. In: International Conference on Robotics and Automation (1990)
12. Leake, D.B. (ed.): Case-Based Reasoning: Experiences, Lessons, and Future Directions. AAAI Press, Menlo Park (1996)
13. Atkeson, C.G., Moore, A.W., Schaal, S.: Locally weighted learning for control. Artificial Intelligence Review 11(1-5), 75–113 (1997)
14. Gabel, T., Riedmiller, M.: Cbr for state value function approximation in reinforcement learning. In: 6th International Conference on Case-Based Reasoning (2005)
15. Atkeson, C., Santamaria, J.: A comparison of direct and model-based reinforcement learning. In: International Conference on Robotics and Automation (1997)
16. Stone, P., Dresner, K., Fidelman, P., Kohl, N., Kuhlmann, G., Sridharan, M., Stronger, D.: The UT Austin Villa 2005 RoboCup four-legged team. Technical Report UT-AI-TR-05-325, The University of Texas at Austin, Department of Computer Sciences, AI Laboratory (November 2005)
17. Puterman, M.L.: Markov Decision Processes. Wiley, NY (1994)
18. Kwok, C., Fox, D.: Reinforcement learning for sensing strategies. In: The IEEE International Conference on Intelligent Robots and Systems (2004)

Precise Extraction of Partially Occluded Objects by Using HLAC Features and SVM

Kazutoki Otake, Kazuhito Murakami, and Tadashi Naruse

Graduate School of Information Science and Technology, Aichi Prefectural University
Kumabari, Nagakute-cho, Aichi 480-1198 Japan
im071006@cis.aichi-pu.ac.jp,
{murakami,naruse}@ist.aichi-pu.ac.jp

Abstract. In the RoboCup competition, robot soccer game, ball and robots are extracted by using color information. If color markers attached on the robot or a ball itself are occluded, especially the occlusion ratio is high, it will be difficult to extract them. This paper proposes a new and high precision method which extracts partially occluded objects based on the statistical features of the pixel and its neighborhoods. Concretely, at first, input image is labeled by using color information and small candidate regions which have similar color to the color markers or the ball are extracted, then each candidate region is classified into partially occluded object or noise by using HLAC features and SVM. We applied our method to the global vision system of RoboCup small size league (SSL) and confirmed that it could extract partially occluded objects, 94.23% for 5 to 8 pixels area and 80.06% for 3 to 4 pixels area, and worked more than 60fps.

1 Introduction

RoboCup, robot soccer game, is an international research project to clarify and promote robot engineering and artificial intelligence by using autonomous robots and started on 1997. The final objective is "By the year 2050, develop a team of fully autonomous humanoid robots that can win against the human world soccer champion team."

To extract robots and a ball, the images of the CCD cameras that are mounted at the ceiling of the hall or on the robot itself are used in general. The former is called "global vision" and the latter is called "local vision", respectively. Figure 1 shows examples of the soccer scene in RoboCup competition. Figure 1(a) shows SSL's robots which utilize a global vision system, and Fig.1(b) shows 4-legged league's robot which utilizes a local vision system.

In the real game, many teams use color markers to extract and classify each robot and ball. There is no restriction, except for the team color marker, to the color and the shape of the color markers, so each team can uses different color/shape markers[1].

U. Visser et al. (Eds.): RoboCup 2007, LNAI 5001, pp. 17–28, 2008.

Figure 2 shows an example image obtained through a CCD camera. Figure 2(a) is an original image and Fig.2(b) is its labeled result. In the small size league, it needs less than 1/60 seconds to process 1 cycle. When all of the color markers are observed completely, the robots and ball can be detected, but if the objects are partially occluded, then it will be difficult to extract. This kind of occlusion occurs in the global vision system and also in the local vision system. Most of the conventional methods which calculate width, height, area, etc. of the candidate objects or regions have not been succeeded in extracting partially occluded objects and in judging noise or target object. As described above, it is necessary for the camera to observe whole of the object without occlusion. Narita et al. have reported that multi-cameras could reduce the occlusion of the ball for RoboCup SSL[2]. However, even though the number of cameras increases, the occlusion would not be 0.

In general, particle filter or Kalman filter etc. are used for the object tracking with occlusion. Sugimura et al. have reported the robust tracking method for the object detection[3]. However, an essential problem remains that the moment when the object just begins to be seen could not be caught by the conventional tracking methods.

This paper proposes a new and high precision method to extract partially occluded objects. The candidate regions that are extracted and labeled by color information are classified into the target object or noise by using higher order local autocorrelation (HLAC) features and support vector machines (SVM). In the following parts, the analysis of extraction errors from the view point of extraction and occlusion are described in section 2, a high speed and robust method is explained in section 3, and the experimental results for the RoboCup vision system are expressed in sections 4 and 5.

(a) Small-size robot

(b) 4-legged robot

Fig. 1. Soccer scenes in RoboCup

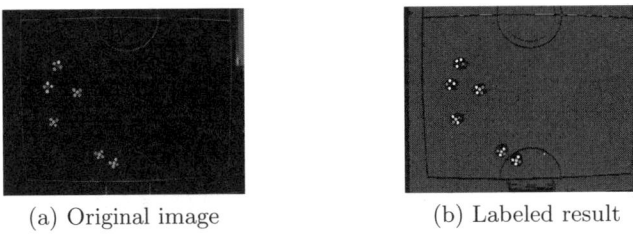

(a) Original image   (b) Labeled result

Fig. 2. An image obtained through a global vision

2 Problems of Conventional Methods

2.1 Extraction Error by Noise

Noise appears due to the aberration of camera's lens or the threshold setting of the parameters such as the width, height, area, and so on, but when the noise has the same values as the target pixel, it will be difficult to remove it.

Usually, since the target is not so small, in general it is larger than the noise, conventional method succeeds in thresholding. If the extracted area becomes small by the occlusion, then the extraction rate goes down.

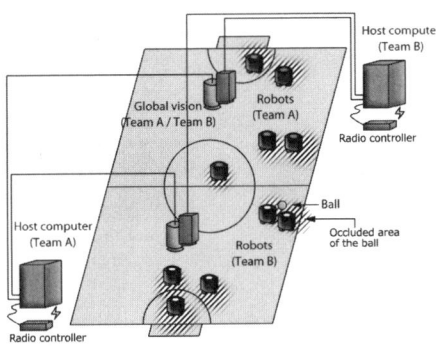

Fig. 3. Occluded area in SSL (Occlusion area is displayed with diagonal lines)

2.2 Extraction Error by Partial Occlusion

Many teams in SSL including our team use a global vision system composed of 2 or more cameras. Occlusion occurs a little in the center of the image, and on the contrary, it becomes more in the area far from the center of the image as shown in Fig.3.

Figure 4 shows a typical situation which causes occlusion. In the RoboCup SSL competition, a golf ball colored in orange is used. Figure 5 shows examples of labeled result for orange regions. Figure 5(a) is a case that only a ball exists and Fig.5(b) is a case that some part of a ball is occluded by a robot.

Figures 5(c) and 5(d) are the cases that some pixels around color marker or a line on the field are detected as pseudo orange region. In the RoboCup competition, it is necessary to classify these similar regions in real time whether it is a real target object or a noise.

Fig. 4. A typical scene of occlusion

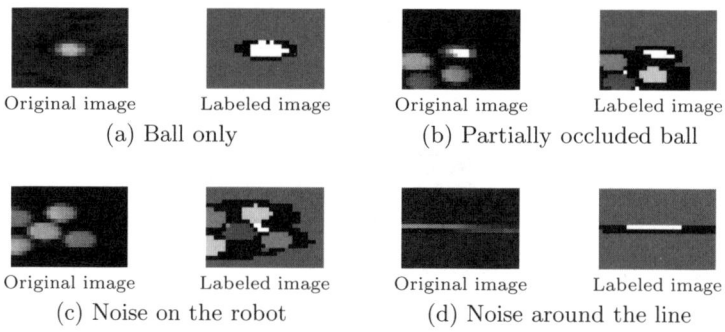

Original image   Labeled image   Original image   Labeled image
(a) Ball only   (b) Partially occluded ball

Original image   Labeled image   Original image   Labeled image
(c) Noise on the robot   (d) Noise around the line

Fig. 5. Examples of orange objects

3   A Method to Extract Partially Occluded Object by HLAC and SVM

3.1   Principle of the Proposed Method

This section describes a method to extract partially occluded object by using higher order local autocorrelations (HLAC)[4] and support vector machines (SVM). Figure 6 shows the concept of the proposed method. First, the system calculates HLAC features by using 35 masks shown in Fig.7 for each pixel which belong to class 1 (i.e. ball) and class2 (i.e. noise), respectively. The number of dimension of HLAC features is $35 \cdot 3 = 105$ because each pixel's value is composed of Y, U and V values. Then, the distributions of HLAC features for class1 and

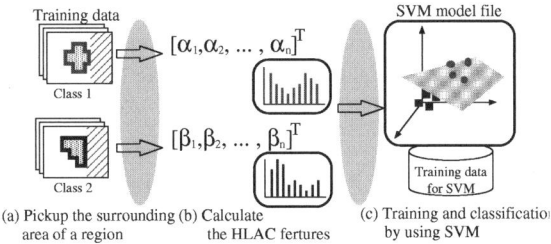

Fig. 6. The flow of the proposed method

class2 are applied to SVM. The combination of HLAC and SVM realizes a robust extraction to the changes of direction (occlusion occurs in any directions) and also to the changes of lighting.

3.2   Calculation of HLAC Features and Normalization

How to expand autocorrelation function to higher orders is presented, for example, in [3]. Let the reference value be r on the image f, then the N-th order autocorrelation $c(a_1, \cdots a_N)$ is defined by the calculation as

$$c(a_1, \cdots a_N) = \int f(r)f(r + a_1) \cdots f(r + a_N)dr, \qquad (1)$$

here $a_1, a_2, ...a_N$ are the reference values around r.

Now, restrict the order N up to $2(N = 0, 1, 2)$ and its displacement is the correlation around the $f(r)$ at a maximum 3 in the 3×3pixels. Then the number of feature values becomes 35 as shown in Fig.7.

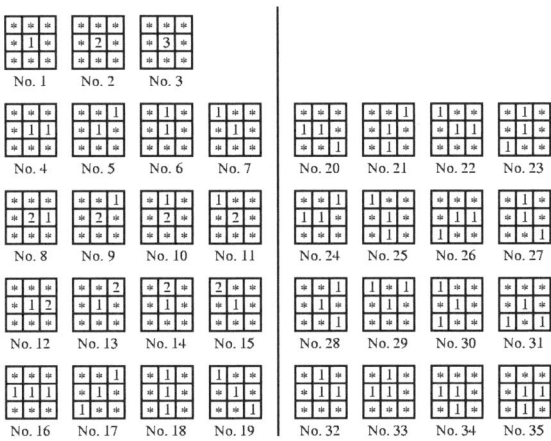

Fig. 7. Templates of higher order local autocorrelation feature that N is limited to 2

As for the feature vector calculated by the sum of products of pixel value, it is necessary to normalize so that the value of each feature vector be the same range. In case of 8-bit quantization for each pixel value, the normalized autocorrelation $c'(a_1, \cdots a_N)$ becomes as follows.

$$c'(a_1, \cdots a_N) = \sum_r \frac{f(r)f(r+a_1)\cdots f(r+a_N)}{255^N} \qquad (2)$$

In addition, feature vectors are separately calculated for each element of the YUV color information, so the feature vector H is expressed as

$$H = \{c'_Y(a_1, \cdots a_N), c'_U(a_1, \cdots a_N), c'_V(a_1, \cdots a_N)\} \qquad (3)$$

An example of H is shown in Fig.8. As shown in Fig.8(a), the feature values corresponding to $N = 0, 1$ (0 to 2 for x-axes) are extremely smaller than others. In contrast, Fig.8(b) shows the normalized values.

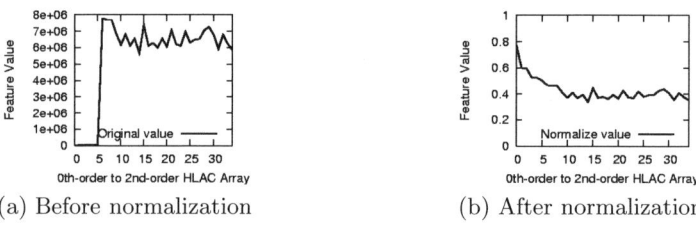

(a) Before normalization   (b) After normalization

Fig. 8. An example of HLAC features

3.3   Training by SVM

It is reported that SVM is effective to the changes of the position and the luminance in the object extraction[5,6]. Feature vectors, H_{ball} for the object to be detected as the target object and H_{noise} for the noise, are used for the training data of SVM. These data are collected as the training data in the beginning and creates a model for SVM[1].

4   Application to RoboCup SSL

4.1   Time Restriction in RoboCup

The flow of the ball detection is shown in Fig.9.

First, labeling and segmentation[2] processes are applied to the color space. The proposed method uses the parameters such as width, height and area features like a conventional method.

---

[1] C-SVC is used for the classifier and RBF kernel is used in LibSVM[7].
[2] Segmentation process by using CMVision2[9].

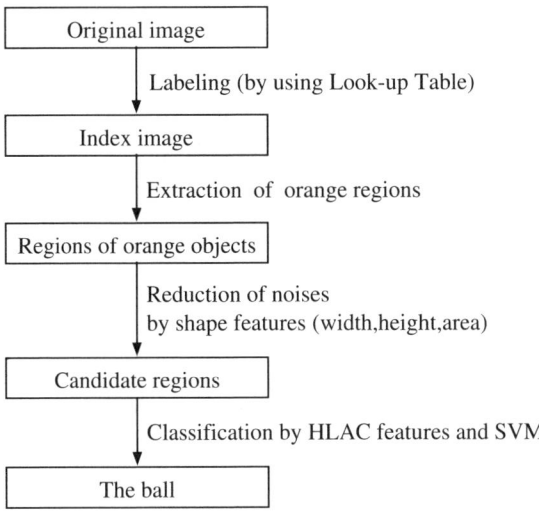

Fig. 9. Ball detection flow

In general, the processing interval is 60fps or more. And the total processing time for each frame is 16.7msec or less. In our conventional method, it took 1.9msec for image processing and 2.1 to 6.1msec for other processing such as strategy, pass generation, and so on. Therefore, our proposed method can use about $16.7 - 6.1 = 10.6$msec or less[8].

4.2 Host Computer System

The specification of our computer is as follows:

- CPU Athron 64 3500+
- 512MB Memory
- Debian Linux Operating system
- GV-VCP/PCI capture boards
  (Bt848 chipset is popular and low-cost board.)

4.3 Characteristics of Robots and Ball in Occlusion

We have used the black color for the robot's body except for the color marker as shown in Fig.10. The height of the robot is 150mm and the ball is 42mm. When the ball is occluded, the color of the neighborhood region of the ball becomes black.

5 Experiment

5.1 Construction of Training Model and Classification

We have prepared training data with occlusion which occurred around periphery of the image, and at the same time, under the condition that the lighting changes in

Fig. 10. A robot and a ball that are used for occlusion experiment

order to rise up the robustness of the proposed method. The occlusion images are obtained by controlling a robot to occlude some part of the ball as shown in Fig.10.

We collected the features of a ball and noise from orange region of size 8 3 pixel, because the size of the ball was observed more than 83 pixel without occlusion under our experimental conditions. The collected training data are shown in Table 1. The distributions of the training data are shown in Fig.11. Figures 11(a), 11(b) and 11(c) are the distributions of the width, height and area of the ball, respectively, and Fig.s11(d), 11(e) and 11(f) are those of the noise.

Table 1. The number of training data of ball and noise from orange region

Picked up region		The number of	The number of
Width×Height [pixel]	Area [pixel]	training data for balls	training data for noise
8x3 or less	1	300	2000
8x3 or less	2	1500	2000
8x3 or less	3,4	1800	2000
8x3 or less	5-8	2000	2000
8x3 or less	9-16	2000	0
8x3 or less	17-24	2000	0

The classifier based on SVM was created by these training data. We evaluated the classifier by using the 10-fold cross validation. From Fig.11, all of noise observed became 8 pixel or less, then the experiment and classification for partially occluded balls were done for under 8 pixels.

The training data was evaluated by using the following expressions.

$$precision = \frac{R}{N} \tag{4}$$

$$recall = \frac{R}{C} \tag{5}$$

$$F - measure = \frac{2 \cdot precision \cdot recall}{precision + recall} \tag{6}$$

here, R is the the number of orange regions of the detected balls (noise is not included), N is the the number of orange regions of the detected balls (noise is included), and C is the the number of orange regions of the balls in training data set.

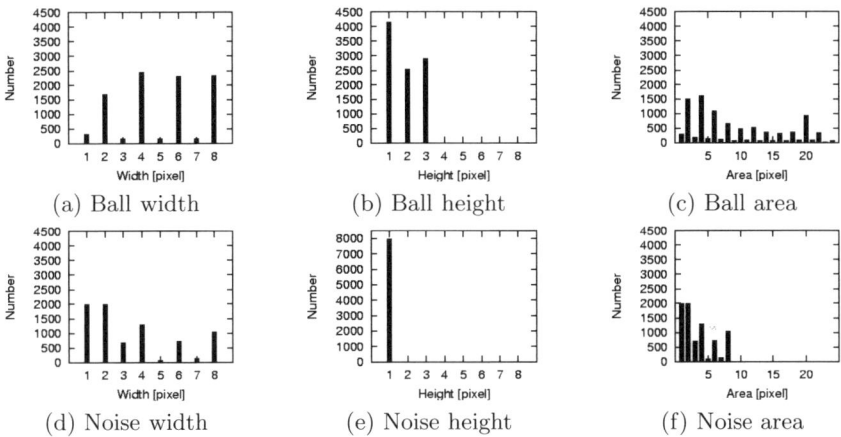

Fig. 11. Distributions of shape features of training data

The ball has been detected only by using width, height and area of the region in a conventional method. Therefore, not only the ball but also the noise is included in the region when the size becomes smaller. The detection rate D of the ball in a conventional method is calculated as $D = \frac{B}{B+N}$ for each area of the region, here, B and N are the numbers of balls and noise observed in the unit time.

The robot shown in Fig.10 has moved at random by using a remote controller, and the numbers of observations of the ball and noise were examined. Table 2 shows this result. The detection rate of a conventional method was calculated by using it. The number of frames in which the orange regions were observed was assumed to be 6000 frames.

5.2 Calculation of Detection Rate in a Practical System

In our practical system, in the unit time for ball detection rate $D^{(n)}$ is calculated from the product of the classification rate $BallClassificationRate(n)$ and the number $B^{(n)}$ of the ball observed by a certain number n of pixels.

$$D(n) = \frac{\sum_{i=n}^{24} \left(B^{(i)} \cdot BallClassificationRate(i) \right)}{\sum_{i=n}^{24} B^{(i)}} \tag{7}$$

Table 2. The numbers of orange regions for ball and noise observed in 6000 frames

Picked up region		The numbers of orange regions	
Width×Height [pixel]	Area [pixel]	ball	noise
8x3 or less	1	62	8879
8x3 or less	2	173	8114
8x3 or less	3,4	304	5388
8x3 or less	5-8	490	7109
8x3 or less	9-16	1518	0
8x3 or less	17-24	2086	0

From the data in Table 2 and classification result, ball's detection rate $D(i)$ ($i = 1, 2, 3, 5, 9$) was calculated and the processing time was measured.

6   Experimental Result

6.1   Classification Result by Each Area

Table 3 expresses the experimental result for the object classification of a conventional method and our proposed method for each area.

6.2   Detection Result in a Practical System

Table 4 shows the detection rate and the processing time in our practical system. The detection rate was calculated based on ball's samples in continuous observed data(6000 frames).

Although our proposed method can be applied to both of the ball detection process and the marker detection process, in the experiment, we applied only to the ball detection. The processing time shown in Table 4 denotes all of the vision processing.

7   Discussions

Table 3 shows the performance of the classification in each area. The proposed method is higher than a conventional method, and especially for the case that the size of the object is small. Furthermore, it is robust for the changes of lighting intensity, concretely it could work even if it changes in 1.1ev(+214%).

Table 4 shows the real detection rate in our particle system. RoboCup system requires more than 97 or 98% extraction rate. Although a conventional method has the limit that the target size should be more than 2 pixels area, and our proposed method has enough potential for all cases.

For the object of size 3 pixels area or more, it is known from Table 4 that the processing time is less than 10.6msec as discussed in 4.1, and it satisfies our system's condition to work in real time.

Table 3. Classification result for each area

Picked up region		Classification result		
Width × Height [pixel]	Area [pixel]	*precision*	*recall*	*F-measure*
8x3 or less	1	59.00%	45.12%	51.13%
8x3 or less	2	71.80%	73.88%	73.88%
8x3 or less	3,4	73.83%	87.43%	80.06%
8x3 or less	5-8	93.05%	95.45%	94.23%
8x3 or less	9 or more	100.00%	100.00%	100.00%

Table 4. Detection result and processing time in our practical system

Picked up region		Conventional method		Proposed method	
Width × Height [pixel]	Minimum area [pixel]	Detection rate	Processing time[ms]	Detection rate	Processing time[ms]
8x3 or less	1	78.94%	1.84	96.49%	120.31
8x3 or less	2	88.00%	1.83	97.09%	14.50
8x3 or less	3	83.03%	1.71	97.98%	9.86
8x3 or less	5	88.00%	1.69	99.31%	9.56
8x3 or less	9	100.00%	1.65	100.00%	8.53

8   Conclusions

This paper described a high speed method to extract partially occluded object by using HLAC and SVM. And it is shown that this method is robust for the changes of luminance. From experimental result, it has clarified to have the effectiveness for the extraction of partially occluded objects and also confirmed that it works in real time. We applied our proposed method to the RoboCup's global vision system and confirmed that it could extract color markers on the robots and a ball in the occluded situations.

Although this method works in real time, there still remains some subjects to be solved. It is required to realize more high speed processing in order to use the rest time for the strategic motion planning, and also it is also necessary to select more effective features to shorten the training process. These are coming subjects.

References

1. Bruce, J., Veloso, M.: Fast and Accurate Vision-Based Pattern Detection and Identification. In: Proc. of ICRA 2003, the 2003 IEEE International Conference on Robotics and Automation, Taiwan (May 2003)
2. Narita, R., Murakami, K., Naruse, T.: A Better Solution of Multi-camera's Layout for RoboCup Small Size League. In: Proc. of the 13th Japan-Korea Joint Workshop on Frontiers of Computer Vision, 2007(FCV 2007), pp. 351–356 (2007)
3. Sugiura, D., Kobayashi, Y., Sato, Y., Sugimoto, A.: People Tracking with Adaptive Environmental Attributes using the History of Human Activity, IPSJ SIG Notes. In: CVIM, vol. 115, pp. 171–178 (November 2006) (in Japanese)
4. Otsu, N., Kurita, T.: A new scheme for practical flexible and intelligent vision systems. In: Proc. of IAPR Workshop on Computer Vision, pp. 431–435 (1988)
5. Pontil, M., Verri, A.: Support vector machines for 3D object recognition. Proc. of IEEE Trans. PAMI 20(6), 637–646 (1998)
6. Okabe, T., Sato, Y.: Support Vector Machines for Object Recognition under Varying Illumination. In: Proc. of IPSJ Transctions on Computer Vision and Image Media, vol. 44, pp. 22–29 (2003) (in Japanese)

7. LIBSVM – A Library for Support Vector Machines,
   http://www.csie.ntu.edu.tw/~cjlin/libsvm/
8. Otake, K., Murakami, K., Naruse, T.: A High Speed and Robust Method to Extract
   Partially Occluded Objects for RoboCup Small Size League. In: Proc. of the 13th
   Japan-Korea Joint Workshop on Frontiers of Computer Vision, 2007(FCV 2007),
   pp. 63–68 (2007)
9. CMVision, http://www-2.cs.cmu.edu/~jbruce/cmvision/

Probabilistic Decision Making in Robot Soccer*

Pablo Guerrero, Javier Ruiz-del-Solar, and Gonzalo Díaz

Department of Electrical Engineering, Universidad de Chile
{pguerrer,jruizd,gonzdiaz}@ing.uchile.cl

Abstract. Decision making is an important issue in robot soccer, which has not been investigated deeply enough by the RoboCup research community. This paper proposes a probabilistic approach to decision making. The proposed methodology is based on the maximization of a game situation score function, which generalizes the concept of accomplishing different game objectives as: passing, scoring a goal, clearing the ball, etc. The methodology includes a quantitative method for evaluating the game situation score. Experimental results in a high-level strategy simulator, which runs our four-legged code in simulated AIBOs' robots, show a noticeable improvement in the scoring effectiveness achieved by a team that uses the proposed approach for making decisions.

1 Introduction

The aim of this paper is to propose a general methodology for taking decisions probabilistically in robot soccer. In a robotic soccer match, a player needs to take several decisions as for example: (i) where to position itself in the field when not having the ball, (ii) when to approach the ball, (iii) when to act as a support player, either supporting an attacker or a defender, (iv) what movements to do with the ball when having it, and (v) when and (vi) to which position to kick the ball. The decisions must take into account the role of the robot (defender, attacker, etc), the state of the game (score), the robot surround (position of teammates, opponents and the ball), and the teammates actions. In addition, decisions should be taken as fast as possible.

Most of the existent work related with decision making in robot soccer has focused in resolving specific tasks such as pass selection, and has not taken enough care of the big picture. The few approaches that consider several tasks at the same time, start their reasoning by considering a lot of reasonable decision criterions, and finally trying to mix them as best as possible. On the contrary, we believe that any strategy must start by defining a clear and general objective to be accomplished. Then, this general objective may be decomposed in more specific ones. In soccer, the general objective is to win the match, which can be also said as: "to score more goals than the opponent". Thus, instead of making a detailed list of possible risks, gains and costs, and then trying to take them all into account in the best way, we are proposing to reason in the opposite way: to clearly define the general objective to achieve, and then

---

* This research was partially supported by FONDECYT (Chile) under Project Number 1061158.

U. Visser et al. (Eds.): RoboCup 2007, LNAI 5001, pp. 29–40, 2008.

to find the more relevant criterions that can lead us to right decisions in order to accomplish this objective.

When the problem is faced in this fashion, it is clear how to balance the specific objectives as passing the ball, shooting to the goal, etc., and a wide spectrum of decisions' classes can be performed. Probabilities are nice to define such an approach, because in a probabilistic framework the natural uncertainties found in the process can be easily considered. The here-proposed methodology considers a score function of a given game situation. Decisions are taken in order to maximize the expected value of this score function. To make the kick decisions probabilistically, Montecarlo-based algorithms are used to integrate the PDFs (Probability Density Functions) of the available kicks over the field space. Another particularity of the proposed system is the way it takes opponents into account: they are not merely seen as possible blockers of the intended actions. Instead, we consider that the opposite team is intending, as much as the own, to score goals. Thus, we evaluate their possibilities with the same deepness that we do with the own: all of our analysis is symmetric for both teams. As a result, the presented approach is able to naturally balance defensive and offensive behaviors, and furthermore, it is able to change this balance according to the present situation. As human players do, robots following our approach will be more averse to risk when facing a defensive situation, and will gradually become more prone to take risks as the situation gets more offensive. Finally, the proposed methodology provides a quantitative method for evaluating the game situation score.

The advantages of the proposed system are the following: (i) the method relays only on the expected scored goal difference, and not in others conventionally taken into account such as pass success or ball possession time length; (ii) as stated in [8], when the space of the possible decisions is explored with a grid, it is possible to balance the accuracy of the decision and the computational cost; (iii) the uncertainty in the kicks result is considered; and (iv) the symmetric analysis of the situations allows a natural balancing between offensive and defensive behaviors. One disadvantage of the proposed method is the assumption of arbitrary models for the calculation of several of the probabilities. However, we believe this disadvantage may be corrected, by redefining if necessary the model of these probabilities, without affecting the core of the proposed system.

This paper is organized as follows. In section 2 is presented some related work. The proposed probabilistic methodology for decision making is described in section 3. In section 4, experimental results are presented. Finally, in section 5 conclusions of this work are given.

2   Related Work

For simulated soccer there have been proposed several interesting approaches that take into account several factors to make decisions ([4][7][8] to name a few). Some of them are based in reward functions, but finally, they use heuristics to mix probabilities (for example it is not clear how to compare the reward of a successful pass with the one of a successful shoot to the goal). Besides, they do not consider the uncertainty in the kicks' result.

When choosing an appropriate kick for an objective, most of the teams consider the time that it takes to be realized, the ball departure angle, and the shoot power, which is reflected on the ball speed after the kick (see for example the Team Description Papers in [2]). This information is usually acquired using statistics of data obtained from the repetition of a particular kick, and calculating the mean values of the distance and the angle of the final ball position for each available punch. There are different ways to choose the kick as a function of these parameters. From the strategic point of view there are differences at the moment of choosing a kick. For instance, the method implemented by the German Team [5] to pass the ball does not only use the information provided by its team partners; it uses in addition some visual information about the position of the receiver. Then it chooses the pass so that the objective is exactly the position of the receiving robot, which has to be warned right on that moment to react, and go back to the initial position for a better control of the ball.

In [1], it is proposed an interesting approach to deal with kicks uncertainty, based on a MonteCarlo sampling. The probabilities of accomplishing some prioritized objectives (passing, self-passing, shooting, and clearing) were estimated for each kick. We have incorporated the idea of the MonteCarlo sampling to our work, but instead of using a prioritized list of objectives for the objective and kick selection, we are proposing the use of a generalized objective which takes into account simultaneously all the listed objectives considered in [1], plus other possible objectives which are very difficult to consider in such an approach, as for example leading passing (passing not directly to the teammate but to a point ahead).

3 Proposed Approach

3.1 Game Segment

A RoboCup soccer match may be split into *game segments*. A game segment is the interval between two kick offs (kick offs occur when the match starts, and after a goal is scored). Every game segment may end in two ways: time out or goal. We can then define the score obtained in the current game segment as:

$$\beta = \begin{cases} -1 & \omega^{g'} \\ 0 & \omega^{t} \\ 1 & \omega^{g} \end{cases} \tag{1}$$

Where $\omega^{g'}$, ω^{t} and ω^{g} are respectively the events: "the opposite team scores", "time is out before anyone scores" and "the own team scores".

3.2 Ball Control Action

A *ball control action* (*BCA*) is what a robot does after catching the ball, and it consists in a relative displacement $\Delta\mathbf{x} = (\Delta x, \Delta y)^{T}$ and rotation $\Delta\theta$ of the robot holding the ball, and a kick $\mathbf{k}$ of the ball:

$$a = (\mathbf{d}, \mathbf{k})^T ; \quad \mathbf{d} = (\Delta \mathbf{x}, \Delta \theta)^T \tag{2}$$

Each game segment may be seen as a succession of BCA's $\{a_k\}$. We have a limited set of kicks $\mathbf{\Omega} = \{\mathbf{k}_l\}$. Let (r_l, θ_l) be the polar coordinates, relative to the kicking robot, to what the ball will arrive, if it is allowed to roll freely, after the kick $\mathbf{k}_l$ is performed. We assume that r_l and θ_l are independent Gaussian random variables with respective means $\mu_{r,l}$ and $\mu_{\theta,l}$, and variances $\sigma_{r,l}^2$ and $\sigma_{\theta,l}^2$. Then, the kick $\mathbf{k}_l$ can be parameterized using: $\mathbf{\Pi}_l = (\mu_{r,l}, \mu_{\theta,l}, \sigma_{r,l}^2, \sigma_{\theta,l}^2)$. The parameters $\mathbf{\Pi}_l$ have to be calculated previously for each of the available kicks. Figure 1.a shows our current available set of kicks and their parameters.

3.3 Score Function

Let us define a *game situation* as a vector $\mathbf{S} = (\mathbf{R}, \mathbf{b})^T$ where $\mathbf{b}$ is the estimated position of the ball and $\mathbf{R} = (\mathbf{x}_1, ..., \mathbf{x}_{N_R}, \mathbf{x}_1', ..., \mathbf{x}_{N_R}')^T$ is a vector containing the estimated poses of all robots, being N_R the number of robots per team. In particular, each robot may have an estimation of $\mathbf{S}$. In our implementation, teammate robots share their own estimated positions, the observations of the ball and of the other robots, and each robot tracks all the mobile objects using an EKF based approach. We propose that any situation of the game may be evaluated in terms of how advantageous it is. We will call this measurement the *Game Situation Score (GSS)*. The *GSS* is defined as:

$$GSS(\mathbf{S}) = E(\beta|\mathbf{S}) = P(\omega^g |\mathbf{S}) - P(\omega^{g'} |\mathbf{S}) \tag{3}$$

We are especially interested in situations when the ball just arrived to a new position, after a BCA. We define $\mathbf{S}_k = (\mathbf{R}_k, \mathbf{b}_k)^T$ as the situation produced by a_k, in the moment when the ball stops rolling. The event "a goal is scored by means of a_k" is defined as ω_k^g or $\omega_k^{g'}$, depending on which team scored. The events ω_{k+}^g and $\omega_{k+}^{g'}$ correspond to a goal scored, by means of a later BCA than a_k, by respectively the own team and the opposite team. Then $P(\omega^g |\mathbf{S}_k)$ is calculated as (the calculation of $P(\omega^{g'} |\mathbf{S})$ is symmetrical):

$$P(\omega^g |\mathbf{S}_k) = P(\omega_k^g |\mathbf{S}_k) + (1 - P(\omega_k^g |\mathbf{S}_k)) P(\omega_{k+}^g |\mathbf{S}_k) \tag{4}$$

It is straightforward from the previous definitions that the *immediate goal probability* $P(\omega_k^g |\mathbf{S}_k)$ is 1 or 0 depending on whether $\mathbf{b}_k$ is inside or outside the opposite goal.

The *future scoring probability* of the own team may be calculated in a recursive form:

$$P\left(\omega_{k+}^{g}\big|\mathbf{S}_{k}\right)=\int P\left(\omega_{k+}^{g}\big|\mathbf{S}_{k+1}\right)P\left(\mathbf{S}_{k+1}\big|\mathbf{S}_{k}\right)d\mathbf{S}_{k+1} \tag{5}$$

It is impractical to calculate the former integral, so we make some simplifications: (i) after the ball arrives to $\mathbf{b}_{k}$, the closest robot of each team, will lead to $\mathbf{b}_{k}$ until one of them catches the ball, (ii) as the pose of the rest of the robots at $k+1$ is unpredictable, we will assume they will remain static, and (iii) a_{k+1}, and thus $\mathbf{b}_{k+1}$, are totally determined by the team of the robot which will perform a_{k+1} and by all the robots' poses. Therefore, $\mathbf{S}_{k+1}$ is only a function of which robot will capture the ball and consequently perform a_{k+1}. Two events are defined: "the closest robot of the own team will catch the ball", called ω_{k+1}^{c}, and "the closest robot of the opposite team will catch the ball", called $\omega_{k+1}^{c'}$. Then, equation (5) can be rewritten as:

$$P\left(\omega_{k+}^{g}\big|\mathbf{S}_{k}\right)\approx P\left(\omega_{k+}^{g}\big|\omega_{k+1}^{c},\mathbf{S}_{k}\right)P\left(\omega_{k+1}^{c}\big|\mathbf{S}_{k}\right)+P\left(\omega_{k+}^{g}\big|\omega_{k+1}^{c'},\mathbf{S}_{k}\right)P\left(\omega_{k+1}^{c'}\big|\mathbf{S}_{k}\right) \tag{6}$$

The *catching probabilities* $P\left(\omega_{k+1}^{c(c')}\big|\mathbf{S}_{k}\right)$ are approximated as (analogous for $\omega_{k+1}^{c'}$):

$$P\left(\omega_{k+1}^{c}\big|\mathbf{S}_{k}\right)=\frac{t_{c}'\left(\mathbf{S}_{k}\right)}{t_{c}'\left(\mathbf{S}_{k}\right)+t_{c}\left(\mathbf{S}_{k}\right)} \tag{7}$$

Where $t_{c}\left(\mathbf{S}_{k}\right)$ and $t_{c}'\left(\mathbf{S}_{k}\right)$ are the amounts of time required to arrive to $\mathbf{b}_{k}$ for the closest robot of respectively the own team and the opposite team:

$$t_{c}\left(\mathbf{S}_{k}\right)=\frac{\left|\mathbf{x}_{i,k}-\mathbf{b}_{k}\right|}{v_{R}}+\frac{\left|\theta_{i,k}-\sphericalangle\left(\mathbf{b}_{k}-\mathbf{x}_{i,k}\right)\right|}{\omega_{R}} \tag{8}$$

Where $\mathbf{x}_{i,k}$ and $\theta_{i,k}$ are respectively the position and orientation of the robot of the own team closest to the ball at time k. Note that the time required for displacing and for rotating are considered in terms of the estimated robot linear speed v_{R} (=40cm/sec) and angular speed ω_{R} (=120°/sec) (these values correspond to AIBO ERS7 robots). The calculation of $t_{c}'\left(\mathbf{S}_{k}\right)$ is analogous.

The *future scoring probabilities* $P\left(\omega_{k+}^{g}\big|\omega_{k+1}^{c(c')},\mathbf{S}_{k}\right)$ can be calculated using (4):

$$P\left(\omega_{k+}^{g}\big|\omega_{k+1}^{c(c')},\mathbf{S}_{k}\right)=P\left(\omega_{k+1}^{g}\big|\omega_{k+1}^{c(c')},\mathbf{S}_{k}\right)+\left(1-P\left(\omega_{k+1}^{g}\big|\omega_{k+1}^{c(c')},\mathbf{S}_{k}\right)\right)P\left(\omega_{k+1+}^{g}\big|\omega_{k+1}^{c(c')},\mathbf{S}_{k}\right) \tag{9}$$

This leads to a possibly infinite recursion, therefore we will approximate all the remaining probabilities as a function of some coarse indicators of how advantageous the resulting situations are. We introduce the *expected free time* (t_{f} or t_{f}') of the robot that catches the ball, as the amount of time that the robot will be able to hold the ball without the direct presence of a rival, and is calculated as (analogous for t_{f}'):

$$t_f = bnd\left(0; t'\left(\mathbf{S}_k\right) - t\left(\mathbf{S}_k\right); \infty\right) \tag{10}$$

With $bnd\left(c; d; e\right)$ defined as the quantity d lower bounded by c and upper bounded by e. We also define the *aligning time* (t_a or t'_a) of the robot that catches the ball as the amount of time that it will need for aligning to its opposite goal. If $\mathbf{g}'$ is the position of the opposite goal , t_a is calculated as (analogous for t'_a):

$$t_a = \frac{\theta_a}{\omega_R} = \frac{\left|\sphericalangle\left(\mathbf{b}_k - \mathbf{x}_{i,k}\right) - \sphericalangle\left(\mathbf{g}' - \mathbf{b}_k\right)\right|}{\omega_R} \tag{11}$$

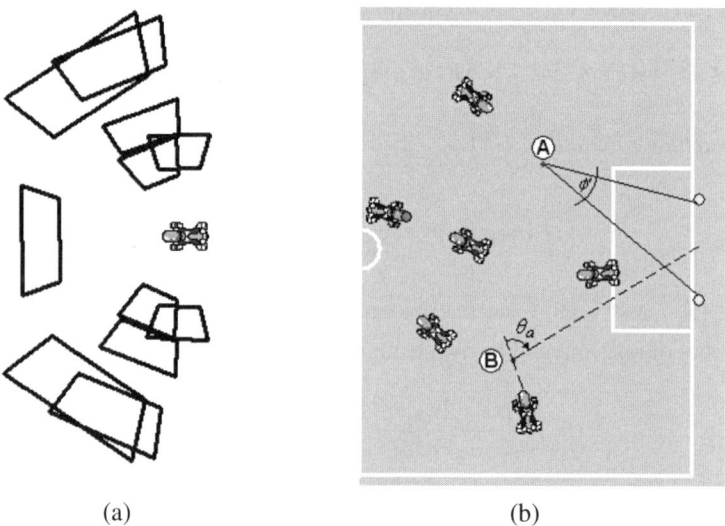

(a)   (b)

Fig. 1. (a) Set of available kicks with their relative means and variances, each plotted polar rectangle is bounded by $\left(\mu_{r,l} \pm \sigma_{r,l}, \mu_{\theta,l} \pm \sigma_{\theta,l}\right)$. (b) illustration of ϕ' and θ_a for two objective points (A and B, respectively).

We approximate $P\left(\omega_{k+1}^g \middle| \omega_{k+1}^c, \mathbf{S}_k\right)$ as a function of the *opening angle* ϕ', which is the angle difference between the two posts of the goal from the point of the ball.

$$P\left(\omega_{k+1}^g \middle| \omega_{k+1}^c, \mathbf{S}_k\right) = bnd\left(0; \frac{t_f - t_a}{\sec}; 3\right) bnd\left(0; \frac{\phi'}{\bar{\sigma}_\theta}; 1\right) u\left(\max_j\left(\mu_{r_j}\right) - \left|\mathbf{b}_k - \mathbf{g}'\right|\right) \tag{12}$$

Where $\bar{\sigma}_\theta$ is the mean of the angle variances of the available kicks, and u is the step function, which will become 1 if it is possible to reach the goal, considering the maximum mean distance reached by an available kick.

The remaining probabilities are even fuzzier, therefore we make use of coarser indicators. We approximate $P\left(\omega_{k+1+}^{g}\,|\,\omega_{k+1}^{c},\mathbf{S}_{k}\right)$ as:

$$P\left(\omega_{k+1+}^{g}\,|\,\omega_{k+1}^{c},\mathbf{S}_{k}\right)=v_{1}\left(t_{f}-t_{a}\right)\frac{\max\left(\mu_{r_{i}}\right)}{\left|\mathbf{b}_{k}-\mathbf{g}'\right|} \tag{13}$$

With a selected value of $v_{1}=0.3Hz$. For the calculation of $P\left(\omega_{k+}^{g}\,|\,\omega_{k+1}^{c},\mathbf{S}_{k}\right)$, we assume that a robot will not score in its own goal. Thus,

$$P\left(\omega_{k+1}^{g}\,|\,\omega_{k+1}^{c},\mathbf{S}_{k}\right)=0 \tag{14}$$

$$\Rightarrow P\left(\omega_{k+}^{g}\,|\,\omega_{k+1}^{c},\mathbf{S}_{k}\right)=P\left(\omega_{k+1+}^{g}\,|\,\omega_{k+1}^{c},\mathbf{S}_{k}\right) \tag{15}$$

The *future crossed score probability* $P\left(\omega_{k+1+}^{g}\,|\,\omega_{k+1}^{c},\mathbf{S}_{k}\right)$ is approximated as:

$$P\left(\omega_{k+1+}^{g}\,|\,\omega_{k+1}^{c},\mathbf{S}_{k}\right)\approx\frac{\tau}{t_{f}'}\frac{\max\left(\mu_{r_{i}}\right)}{\left|\mathbf{b}_{k}-\mathbf{g}'\right|} \tag{16}$$

Where a value of $\tau=0.3\sec$ is found to yield satisfactory results. Summarizing, $GSS\left(\mathbf{S}_{k}\right)$ may be calculated using equations (3), (4), (6), (7), (8), (9), (10), (11), (12), (13), (15), (16). Figure 1.b illustrates some of the variables used in the calculation of the *GSS*.

3.4 Decision Map

In the moment where a robot holds the ball, it has infinite possible BCA's that should be evaluated in order to decide for the best. We make a discretization of this space to be able to explore it. The discretization consists in a polar grid, where the distance is limited by the maximum distance that the ball can be kicked considering the available kicks, and the amount of time that the ball can be held. This grid is called *decision map* and consists in *M objective points* $\mathbf{p}_{m}$. Figure 2 shows some examples of decision maps. Accomplishing the generalized objective is defined as maximizing the expected *GSS* of the final position of the ball. The decision map is used to explore the space of feasible final positions of the ball after a BCA.

3.5 Objective and Ball Control Action Selection

If we leave $\mathbf{R}$ fixed, *GSS* may be seen as a function of the ball position $\mathbf{b}$, $GSS_{\mathbf{R}}\left(\mathbf{b}\right)$. Then, for each objective point $\mathbf{p}_{m}$ in the decision map, its *ideal score* π_{m} is calculated as:

$$\tilde{\pi}_{m}=GSS_{\mathbf{R}}\left(rep\left(\mathbf{p}_{m}\right)\right) \tag{17}$$

If $\mathbf{p}_m$ is out of the field, the ball will be replaced by a human referee in an arbitrary point (see [3] for details). Thus $rep(\mathbf{p}_m)$ is the expected ball replacement position if $\mathbf{p}_m$ is out of the field, and in other case it is equal to $\mathbf{p}_m$.

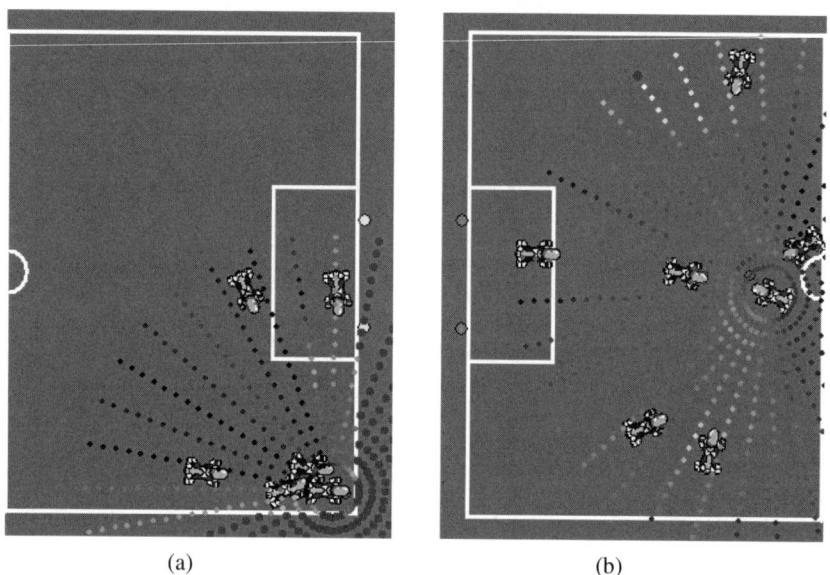

(a)   (b)

Fig. 2. Examples of decision maps and taken decisions, using the developed high-level strategy's simulator. The polar grid is around the red robot that holds the ball. Lighter points correspond to higher scores in the decision map. The big red points correspond to the selected points. (a) Defensive situation, the red robot holding the ball is blocked by two blue robots, thus points out of the field are selected (even preferring them over a possible but risky pass to the goalie), because its partner will be very close to the ball after the referee replace it. (b) Offensive situation, where a leading pass is selected, preferring it over a direct pass.

Taking into account objective points out of the field, the *rep* function has the nice effect, often observed in human players, that in some situations the robot may decide to kick the ball out of the field (see a simulated example en figure 2.a). Let us define the *filtered score* of the objective point $\mathbf{p}_m$ as:

$$\bar{\pi}_m = E\left(GSS_{\mathbf{R}}\left(rep(\mathbf{b})\right)\big|\mathbf{p}_m\right) \qquad (18)$$

Note that $\bar{\pi}_m \neq \tilde{\pi}_m$ since there is an uncertainty in the final position of the ball after performing any kick. To consider this uncertainty, $\bar{\pi}_m$ is calculated as the result of applying a Gaussian low-pass filter over each polar coordinate to $\tilde{\pi}_m$. Consequently, smooth maxima of $\tilde{\pi}_m$ are preferred over sharp ones.

For the sake of simplicity, to calculate $\tilde{\pi}_m$ and $\bar{\pi}_m$ we use $\mathbf{R}$ as the estimation of the poses of all the robots in the moment when the decision is taken. However, $\mathbf{R}$ will probably vary from the moment when the robot makes de decision of where to kick the ball, to the moment when the ball finally arrives to its final position **b**. We assume that the variation of **R** when time passes will always diminish the maxima of $\bar{\pi}_m$, which is a reasonable assumption since as time passes by, other robots may block the way from the robot holding the ball to any given objective point. Thus, for each objective point $\mathbf{p}_m$ in the decision map, we select the index $l(\mathbf{p}_m)$ of the required kick $\mathbf{k}_{l(\mathbf{p}_m)}$ as:

$$l(\mathbf{p}_m) = \arg\min_l \left(t_d \left(\mathbf{k}_l, \mathbf{p}_m, \mathbf{R} \right) \right) \tag{19}$$

Where $t_d(\mathbf{k}_l, \mathbf{p}_m, \mathbf{R})$ is the *required dribbling time* for kicking to the objective point $\mathbf{p}_m$, using the kick $\mathbf{k}_l$, and given the robots (teammates and opponents) poses **R**, and is calculated as:

$$t_d(\mathbf{k}_l, \mathbf{p}_m, \mathbf{R}) = \frac{|\Delta\mathbf{x}_{m,l}|}{v_R} + \frac{|\Delta\theta_{m,l}|}{\omega_R} \tag{20}$$

With $\Delta\mathbf{x}_{m,l}$ and $\Delta\theta_{m,l}$ being respectively the required displacement and rotation of the robot to perform $\mathbf{k}_l$ and reach $\mathbf{p}_m$, if the kick results in its expected values $\mu_{r,l}$, $\mu_{\theta,l}$. If the way from the robot to $\mathbf{p}_m$ is free, $\Delta\mathbf{x}_{m,l}$ just aims to adjust the distance to $\mathbf{p}_m$ (the robot moves in the axis between it and $\mathbf{p}_m$). If the way to $\mathbf{p}_m$ is blocked, $\Delta\mathbf{x}_{m,l}$ also considers an obstacle-avoiding component, which means that the robot will move to the closer free axis to $\mathbf{p}_m$, to the point at a distance $\mu_{r,l}$ of $\mathbf{p}_m$. In both cases, $\Delta\theta_{m,l}$ is calculated to align the robot with the needed angle to kick to $\mathbf{p}_m$ using $\mathbf{k}_l$. Once $l(\mathbf{p}_m)$ is selected, the minimum dribbling time, $t_d\left(\mathbf{k}_{l(\mathbf{p}_m)}, \mathbf{p}_m, \mathbf{R}\right)$, is used to punish the *final score* π_m of the objective point $\mathbf{p}_m$.

$$\pi_m = \begin{cases} \bar{\pi}_m - v_2 t_d\left(\mathbf{k}_{l(\mathbf{p}_m)}, \mathbf{p}_m, \mathbf{R}\right) & t_d\left(\mathbf{k}_{l(\mathbf{p}_m)}, \mathbf{p}_m, \mathbf{R}\right) < 3 \\ -1 & \sim \end{cases} \tag{21}$$

With a selected value of $v_2 = 0.12 Hz$. The condition in (21) ensures that only feasible points are considered (the robot is allowed to hold the ball for a maximum of 3 seconds [3]). The selected objective point $\mathbf{p}_m$ is selected as the one that maximizes π_m. Figure 2 shows some examples of the calculation of π_m in determined situations.

4 Results

As we have defined the decision making problem –in terms of maximizing the expected score advantage obtained– results should show that a team using the presented decision making framework is able to beat, getting as much score advantage as possible, another team using another decision making framework. The complete benefits of the system should be noticeable in a standard 4 versus 4 robots match. To test the system and to be able to present comprehensive results, we have developed a high-level strategy simulator, *UChile HL-SIM*, which runs our four-legged code in simulated AIBO's robots. Differing from our realistic simulator, UChilSim [6], UChile HL-SIM is not focused in realistic 3D visualization of scenes, neither in realistic dynamic interactions simulation, but it is intended for debugging specifically high-level strategy and behaviors. For that purpose, each simulated robot runs our strategy and actuation code, and the simulator brings them error-free perception and world modeling information. The result of the intended displacements of the robot is also simulated as error-free. Dynamic interactions between objects (ball, robots, and goals) are modeled in an idealized but comprehensive fashion (simplified 2D geometry). In order to provide a normal game flow, refereeing is also simulated, taking into account the RoboCup 2006 Four Legged League Competition Rules [3]. Figure 3 shows a screenshot of UChile HL-SIM.

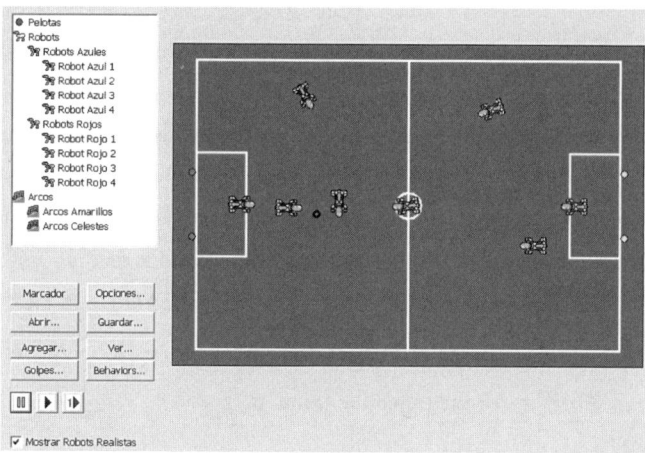

Fig. 3. UChile HL-SIM: High Level simulator used for testing the proposed strategy

For testing and validation purposes, we tested the described probabilistic-based decision making strategy, in 10 simulated matches between a team which uses this new strategy against a team which uses the decision making system proposed in [1] (probabilistic kick selection). It should be stressed that in both cases the only difference in the robot control software (UChile) is the strategy module. The matches were always won by the team running the proposed approach with an average goal difference of 8.5 (see Table 1 for details on the results). In the simulated matches, it was evident how some of the described improvements, as leading passes and clearing outside the field, appeared.

Table 1. Detailed results of the simulated matches. The score of the team running the proposed strategy goes first.

Match	Score	Goal Difference
1	12 - 5	7
2	14 - 3	11
3	6 - 4	2
4	7 - 3	4
5	9 - 2	7
6	15 - 2	13
7	14 - 1	13
8	16 - 3	13
9	9 - 3	6
10	10 - 1	9
Average	**11.2 - 2.7**	**8.5**

5   Conclusions

We have presented a novel approach for general decision making in robot soccer, based on the definition of a game situation score function, and the consequent discrimination of more specific objectives as passing and shooting to the goal.

The main advantage of the proposed system is that it relays only on the scored goals probability, and not in others conventionally taken into account such as pass success or ball possession time length. Additional advantages are the possibility of balancing the accuracy of the decision and the computational cost, by modifying the decision map resolution, and the consideration of the kicks' result uncertainty. The assumption of arbitrary models for the calculation of some of the probabilities should be corrected in future works, for example by using a machine learning approach.

The presented approach takes into account the uncertainty in the actions' results (kicks PDF's), but it does not take into account the uncertainty in the perception of the situations (vision, objects tracking and localization). We are planning to extend our work to make it able to consider the perceptual uncertainty.

The presented high-level strategy simulator is very well suited for testing high-level strategy and behaviors. We are planning to extend its capabilities in order to learn the parameters and morphology of the decision-making's algorithms inside the behaviors of different levels.

The preliminary results encourage us to continue developing our system. In particular, more factors may be included to better estimate some probabilities, but always keeping the conceptually hierarchized approach. On the other hand, some of the parameters used for calculating probabilities may be learned during a game, in order to adapt the strategy to the opponent characteristics.

References

1. Dodds, R., Vallejos, P., Ruiz-del-Solar, J.: Probabilistic kick selection in robot soccer. In: 3rd IEEE Latin American Robotics Symposium – LARS 2006 (CD Proceedings), Santiago, Chile, October 26 - 27 (2006)

2. Lakemeyer, G., Sklar, E., Sorrenti, D., Takahashi, T.: RoboCup 2006: Robot Soccer World Cup X. LNCS (LNAI). Springer, Heidelberg (to appear, 2007)
3. RoboCup Technical Committee. RoboCup Four-Legged League Rule Book (2006), http://www.tzi.de/4legged/bin/view/Website/WebHome
4. Stone, P., McAllester, D.: An Architecture for Action Selection in Robotic Soccer. In: Proceedings of the Fifth International Conference on Autonomous Agents, pp. 316–323 (2001)
5. Röfer, T., et al.: German Team 2006 Team Description Paper. In: Lakemeyer, G., Sklar, E., Sorrenti, D.G., Takahashi, T. (eds.) RoboCup 2006: Robot Soccer World Cup X. LNCS (LNAI), vol. 4434, Springer, Heidelberg (2007)
6. Zagal, J., Ruiz-del-Solar, J.: UCHILSIM: A Dynamically and Visually Realistic Simulator for the RoboCup Four Legged League. In: Nardi, D., Riedmiller, M., Sammut, C., Santos-Victor, J. (eds.) RoboCup 2004. LNCS (LNAI), vol. 3276, pp. 34–45. Springer, Heidelberg (2005)
7. Reis, L., Lau, N.: FC Portugal Team Description: RoboCup 2000 Simulation League Champion. In: Stone, P., Balch, T., Kraetzschmar, G.K. (eds.) RoboCup 2000. LNCS (LNAI), vol. 2019, pp. 29–40. Springer, Heidelberg (2001)
8. Kyrylov, V.: Balancing Gains, Risks, Costs, and Real-Time Constraints in the Ball Passing Algorithm for the Robotic Soccer. In: Lakemeyer, G., Sklar, E., Sorrenti, D.G., Takahashi, T. (eds.) RoboCup 2006: Robot Soccer World Cup X. LNCS (LNAI), vol. 4434, Springer, Heidelberg (2007)
9. Vallejos, P., Ruiz-del-Solar, J., Duvost, A.: Cooperative Strategy using Dynamic Role Assignment and Potential Fields Path Planning. In: 1st IEEE Latin American Robotics Symposium – LARS 2004, México City, México, October 28 – 29, pp. 48–53 (2004)

Multi-robot Cooperative Localization through Collaborative Visual Object Tracking

Zhibin Liu, Mingguo Zhao, Zongying Shi, and Wenli Xu

Department of Automation, Tsinghua University, Beijing, China
liu-zb04@mails.tsinghua.edu.cn

Abstract. In this paper we present an approach for a team of robots to cooperatively improve their self localization through collaboratively tracking a moving object. At first, we use a Bayes net model to describe the multi-robot self localization and object tracking problem. Then, by exploring the independencies between different parts of the joint state space of the complex system, we show how the posterior estimation of the joint state can be factorized and the moving object can serve as a *bridge* for information exchange between the robots for realizing cooperative localization. Based on this, a particle filtering method for the joint state estimation problem is proposed. And, finally, in order to improve computational efficiency and achieve real-time implementation, we present a method for decoupling and distributing the joint state estimation onto different robots. The approach has been implemented on our four-legged AIBO robots and tested through different scenarios in RoboCup domain showing that the performance of localization can indeed be improved significantly.

1 Introduction

Autonomous robots need to know their own positions within the environment, and the positions of other robots and moving objects in order to complete their tasks individually or in a cooperative way. However, it is not an easy job to accurately estimate the robot's own position as well as the state of the moving objects, because the information that robots receive through their sensors is inherently uncertain, and the control over their actuators is also inaccurate. Additionally, the estimating problem is made more difficult when there are unmodeled interactions or collisions between the robots or the robot haven't seen any distinct landmarks for a long time, which are especially typical in RoboCup domain.

During the soccer games, ball is the focus of robot's attention. Searching for ball, chasing and dribbling the ball and seeking for opportunities to kick a goal are usually the most important tasks of the robots. So, it is often the cases that there are few or no distinct landmarks in robot's sight. As a consequence, odometry errors accumulate as the time goes by without compensation, and the accuracy and reliability of localization result is seriously affected. However, just as mentioned above, the ball is usually in the sight of the robots. If there are some ways to improving the robots' self localization based on the ball information, much better performance can be expected.

U. Visser et al. (Eds.): RoboCup 2007, LNAI 5001, pp. 41–52, 2008.
© Springer-Verlag Berlin Heidelberg 2008

Considering all of the factors mentioned above, we enable the robots to share information and improve their self localization cooperatively by making them track the moving objects collaboratively and then refine their self localization results based on the common knowledge of the objects. We implement this idea on a team of four-legged AIBO robots to collaboratively track and estimate the state of a ball and use the ball information to improve their self localization simultaneously.

This paper is organized as follows. After introducing the related works in the next section, we present our multi-robot cooperative localization and ball tracking method in Section 3. Experimental results are given in Section 4, followed by conclusions drawn in Section 5.

2 Related Work

In Recent years, multi-robot cooperative localization has received increasing attention in robotics community. Most of the works on this problem are based on the assumption that the robots have abilities to detect and identify each other and estimate their relative positions [1, 2, 3, 4, 5]. They usually requires sophisticated image processing methods or adding artificial marks onto the robot platform. However, these may not be granted in many cases, especially in RoboCup competitions. Because adding distinctly colored marks to the AIBO robots is not allowed by rules, so it is quite difficult to identify the robots or accurately estimate their relative positions, due to the irregular and complex shape of the robot.

To our knowledge, the first work using moving objects' information to improve the robots' self localization is [6], in which Schmitt et al presented a method for enabling a team of robot to estimate their joint positions in a known environment and track the positions of autonomously moving objects (e.g., the ball). By using the ball's position estimations received from the other robots to correct the robot's own pose, the state estimators of different robots can cooperate to increase the accuracy and reliability of the estimation process. But this method is based on Kalman filtering, which is inefficient to track multiple ball hypotheses in face of false positive ball detection and sensor noises. In [7], Kwok and Fox presented a Rao-Blackwellised particle filtering method for estimating the robot's self location as well as the ball state. It provides a powerful model for multiple model object tracking and also allows the robot to infer where it is by observing the ball. However, the cooperative localization or object tracking problem are not discussed in their work. In another most recent work [8], Göhring presented an approach to estimate the position of objects tracked by a team of mobile robots by using the spatial relation of the objects respect to stationary landmarks detected in the same camera images, and then use these objects for better self localization. Though the objects' position estimation resulted by this method is robust to the localization errors of the robots, it requires that each robot can detect the ball as well as some landmarks at the same time. Moreover, only the static object model is considered in their work.

3 Multi-robot Cooperative Localization through Collaborative Object Tracking Using Particle Filters

In this section, we will first describe the multi-robot localization and object (ball) tracking problem using Bayes net. Then, through formal analysis, we will show how this joint state estimation problem can be factorized and tackled using particle filters. Finally, we conclude that the moving ball can serves as a bridge to realize cooperative localization, and an efficient distributed implementation method is presented.

3.1 Problem Description Using Bayes Net

Without loss of generality, we consider a system consisting of a pair of robots and a ball. Let $<b_k, r_k^1, r_k^2>$ denote the state of the system at time k. $b_k =< x_b, y_b, \dot{x}_b, \dot{y}_b, m_b >$ denotes the state of the ball in global coordinates, where $x_b, y_b, \dot{x}_b, \dot{y}_b$ represent ball location and velocity and $m_b \in \{0,1,2\}$ indicates the interaction model of the ball and robots. $m_b = 0$ means the ball is not grabbed by any teammate of the robots, while $m_b = 1$ or 2 indicates that the ball is grabbed by robot 1 or 2 respectively. $r_k^j =< x_r^j, y_r^j, \theta_r^j >$, $j = 1,2$, is the robot location and orientation on the field. Moreover, denote the observations of the ball and landmarks made by robot j as z_k^j, which is provided in relative bearing and distance.

A graphical model description of the state estimation problem of the system is given in Fig.1, where the nodes represent different random variables and the arrows indicate dependencies between these variables. The model shows the following relationships:

1) Robot-j's location at time k, r_k^j, only depends on the previous location r_{k-1}^j and the robot motion control u_{k-1}^j.
2) The observations z_k^j consist of $z_k^{j,L}$ and $z_k^{j,B}$, which describe landmark observations and ball observations respectively. $z_k^{j,L}$ only depend on the current robot location r_k^j (since the map of field is given); relative ball observations $z_k^{j,B}$ only depend on the current ball and robot positions.
3) The location, velocity and interaction model of the ball b_k typically depend on the previous ball state b_{k-1}, the actions of all robots, u_{k-1}^1, u_{k-1}^2, and the robots location r_k^1, r_k^2. However, just as the dashed arrows indicate, the existence of the relationship between robot location, motion control and ball state depends on which robot grab the ball, i.e. the component m_b in b_k. For example, if $m_b = 1$, i.e. ball is grabbed by robot 1, then the ball location is tightly attached to the robot location r_k^1 and the arrow from r_k^1 to b_k exists.

3.2 Factorizing the Joint State Space Posterior of Multi-robot Cooperative Localization and Object Tracking

Since the dependencies between different parts of the joint state space are defined based on Bayes net description, we can address the problem of filtering, which aims to compute the posterior over the joint state vector $<b_k, r_k^1, r_k^2>$ conditioned on all sensor measurements obtained so far, i.e. to compute:

$$p(b_k, r_k^1, r_k^2 \mid z_{1:k}^1, u_{0:k-1}^1, z_{1:k}^2, u_{0:k-1}^2) \qquad (1)$$

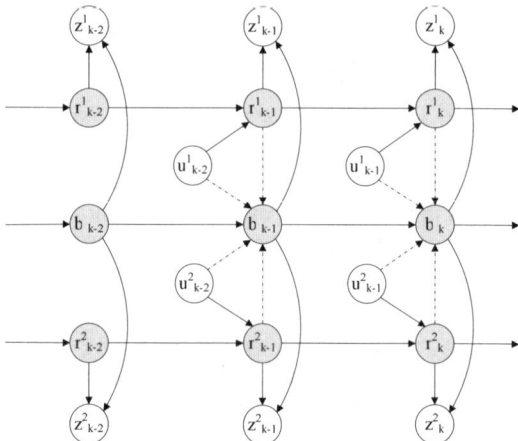

Fig. 1. Bayes net for multi-robot localization and ball tracking. The nodes in this graph represent the different parts of the dynamic system at consecutive time instances, and the edges represent dependencies between the individual parts of the state space. Filled circles indicate system state variable nodes, while the other circles stand for observations and motion control.

Based on the posterior estimation resulted from previous step, (1) can be written in a recursive form:

$$p(b_k, r_k^1, r_k^2 \mid z_{1:k}^1, u_{0:k-1}^1, z_{1:k}^2, u_{0:k-1}^2)$$
$$= \iiint_{b_{k-1}, r_{k-1}^1, r_{k-1}^2} p(b_k, r_k^1, r_k^2 \mid b_{k-1}, r_{k-1}^1, r_{k-1}^2, z_k^1, u_{k-1}^1, z_k^2, u_{k-1}^2) \cdot$$
$$p(b_{k-1}, r_{k-1}^1, r_{k-1}^2 \mid z_{1:k-1}^1, u_{0:k-2}^1, z_{1:k-1}^2, u_{0:k-2}^2) db_{k-1} dr_{k-1}^1 dr_{k-1}^2 \qquad (2)$$

The second term in (2) is the previous posterior, and the first term can be further factorized by employing the dependencies and independencies described in Bayes net model presented above. First, it can be factorized as:

$$p(b_k, r_k^1, r_k^2 \mid b_{k-1}, r_{k-1}^1, r_{k-1}^2, z_k^1, u_{k-1}^1, z_k^2, u_{k-1}^2)$$
$$= p(b_k \mid r_k^1, r_k^2, b_{k-1}, r_{k-1}^1, r_{k-1}^2, z_k^1, u_{k-1}^1, z_k^2, u_{k-1}^2) p(r_k^1, r_k^2 \mid b_{k-1}, r_{k-1}^1, r_{k-1}^2, z_k^1, u_{k-1}^1, z_k^2, u_{k-1}^2) \quad (3)$$

Since when $r_k^1, r_k^2, b_{k-1}, u_{k-1}^1, u_{k-1}^2$ are given b_k can be determined, (3) can be written as:

$$p(b_k, r_k^1, r_k^2 \mid b_{k-1}, r_{k-1}^1, r_{k-1}^2, z_k^1, u_{k-1}^1, z_k^2, u_{k-1}^2)$$
$$= p(b_k \mid r_k^1, r_k^2, b_{k-1}, z_k^1, u_{k-1}^1, z_k^2, u_{k-1}^2) p(r_k^1, r_k^2 \mid b_{k-1}, r_{k-1}^1, r_{k-1}^2, z_k^1, u_{k-1}^1, z_k^2, u_{k-1}^2) \quad (4)$$

Then, according to Bayes rule, we have:

$$p(b_k, r_k^1, r_k^2 \mid b_{k-1}, r_{k-1}^1, r_{k-1}^2, z_k^1, u_{k-1}^1, z_k^2, u_{k-1}^2)$$
$$\propto p(b_k \mid r_k^1, r_k^2, b_{k-1}, z_k^1, u_{k-1}^1, z_k^2, u_{k-1}^2) p(z_k^1, z_k^2 \mid r_k^1, r_k^2, b_{k-1}, r_{k-1}^1, r_{k-1}^2, u_{k-1}^1, u_{k-1}^2) \cdot$$
$$p(r_k^1, r_k^2 \mid b_{k-1}, r_{k-1}^1, r_{k-1}^2, u_{k-1}^1, u_{k-1}^2) \quad (5)$$

Since r_k^1 only depends on r_{k-1}^1, u_{k-1}^1, and r_k^2 only depends on r_{k-1}^2, u_{k-1}^2, the rightmost term in (5) can be factorized as:

$$p(r_k^1, r_k^2 \mid b_{k-1}, r_{k-1}^1, r_{k-1}^2, u_{k-1}^1, u_{k-1}^2) = p(r_k^1 \mid r_{k-1}^1, u_{k-1}^1) p(r_k^2 \mid r_{k-1}^2, u_{k-1}^2) \tag{6}$$

Exploiting the dependencies in the graph model, we know that z_k^1, z_k^2 are conditional independent from r_{k-1}^1, r_{k-1}^2, so the second term in (5) can be written as:

$$p(z_k^1, z_k^2 \mid r_k^1, r_k^2, b_{k-1}, r_{k-1}^1, r_{k-1}^2, u_{k-1}^1, u_{k-1}^2) = p(z_k^1, z_k^2 \mid r_k^1, r_k^2, b_{k-1}, u_{k-1}^1, u_{k-1}^2) \tag{7}$$

Substituting (6) and (7) into (5), we have:

$$
\begin{aligned}
& p(b_k, r_k^1, r_k^2 \mid b_{k-1}, r_{k-1}^1, r_{k-1}^2, z_k^1, u_{k-1}^1, z_k^2, u_{k-1}^2) \\
& \propto p(b_k \mid r_k^1, r_k^2, b_{k-1}, z_k^1, u_{k-1}^1, z_k^2, u_{k-1}^2) p(z_k^1, z_k^2 \mid r_k^1, r_k^2, b_{k-1}, u_{k-1}^1, u_{k-1}^2) \cdot \\
& \qquad\qquad\qquad\qquad\qquad p(r_k^1 \mid r_{k-1}^1, u_{k-1}^1) p(r_k^2 \mid r_{k-1}^2, u_{k-1}^2) \\
& = p(b_k, z_k^1, z_k^2 \mid r_k^1, r_k^2, b_{k-1}, u_{k-1}^1, u_{k-1}^2) p(r_k^1 \mid r_{k-1}^1, u_{k-1}^1) p(r_k^2 \mid r_{k-1}^2, u_{k-1}^2) \\
& = p(z_k^1, z_k^2 \mid r_k^1, r_k^2, b_k) p(b_k \mid r_k^1, r_k^2, b_{k-1}, u_{k-1}^1, u_{k-1}^2) p(r_k^1 \mid r_{k-1}^1, u_{k-1}^1) p(r_k^2 \mid r_{k-1}^2, u_{k-1}^2) \\
& = p(z_k^1 \mid r_k^1, b_k) p(z_k^2 \mid r_k^2, b_k) p(b_k \mid r_k^1, r_k^2, b_{k-1}, u_{k-1}^1, u_{k-1}^2) p(r_k^1 \mid r_{k-1}^1, u_{k-1}^1) p(r_k^2 \mid r_{k-1}^2, u_{k-1}^2)
\end{aligned}
\tag{8}
$$

Substituting (8) into (2) we get:

$$
\begin{aligned}
& p(b_k, r_k^1, r_k^2 \mid z_{1:k}^1, u_{0:k-1}^1, z_{1:k}^2, u_{0:k-1}^2) \\
& \propto p(z_k^1 \mid r_k^1, b_k) p(z_k^2 \mid r_k^2, b_k) \iiint\limits_{b_{k-1}, r_{k-1}^1, r_{k-1}^2} p(b_k \mid r_k^1, r_k^2, b_{k-1}, u_{k-1}^1, u_{k-1}^2) p(r_k^1 \mid r_{k-1}^1, u_{k-1}^1) \cdot \\
& \qquad p(r_k^2 \mid r_{k-1}^2, u_{k-1}^2) p(b_{k-1}, r_{k-1}^1, r_{k-1}^2 \mid z_{1:k-1}^1, u_{0:k-2}^1, z_{1:k-1}^2, u_{0:k-2}^2) db_{k-1} dr_{k-1}^1 dr_{k-1}^2
\end{aligned}
\tag{9}
$$

It is clearly shown in equation (8) that, the variable b_k (ball) serves as a linkage between the states of the robots, r_k^1 and r_k^2, which allows the information flow to travel from one robot to another and vice versa to achieve cooperative localization.

3.3 Particle Filtering for Joint Estimation

To implement the idea presented in the previous subsection, we have to specify the representation of the posterior distribution. We utilize particle filtering, which represent posteriors by sets of weighted samples, or particles:

$$S_k = \{< s_k^{(i)}, w_k^{(i)} > \mid 1 \le i \le N\}$$

where each particle $s_k^{(i)} = < b_k^{(i)}, r_k^{1(i)}, r_k^{2(i)} >$ and N is the total number of samples. The task is to generate samples distributed according to (1) based on the samples drawn from the posterior at k-1, denoted by S_{k-1}. We generate the different components of $s_k^{(i)}$ stepwise according to (8). In the first step, a sample $s_{k-1}^{(i)} = < b_{k-1}^{(i)}, r_{k-1}^{1(i)}, r_{k-1}^{2(i)} >$ is drawn from S_{k-1}, and then we draw new robot pose $r_k^{1(i)}$ and $r_k^{2(i)}$ for robot 1 and robot 2 respectively, according to:

$$r_k^{1(i)} \sim p(r_k^{1(i)} \mid r_{k-1}^{1(i)}, u_{k-1}^1) \qquad (10)$$

$$r_k^{2(i)} \sim p(r_k^{2(i)} \mid r_{k-1}^{2(i)}, u_{k-1}^2) \qquad (11)$$

This gives us $s_k^{(i)} = < \_, r_k^{1(i)}, r_k^{2(i)} >$, where $\_$ denotes uninitialized value. Then, the sample's ball state $b_k^{(i)}$ is estimated:

$$b_k^{(i)} \sim p(b_k^{(i)} \mid r_k^{1(i)}, r_k^{2(i)}, b_{k-1}^{(i)}, u_{k-1}^1, u_{k-1}^2) \qquad (12)$$

Finally, the importance weight of the sample $w_k^{(i)}$ is calculated as:

$$w_k^{(i)} = \eta \cdot p(z_k^1 \mid r_k^{1(i)}, b_k^{(i)}) p(z_k^2 \mid r_k^{2(i)}, b_k^{(i)}) \qquad (13)$$

where η is a normalizing factor which ensures all of the importance weights sum up to 1. Note that, since the observations z_k^j are composed of landmarks detection $z_k^{j,L}$ and ball detection $z_k^{j,B}$, equation (13) can be further factorized as:

$$
\begin{aligned}
w_k^{(i)} &= \eta \cdot p(z_k^{1,L}, z_k^{1,B} \mid r_k^{1(i)}, b_k^{(i)}) p(z_k^{2,L}, z_k^{2,B} \mid r_k^{2(i)}, b_k^{(i)}) \\
&= \eta \cdot p(z_k^{1,L} \mid r_k^{1(i)}) p(z_k^{1,B} \mid r_k^{1(i)}, b_k^{(i)}) p(z_k^{2,L} \mid r_k^{2(i)}) p(z_k^{2,B} \mid r_k^{2(i)}, b_k^{(i)})
\end{aligned} \qquad (14)
$$

where the facts that, when the robots' pose $r_k^{j(i)}$ and ball state $b_k^{(i)}$ are given the landmarks detection and ball detection are independent, and the landmark observation only depends on the robot location (as the map of the environment is already known), are used.

3.4 Distributed Implementation

There are different ways to implement our multi-robot cooperative localization and ball tracking method. The most intuitive one is to make every robot maintain and estimate the full joint state vector $< b_k, r_k^1, r_k^2 >$. But, unfortunately, it requires a large amount of particles to achieve satisfying estimation result, due to the high dimension of the joint state space. As the members of the robots increase, this problem becomes more serious. It will be computationally too demanding for the AIBO robots.

Here we present a distributed method in which each robot only have to estimate its own self location and ball state, then through communication the information are shared and cooperative localization and ball tracking is achieved.

Introducing two affiliated factors b_k^1 and b_k^2, which correspond to the ball estimation made by robot 1 and 2 respectively, the third term in equation (9) can be transformed as:

$$p(b_k \mid r_k^1, r_k^2, b_{k-1}, u_{k-1}^1, u_{k-1}^2)$$

$$= \iint_{b_k^1, b_k^2} p(b_k, b_k^1, b_k^2 \mid r_k^1, r_k^2, b_{k-1}, u_{k-1}^1, u_{k-1}^2) db_k^1 db_k^2$$

$$= \iint_{b_k^1, b_k^2} p(b_k \mid b_k^1, b_k^2) p(b_k^1, b_k^2 \mid r_k^1, r_k^2, b_{k-1}, u_{k-1}^1, u_{k-1}^2) db_k^1 db_k^2$$

$$= \iint_{b_k^1, b_k^2} p(b_k \mid b_k^1, b_k^2) p(b_k^1 \mid r_k^1, b_{k-1}, u_{k-1}^1) p(b_k^2 \mid r_k^2, b_{k-1}, u_{k-1}^2) db_k^1 db_k^2 \qquad (15)$$

This is attractive, since it allows each robot to estimate the ball state individually and then through an information fusion process the *team* ball estimation b_k is obtained.

Now, we present our method for performing the joint estimation in a distributed form: first each robot only estimate the joint state vector $< b_k^j, r_k^j >$, i.e. the individual ball state and self location, based on its own observations; then they send their estimation results to their teammates as well as receive the information coming from their teammates; whereafter, the team ball state b_k is estimated and the partial joint state $< b_k, r_k^j >$ maintained by each robot is finally updated.

Additionally, we enable each robot to use two kinds of ball model: egocentric ball model and global ball model. The egocentric ball model represents the ball state in robot-centric coordinate. It is more robust against global localization errors, and its uncertainty is much smaller than global ball state so that fewer particles are needed to represent its probabilistic distribution. The global ball model represents the ball state in global allocentric reference coordinate, which is used for communicating information to other robots. By associating egocentric ball state with robot's self location, the global ball state can be calculated. And the global ball estimation resulting from all robots are fused to get the *team* ball estimation. It is none other but this team ball estimation that enables the robots to act harmoniously and position themselves strategically on the field, and further to improve their self localization cooperatively.

Suppose, for any robot j, we use $n_{b,L}$ particles $^l b_k^{j(i)}$, $i \in [1, n_{b,L}]$ to represent the probabilistic distribution of egocentric ball state, and n_r particles $r_k^{j(\tau)}$, $\tau \in [1, n_r]$ for self localization. The procedure of the cooperative localization and ball tracking algorithm running on each robot j is as follows:

1) **Predict self location:** generate robot pose $r_k^{j(\tau)} \sim p(r_k^{j(\tau)} \mid r_{k-1}^{j(\tau)}, u_{k-1}^j)$;

2) **Update self localization using landmark measurement:** if any landmark is detected, the weights $w_k^{j(\tau)}$ of the samples $r_k^{j(\tau)}$ are calculated as $w_k^{j(\tau)} = p(z_k^{j,L} \mid r_k^{j(\tau)}) \cdot w_{k-1}^{j(\tau)}$, if the sum of the weights is smaller than a given threshold, substitute the low-weight samples by new samples randomly drawn according to the observations (similar to the sensor resetting method presented in [9]); else, if no landmark is detected, go to next step;

3) **Predict egocentric ball state:** if the ball is grabbed by robot j, the relative position of the ball $^l x_b^{j(i)}$, $^l y_b^{j(i)}$ in all particles $^l b_k^{j(i)}$ are set to zero; else, if the

ball is not grabbed by robot j, the state of the ball is predicted as $^lb_k^{j(i)} \sim p(^lb_k^{j(i)} \mid {}^lb_{k-1}^{j(i)}, u_{k-1}^1)$, and the weights of these particles are set to be equal;

4) **Egocentric ball update:** *a)* if the ball is neither seen nor grabbed by the robot, go to the final step; *b)* if the ball is grabbed, go to next step; *c)* if the ball is seen, update the weights of egocentric ball particles as $_b^lw_k^{j(i)} = p(z_k^{j,B} \mid {}^lb_k^{j(i)}) \cdot {}_b^lw_{k-1}^{j(i)}$, and then normalize these weights;

5) **Generate robot pose hypotheses:** calculate robot pose hypotheses by clustering the particles $r_k^{j(\tau)}$ of the self location, and then pick out n_h three robot pose hypotheses with the highest probabilities;

6) **Generate ball particles in the global coordinate:** associate the $n_{b,L}$ egocentric ball particles $^lb_k^{j(i)}$ with each of the robot pose hypotheses resulting from the last step to generate $n_h \times n_{b,L}$ particles in global coordinate $^gb_k^{j(i)}$; calculate the weights $_b^gw_k^{j(i)}$ of $^gb_k^{j(i)}$ by multiplying the weights of the egocentric ball particles and the probability of the robot pose hypotheses;

7) **Subsample global ball particles to obtain representative particles:** in this step we follow the method presented in [10], i.e. first the soccer field is recursively split into cells to form a quad-tree with a maximum depth of d_{max}; then for each cell a representative particle is calculated as the weighted average of the particles contained in that cell, and the weight of the representative particle is the sum-weight of the involved particles; finally, the n_{rep} representative particles with the highest weights are chosen with their weights normalized;

8) **Send/receive representative ball particles to/from teammates:** representative global ball particles resulting from the last step are sent to/received from the teammates through wireless communication;

9) **Calculate the entropy of robot pose estimation:** based on the particles and weights resulting from step 1) and 2), the entropy of the robot pose is calculated as a metric of the underlying uncertainty in the pose estimation;

10) **Estimate team ball location hypotheses based on the fused information:** *a)* if the entropy of robot pose is within a certain range, go to the final step; *b)* otherwise, if the entropy is higher than the given threshold, the robot will calculate the global ball position hypotheses by utilizing the received representative particles together with its own representative particles; these $n_{rep} \times n$ particles (n is the total number of robots) are classified into clusters following a clustering method similar to step 5), and the location hypothesis with the highest probability b_k^t is selected out;

11) **Update self localization using team ball estimation:** update the particle set representing the robot pose (resulting from step 1) and 2)) by calculating the weights as $w_k^{j(\tau)} \leftarrow p(z_k^{j,B} \mid r_k^{j(\tau)}, b_k^t) \cdot w_k^{j(\tau)}$; if the sum-weight of particles is smaller than a given threshold, substitute some of the lowest-weight samples by new samples drawn according to the team ball location b_k^t and ball observation $z_k^{j,B}$ (similar to the method used in Step 2));

12) **Particle weight normalization:** at this final step of iteration, the weights of the particles representing the robot pose $w_k^{j(\tau)}$ are normalized ensuring them sum up to 1.

4   Experiments and Results

To verify the effectiveness of the multi-robot cooperative localization and ball tracking method, we conduct experiments on Sony AIBO robots on the field of RoboCup Soccer 4-legged League. Our method is compared with the reference method presented in [11], which has been adopted by more than 6 different teams in RoboCup Soccer 4-legged League and its source code is publicly available.

We set up two scenarios in our experiments, both of which went on RoboCup 2006 Four-legged League soccer field. Throughout all of the experiments, the rule that '*the robot should not carry ball for longer than 3 seconds at one time*' is obeyed. The parameters in the algorithm presented in the previous section are set as: $n_{b,L} = 40$, $n_r = 100$, $n_h = 3$, $d_{\max} = 6$, $n_{rep} = 12$.

4.1   Scenario A: 1 Team of 4 Robots, RoboCup 2006 Field

In our first test scenario, a team of 4 robots are placed on the field, without opponents or other obstacles. This scenario represents a "best case" scenario to evaluate the performance of the two localization methods, because there is no collision between the robots, and the chances that the ball be occluded from the sight of the robots are smaller. During the experiment, robots NO.1~ NO.3 are expected to stay at the fixed points on the field (shown as the small solid black squares in Fig. 2(a)). They concentrate on tracking the ball, but also have to periodically distract their attention from it in order to see the landmarks and localize themselves. The positions of these three robots keep not changed, but their orientations can be adjusted by themselves so as to face directly to the moving ball and keep tracking of it. Robot NO.4 (its localization results are examined) can walk freely, chase the ball and carry ball to go toward 5 appointed locations (the small solid red squares labeled L#1~L#5 in Fig. 2(a)) sequentially. When the robot gets quite near to an appointed location, the experimenter would tap the back button on the robot manually so as to conduct it to change its destination and go to the next appointed location.

We compare the performance of the reference method with our cooperative localization method by running them in parallel on the robots and making them process exactly the same sensor data. The entropy [12, 13] of robot NO.4's pose estimations resulting from the two methods are automatically recorded in a log file one time per second by the robot. The ground truth of robot positions are obtained as follows: every ten seconds, the robot and localization algorithms pause; the real position of the robot is measured manually with the current localization results of the two methods recorded; then, by tapping the head button of the robot manually, it continue to move.

Fig.2 depicts the results for this scenario. At beginning, robot NO.4 was placed at the start point (small red solid circle in Fig. 2(a)), then every ten seconds its real position was recorded (magenta ☆ in Fig. 2(a)). The estimated positions of our

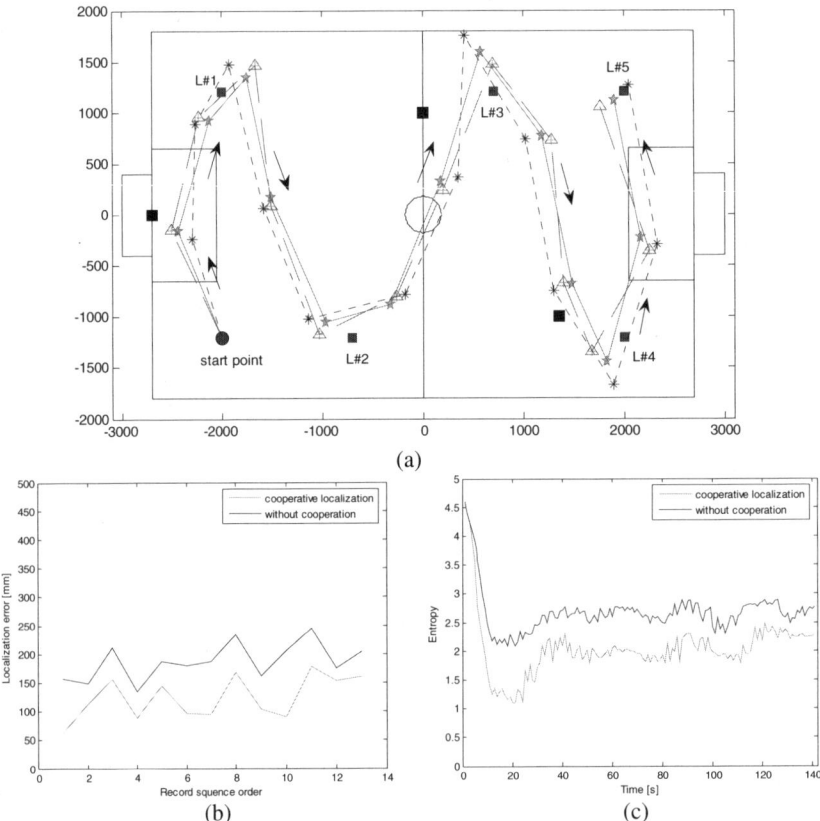

Fig. 2. Results for scenario A: (a) robot's real positions and the estimated positions of the two methods. (b) Localization errors. (c) Entropy of pose estimations at different time instances.

method and reference method are shown by green △ and blue * respectively. Note that, the colored lines linking the recorded positions in Fig. 2(a) do not stand for the trajectories, but only show the sequential order of the positions.

The localization errors, which are measured by the distance between the real positions and estimated positions, are shown in Fig. 2(b). In Fig. 2(c), the entropy of pose estimation resulting from the two methods is visualized. It is clear that both the localization errors and entropy of our cooperative localization method are significantly smaller than that of the reference method.

4.2 Scenario *B*: Real Game, 2 Teams of 6 Robots, RoboCup 2006 Field

This scenario aims to deal with the real game situation: two teams of robots play competitively on a standard RoboCup 2006 field. Through this scenario, we can examine that to what level our cooperative localization method can promote the performance of robot's self localization.

Since we only have 6 AIBO robots at hand, we can only assign 3 members for each team. Moreover, because it is a real soccer game, the ground truth of robots' positions are difficult to measure manually. But, the entropy of robot pose is a useful metric to measure the uncertainty of the robot's state. So, in this scenario, we evaluate the performance of the two methods by focusing on comparing the resulted entropy.

The experiment lasted for 5 minutes. Each team has 3 robots: goalie, defender and attacker. Since the attacker is the most active role in the team, it has more chances to collide with opponent robots when chasing the ball or seeking for opportunity to shoot. So, its self localization results can to some extent provide a "worst case" scenario for localization algorithms' performance. Therefore, we recorded the red attacker's pose estimation entropy during the game. The entropy was written into a log file by the robot at a rate of one record per second. Fig.3 depicts the results.

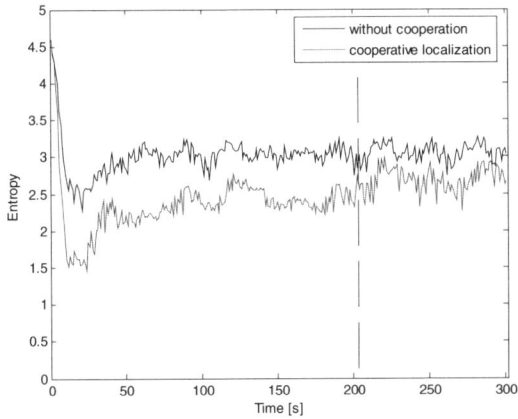

Fig. 3. Entropy of pose estimation at different time instances in Scenario B

It is clear that our cooperative localization method outperforms the reference method again. And, by examining the result carefully, we found that the difference lying in the performances of the two methods becomes less significant after the time instance labeled by the dashed line in Fig.3. This is due to the fact that there were more collisions between the attacker and the opponent robots and the ball was usually occluded by the robots. There were fewer chances for the red attacker's teammates to see the ball and provide accurate *team ball* estimation. So, the improvement made by utilizing ball information to promote self localization was affected, and became less significant. This is reasonable and consistent with our common knowledge.

5 Conclusion

In this paper we presented a probabilistic method for multi-robot cooperative localization and object tracking. By viewing the object and robots as a whole system, a Bayes net model is established to describe the joint state estimation problem. Then, through exploring the independences between different parts of the state space, we show how the posterior estimation of the joint state can be factorized and tackled using a particle filtering method. Finally, in order to improve computational efficiency and achieve real-time implementation, we distributed the joint state estimation task to different robots: first, each of the robot estimate their self location and ball state based on their own sensor data; then, by exchanging information between the robots the ball state estimation is refined; at last, each robot use the refined ball state estimation to correct their self localization.

By utilizing the proposed method, the state estimation modules of different robots can cooperate to increase the accuracy and reliability of their self localization and ball state estimation. It is capable of dealing with multiple hypotheses lying in the state of both the ball and robots. The experimental results show that the proposed method is effective and can evidently improve the robots' self localization in RoboCup domain.

References

1. Fox, D., Burgard, W., Kruppa, H., Thrun, S.: A Probabilistic Approach to Collaborative Multi-Robot Localization. Autonomous Robots 8, 325–344 (2000)
2. Kurazume, R., Hirose, S.: An Experimental Study of a Cooperative Positioning System. Autonomous Robots 8, 43–52 (2000)
3. Roumeliotis, S.I., Bekey, G.A.: Distributed Multirobot Localization. IEEE Transactions on Robotics and Automation 18(5), 781–795 (2002)
4. Madhavan, R., Fregene, K., Parker, L.E.: Distributed Cooperative Outdoor Multirobot Localization and Mapping. Autonomous Robots 17, 23–39 (2004)
5. Howard, A.: Multi-robot Simultaneous Localization and Mapping using Particle Filters. International Journal of Robotics Research 25, 1243–1256 (2006)
6. Schmitt, T., Hanek, R., Beetz, M., et al.: Cooperative Probabilistic State Estimation for Vision-based Autonomous Mobile Robots. IEEE Transactions on Robotics and Automation 18(5), 670–684 (2002)
7. Kwok, C., Fox, D.: Map-based Multiple Model Tracking of a Moving Object. In: Nardi, D., Riedmiller, M., Sammut, C., Santos-Victor, J. (eds.) RoboCup 2004. LNCS (LNAI), vol. 3276, pp. 18–33. Springer, Heidelberg (2005)
8. Göhring, D., Burkhard, H.D.: Multi Robot Object Tracking and Self Localization Using Visual Percept Relations. In: Proc. of the 2006 IEEE International Conference on Intelligent Robots and Systems (IROS 2006), Beijing, China (2006)
9. Lenser, S., Veloso, M.: Sensor Resetting Localization for Poorly Modelled Mobile Robots. In: Proc. of the IEEE International Conference on Robotics and Automation (ICRA 2000), San Francisco (2000)
10. Nisticò, W., Hebbel, M., Kerkhof, T., Zarges, C.: Cooperative Visual Tracking in a Team of Autonomous Mobile Robots. In: 10th International Workshop on RoboCup 2006 (Robot World Cup Soccer Games and Conferences) (2006)
11. Röfer, T., et al.: GermanTeam RoboCup 2005. Technical report (2005), http://www.germanteam.org/GT2005.pdf
12. Fox, D., Burgard, W., Thrun, S.: Active Markov Localization for Mobile Robots. Robotics and Autonomous Systems 25(3-4), 195–207 (1998)
13. Hoffmann, J., Spranger, M., Gohring, D., Jungel, M., Burkhard, H.: Further studies on the use of negative information in mobile robot localization. In: Proc. of the IEEE International Conference on Robotics and Automation (ICRA 2006), Orlando,Florida (2006)

Cooperative Object Localization Using Line-Based Percept Communication

Daniel Göhring

Institut für Informatik
LFG Künstliche Intelligenz
Humboldt-Universität zu Berlin, Germany
goehring@informatik.hu-berlin.de

Abstract. In this paper we present a novel approach to estimate the position of objects tracked by a team of robots. Moving objects are commonly modeled in an egocentric frame of reference, because this is sufficient for most robot tasks as following an object, and it is independent of the robots localization within its environment. But for multiple robots, to communicate and to cooperate the robots have to agree on an allocentric frame of reference. Instead of transforming egocentric models into allocentric ones by using self localization information, we will show how relations between different objects within the same camera image can be used as a basis for estimating an object's position. The spacial relation of objects with respect to stationary objects yields several advantages: a) Errors in feature detections are correlated. The error of relative positions of objects within a single camera frame is comparably small. b) The information is independent of robot localization and odometry. c) Object relations can help to detect inconsistent sensor data. We present experimental evidence that shows how two non-localized robots are capable to infer the position of an object by communication on a RoboCup Four-Legged soccer field.

1 Introduction

For a mobile robot to perform a task, it is important to model its environment, its own position within the environment and the position of surrounding objects, which can be other robots as well. This task is made more difficult when the environment is only partially observable. The task is characterized by extracting information from the sensor data and by finding a suitable internal representation (model).

In hybrid architectures [1], basic behaviors or skills, such as, e.g., following a ball, are often based directly on sensor data, e.g., the ball percept. Maintaining an object model becomes important if sensing resources are limited and a short term memory is required to provide an estimate of the object's location in the absence of sensor readings.

Modeling objects and localization is often decoupled to reduce the computational burden. In this loosely-coupled system, information is passed from localization to object tracking. The effect of this loose coupling is that the quality of

U. Visser et al. (Eds.): RoboCup 2007, LNAI 5001, pp. 53–64, 2008.
© Springer-Verlag Berlin Heidelberg 2008

the localization of an object in a map is determined not only by the uncertainty associated with the object being tracked, but also by the uncertainty of the observer's localization. In other words, the localization error of the object is the combined error of allocentric robot localization and the object localization error in the robot coordinate frame.

For this reason, robots often use an egocentric model of objects relevant to the task at hand, thus making the robot more robust against global localization errors. A global model is used for communicating information to other robots [11] or to commonly model a ball by many agents with Kalman filtering [2]. In all cases, the global model inherits the localization error of the observer.

We address this problem by modeling objects in allocentric coordinates from the start. Furthermore in RoboCup one can see a removal of more and more uniquely identifiable landmarks during the last years. The number beacons in the Four-Legged League has decreased from six to two beacons within four years. Therefore in this paper we focus on using *object to field line relations*.

In feature based belief modeling, features are extracted from the raw sensor data. We call such features *percepts* and they correspond directly to objects in the environment detectable in the camera images. In a typical camera image of a RoboCup environment, the image processing could, for example, extract the following percepts: *ball, line point, so called edgel, opponent player*, and *goal*. A *edgel* describes in our case the detection of a point that lies on a field line. Here it contains the position of that point relative to the robot in 2D space and the normal vector angle of the field line in this point, relative to the robot. Usually percepts are considered to be independent of each other to simplify computation, even if they are used for the same purpose, such as localization. Using the distance of features detected within a single camera image to improve Monte-Carlo Localization was proposed by [6]. The idea of using object relations has already been used in various map buildings tasks [12]. Using the spacial ordering of landmarks in the image for self localization was introduced by [14].

When modeling objects in relative coordinates, using only the respective percept is often sufficient. However, information that could help localize the object within the environment is not utilized. That is, if the ball was detected in the image right next to a goal, this helpful information is not used to estimate its position in global coordinates.

We show how using the object relations derived from percepts that were extracted from the same image yields several advantages:

Sensing errors. As the object of interest and the reference object are detected in the same image, the sensing error caused by joint slackness, robot motion, etc. becomes irrelevant as only the relation of the objects within the camera image matters.

Global localization. The object can be localized directly within the environment, independent of the quality of current robot localization.

Communication. Using object relations offers an efficient way of communicating sensing information, which can then be used by other robots to update their belief by sensor fusion.

Fig. 1. As testbed served the play field of the Sony 4-Legged League. Flags, goals, lines and the ball can be found on the field at fixed positions as shown.

1.1   Outline

We will show how relations between objects in camera images can be used for estimating the object's position within a given map and in which way different types of information can be used for this task. Particularly we want to analyze how information from non-uniquely identifiable objects as field lines can be incorporated. We will present experimental results using a Monte-Carlo Particle Filter to track the ball. Furthermore, we will show how communication between agents can be used to combine incomplete knowledge from individual agents about object positions, allowing the robot to infer the object's position from this combined data.

Our experiments were conducted on the color coded field of the *Sony Four Legged League* using the Sony Aibo ERS-7, which has a camera resolution of $208 * 160$ pixels YUV and an opening angle of only 55°.

2   Object Relation Information

In a RoboCup game, the robots permanently scan their environment for landmarks as there are flags, goals, the ball and field lines. We abstract from the algorithms which recognize the ball and the landmarks in the image as they are part of the image processing routines. In the next section we will give a brief overview over the information to be gained from each of the percepts, which is already described in more detail in [4].

2.1   Information Gained by Percepts

While describing percepts the robot receives, we want to distinguish uniquely identifiable objects from those which can not be uniquely identified. Fig. 2 gives an example of possible percepts the robot can perceive.

Fig. 2. Examples for what the robot can perceive: a) Flag and the ball, b) goal and the ball, c) a field-line and the ball

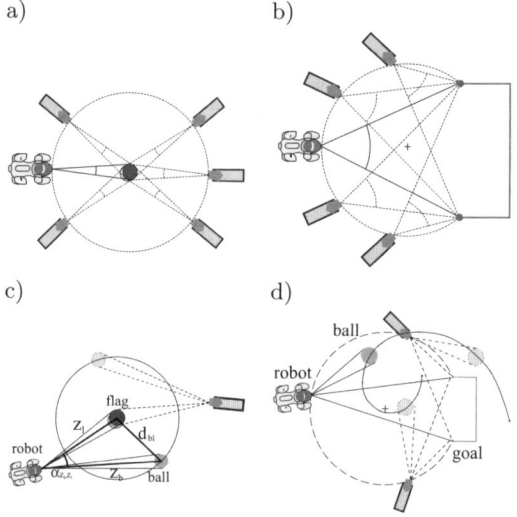

Fig. 3. Single percept: a) When a flag is seen, a circle containing all possible robot positions remains, b) The circle shows all possible positions for a seen goal. Light grey robot shapes represent possible robot positions; Two percepts in one image c) When seeing a flag and a ball in one image, the distance d_{bl} of the ball to the flag can be calculated; for all possible ball positions a circle remains, d) same situation for a seen goal and a ball, the spiral arc represents all possible ball positions.

Unique Objects. When seeing a two-colored flag, a robot actually perceives the left and right border of the flag, which enables it to calculate the distance and the angle to the flag (fig. 3 a). In the given approach this information is not being used for self localization but for calculating the distance from other objects as the ball to the flag. If a goal is detected, the robot can measure the angle between the left and the right goal-post. For a given goal-post angle the robot can calculate its distance and angle to a hypothetical circle center, whereas the circle includes the two outer points of the goal-posts and the point of the robot camera (fig. 3 b). If a ball is perceived, the distance to the ball and its direction relative to the robot can be calculated. So far all percepts we described are more or less unique, i.e., every percept can be assigned to a certain object in the robot's environment.

Table 1. Percept Statistics (Example)

Percentage of Percept Occurance in Images			
Ball	Flag	Goal	Line
35	52	22	59
Only Ball	Ball and Flag	Ball and Goal	Ball and Line
3	24	8	28

a)　　　　　　　　　　b)

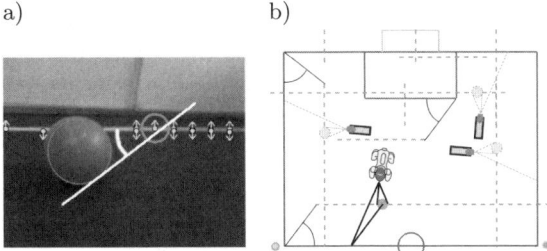

Fig. 4. a) A line-point (small circle) is seen together with a ball. The edgel data contains the position of the line on the field and the normal vector of the corresponding line (small arrows). Therefrom the ball distance to the line and the angle from the line-point to the ball can be calculated; b) Grey dotted lines represent all remaining possible ball positions on the field, when the ball-line percept is known; for better understanding the real robot position is drawn in detail, the other (schematic) robot drawings represent other possible robot and ball positions on the field.

Now we want to describe what kind of information can be gathered from field lines as an example for non-unique objects.

Non-unique Objects. On a soccer field, line information is a useful feature to reason about the robot position or about object positions. As can be seen in table 1, field lines are very often present in robot images - often together with other percepts as flags, goals or the ball. Now we will analyze which information line data can bring. We will investigate this question for the case, in which a ball and a line are seen simultaneously. When our robot perceives a line, it actually perceives one or more points of the line, together with the normal vector of the line, as fig. 4 a) shows. When a ball is seen in the same image as well, the robot can calculate a shape, containing all possible ball positions on the field. When, e.g., a ball is seen 10 cm away from an edgel in an angle of $45°$, all points on the field are possible ball positions, which lie in 10 cm and an $45°$ angle from any edgel of the field, see fig. 4 b). Or very easy speaking: when the robot sees a ball directly on a field line, then every point on any field line could be a possible ball position.

When there is more than just one line percept in the image, many approaches exist to combine different edgels to one or more different lines [10]. Every edgel

a)   b)

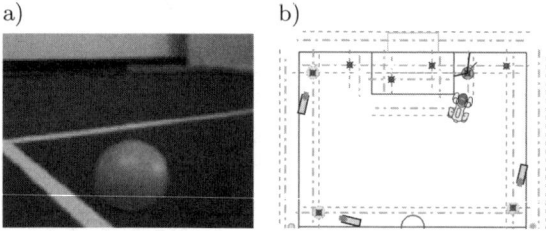

Fig. 5. a) The robot sees a ball next to two different lines; b) assuming, that the robot perceived for each of both lines an edgel percept, resulting in two solution sets (weak grey and strong orange dotted lines). One can calculate the remaining possible ball positions by cutting both solution spaces. The cut is then reduced by all solutions which would result in a wrong angle between the line percepts, related to the ball. Crossed circles (red) represent all remaining possible ball positions on the field.

can be treated as different evidence for modeling the object's position. This is especially interesting in situations where line crossings or other alignments of different lines occur at once. Being able to relate an object's position to different lines constraints the solution space for the remaining ball positions drastically, as fig. 5 b) shows. Every ball-edgel pair enables the robot to calculate possible ball positions on the field (the solution space) as in fig. 4 b). When seeing two or more of these ball-line pairs, the resulting ball positions can be calculated as the cut operation of all these solution spaces. The remaining solution space can be reduced even more, because the angle between the different edgels related to the ball is also measurable from the image (fig. 5 b)).

2.2   Dependencies between Percepts / Sensor Model

In this section we want to analyze the correlation between errors of different percepts within one image. For the sensor model, we measure the standard deviation σ^l by letting a robot take multiple images of certain scenes: a ball, a flag, a goal, a line and combinations of it. The standard deviation of distance differences and respectively angle differences of objects in the image relative to each other were measured as well. The robot is walking on the spot to keep the distance within the environment constant an to get noisy sensor data as during real robot motions. We found out that the angle errors of different percepts within the same image are strongly correlated which can be seen in fig. 6 in case of a ball and a flag.

3   Multi-agent Modeling

Now we want to describe a possible implementation of this approach. As the sensor data of our Aibo ERS-7 robot are not very accurate, we have to cope with a lot of sensor noise. Furthermore, the probabilistic distribution is not always unimodal, e.g., in cases where the observations lead to more than one

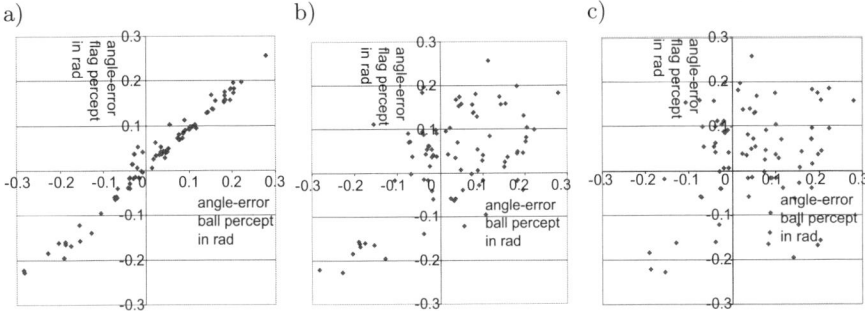

Fig. 6. The diagrams show the measured angle error to a ball and to a flag. The ball is located at a distance of 1.5m, the flag at 2.0m. a) flag and ball are seen in the same image, the angle errors between both are strongly correlated; b) the ball is seen 0.03 seconds earlier than the flag, lower correlation; c) the ball is seen 0.2 seconds earlier than the flag, almost no correlation between the angle errors.

solution for possible ball positions. This is why a simple Kalman filter would not be sufficient [7]. We chose an implementation using a Monte-Carlo Particle Filter because of its ability to model multimodal distributions and its robustness to sensor noise. Other approaches as Multi Hypothesis Tracking or Grid Based algorithms might work also [5]. As we cope with static situations this time only, we could abstract from network communication time and the delay after which percept relations were received.

3.1   Monte-Carlo Filter For Multi Agent Object Localization

Markov localization methods, in particular Monte-Carlo Localization (MCL), have proven their power in numerous robot navigation tasks, e.g., in office environments [3], in the museum tour guide Minerva [13], in the highly dynamic RoboCup environment [8], and outdoor applications in less structured environments [9]. MCL is widely used in RoboCup for object and self localization [7] because of its ability to model arbitrary distributions and its robustness towards noisy input data. The probability distribution is represented by a set of samples, called particle set. Each particle represents a pose hypothesis. The current belief of the object's position is modeled by the particle density, i.e., by knowing the particle distribution the robot can approximate its belief about the object state. Thereby the belief function $Bel(s_t)$ describes the probability for the object state s_t at a given time t. Using the Markov assumption and Bayes law, the belief function $Bel(s_t)$ depends only on the previous belief $Bel(s_{t-1})$, the last robot action u_{t-1} and the current observation z_t:

$$Bel^-(s_t) \longleftarrow \int \underbrace{p(s_t|s_{t-1}u_{t-1})}_{\text{process model}} Bel(s_{t-1})ds_{t-1} \tag{1}$$

$$Bel(s_t) \longleftarrow \eta \underbrace{p(z_t|s_t)}_{\text{sensor model}} Bel^-(s_t) \tag{2}$$

whereas η is a normalizing factor. Equation (1) shows how the *a priori* belief Bel^- is calculated from the previous Belief $Bel^-(s_{t-1})$. It is the belief prior the sensor data, therefore called prediction. As our robots do not perform any actions with the ball and as the situation is static, our propagation step becomes very simple or can be left out. In (2) the a-priori belief is updated by sensor data z_t, therefore called update step. Our update information is information about object relations as described in section 2.1. The data from fig. 6 can serve as a sensor model, telling the filter how accurate the sensor data are. The particles are distributed equally at the beginning, then the filtering process begins.

3.2 Monte-Carlo Localization, Implementation

Our hypotheses space for object localization has two dimensions for the position q on the field. Each particle s^i can be described as a state vector $\overrightarrow{s}^i$

$$\overrightarrow{s}^i = \begin{pmatrix} q^i_{x_t} \\ q^i_{y_t} \end{pmatrix} \tag{3}$$

and its likelihood p^i.

The likelihood of a particle p^i can be calculated as the product of all likelihoods of all gathered evidence [12]. From every given sensor data, e.g., a landmark l and a ball (with its distances and angles relative to the robot) we calculate the resulting possible ball positions relative to the landmark l. The resulting arc will be denoted as ξ^l. We showed in 2.1 that ξ^l has a circular form, when l is a flag, a spiral form, when l is a goal or a set of lines, when l is an edgel. The shortest distance δ^l from each particle $\overrightarrow{s}^i$ to ξ^l is our argument for a Gaussian likelihood function $\mathcal{N}(\delta, \mu, \sigma)$. The parameters of the Gaussian where derived experimentally. The sensor model being assumed to be Gaussian showed to be a good approximation in experiments. The likelihood is being calculated for all seen landmarks l and then multiplied:

$$p^i = \prod_{l \in L'} \mathcal{N}(\delta^l, 0, \sigma) \tag{4}$$

In cases without new evidence all particles get the same likelihood. After likelihood calculation, particles are resampled.

Multi Agent Modeling. Percept relations from every robot are communicated to every other robot. The receiving robot uses the communicated percept relations the same way it uses its own for likelihood calculation of each particle

- By communicating percept relations rather than particles, every robot can incorporate the communicated sensor data to calculate the likelihood of its particle set.

4   Experimental Results

As a test platform served the Aibo ERS-7. In the first reference algorithm, to which we compare our approach, two robots try to localize and to model the ball in an egocentric model. As a result each robot maintains a particle distribution for possible ball positions, resulting from self localization belief and the locally modeled ball positions. In our situation neither robot is able to accurately determine the ball position. Then the two robots communicate their particle distribution to each other. After communication each robot creates a new particle cloud as a combination of its own belief and the communicated belief (communicated particle distribution). We want to check how this algorithm performs in contrast to our presented algorithm in situations, where self localization is not possible, e.g., when every robot can only see one landmark and the ball.

In our first experiment, we placed both robots in front of different landmarks, one in front of a goal and one in front of a line with partially overlapping fields of view, such that both robots could see the ball (fig. 7 and 8).

The robots cannot accurately model the ball position when just communicating particle distributions, whereas by communicating percept relations the modeled

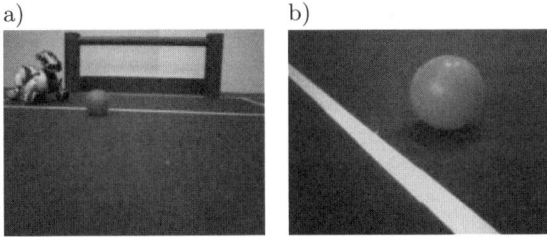

Fig. 7. Experiment A: views from two robots: a) robot A seeing a ball and a goal; b) robot B seeing a ball and a line

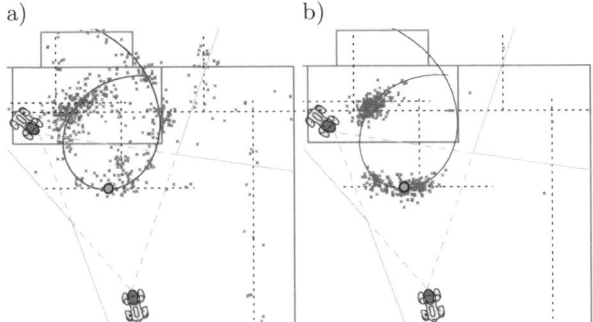

Fig. 8. Experiment A: the modeled ball position. a) both robots try to localize and have an egocentric ball model. After interchanging their particle distribution, the particle cloud does not convergence to a confined area; b) robots interchange the percept relations (ball-line and ball-goal), then updating and resampling the particle distribution. The distribution converges quickly to two small areas.

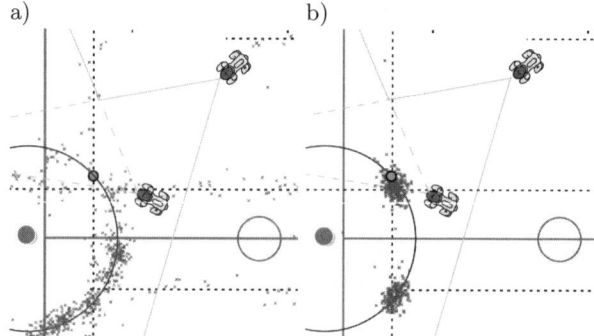

Fig. 9. Experiment B: The upper robot can see the ball and a line, the lower robot can see the flag only, because it is too far away to see the line. a) communicating particles does not lead to a convergence of the particles; b) communicating percept relations leads to convergence of the particle cloud to two small areas.

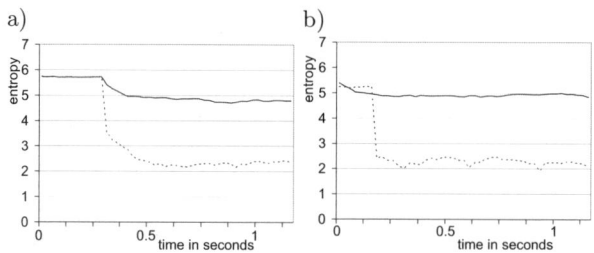

Fig. 10. Entropies over time for the experiments A and B. a) Experiment A: the dotted line represents the entropy for communicating percept relations, the continuous line represents the particle entropy for communicating percept distributions; b) Experiment B: dotted line represents ball particle entropy when communicating percept relations, continuous line for communicating particle distributions.

position converges to two small areas (fig. 8). Entropy measurement shows this quantitatively in fig. 10 a) - the entropy is much smaller, when percept relations are communicated. In Experiment B (fig. 9) one robot sees a flag, the other robot sees a line and both can see the ball. Again the robots try to localize and model the ball position egocentricly. Then they transform the egocentricly modeled ball particles into allocentric coordinates and communicate the particle distribution to each other. Simple particle communication does not lead to a convergence of the resulting particle distribution, whereas communicating percept relations leads to a convergence to a confined area (fig. 9 b)). Also entropy is much smaller again, when communicating percept relations 10 b).

5   Conclusion

Object relations, especially line information, in robot images can be used to localize objects in allocentric coordinates, e.g., if a ball is detected in an image next to a goal, the robot can infer something about where the ball is on the field. Percept relations can also help to detect image processing errors. Without having to be localized at all, it can accurately estimate the position of an object within a map of its environment using nothing but object relations. Furthermore, we were able to show how the process of object localization can be sped up by communicating object relations to other robots. Two non-localized robots are thus able to both localize an object using their sensory input in conjunction with communicated object relations.

Future Work. Future work will investigate, how the presented approach can be extended to moving objects, letting the robot infer not only about the position but also about the speed. Another interesting question would be, how redundant computation that is done by every agent can be distributed among the different robots while staying robust against system failures of different robots.

Acknowledgments

Program code used was developed by the GermanTeam. Source code is available for download at http://www.germanteam.org

References

1. Arkin, R.C.: Behavior-Based Robotics. MIT Press, Cambridge (1998)
2. Dietl, M., Gutmann, J., Nebel, B.: Cooperative sensing in dynamic environments. In: IEEE/RSJ International Conference on Intelligent Robots and Systems (IROS 2001), Maui, Hawaii (2001)
3. Fox, D., Burgard, W., Dellaert, F., Thrun, S.: Monte carlo localization: Efficient position estimation for mobile robots. In: Proceedings of the Sixteenth National Conference on Artificial Intelligence and Eleventh Conference on Innovative Applications of Artificial Intelligence (AAAI), pp. 343–349. The AAAI Press/The MIT Press (1999)
4. Göhring, D., Burkhard, H.-D.: Multi robot object tracking and self localization using visual percept relations. In: Proceedings of IEEE/RSJ International Conference of Intelligent Robots and Systems (IROS), pp. 31–36. IEEE (2006)
5. Gutmann, J.-S., Fox, D.: An experimental comparison of localization methods continued. In: Proceedings of the 2002 IEEE/RSJ International Conference on Intelligent Robots and Systems (IROS). IEEE (2002)
6. Kaplan, K., Celik, B., Mericli, T., Mericli, C., Akin, L.: Practical extensions to vision-based monte carlo localization methods for robot soccer domain. In: Bredenfeld, A., Jacoff, A., Noda, I., Takahashi, Y. (eds.) RoboCup 2005. LNCS (LNAI), vol. 4020. Springer, Heidelberg (to appear, 2006)
7. Kwok, C., Fox, D.: Map-based multiple model tracking of a moving object. In: Nardi, D., Riedmiller, M., Sammut, C., Santos-Victor, J. (eds.) RoboCup 2004. LNCS (LNAI), vol. 3276, pp. 18–33. Springer, Heidelberg (2005)
8. Lenser, S., Bruce, J., Veloso, M.: CMPack: A complete software system for autonomous legged soccer robots. In: AGENTS 2001: Proceedings of the fifth international conference on Autonomous agents, pp. 204–211. ACM Press (2001)

9. Montemerlo, M., Thrun, S.: Simultaneous localization and mapping with unknown data association using FastSLAM. In: Proceedings of the 2003 IEEE International Conference on Robotics and Automation (ICRA), pp. 1985–1991. IEEE (2003)
10. Nguyen, V., Martinelli, A., Tomatis, N., Siegwart, R.: A comparison of line extraction algorithms using 2d laser rangefinder for indoor mobile robotics. In: Proceedings of the IEEE/RSJ Intenational Conference on Intelligent Robots and Systems, IROS, Edmonton, Canada. IEEE (2005)
11. Schmitt, T., Hanek, R., Beetz, M., Buck, S., Radig, B.: Cooperative probabilistic state estimation for vision-based autonomous mobile robots. IEEE Transactions on Robotics and Automation 18(5), 670–684 (2002)
12. Thrun, S., Burgard, W., Fox, D.: Probabilistic Robotics. MIT Press, Cambridge (2005)
13. Thrun, S., Fox, D., Burgard, W.: Monte carlo localization with mixture proposal distribution. In: Proceedings of the National Conference on Artificial Intelligence (AAAI), pp. 859–865 (2000)
14. Wagner, T., Huebner, K.: An egocentric qualitative spatial knowledge representation based on ordering information for physical robot navigation. In: ECAI 2004, Workshop on Issues in Designing Physical Agents for Dynamic Real-Time Environments (2004)

Adaptive Recognition of Color-Coded Objects in Indoor and Outdoor Environments

Yasutake Takahashi[1], Walter Nowak[2], and Thomas Wisspeintner[2]

[1] Adaptive Machine Systems, Graduate School of Engineering, Osaka University,
Yamadaoka 2-1, Suita, Osaka, 565-0871, Japan
yasutake@ams.eng.osaka-u.ac.jp
http://www.er.ams.eng.osaka-u.ac.jp
[2] Fraunhofer Institute IAIS,
Schloss Birlinghoven, D-53754 Sankt Augustin, Germany
{walter.nowak,thomas.wisspeintner}@iais.fraunhofer.de
http://www.iais.fraunhofer.de

Abstract. To achieve robust color perception under varying light conditions in indoor and outdoor environments, we propose a three-step method consisting of adaptive camera parameter control, image segmentation and color classification. A controller for the intrinsic camera parameters is used to improve color stability in the YUV space. Segmentation is done to detect spatially coherent regions of uniform color belonging to objects in the image. Then, a probabilistic classification method is applied to label the colors by use of a Gaussian color distribution model. Experiments under combination of artificial and natural illuminations indoors and outdoors have been carried out. The results show the feasibility of this approach as well as the problems that occur under these highly diverse light situations. In particular we investigate the application in a RoboCup soccer scenario pointing toward future outdoor use.

Keywords: Color constancy, adaptive camera parameter control, segmentation, color classification, outdoor color vision.

1 Introduction

Computer vision has been for long identified to provide rich information about the environment for mobile robots. One of the major challenges in interpreting camera images is to cope with influences from illumination changes. In particular color information, which humans easily can classify, may appear very differently in the camera image. This is even more the case when the robot is supposed to work in indoor and outdoor environments.

In the context of RoboCup several soccer leagues use color coded environments in well defined light conditions. One long-term goal of RoboCup is to remedy these artificial regulations and cope with natural light and objects. Yet up to now, most RoboCup teams use manually calibrated color tables and fixed camera parameters which have to be tuned right before the games. This tedious

U. Visser et al. (Eds.): RoboCup 2007, LNAI 5001, pp. 65–76, 2008.

procedure only works when the environment does not undergo severe changes like direct sunlight and clouds during play.

We propose a combination of several techniques to approach this problem. First of all we continuously control the intrinsic camera parameters aiming for best possible color constancy. A first segmentation step based on Markov Random Fields leads to regions of uniform colors, which are then probabilistically classified to a set of discrete colors. Experiments convey that this method provides robust color classification under a variety of illumination conditions.

The paper is structured as follows. In chapter 2 we give an overview on related work regarding color constancy and color classification. Our approach is described in chapter 3 including camera parameter control, color segmentation and classification. In Chapter 4 we evaluate these steps by providing experimental results. Chapter 5 concludes with a discussion of the results and future work.

2   Related Work

A vast body of research has been done in the field of color constancy. Here the focus traditionally lies on the identification of illumination-independent descriptors for surfaces in a scene [1]. This includes the two tasks of determining the illuminant of a scene and mapping color values to a set of descriptors. An important instance of this general problem is the correction of colors in one image to match another image with some other illumination [2]. The available algorithms can be roughly divided into physics-based methods which try to model and explain the underlying physical processes, such as the dichromatic reflectance model, and statistics-based methods. These try to correlate distributions of colors under different illuminations, usually requiring enough colors to be present in the image. Examples are the diagonal method, gray-world methods, max-RGB [3], gamut mapping [4] as well as machine learning methods [5]. Another option is the use of chromaticity (normalized) color spaces such as YUV or HSI, where the brightness of each color is stored explicitly. It has been shown in [4] that methods using only chromaticity show similar performance as in full RGB space, but are more stable to shadows. Since the brightness also has an influence on the color value in the image due to camera characteristics and limitations, in our paper we go beyond those approaches by implementing an online method to keep the brightness in the image stable.

As several authors [6] [7] [8] point out, such algorithms have to deal with very big differences in the appereance of one color. Color regions may overlap, and the values of a set of colors change in various and highly nonlinear ways. This is particulary the case when the type of light changes, e.g. from natural to artificial light.

The application of mobile robots enables the use of online methods, such as online adaptation of camera parameters. The problem here lies in the nonlinear control and calculation of the control error. The concrete meaning of a camera parameter can vary much between different cameras, and is often not exactly specified. Even worse the relation between parameter value and effect is usually very non-linear. One possible remedy is to apply learning methods as in [9].

One approach for determining the required control errors is the use of reference colors. For example white can be used to set the camera's white balance parameter. Catadioptric camera systems are used by many RoboCup teams. Here a small colored ring can be laid around the camera objective so that it is always visible in the image and does not hide the view of the field itself.

Another approach is the use of semantic knowledge about the environment. This can be especially well applied in RoboCup environments due to the known field specifications. [10] and [11] first compute the pose of the robot on the field using mainly black-and-white information, then calculate the position of colored objects and finally adapt to the observed colors dynamically. [12] apply knowledge about the field and borders of objects. A comparable approach was done by [13] to recognize roads, assuming that a road is mostly flat and the car is driving on one. In our approach we avoid using such context information to account for broader application scenarios. Alternatively [14] use a three-step method to identify pixels usable as white reference; in contrast to our work they only control the white balance parameter and require white colors to be present in the environment.

Several papers investigate the benefits of first doing a segmentation or edge-detection step, and then classify the colors of whole segments. [15] and [16] use such methods with the main aims of improved color recognition and fast processing time. In [17] it was found that among different alternatives the method of choosing an unsure color ("maybe-color") to fit to its surrounding ones gave the best results.

Color classification by modeling color distributions as Gaussians was used by [18] and [19]. [20] shows that a discrete set of illumination conditions (bright, intermediate, dark) already improves the classification result significantly. In our paper we give further evidence for such a differentiation, as well as highlighting the benefits resulting from a continuous adaption.

3   Process for Robust Color Perception

The proposed method consists of the following major processing steps:

1. segment vertical lines into regions based on spatial uniformity of color
2. calculate mean color value for each segment
3. classify each segment to a set of color representatives
4. control camera parameters using reference colors

These steps will be described in more detail in the following subsections.

3.1   Image Segmentation

We adopt a boundary-based Markov Random Field method for line-based segmentation of an image. Markov Random Fields have been proposed as model

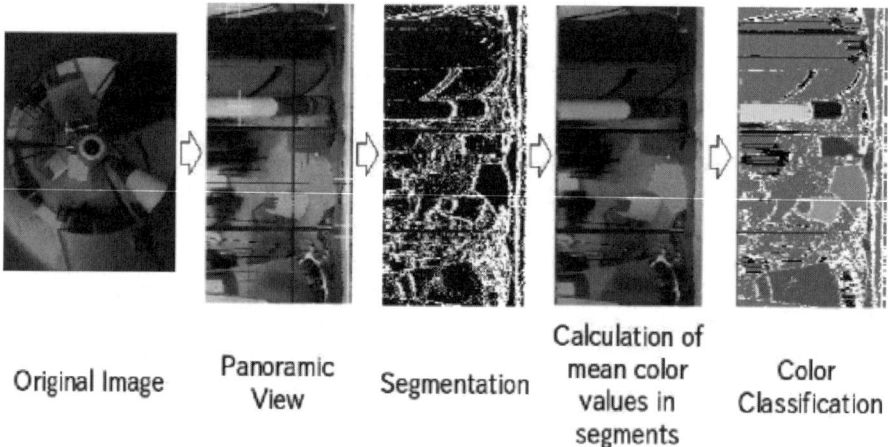

| Original Image | Panoramic View | Segmentation | Calculation of mean color values in segments | Color Classification |

Fig. 1. Image processing steps

for the visual field in the brain. Many variations of Markov Random Field have been developed, some of them have been already applied to the task of image segmentation. This method provides a sophisticated way to segment an image into spatially uniform regions. Here, we introduce the idea of the boundary-based Markov Random Field briefly.

First, we define an energy function $E(f, l|d)$ as follows:

$$E(f, l|d) = \frac{1}{2} \sum_i (f_i - d_i)^2 + \lambda \sum_i (1 - l_i)(f_{i+1} - f_i)^2 + \theta \sum_i l_i \qquad (1)$$

where d is an intensity process vector representing the observed image line. Each intensity value d_i is supposed to include some noise. f is the estimated value vector. l is called line process. l_i represents the discontinuity (edge) between the ith pixel and pixel $i + 1$. It is 1 if it is a boundary, and 0 otherwise.

The first term of equation 1 is for data fitting and tries to minimize the error of estimation. The second term is for smoothness in space. While there is no boundary specified by the line process l_i it tries to minimize the difference between conjunct pixels f_i and f_{i+1}. When the line process l_i is 1, i.e. there is a boundary, then no constraint between the conjunct pixels is introduced. The third term of 1 is a constraint on the number of boundaries. This means there should be less boundaries in the image than number of pixels.

In order to minimize the energy function (1), we use a hill-climbing method and introduce derivatives of f_i and l_i:

$$\frac{\partial f_i}{\partial t} = \lambda\{(1 - l_{i-1})(f_{i-1} - f_i) + (1 - l_i)(f_{i+1} - f_i)\} - (f_i - d_i) \qquad (2)$$

$$\frac{\partial l_i}{\partial t} = -l_i + H(\frac{\lambda}{2}(f_{i+1} - f_i)^2 - \theta) \qquad (3)$$

where $H(\cdot)$ is a step function. Each parameter is updated with the above derivative iteratively until it reaches convergence.

After obtaining the segmentation of the image, the mean value of each segment is calculated and used for color classification.

3.2 Color Classification

A probabilistic classification method based on Mahalanobis distances is applied to label colors. A Gaussian model of the color distribution for each color, consisting of mean vector and covariance matrix in YUV space must be provided beforehand. The mean color value of each segment is used to calculate the Mahalanobis distances with respect to the color distribution models. To illustrate this, one reference color is assumed to be a distribution with mean $\mu = (\mu_y, \mu_u, \mu_v)$ and covariance matrix Σ. The Mahalanobis distance between a color value $x = (x_y, x_u, x_v)$ and this distribution is defined as:

$$D_M(x) = \sqrt{(x - \mu)^T \Sigma^{-1}(x - \mu)} \qquad (4)$$

Each segment is associated to the reference color with the minimal Mahalanobis distance to the segment's mean value, provided this is below a predefined threshold. This threshold value offers a way to tune the ratio between unidentified pixels and false positive ones.

3.3 Camera Parameter Control

To achieve color constancy under different light conditions we use a set of PID controllers to modify relevant intrinsic camera parameters. To compensate for intensity changes of the illumination, Gain and Iris are being controlled, using the mean Y value of a white reference color visible in the camera image.

To account for changes of the type of illumination the two White Balance channels are being controlled by using the mean U and V values of the white reference color. Furthermore, an additional red reference color is used to control Saturation in the same way using the calculated mean saturation value.

The parameters of each PID controller can be tuned by analysis of the step response switching from dark to bright illumination, from bright to dark and between different types of illumination.

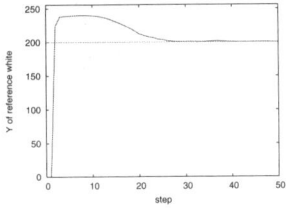

Fig. 2. Step response with optimized control parameters for the brightness controller

Fig. 2 shows a step response with optimized control parameters for the Brightness controller, switching illumination from dark to bright. In this example, the controller only operates on every fifth step.

After the PID parameter optimization, a proper desired value for each controller has to be determined. This can be done by qualitative analysis of the YUV color distribution. In a distribution optimal for color classification the colors should be widely spread in the color space. On the other hand colors should not be over-saturated, i.e. the distribution should not reach the borders of the YUV cube. Furthermore the center of the distribution should lie in the center of the given YUV space.

4   Experiments

To evaluate the performance of our approach, we conducted several experiments in indoor and outdoor environments under different light conditions.

As basis for our experiments we used a VolksBot robot[21] with a catadioptric camera system. A variant is used in the AIS/BIT RoboCup MSL team. Processing was done on an onboard laptop with a Pentium M 1.8 GHz processor. The complete vision processing takes less then 20 ms for one image, depending on the number and sizes of recognized color regions. Thus the algorithm works in real-time.

The vision system consists of a Sony DSW 500 camera looking into a hyperbolic mirror, thus producing 360 degree panoramic YUV images. A ring of white and red paper is fixed around the camera lens, see Fig. 1 left. This ring provides

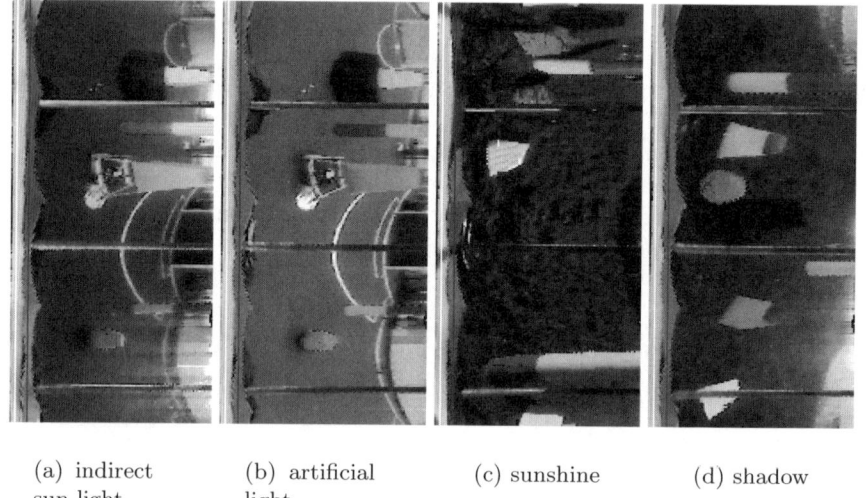

(a) indirect   (b) artificial   (c) sunshine   (d) shadow
sun light   light

Fig. 3. Captured panoramic images with PID controller under four light conditions

the reference colors used by the camera parameter controller without interfering with the view of the scene itself.

The colored objects used for color classification are mainly taken from the RoboCup scenario, in particular blue and yellow goals, a green field with white lines, cyan and magenta markers, a red ball and black robots. For outdoor tests we used a subset of these.

To account for a broad range of light conditions, we regard the following situations:

1. Indoor: only artificial light of one light source (630 Lux)
2. Indoor: mixed artificial and indirect sun light (1370 Lux)
3. Indoor: only indirect sun light (500 Lux)
4. Outdoor: camera and objects in direct sun light (97,000 Lux)
5. Outdoor: camera and objects in shadow (2,550 Lux)

Fig. 3 shows the camera images under these different light conditions.

4.1 Color Constancy

Fig. 4 shows the merged distributions of YUV values obtained from the color objects under the three indoor light conditions. The upper left image shows the 3D-view and the upper right image shows the 2D-projection on the UV-plane using PID control of the camera parameters. The lower left and right images

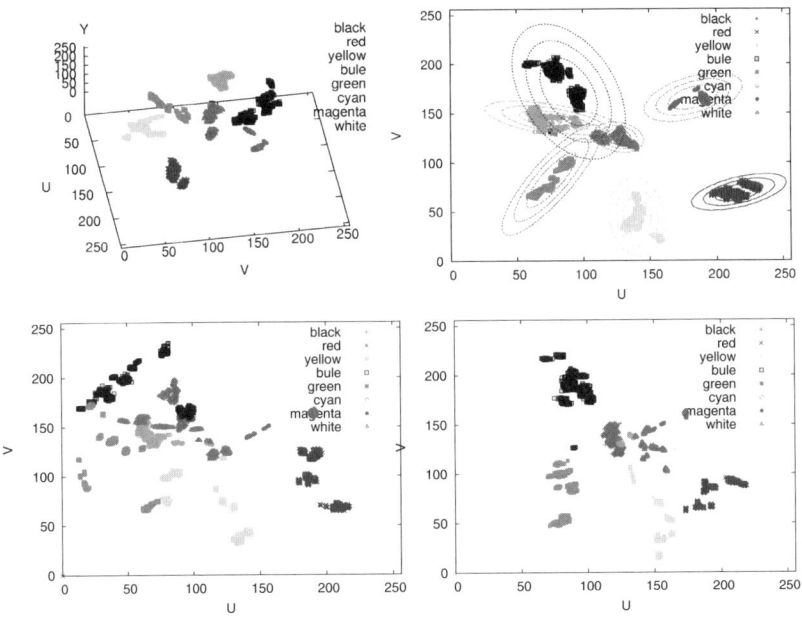

Fig. 4. Plots of YUV color distribution indoors

Table 1. Means $\mu_{y,u,v}$ and standard deviations $\sigma_{y,u,v}$ of typical colors in YUV space under various light condition in indoor/outdoor environment

	Indoor									Outdoor		
	PID			No PID			Embedded			PID		
red μ	127.9	216.8	69.0	133.2	179.6	87.9	162.2	194.1	81.1	134.9	213.6	60.6
red σ	15.6	8.9	4.6	44.6	53.2	33.2	17.7	12.6	10.8	8.8	6.3	10.1
yellow μ	187.0	140.0	44.0	206.7	118.4	81.6	219.1	148.9	51.9	189.5	163.5	24.1
yellow σ	26.0	5.2	9.8	34.8	21.3	34.1	20.5	10.1	30.9	16.3	7.3	11.5
blue μ	63.1	89.2	172.1	98.6	64.3	191.6	101.4	89.8	192.0	83.2	84.1	186.1
blue σ	24.1	5.2	9.8	43.8	26.8	23.1	32.2	8.7	14.0	22.8	21.6	23.8

illustrate the distributions with fixed camera parameters and embedded camera control in the UV-plane respectively.

It is shown how the color drift is greatly reduced when applying the PID controller, while the colors drift heavily for the other two approaches. Not using PID control, the drift can be so big that the color distributions overlap, making it impossible to deduce from one YUV value a unique color class.

It should be noted that also with the PID control the colors significantly drift depending on changes of direction of illumination, changes of intensity, changes of the ratio of different kinds of illumination or reflections. Still, the PID control provides better stability and spreading of the distributions compared to the other approaches evaluated. The system provides highest color constancy, when both, the object and the reference colors rings are exposed to equal light conditions.

The biggest change in color value occurs without any parameter control. It is interesting that not only the brightness Y, but also U and V change when illumination intensity decreases. This indicates that a simple brightness normalization is not enough to identify colors robustly, giving reason to also control the saturation value of the camera.

Table 1 lists mean values and standard deviations for three object colors under diverse light conditions with different control methods for the indoor and outdoor experiments. The table only shows the standard deviation in the direction of Y, U and V axes. Comparing the standard deviations of the different approaches for a certain color, like e.g. red, the lower drift of the PID control method can be confirmed. It is apparent that the standard deviation with PID control is nearly always smaller than for the others.

The conditions change drastically when going from indoor to natural light conditions outdoors. The image in Fig. 5 shows the YUV distributions of the object colors and their projection into the UV-plane for the outdoor experiment in direct sunshine and shadow. The reason for the observable higher color drift lies in the fact of having a huge intensity range from 2550 to 97,000 Lux between shadow and direct sunlight.

Especially in the experiment undertaken in direct sunlight these extreme illumination ranges occur in a single scene, having the same objects partly exposed to direct sunshine and partly lying in its own shadow. Furthermore, the drift in

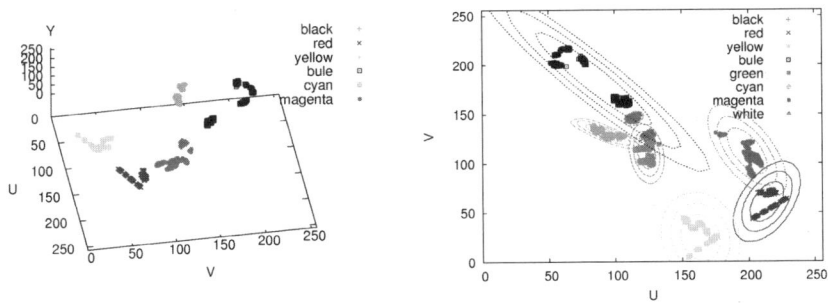

Fig. 5. YUV color distributions in outdoor environment

color space is highly depending on the pose of the objects relative to the light source and to the camera. Also surface properties of the objects have a bigger influence here. Related to this huge illumination range, one can also see the need for the saturation control, as saturation of an object color decreases for dark and bright situations significantly.

The red color for example has a much lower saturation V value when the camera is outdoors. We assume that not the kind of illumination, but the high intensity and the limited color range of the camera sensor is responsible for this effect. The color is much brighter outdoors; since the YUV space is of conical shape, this results in a lower range of possible saturation values.

Still, the color distributions do not overlap, which indicates that a proper classification of colors should be possible. This will be evaluated in the next section.

4.2 Classification Results

First we have a look at the mean values and standard deviations of the reference color distributions, since these form the basis for the color classification step. In Fig. 4 upper right and Fig. 5 right, these regions are drawn as ellipses around the distribution of the respective colors. The images show the projection of the 3-dimensional ellipsoids on the UV-plane. The drawn ellipses represent the borders of 2-σ, 3-σ and 4-σ areas. Since the ellipsoids differ in the Y-values they cover, they do actually not overlap in the way the image of their projections may suggest.

The drawing of the ellipsoids indicates what threshold to use to retrieve a binary classification result. Since the majority of already measured color pixels should be included, at least $3\,\sigma$ seems reasonable. For a more robust identification towards unexpected light variations a higher value could be useful. But as this can result in more false positive classifications, a compromise must be found. For our classification experiments we chose a threshold of $3\,\sigma$.

Fig. 6 shows the classification results in multiple light conditions. In general for all situations the classification algorithm shows a good performance. In

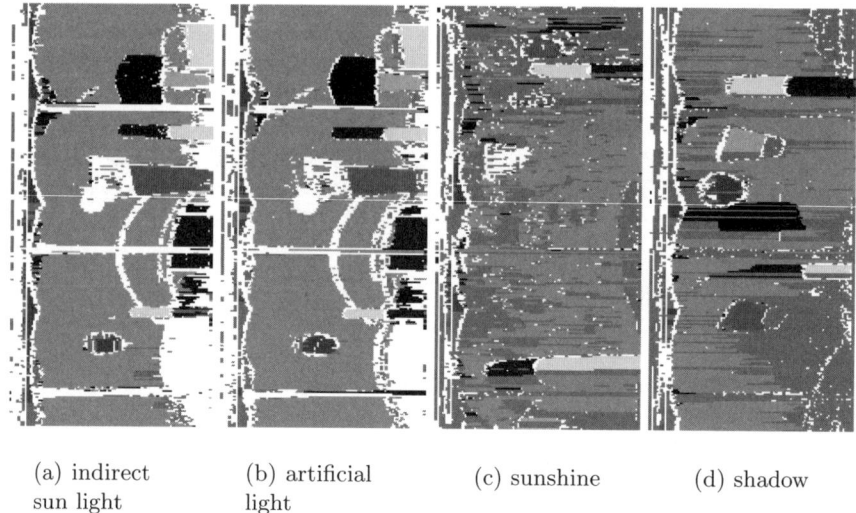

| (a) indirect sun light | (b) artificial light | (c) sunshine | (d) shadow |

Fig. 6. Classification results with PID controller in indoor and outdoor environments

the indoor environment all objects are recognized with their correct colors, and only very few false positive classifications exist. In the outdoor environment the method has problems with very dark pixels resulting from the high differences in intensities due to sunlight and shadow.

5 Conclusion

We have presented a robust color perception method including PID controller of camera parameters, segmentation by Markov Random Field, and classification based on Mahalanobis distance. The PID controller provided enough color constancy to be able to fuse the distribution under different light conditions and to generate reference color models for indoor and outdoor. These reference color models have shown to provide a robust basis for color classification under a variety of different light conditions. The big difference of color distribution in indoor and outdoor suggest the use of separate reference models for these two cases.

The vast illumination range occurring outdoors within one image has shown the physical limitations of the camera. Future work will investigate possible use of attention based mechanisms to choose from different parameter sets for different light situations.

References

1. Maloney, L., Wandell, B.: Color constancy: A method for recovering surface spectral reflectance. Journal of the Optical Society of America A 3, 29–33 (1986)
2. Forsyth, D.: A novel algorithm for color constancy. Int. Journal of Computer Vision 5(1), 5–36 (1990)

3. Barnard, K., Funt, B., Cardei, V.: A comparison of computational colour constancy algorithms; part i: Methodology and experiments with synthesized data. IEEE Transactions on Image Processing 11(9), 972–984 (2002)
4. Finlayson, G., Hordley, S.: Improving gamut mapping color constancy. IEEE Transactions on Image Processing 9(10), 1774–1783 (2000)
5. Austermeier, H., Hartmann, G., Hilker, R.: Colour-calibration of a robot vision system using self-organising feature maps. In: Vorbrüggen, J.C., von Seelen, W., Sendhoff, B. (eds.) ICANN 1996. LNCS, vol. 1112, pp. 257–262. Springer, Heidelberg (1996)
6. Mayer, G., Utz, H., Kraetzschmar, G.: Playing robot soccer under natural light: A case study. In: Polani, D., Browning, B., Bonarini, A., Yoshida, K. (eds.) RoboCup 2003. LNCS (LNAI), vol. 3020, pp. 238–249. Springer, Heidelberg (2004)
7. Austin, D., Barnes, N.: Red is the new black - or is it? In: Proceedings of the 2003 Australian Conference on Robotics and Automation, Brisbane, Australia (2003)
8. Funt, B., Barnard, K., Martin, L.: Is machine colour constancy good enough? In: Burkhardt, H., Neumann, B. (eds.) ECCV 1998. LNCS, vol. 1407, Springer, Heidelberg (1998)
9. Grillo, E., Matteucci, M., Sorrenti, D.G.: Getting the most from your color camera in a color-coded world. In: Nardi, D., Riedmiller, M., Sammut, C., Santos-Victor, J. (eds.) RoboCup 2004. LNCS (LNAI), vol. 3276, pp. 221–235. Springer, Heidelberg (2005)
10. Gunnarsson, K., Wiesel, F., Rojas, R.: The color and the shape: Automatic online color calibration for autonomous robots. In: Bredenfeld, A., Jacoff, A., Noda, I., Takahashi, Y. (eds.) RoboCup 2005. LNCS (LNAI), vol. 4020, pp. 347–358. Springer, Heidelberg (2006)
11. Heinemann, P., Sehnke, F., Streicher, F., Zell, A.: Towards a calibration-free robot: The act algorithm for automatic online color training. In: Lakemeyer, G., Sklar, E., Sorrenti, D.G., Takahashi, T. (eds.) RoboCup 2006: Robot Soccer World Cup X. LNCS (LNAI), vol. 4434. Springer, Heidelberg (2007)
12. Jüngel, M., Hoffmann, J., Lötzsch, M.: A real-time auto-adjusting vision system for robotic soccer. In: Polani, D., Browning, B., Bonarini, A., Yoshida, K. (eds.) RoboCup 2003. LNCS (LNAI), vol. 3020, pp. 214–225. Springer, Heidelberg (2004)
13. Thrun, S., Montemerlo, M., Dahlkamp, H., Stavens, D., Aron, A., Diebel, J., Fong, P., Gale, J., Halpenny, M., Hoffmann, G., Lau, K., Oakley, C., Palatucci, M., Pratt, V., Stang, P., Strohband, S., Dupont, C., Jendrossek, L.E., Koelen, C., Markey, C., Rummel, C., van Niekerk, J., Jensen, E., Alessandrini, P., Bradski, G., Davies, B., Ettinger, S., Kaehler, A., Nefian, A., Mahoney, P.: Stanley, the robot that won the darpa grand challenge. Journal of Field Robotics 23(9), 661–692 (2006)
14. Chikane, V., Fuh, C.S.: Automatic white balance for digital still cameras. Journal for Information Science and Engineering 22, 497–509 (2006)
15. de Cabrol, A., Bonnin, P., Costis, T., Hugel, V., Blazevic, P., Bouchefra, K.: A new video rate region color segmentation and classification for sony legged robocup application. In: Bredenfeld, A., Jacoff, A., Noda, I., Takahashi, Y. (eds.) RoboCup 2005. LNCS (LNAI), vol. 4020, pp. 436–443. Springer, Heidelberg (2006)
16. Lovell, N.: Illumination independent object recognition. In: Bredenfeld, A., Jacoff, A., Noda, I., Takahashi, Y. (eds.) RoboCup 2005. LNCS (LNAI), vol. 4020, pp. 384–395. Springer, Heidelberg (2006)
17. Stanton, C., Williams, M.A.: A novel and practical approach towards color constancy for mobile robots using overlapping color space signatures. In: Bredenfeld, A., Jacoff, A., Noda, I., Takahashi, Y. (eds.) RoboCup 2005. LNCS (LNAI), vol. 4020, pp. 444–451. Springer, Heidelberg (2006)

18. Kikuchi, T., Umeda, K., Ueda, R., Jitsukawa, Y., Osumi, H., Arai, T.: Improvement of color recognition using colored objects. In: Bredenfeld, A., Jacoff, A., Noda, I., Takahashi, Y. (eds.) RoboCup 2005. LNCS (LNAI), vol. 4020, pp. 537–544. Springer, Heidelberg (2006)
19. Anzani, F., Bosisio, D., Matteucci, M., Sorrenti, D.G.: On-line color calibration in non-stationary environments. In: Bredenfeld, A., Jacoff, A., Noda, I., Takahashi, Y. (eds.) RoboCup 2005. LNCS (LNAI), vol. 4020, pp. 396–407. Springer, Heidelberg (2006)
20. Sridharan, M., Stone, P.: Towards illumination invariance in the legged league. In: Nardi, D., Riedmiller, M., Sammut, C., Santos-Victor, J. (eds.) RoboCup 2004. LNCS (LNAI), vol. 3276, pp. 196–208. Springer, Heidelberg (2005)
21. Wisspeintner, T., Nowak, W., Bredenfeld, A.: Volksbot - a flexible component-based mobile robot system. In: Bredenfeld, A., Jacoff, A., Noda, I., Takahashi, Y. (eds.) RoboCup 2005. LNCS (LNAI), vol. 4020, pp. 716–723. Springer, Heidelberg (2006)

3D Tracking by Catadioptric Vision Based on Particle Filters*

Matteo Taiana[1,2], José Gaspar[1], Jacinto Nascimento[1], Alexandre Bernardino[1], and Pedro Lima[1]

[1] IST, Instituto de Sistemas e Robótica – Lisboa, Portugal
{mtajana,jag,jan,alex,pal}@isr.ist.utl.pt
[2] Politecnico di Milano, Italy

Abstract. This paper presents a robust tracking system for autonomous robots equipped with omnidirectional cameras. The proposed method uses a 3D shape and color-based object model. This allows to tackle difficulties that arise when the tracked object is placed above the ground plane floor. Tracking under these conditions has two major difficulties: first, observation with omnidirectional sensors largely deforms the target's shape; second, the object of interest embedded in a dynamic scenario may suffer from occlusion, overlap and ambiguities. To surmount these difficulties, we use a *3D particle filter* to represent the target's state space: position and velocity with respect to the robot. To compute the likelihood of each particle the following features are taken into account: *i*) image color; *ii*) mismatch between target's color and background color. We test the accuracy of the algorithm in a RoboCup Middle Size League scenario, both with static and moving targets.

1 Introduction

In order to carry out complex tasks (e.g. playing football) robots need to extract sufficient information from the environment they operate in. Catadioptric sensors are widely used in robotics, especially for self localization and navigation [8],[1], as they gather information from a large portion of the space surrounding a robot. One drawback is that images are affected by strong distortion and perspective effects, which may force the use of non-standard algorithms for target detection and tracking.

Automated tracking is still an open problem, e.g., surveillance applications [2], sports [5,10] or smart rooms [6]. In general, tracking visual features in complex and cluttered environments is fraught with uncertainty. It is therefore crucial to adopt principled probabilistic models. Over the past few years, particle filters, also known as sequential Monte Carlo (MC), proved to be effective in image

* This work was supported by Fundação para a Ciência e a Tecnologia (ISR/IST pluriannual funding) through the POS-Conhecimento Program that includes FEDER funds. We would like to thank Dr. Luis Montesano and Dr. Alessio Del Bue for the helpful discussions.

U. Visser et al. (Eds.): RoboCup 2007, LNAI 5001, pp. 77–88, 2008.

processing tracking techniques, e.g., [11,12,13]. The strength of these methods lies in their simplicity and flexibility on nonlinear and non-Gaussian settings [7].

We use a 3D particle-filter [9,11] tracker in which the hypotheses are 3D positions and velocities of the object, and whose likelihood is a function of object color and shape. From one image frame to the next, the hypotheses are moved according to an appropriate motion model. Then, for each particle, a likelihood is computed, in order to estimate the object state. To calculate the likelihood of a particle we first project the contour of the object it represents on the image plane (as a function of the object 3D shape, position and orientation) using an approximated model for the catadioptric system, the Unified Projection Model [15]. The likelihood is then calculated as a function of three color histograms: one represents the object color model and is computed in a training phase with several examples taken from distinct locations and illumination conditions; the other two histograms represent the inner and outer boundaries of the projected contour, and are computed at every frame for all particles. The idea is to assign a high likelihood to the contours for which the inner pixels have a color similar to the object, and are sufficiently distinct from outside ones.

A work closely related to this is described in [14], although in that case the tracking of RoboCup Middle Size League (MSL) balls is accomplished on the image plane. Tracking the 3D trajectory of a ball has become relevant in the RoboCup MSL scenario, as robots are now provided with the ability to kick the ball off the ground. Tracking the position of an object in 3D space instead of on the image plane has two main advantages: (i) the motion model used by the tracker can be the actual motion model of the object, while in image tracking the motion model should describe movements of the projection of the object on the image plane and, because of the aforementioned distortion, a good model can be difficult to formulate and use; (ii) with 3D tracking the actual position of the tracked object is directly available, while a further non-trivial step is needed for a system based on an image tracker to provide it.

The paper is organized as follows. In Section 2 we describe the catadioptric sensor and the used projection model. The particle filter is described in Section 3, and customized to our particular problem in 4. The experimental results are shown in Section 5 and, finally, Section 6 concludes the paper and presents ideas for future work.

2   Catadioptric Imaging System

In this section we describe the imaging system, its projection model and the used calibration method. Our catadioptric vision system, see Fig.1a, combines a camera looking upright to a convex mirror, having omnidirectional view in the azimuth direction [16]. The system is designed to have a wide-angle and a constant-resolution view of the ground plane [17,18]. The system has the constant-resolution property at one reference plane, the ground plane, and has only approximately constant-resolution at planes parallel to the reference one. As compared to perspective cameras, the constant-resolution design is a good compromise between approximating ubiquitous constant-resolution and enlarging the field of view. Note that perspective cameras can have constant-resolution for all planes orthogonal to the optical axis only for narrow view fields. Large fields-of-view imply using small focal length lenses which introduce large radial distortions.

Let the projection model, $\mathcal{P}$ of the constant-resolution system represent the transformation of a 3D point, $[X\ Y\ Z]^T$ into the 2D coordinates of its projection on the image plane, $[u\ v]^T$, considering the parameters θ:

$$[u\ v]^T = \mathcal{P}\left([X\ Y\ Z]^T; \theta\right). \tag{1}$$

$\mathcal{P}$ is trivial for the ground-plane, as it is just a scale factor between pixels and meters. Deriving $\mathcal{P}$ for the complete 3D field of view is complex as it involves using the actual mirror shape [18]. Here we assume that the system approximates a single projection center system, considering that the mirror size is small when compared to the distances to the imaged-objects. Hence, we can use a standard model for catadioptric omnidirectional cameras, namely the Unified Projection Model (UPM) pioneered by Geyer and Daniilidis [15].

The UPM represents all omnidirectional cameras with a single center of projection [15]. It is simpler than the model which takes into account the actual shape of the mirror and gives good enough approximations for our purposes.

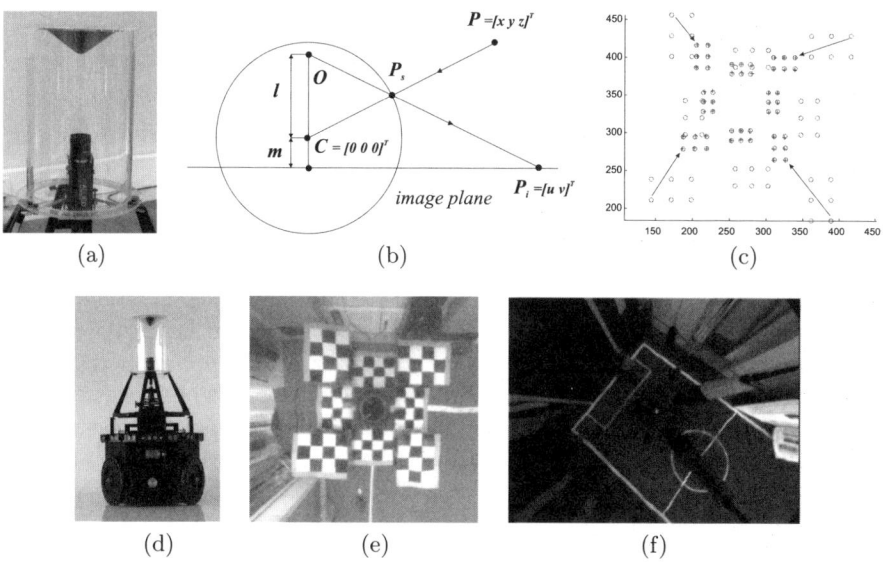

Fig. 1. (a) Catadioptric camera. (b) The Unified Projection Model. (c) Calibration result: observed image points (crosses), 3D points projected using initial projection parameters (dark gray circles) and using calibrated parameters (light gray circles) - the arrows show the calibration effect at four points. (d) The OmniISocRob robotic platform. (e) Image used for calibration. (f) Sample image taken in a RoboCup MSL scenario.

The model consists of a two-step mapping via a unit-radius sphere: (i) project a 3D world point, $P = [x \ y \ z]^T$ to a point P_s on the sphere surface, such that the projection is normal to the sphere surface; (ii) project to a point on the image plane, $P_i = [u \ v]^T$ from a point, O on the vertical axis of the sphere, through the point P_s. This mapping is graphically illustrated in Fig.1b. The mapping is mathematically defined by:

$$\begin{bmatrix} u \\ v \end{bmatrix} = \frac{l+m}{l\sqrt{x^2+y^2+z^2}-z} \begin{bmatrix} s_u & 0 \\ 0 & s_v \end{bmatrix} \begin{bmatrix} x \\ y \end{bmatrix} + \begin{bmatrix} u_0 \\ v_0 \end{bmatrix} \tag{2}$$

where (l, m) parameters describe the type of camera, (s_u, s_v, u_0, v_0) represent pixel scaling and offsetting in the image plane, and $[x \ y \ z]^T$ is a 3D point in the camera coordinate system, whose relationship to world coordinates is given by the 3D rigid transformation, $[x \ y \ z]^T = R[X \ Y \ Z]^T + [x_0 \ y_0 \ z_0]^T$.

To calibrate the model we use a set of known non-coplanar 3D points $[X_i \ Y_i \ Z_i]^T$ and measure their images $[u_i \ v_i]^T$. Then, we minimize the mean squared error between the measurements and the projection with the parametric model $\mathcal{P}$:

$$\theta^* = \arg_\theta \min \sum_i \left\| [u_i \ v_i]^T - \mathcal{P}\left([X_i \ Y_i \ Z_i]^T; \theta\right) \right\|^2 \tag{3}$$

where θ contains the 3D rigid transformation from world to camera coordinates, pixels scaling and offsetting, and the camera type parameters (l, m). We set the calibration patterns coordinate system in accordance with the robot frame by aligning the patterns with the center of the robot, see Fig.1e.

3   3D Tracking with Particle Filters

In this section we introduce the methods employed for 3D target tracking with particle filters. We are interested in computing, at each time $t \in \mathbb{N}$, an estimate of the 3D pose of a target. We represent this information as a "state-vector" defined by a random variable $\mathbf{x}_t \in \mathbb{R}^{n_\mathbf{X}}$ whose distribution in unknown (non-Gaussian); $n_\mathbf{X}$ is the dimension of the state vector. In the present work we are mostly interested in tracking balls and cylindrical robots, whose orientation is not important for tracking. However, the formulation is general and can easily incorporate other dimensions in the state-vector, e.g. target orientation and spin.

Let $\mathbf{x}_t = [x, y, z, \dot{x}, \dot{y}, \dot{z}]^T$, with (x,y,z), $(\dot{x},\dot{y},\dot{z})$ the 3D cartesian position and linear velocities in a robot centered coordinate system. The state sequence $\{\mathbf{x}_t; t \in \mathbb{N}\}$ represents the state evolution along time and is assumed to be an unobserved Markov process with some initial distribution $p(\mathbf{x}_0)$ and a transition distribution $p(\mathbf{x}_t \mid \mathbf{x}_{t-1})$.

The observations taken from the images are represented by the random variable $\{\mathbf{y}_t; t \in \mathbb{N}\}$, $\mathbf{y}_t \in \mathbb{R}^{n_\mathbf{Y}}$, and are assumed to be conditionally independent given the process $\{\mathbf{x}_t; t \in \mathbb{N}\}$ with marginal distribution $p(\mathbf{y}_t \mid \mathbf{x}_t)$, where $n_\mathbf{Y}$ is the dimension of the observation vector.

In a statistical setting, the problem is posed as the estimation of the *posteriori* distribution of the state given all observations $p(\mathbf{x}_t \mid \mathbf{y}_{1:t})$. Under the Markov assumption, we have:

$$p(\mathbf{x}_t \mid \mathbf{y}_{1:t}) \propto p(\mathbf{y}_t \mid \mathbf{x}_t) \int p(\mathbf{x}_t \mid \mathbf{x}_{t-1}) \, p(\mathbf{x}_{t-1} \mid \mathbf{y}_{1:t-1}) d\mathbf{x}_{t-1} \qquad (4)$$

The previous expression tells us that the *posteriori* distribution can be computed recursively, using the previous estimate, $p(\mathbf{x}_{t-1} \mid \mathbf{y}_{1:t-1})$, the motion-model, $p(\mathbf{x}_t \mid \mathbf{x}_{t-1})$ and the observation model, $p(\mathbf{y}_t \mid \mathbf{x}_t)$.

To address this problem we use particle filtering methods. Particle filtering is a Bayesian method in which the probability distribution of an unknown state is represented by a set of M weighted particles $\{\mathbf{x}_t^{(i)}, w_t^{(i)}\}_{i=1}^M$ [11]:

$$p(\mathbf{x}_t \mid \mathbf{y}_{1:t}) \approx \sum_{i=1}^M w_t^{(i)} \delta(\mathbf{x}_t - \mathbf{x}_t^{(i)}) \qquad (5)$$

where $\delta(\cdot)$ is the dirac delta function. Based on the discrete approximation of $p(\mathbf{x}_t \mid \mathbf{y}_{1:t})$, different estimates of the best state at time t are possible to be devised. For instance we may use the Monte Carlo approximation of the expectation, $\hat{\mathbf{x}} \doteq \frac{1}{M} \sum_{i=1}^M w_t^{(i)} \mathbf{x}_t^{(i)} \approx \mathbb{E}(\mathbf{x}_t \mid \mathbf{y}_{1:t})$, or the maximum likelihood estimate, $\hat{\mathbf{x}}_{ML} \doteq \mathrm{argmax}_{\mathbf{x}_t} \sum_{i=1}^M w_t^{(i)} \delta(\mathbf{x}_t - \mathbf{x}_t^{(i)})$.

To compute the approximation to the *posteriori distribution*, a typical tracking algorithm works cyclically in three stages:

1. *Prediction* - computes an approximation of $p(\mathbf{x}_t \mid \mathbf{y}_{1:t-1})$, by moving each particle according to the motion model;
2. *Update* - each particle's weight i is updated using its likelihood $p(\mathbf{y}_t \mid \mathbf{x}_t^{(i)})$:

$$w_t^{(i)} \propto w_{t-1}^{(i)} p(\mathbf{y}_t \mid \mathbf{x}_t^{(i)}) \qquad (6)$$

3. *Resampling* - the particles with a high weight are replicated and the ones with a low weight are forgotten.

For this purpose, we need to model probabilistically both the motion dynamics, $p(\mathbf{x}_t \mid \mathbf{x}_{t-1})$, and the computation of each particle's likelihood $p(\mathbf{y}_t \mid \mathbf{x}_t^{(i)})$.

3.1   The Motion Dynamics

In the system proposed herein we assume motion dynamics follow a standard autoregressive dynamic model:

$$\mathbf{x}_t = A\mathbf{x}_{t-1} + \mathbf{w}_t, \qquad (7)$$

where $\mathbf{w}_t \sim \mathcal{N}(0, Q)$. The matrices A, Q, could be learned from a set of representative correct tracks, obtained previously (e.g., see [3]), however, we choose pre-defined values for these two matrices (see Sections 4 and 5). Since the coordinates in the model are real-world coordinates, the motion model for a tracked object can be chosen in a principled way, both by using realistic models (constant velocity, constant acceleration, etc.) and by defining the covariance of the noise terms in intuitive metric units.

3.2   Observation Model

Each state vector $\mathbf{x}_t$ represents a target pose hypothesis. According to target shape, we compute sets of N points in the 3D inner and outer object boundaries: $\{\mathbf{D}_{in}^n\}$ and $\{\mathbf{D}_{out}^n\}$, $n = 1 \cdots N$. These points must be carefully chosen so that their projection in the image plane, using the projection model of section 2, falls in the 2D inside and outside boundaries of the image contour. Then, we obtain sets of 2D points $\{\mathbf{d}_{in}^n\}$ and $\{\mathbf{d}_{out}^n\}$. Each point $\mathbf{d}$ in the image is represented by its color vector in the HSI representation. For the inner and outer boundary point sets, we will compute HSI histograms, with $B = B_h\,B_s\,B_i$ bins.

Let us denote $b_t(\mathbf{d}) \in \{1, \ldots, B\}$ the bin index associated with the color vector at pixel location $\mathbf{d}$ and frame t. Then the histogram of the color distribution of a generic set of points can be computed by a kernel density estimate $\mathbf{H} \doteq \{h(b)\}_{b=1,\ldots,B}$ of the color distribution at frame t, where each histogram bin is given as in [4]

$$h(b) = \beta \sum_n \delta[b_t(\mathbf{d}^n) - b] \tag{8}$$

where δ is the Kronecker delta function, β is a normalization constant which ensures h to be a probability distribution $\sum_{b=1}^{B} h(b) = 1$.

To compute the similarity between two histograms we apply the Bhattacharyya similarity metric, as in [13]:

$$S\left(\mathbf{H}^1, \mathbf{H}^2\right) = \sum_{b=1}^{B} \sqrt{h^1(b) \cdot h^2(b)} \tag{9}$$

The likelihood of the hypothesis is computed, as a function of two similarities: the similarity between the object color model and the color measured in the inside image boundary, and the similarity between the colors measured in the image inside and outside the contour.

Defining a reference color model for the object as $\mathbf{H}^{\mathrm{model}}$, $\mathbf{H}^{\mathrm{inner}}$ as the inner boundary points color histogram, and $\mathbf{H}^{\mathrm{outer}}$ the outer boundary histogram, we will measure their similarity, using (9). The data likelihood should favor candidate color histograms which are close to the reference histogram and are sufficiently distinct from the background. Therefore we use:

$$p(\mathbf{y}_t \mid \mathbf{x}_t^{(i)}) = \mathrm{pos}\left[S(\mathbf{H}^{\mathrm{model}}, \mathbf{H}^{\mathrm{inner}}) - kS(\mathbf{H}^{\mathrm{outer}}, \mathbf{H}^{\mathrm{inner}})\right] \tag{10}$$

where the pos($\cdot$) function truncates to zero the negative values. This allow us to cope with the detection of the object (first term) and the detection from the background (second term).

4   Implementation of the RoboCup MSL 3D Tracker

The present approach is tested for a ball and robot tracking task, in a typical RoboCup MSL environment. The color model for each object was built collecting a set of images in which the object is present, and calculating the HSI color histogram on the (hand labeled) pixels belonging to the specific object. For target dynamics, we have chosen a constant velocity model, in which the motion equations correspond to a uniform acceleration during one sample time:

$$x_t = Ax_{t-1} + Ba_{t-1}, \quad A = \begin{bmatrix} \mathbf{I} & (\Delta t)\mathbf{I} \\ \mathbf{0} & \mathbf{I} \end{bmatrix}, \quad B = \begin{bmatrix} (\frac{\Delta t^2}{2})\mathbf{I} \\ (\Delta t)\mathbf{I} \end{bmatrix} \quad (11)$$

where I is the 3×3 identity matrix and a_t is a 3×1 white zero mean random vector corresponding to an acceleration disturbance. We have set $\Delta t = 1$ for all the experiments, whereas the covariance matrix of the random acceleration vector was fixed at:

$$cov(a_t) = \sigma^2 \mathbf{I}, \quad \sigma = 90 \text{mm/frame}^2 \quad (12)$$

The observation model requires the definition of adequate points in the 3D object inner and outer boundaries, as described in Section 3.2. Our idea was to determine which points of the 3D model would be projected on the object's contour on the image (see Fig.2) and then create the two sets of 2D boundary points by projecting the selected 3D points for a smaller and a larger model of the object (see the close-up's in Figures 3 and 5: the projected contours are drawn in white, while internal and external points are drawn in black). For the ball, for instance, the 3D contour points lie on the intersection between the sphere modelling it and the plane orthogonal to the line which passes through the virtual projection center and the center of the sphere. With this model, it is possible to adjust the number of points describing the 2D contour, obtaining faster processing times (less points) or more robustness (more points).

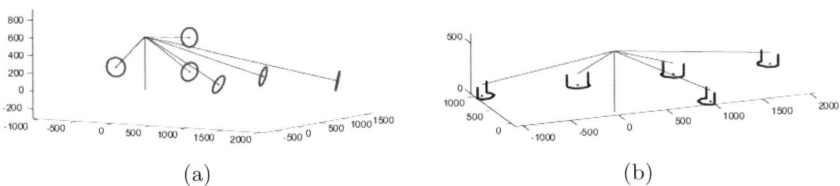

$$(a) \qquad\qquad\qquad\qquad\qquad\qquad (b)$$

Fig. 2. 3D plot of the 3D points projected to obtain the 2D contour points for balls (a) and robots (b), at different positions

5   Experimental Results

We ran several experiments to assess the accuracy and precision of the proposed tracking method: we tracked a ball rolling down a ramp, a ball bouncing on the floor and a robot maneuvering. We furthermore ran an experiment placing a still ball at different positions around the robot and measuring the error with respect to the ground truth.

5.1   Ball Tracking – Ramp and Bouncing

In the first experiment we tracked a ball rolling down a two-rail ramp. The projection of the ball on the image plane changes dramatically in size after some frames (see Fig.3a), due to the nature of the catadioptric system used. The images are affected by both motion blur and heavy sensor noise (see Fig.3b).

This image sequence was acquired with a frame rate of 20fps and we used 10000 particles in the tracker. The initial position for the particles was obtained sampling a 3D Normal distribution, the mean and standard deviation of which were manually set. The initial velocitiy was also manually set, equal for every particle. To sample the pixels in order to build the inner and outer color histograms for each hypothesis we projected a sphere with respectively 0.9 and 1.1 the radius of the actual ball. For each projection we used 50 points, uniformly distributed on each 3D contour. The parameter k was set to 1.5 for all experiments, meaning that we wanted the difference between inner color and outer color to be more discriminative than the similarity between inner color and model color. We repeated the tracking 10 times on the same image sequence.

The results of the tracking are visible in Fig. 4a.

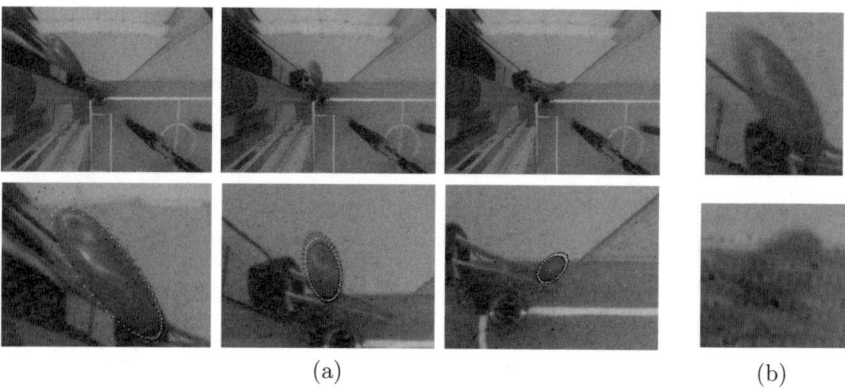

(a)   (b)

Fig. 3. Ball rolling down a ramp. (a) frames 1, 11 and 21 of the sequence and three corresponding close-ups of the tracked ball with the contour of the best hypothesis drawn in white (bottom row). The pixels marked in black are the ones used to build the color histograms. (b) Close-ups of the ball showing motion blur and noise.

In the second experiment we tracked a ball bouncing on the floor. The experimental setting was exactly the same as for the first experiment described, but for the frame rate, which in this case was of 25fps. The image sequence begins with the ball about to hit an obstacle on the ground, while moving horizontally. The collision triggers a series of parabolic movements for the ball, which is tracked until it hits the ground for the fourth time. We repeated the tracking process ten times on the same image sequence. The results are visible in Fig. 4b.

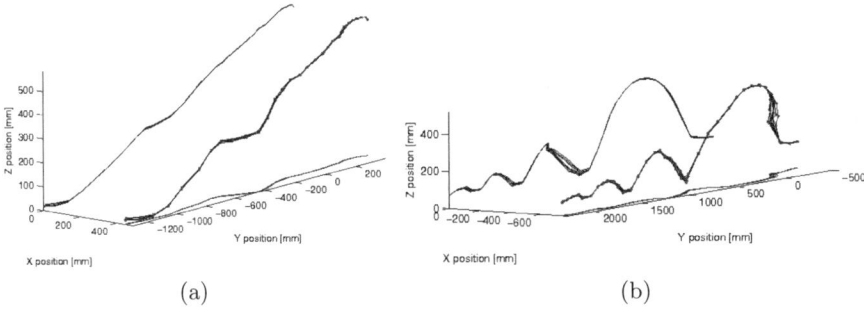

$$(a) \qquad\qquad\qquad (b)$$

Fig. 4. (a) Ball rolling down the ramp: plot of the tracked paths resulting from 10 runs of the algorithm performed on the same image sequence. The 10 dotted liness represent the 10 3D estimated trajectories of the ball, the 20 solid lines are the projection of these trajectories on the ground and lateral plane. (b) Ball jumping: same kind of plot for the estimated trajectories of the ball in the jump image sequence.

5.2 Robot Tracking

In this experiment we tracked a robot moving along a straight line, turning by 90 degrees and continuing its motion along the new direction (Fig. 5a). In this experiment the setting differs from the previous two: the vertical position and speed of the tracked object were constrained to be null. The injected velocity noise was, thereafter, distributed as a 2D Normal. To sample the pixels in order to build the inner and outer color histogram for each hypothesis we projected the contour of an 8-sided-prism with respectively 0.75 and 1.25 the size of the actual robot. The size difference between the projected models and the actual one is greater than in the case of the ball due to the fact that the model for the robot does not exactly fit its actual shape. For each projection, 120 points were used. We repeated the tracking 10 times and results are shown in Fig. 5b.

5.3 Error Evaluation

In this experiment we placed a ball at various positions around a robot and confronted the positions measured with our system against the ground truth. The positions were either in front of the robot, on its right, on its left or behind it, and either on the floor or at a height of 340mm.

Both the ground truth and the estimated ball positions are shown in Figure 6. It is noticeable some bias mainly in the vertical direction, due to miscalibration of the experimental setup. However, we are mostly interested in evaluating errors arising in the measurement process. Distance to the camera is an important parameter in this case, because target size varies significantly. Therefore, we have performed a more thorough error analysis evaluating its characteristics as a function of distance to the camera.

In Figure 7, we plot the measured error in spherical coordinates (ρ–distance, ϕ–elevation, θ–azimuth), as a function of distance to camera's virtual projection

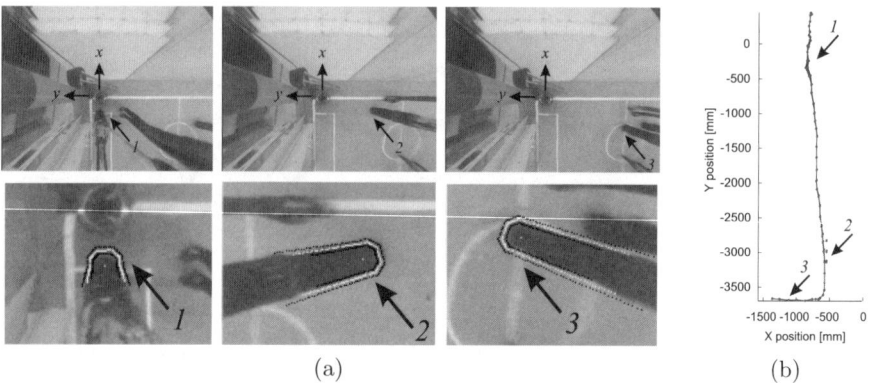

Fig. 5. Robot tracking. (a) three different frames of the sequence (top) and three close-ups of the tracked robot with the contour of the best hypothesis (white) and regions used to build the color histograms (black) drawn (bottom). (b) reconstructed robot trajectory - a view from the top of 10 estimated trajectories.

center. The first plot shows that radial error characteristics are mostly constant along distance, with a systematic error (bias) of about $46mm$, and standard deviation of about $52mm$. This last value is the one we should retain for characterizing the precision of the measurement process. The second plot shows the elevation error, where it is evident a distance dependent systematic error. This has its source on a bad approximation of the projection model for distances close to the cameras. Finally, the third plot shows the azimuthal error. It can be observed that there is a larger random error component at distances close to the camera, but this just a consequence of the fact that equal position errors at closer distances produce larger angular errors. Therefore we conclude that the precision of the observation model is in average of $52mm$ and do not depend significantly on distance to the camera in the tested range.

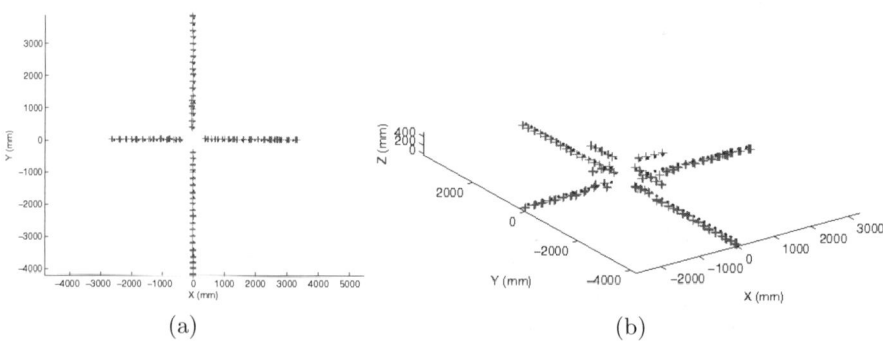

Fig. 6. Two views of ground truth (dots) and measurements (crosses)

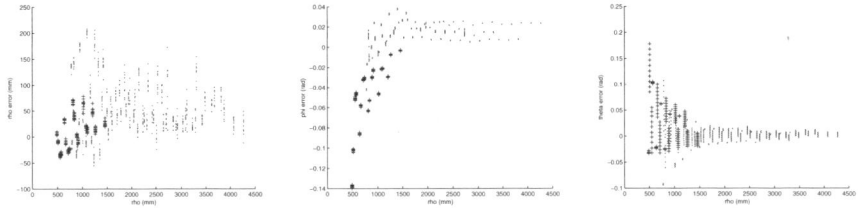

Fig. 7. ρ, ϕ, θ error for balls laying on the ground (dots) and flying at 340mm (crosses)

6   Conclusions

In this paper we have presented a tracking system for MSL Robots equipped with omnidirectional cameras, capable of tracking both targets on or above the floor, to consider the possibility of flying balls. The tracker uses a 3D shape and color model of the targets, and uses particle filtering methods to estimate their 3D location with respect to the Robot. Each hypothesis in the filter represents the 3D pose of an object. Even though in this paper we only consider targets with simple shapes (spheres and cylinders for representing balls and robots), the proposed model is general and copes with arbitrary shapes.

The main advantage of our approach is the use of a full 3D model in which the targets' motion dynamics is naturally expressed. Previous image based (2D) tracking methods require non-linear motion models (image projection is often non-linear) or must rely on approximations. This non-linearity becomes even more drastic in the case of omnicamera systems when targets are not lying on the floor. An additional advantage of the proposed method is related to a direct computation of 3D pose, whereas 2D models compute an image based pose that still must be mapped to world coordinates.

We have performed extensive experiments with real robots in a RoboCup MSL scenario. This paper showed the performance of our method in tracking maneuvering robots, rolling and jumping balls, demonstrating its ability to deal with off-the-floor targets and sudden trajectory changes. Additionally, we evaluated the precision of the system in static scenarios with ground truth measurements.

Since it is becoming more frequent to have robots kicking balls off the floor, the presented method constitutes a solution to improve ball position estimation, which, in the case of the goal-keeper, may be of fundamental importance.

References

1. Gaspar, J., Winters, N., Santos-Victor, J.: Vision-based Navigation and Environmental Representations with an Omnidirectional Camera. IEEE Transactions on Robotics and Automation 16(6) (December 2000)
2. Koller, D., Weber, J., Malik, J.: Robust Multiple Cars Tracking with Occlusion Reasoning. In: European Conf. on Computer Vision, pp. 186–196 (1994)

3. Gilles Celeux, J., Nascimento, J.S.: Learning switching dynamic models for objects tracking. Pattern Recognition 37(9), 1841–1853 (2004)
4. Comaniciu, D., Ramesh, V., Meer, P.: Real-Time Tracking of Non-Rigid Objects using Mean Shift. In: CVPR, pp. 142–151 (2000)
5. Misu, T., Naemura, M.: Robust Tracking of Soccer Players based on Data Fusion. In: IEEE 16th Int. Conf. on Patern Recognition, vol. 1, pp. 556–561 (2002)
6. Intille, S.S., Davis, J.W., Bobick, A.F.: Real-Time Closed-World Tracking. In: IEEE Conf. on Compuiter Vision Pattern Recognition, pp. 697–703 (1997)
7. Khan, Z., Balch, T., Dellaert, F.: MCMC Data Association and Sparse Factorization Updating for Real Time Multitarget Tracking with Merged and Multiple Measurements. IEEE Trans. on PAMI 28(12), 1960–1972 (2006)
8. Lima, P., Bonarini, A., Machado, C., Marchese, F., Ribeiro, F., Sorrenti, D.: Omnidirectional catadioptric vision for soccer robots. Robotics and Autonomous Systems 36(2–3), 87–102 (2001)
9. Thrun, S., Burgard, W., Fox, D.: Probabilistic Robotics. MIT Press, Cambridge (2005)
10. Okuma, K., Taleghani, A., de Freitas, N., Little, J.J., Lowe, D.G.: A Boosted Particle Filter: Multitarget Detection and Tracking. In: European Conf. on Computer Vision, pp. 28–39 (2004)
11. Doucet, A., de Freitas, N., Gordon, N.: Sequential Monte Carlo Methods. In: Gordon (ed.) Practice. Springer (2001)
12. Isard, M., Blake, A.: Condensation: conditional density propagation for visual tracking. Int. Journal of Computer Vision 28(1), 5–28 (1998)
13. Perez, P., Hue, C., Vermaak, J., Gangnet, M.: Color-Based Probabilistic Tracking using Unscented Particle Filter. In: CVPR (2002)
14. Olufs, S., Adolf, F., Hartanto, R., Plöger, P.: Towards probabilistic shape vision in robocup: A practical approach. In: RoboCup International Symposium (2006)
15. Geyer, C., Daniilidis, K.: A unifying theory for central panoramic systems and practical applications. In: Vernon, D. (ed.) ECCV 2000, Part II. LNCS, vol. 1843, pp. 445–461. Springer, Heidelberg (2000)
16. Benosman, R., Kang, S.B. (eds.): Panoramic Vision. Springer, Heidelberg (2001)
17. Hicks, R., Bajcsy, R.: Catadioptric sensors that approximate wide-angle perspective projections. In: CVPR, pp. 545–551 (2000)
18. Gaspar, J., Deccó Jr., C., J.O., Santos-Victor, J.: Constant resolution omnidirectional cameras. In: 3rd IEEE Workshop on Omni-directional Vision, pp. 27–34 (2002)

Improving Vision-Based Distance Measurements Using Reference Objects

Matthias Jüngel, Heinrich Mellmann, and Michael Spranger

Humboldt-Universität zu Berlin, Künstliche Intelligenz
Unter den Linden 6, 10099 Berlin, Germany
{juengel,mellmann,spranger}@informatik.hu-berlin.de
http://www.aiboteamhumboldt.com/

Abstract. Robots perceiving their environment using cameras usually need a good representation of how the camera is aligned to the body and how the camera is rotated relative to the ground. This is especially important for bearing-based distance measurements. In this paper we show how to use reference objects to improve vision-based distance measurements to objects of unknown size. Several methods for different kinds of reference objects are introduced. These are objects of known size (like a ball), objects extending over the horizon (like goals and beacons), and objects with known shape on the ground (like field lines). We give a detailed description how to determine the rotation of the robot's camera relative to the ground, provide an error-estimation for all methods and describe the experiments we performed on an Aibo robot.

Keywords: RoboCup, humanoid robots, Aibo, camera matrix, reference objects.

1 Introduction

A main task in robotic vision is to determine the spatial relations between the robot and the objects that surround it. Usually the robot needs to know the angle and the distance to certain objects in order to localize, navigate or do some high-level planning. To determine the distance to an object is easy when the size of the object and the focal length of the camera are known. To determine the distance to an object of unknown size is possible using the knowledge about the height of the camera and the bearing to the point where the object meets the ground. This bearing is given by the position of the object in the image and the known orientation of the camera relative to the ground. Unfortunately this orientation is not known in a lot of cases. The calculation of the kinematic chain of a legged robot from the ground to the camera is usually difficult as the exact contact points of the robot and the ground are hard to determine. Additionally inaccuracies in the joint angle sensors sum up the longer the kinematic chain is. But also for wheeled robots the orientation of the camera relative to the ground can be unknown, especially when there is a suspension for the wheels. In this paper we show how to determine the orientation of the camera

U. Visser et al. (Eds.): RoboCup 2007, LNAI 5001, pp. 89–100, 2008.

using reference objects in the image and how to calculate the distance to objects of unknown size. This work was inspired by our experience in RoboCup where using the field lines to localize a Sony Aibo was inaccurate due to large errors in the orientation and position of the camera which are calculated based on the sensor readings of the joint angles of the robot and assumptions on the contact points of the legs with the ground.

1.1   Related Work

There has been extensive work on the calibration of camera parameters. Typically authors try to infer intrinisic and extrinsic parameters of cameras using specially crafted calibration objects. A lot of work has been put in to reduce the complexity of this objects, i.e. their dimensionality or rigidness of pattern [1,2] or even allow completely other objects for the parameter estimation [3]. RoboCup teams have developed mechanisms to reduce the calibration time after transport of robots [4] or to calibrate ceiling cameras [5]. A lot of these methods involve off-line optimization of the estimated parameters regarding projection errors. In contrast to these methods our approach focuses on determining the camera pose relative to the ground during the operation of the robot. While intrinsic parameters do not change during operation, the extrinsic parameters of the camera are usually hard to determine using proprioception in a highly dynamic environment like RoboCup. We describe how information from the camera images can be used to determine the orientation of the camera and how additional information from the joint sensors can be incorporated.

1.2   Outline

This paper is divided into several parts. In section 2 we motivate our work by giving an error estimation for the bearing based distance measurement approach. In section 3 we describe several methods that determine the camera matrix by means of visual information in order to determine distances to other objects. In section 4 we examine the robustness of these methods concerning errors. Section 5 presents the results of some experiments which were conducted with an AIBO.

2   Motivation

A simplified version of the bearing based distance estimation approach of objects can be seen in figure 1. The model was used to estimate the significance of any correction approach in advance. From this simple mathematical model conclusions about the influence of measurement errors of the rotation angle φ and the estimated height h_{camera} on the calculated distance d_{object} were drawn.

The basic bearing based distance approach depicted in figure 1 calculates d_{object} from known h_{camera}, h_{object} and φ. From

$$d = \tan\left(\varphi\right) \cdot h_{camera} \quad \text{and} \quad d_{rest} = \tan\left(\varphi\right) \cdot h_{object}$$

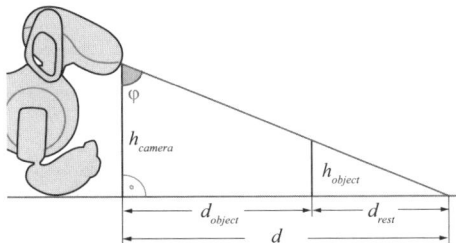

Fig. 1. Simple bearing based distance estimation model

follows with $d_{object} = d - d_{rest}$ that

$$d_{object} = \tan(\varphi) \cdot (h_{camera} - h_{object})$$

With known h_{camera} and h_{object}, d_{object} can be seen as a function depending on φ only, i.e. $d_{object} = d_{object}(\varphi)$. It can be immediately seen that it is also possible to infer the correct bearing φ from known h_{camera}, h_{object} and d_{object}. This simple model is only valid when $h_{camera} > h_{object}$ and $\varphi < \frac{\pi}{2}$. It allows to show the effect of estimation errors of φ on the estimated distance d_{object} of an object of height h_{object}. For an ex ante study suitable values for h_{camera} and h_{object} where chosen from the context of *RoboCup*. The error d_{error} is calculated by

$$d_{error}(\Delta\varphi) = |d_{object}(\varphi + \Delta\varphi) - d_{object}(\varphi)|$$

From the formulas provided it can be seen that even small changes of φ can result in big errors for the estimated distance d_{object}, which is shown in figure 2a) for fixed h_{camera} and h_{object}. For positive $\Delta\varphi$ the error is rising exponentially. Figure 2b) illustrates that this error rises with the growing correct distance of the object.

3   Using Reference Objects for Improving Distance Measurements

A lot of objects in the environment of a robot can be used as reference objects for distance calculation. In this paper we focus on the calculation of distances based on the height of the observing camera and its direction of view. As shown in section 2 this method is very prone to errors in the angle between the optical axis of the camera and the ground. We show several methods to estimate the position and orientation of the camera relative to the ground using different classes of objects:

- objects with known size (e.g. the ball)
- objects with known height, higher than the camera of the robot (e.g goals and beacons)
- objects with known outline on the ground (e.g. goals and field lines)

a)   b)

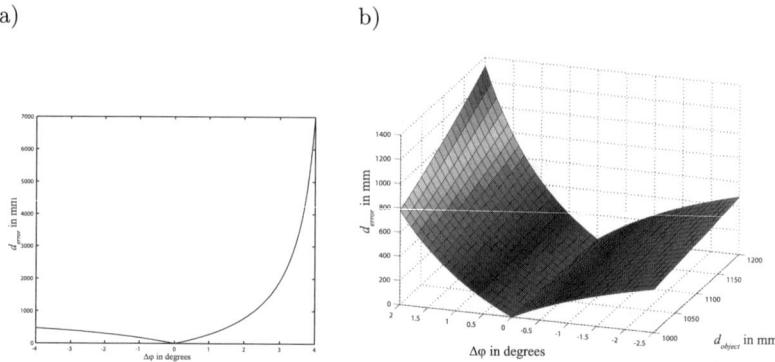

Fig. 2. Bearing based distance estimation error for fixed $h_{camera} = 160mm$ (which is a suitable camera height approximation for the Sony Aibo Robot) and an object height $h_{object} = 90mm$ (height of the *RoboCup* ball in the 4-legged-league) a) Shows the effect of variations of φ (calculated from correct distance of $d_{object} = 1000mm$). Please note that the error gets as big as $7000mm$ for a variation of φ by $4degrees$. b) Shows the same effect as a) in a range for the object distance d_{object} from $1000mm$ to $1200mm$. For bigger distances the error rises dramatically.

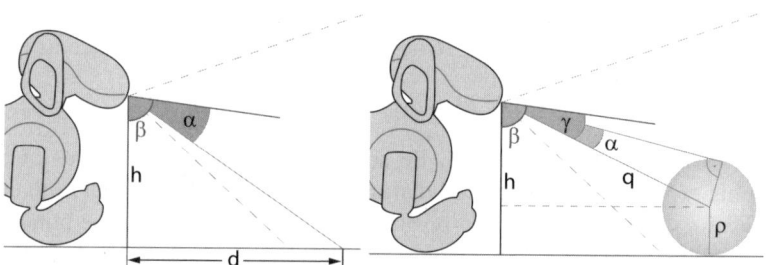

Fig. 3. Relation between the tilt of the camera and the ball used as reference object

All examples given in brackets are usually to be seen during a typical RoboCup game. Thus in almost every image at least one reference object can be used. The following subsections describe the different methods for all classes of reference objects. Given that the camera is not rotated on the optical axis we can limit our following considerations to a two-dimensional model, as shown in figure 3 (left).

3.1 Objects of Known Size

An easy approach in order to determine the camera tilt is to consider reference objects, whose distance can be determined based on their size. If the distance to a point is given, the tilt can be calculated as follows:

$$\beta = \arccos\left(\frac{q}{h}\right) - \alpha.$$

This formula can be deduced from figure 3 (left).

Figure 3 (right) illustrates the relation between the camera tilt and the ball as reference object[1]. Here, two simple formulas can be deduced as follows:

$$\frac{\rho}{q} = \sin(\alpha) \quad \text{and} \quad \frac{h - \rho}{q} = \cos(\beta - \gamma)$$

it can be deduced:

$$\beta = \arccos\left(\frac{h - \rho}{\rho} \cdot \sin(\alpha)\right) + \gamma.$$

This formula allows us to calculate the camera tilt using only the size of the ball without the need of any other sensor information.

3.2 Objects with Known Shape on Ground

If the height and the tilt of the camera are known, the image captured by the camera can be projected to the ground. If the used camera tilt corresponds with the real tilt, the outline of ground-based objects should appear without distortion in this projection. Should there be distortions (e.g. there is not a right angle between the field lines), this is a hint on the fact that the used tilt is incorrect. Thus it is an obvious idea to determine the camera tilt so that the projected field lines are perpendicular to each other.

This idea can be formulated as follows. Let p_1, p_2 and p_3 be the defining points of a corner in the image, p_1 being the vertex. The points are projected to the ground plane by means of the camera matrix $M(\alpha)$, α being the camera tilt. The resulting points are denoted $P_i(\alpha)$. For the angle φ, which is defined by these points, it holds:

$$\cos\varphi = \frac{\langle P_1(\alpha) - P_2(\alpha), P_1(\alpha) - P_3(\alpha)\rangle}{||P_1(\alpha) - P_2(\alpha)|| \cdot ||P_1(\alpha) - P_3(\alpha)||}.$$

However, it is known that $\varphi = \frac{\pi}{2}$ and hence $\cos\varphi = 0$, so that the formula for α is the following:

$$\langle P_1(\alpha) - P_2(\alpha), P_1(\alpha) - P_3(\alpha)\rangle = 0.$$

In general, this equation has an infinite number of solutions. However, in specific cases, as e.g. in the case of AIBO, there is often only one admissible solution due to the limits of the joints. By means of standard methods as Gradient Descent, this solution can be easily found.

This method works best if the corner is viewed on from the outside or the inside. However, if the robot is situated on one of the defining lines, the angle is not distorted by the wrong camera tilt any more and the method fails.

3.3 Tilt Correction Using Objects Higher than Eye-Level

The examination of the horizon yields another approach for the correction of the camera matrix. In many cases the horizon can be measured by means of objects

---

[1] The advantage of taking the ball as reference object is that it is easy to determine its size, as the ball looks equal from every direction. Furthermore, it can be seen on numerous images, being the central object of the game.

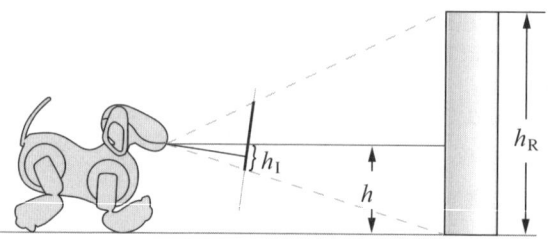 

Fig. 4. (left) A landmark is projected on the image plane. The knowledge of the real height of landmark can be used to determine the height of the horizon in the image. (right) An image captured by the robot, containing the recognized goal and the calculated horizon.

that are higher than the focal point of the camera. For example, if the robot sees a landmark with a known real height h_R and if its height in the image h_I is known as well, it is easy to determine the height of the horizon in the image, as it equals $\frac{h \cdot h_I}{h_R}$, as can be seen in the figure 4. By definition, the line of the horizon goes through the center of the image, if and only if the camera tilt is exactly $\frac{\pi}{2}$. Thus the camera tilt can be determined as follows, in accordance to section 3.1

$$\beta = \frac{\pi}{2} - \alpha.$$

3.4 Roll Correction Using Objects Higher than Eye-Level

In the methods outlined in the sections 3.1, 3.2 and 3.3 we assume that the camera is not rotated on the optical axis (i. e. roll = 0).

Not only does this rotation have an effect on the calculation of the tilt; it also influences the following calculation of the distance to respective objects, if these are not located in the center of the image.

The effects of the rotation on the tilt are examined in detail in section 4.2[2]. In order to calculate the roll we can use the inclination of objects in the image. For example, in the case of a landmark of the 4-Legged League, the horizon is always perpendicular to it. Another method to calculate the slope of the horizon is to determine the height of the horizon by means of several objects, e.g. two goal posts as shown in figure 4 (left). The roll can be easily calculated with the slope of the horizon. If we think of the horizon as a line in the image, the roll of the camera is the gradient of the straight line.

3.5 Using Knowledge about the Kinematic Chain

In some cases, the kinematic chain is known so that the position of the camera can be deduced by means of the joint data. For example, this holds true for AIBO.

---

[2] The effects on the rotation of the camera on the distance become obvious in the section 2.

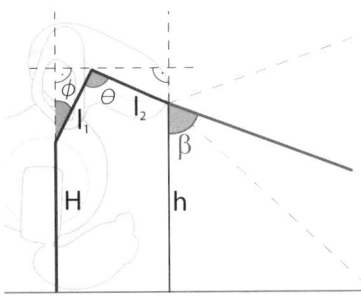

Fig. 5. Relation between the height of the camera and the angle of the neck-joint

In this case the whole kinematic chain can be determined via the joint data. However, the results are partly rather inaccurate. This is due to the fact that the contact points of the robot and the ground cannot be determined precisely. Furthermore, some of the joint sensors provide inaccurate data. In this section we want to examine how the knowledge of the kinematic chain can be combined with the outlined methods in order to achieve better results.

All the outlined methods provide us with the relations, or dependencies, between the different parameters of the camera, as e.g. the tilt and the height, which result from the respective observations. The kinematic chain also yields information on the relations of these parameters. Thus it is evident to try and determine the interesting parameters so that all given relations are fulfilled.

In many cases, there are not enough independent relations to determine all parameters. However, it is possible to write all camera parameters as a function of the joint angles. In turn, we can consider some of the joint angles as parameters and optimize them.

As an example, we use the method outlined in section 3.1 in order to correct the angle of the neck joint in the case of AIBO.

Application for Aibo Robot. According to our findings, AIBO's neck tilt is one of the most substantial error sources. This joint particularly has an effect on the determination of the camera's height. In the method outlined in section 3.1 the height of the camera is implied in the calculations so that this error also affects the results.

In order to completely eliminate the influence of the neck joint we have to make use of our knowledge of the relation between the neck joint and the height of the camera. This relation is depicted in figure 5. The interdependence of the height and the neck tilt can be formulated as follows:

$$h = H + l_1 \cdot \cos(\phi) - l_2 \cdot \sin\left(\frac{\pi}{2} - \theta - \phi\right)$$

and for the camera tilt β it holds $\beta = \theta - \phi$. Applying this to the formula outlined in section 3.1 the following function can be defined:

$$f(\phi) := \left(\frac{h(\phi) - \rho}{\rho} \cdot \sin(\alpha)\right) - \cos(\beta(\phi))$$

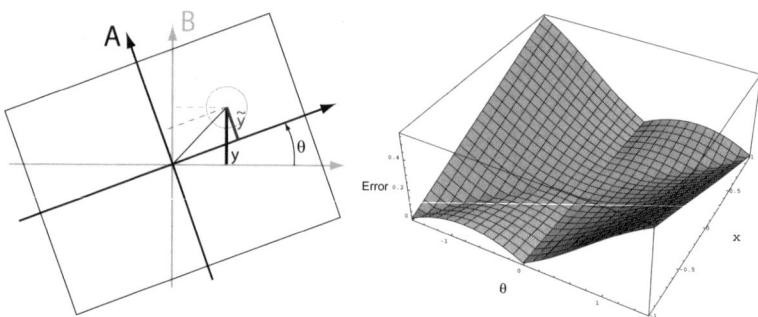

Fig. 6. (left) Camera rotated on its optical axes. (A) is the real coordinate system of the camera, (B) is the not-rotated coordinate system. The coordinate y is necessary for the calculation of the distance to an object. However, the coordinate $\tilde{y}$ measured in the image differs from y in case $\theta \neq 0$. (right) Error $|\tilde{\beta} - \beta|$ caused by ignoring the camera roll θ. The y-position is assumed as $y = 1mm$ (nearly the maximal y-position on the Aibo ERS7 camera chip) and the focal length as $f = 3.5mm$, θ and x-position are varied.

The angle ϕ can be determined as root of the function f. Thus the sensor data of the neck joint does not affect the determination of distances to other objects.

4   Error Estimation

In this section we want to analyze the effects of errors on the above mentioned methods in order to evaluate the quality of the results.

4.1   Errors of the Horizon-Based Methods

In many cases, the height of the robot is not known, e.g. if AIBO is walking. The method outlined in section 3.3 is particularly robust concerning this kind of errors. Let the error of the robot's height be h_e, the resulting error β_e of the roll angle is

$$\tan{(\beta_e)} = \frac{h_e}{d},$$

whereas d is the distance to the object that is used to measure the horizon. In the case of AIBO this would result in an error of $\beta_e = 0.03$, if the error of the height is $h_e = 3cm$ and the distance between the robot and the goal is $d = 1m$. The method becomes more robust with increasing distance to the reference object.

4.2   Estimating Errors Caused by Unknown Camera Roll

The camera tilt is essential for the determination of the distance to other objects. This is why all outlined methods deal with the correction of the camera tilt. Actually, there are cases in which the roll angle has a major influence on the result. The methods in the sections 3.1, 3.2 and 3.3 the roll angle is ignored. Thus we want to examine the effect of this on the results. The error estimation is only performed for the method using the ball as reference object, however for the other methods it can be done in the same way.

We consider all coordinates concerning the center of the image. Let $p = (x, y)^T$ be the center of the not-rotated image and θ the rotation of the camera on the optical axis as shown in figure 6 (left). We need the y-position of the ball's center in order to correct the tilt. After the application of the rotation we get the position of the ball's center as it would be detected in the image, in particular the measured height of the ball's center is then given by

$$\tilde{y} = x \cdot \sin \theta + y \cdot \cos \theta.$$

Thus it is obvious that the extent of the rotation's influence depends on the distance between the center of the image and the ball. Figure 6 (left) illustrates above treatments.

With the notation used in section 3.1 we can denote

$$\beta = \arccos \left(\frac{h - \rho}{\rho} \cdot \sin (\alpha) \right) + \arctan \frac{\tilde{y}}{f}$$

whereas f is the focal length of the camera. Figure 6 (right) illustrates the errors in case $\theta \neq 0$. As the figure shows, the error can be neglected if the angle θ is near zero. Thus, the method will yield acceptable results even though the roll of the camera is ignored if the roll is small enough.

5   Experiments

A number of experiments have been made with AIBO in order to evaluate the outlined methods under real conditions.

5.1   Projection Experiments

A good method to evaluate the accuracy of the camera matrix is to project images to the ground plane. In this subsection we describe two experiments using this methods. The first experiment evaluates the camera matrix obtained using the goal in images. In the second experiment a corner of field lines is used to correct the robots neck tilt joint.

Testing Accuracy of Horizon-Based Tilt and Roll Estimation. This experiment was performed in order to test the accuracy of the horizon based methods outlined in the section 3.3 and 3.4.

In the setup of this experiment the robot is situated in the center of the field and directed towards the blue goal. There is a calibration grid right in front of the robot.

During the Experiment the robot runs on the spot, the camera is directed towards the goal. The camera matrix is calculated and the correction is applied

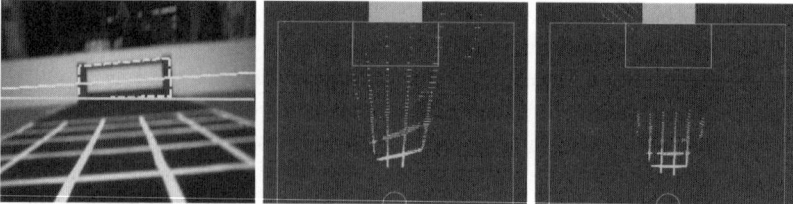

Fig. 7. (left) the scene from the view of the Aibo robot, (center) projection of the grid by means of the camera matrix calculated from the joint data, (right) projection of the grid by means of the corrected matrix

by calculating the tilt and roll angles with the help of the goal in the image (according to the method outlined in section 3.3). Figure 7 (left) shows the situation viewed by the robot. The image captured by the camera is projected to the ground plane by means of both matrices (the one calculated by the means of the kinematic chain and the corrected one).

Distortions occur if the values of the camera matrix do not correspond to reality, i.e. the lengths of the edges are not equal any more and the edges do not form a straight angle. All these effects can be increasingly observed in the case of a camera matrix that is calculated by means of joint data (figure 7 (center)). There are no distortions in the case of the corrected matrix, as can be seen in figure 7 (right).

Testing Field Line Corner Based Tilt Estimation. This experiment consisted of two parts. In the first part the robot was standing and looking straight ahead at a corner of field lines. The robot's hind legs were lifted manually by approximately 10cm resulting in a body tilt of up to 30 degrees. In the second experiment the robot was running on the same spot again looking at a corner of field lines. The running motion caused inaccuracies in the measurement of the neck tilt angle. The body tilt was estimated by the approach described in section 3.2. Both experiments have in common, that the distance to the corner does not change. To visualize the result of this estimation the images of the corner were projected to the ground using the camera matrix obtained from the readings of joint values and using the corrected camera matrix (see figure 8). The distance to the projected vertex of the corner using the corrected camera matrix was almost constant over time. Using the uncorrected camera matrix resulted in a large error in the distance and the angle of the projection of the corner. Thus the method was able to significantly increase the accuracy of bearing based distance measurements.

5.2   Distance Experiment

In this experiment we calculate the distance to another robot with the help of the bearing-based approach. Here, the parameters of the camera are corrected with different methods, which gives us the opportunity to compare them.

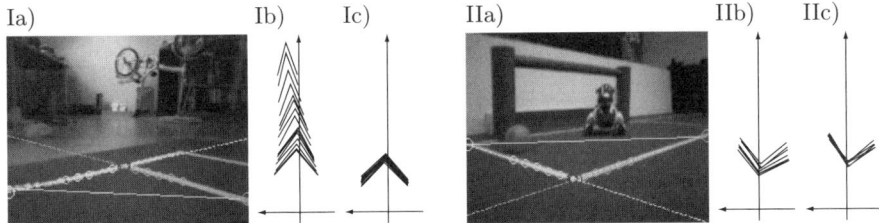

Fig. 8. This figure illustrates the correction of camera tilt by the means of corners of the field lines. The situation from the view of an Aibo and the perceptions of a corner projected to the ground are shown. In the first experiment the hind legs of the Robot were lifted manually, thus the resulting offset in the tilt angle can not be calculated from the joint data only. The figures Ib) and Ic) show the projections of the corner based on the joint data only, and using the corrected neck tilt respectively. IIb) and IIc) illustrate the not-corrected and corrected projections of the corner that was seen while by robot while walking on a spot.

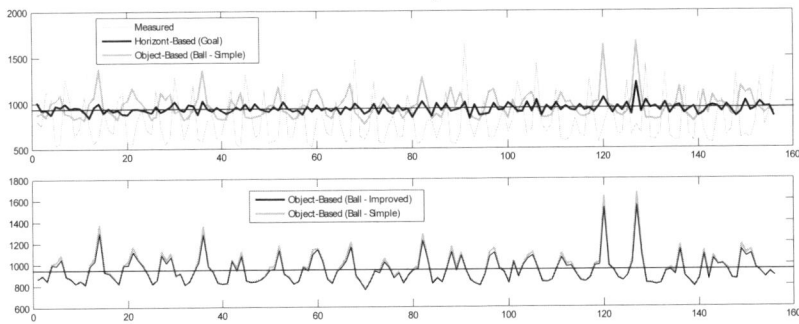

Fig. 9. (top) The distance determined by means of the camera tilt calculated from the joint data is shown in comparison to the distance determined with the help of the method using the size of the ball (outlined in section 3.1) and the method based on the horizon (outlined in section 3.3). (bottom) comparison between the results of the method using the ball as reference object and the combination of this method with the knowledge of the kinematic chain as described in section 3.5.

The setup is the same as in the experiment described above. However, there is no calibration grid in front of the robot. In addition, there are a ball and another robot in the robot's visual field.

In this experiment we correct the camera matrix with the help of the ball (directly and indirectly) and the goal, respectively. In order to compare the results we determine the distance to the other robot with the different corrected camera matrices, respectively. As the observing robot runs on the spot and the other robot does not move, the distance between them is constant. However, the calculated distance varies, due to errors. Figure 9 summarizes and compares the different results within the time scope of about 30 seconds.

The deviations in the not-corrected case are particularly very high. The best results were achieved by applying the horizon-based method. The two methods using the ball as reference object provide nearly identical results that are feasible.

6 Conclusion

We showed several methods to determine the camera pose of a robot relative to the ground using reference objects. This can help to improve bearing-based distance measurements significantly. Our work is relevant for all kinds of robots with long kinematic chains or unknown contact points to the ground as for these robots it is hard to determine the orientation of the camera using proprioception. As we provided methods for different kinds of reference objects there is a hight probability for a robot to see a suitable reference object. Experiments on Aibo showed that the methods work in practice.

References

1. Zhang, Z.: Camera calibration with one-dimensional objects (2002)
2. He, X., Zhang, H., Hur, N., Kim, J., Wu, Q., Kim, T.: Estimation of internal and external parameters for camera calibration using 1d pattern. In: AVSS 2006: Proceedings of the IEEE International Conference on Video and Signal Based Surveillance, Washington, DC, USA, p. 93. IEEE Computer Society, Los Alamitos (2006)
3. Junejo, I., Foroosh, H.: Robust auto-calibration from pedestrians. In: AVSS 2006: Proceedings of the IEEE International Conference on Video and Signal Based Surveillance, Washington, DC, USA, p. 92. IEEE Computer Society, Los Alamitos (2006)
4. Heinemann, P., Sehnke, F., F.S., Zell, A.: Automatic calibration of camera to world mapping in robocup using evolutionary algorithms (2006)
5. Benosman, R., Douret, J., Devars, J.: A simple and accurate camera calibration for the f180 robocup league. In: Birk, A., Coradeschi, S., Tadokoro, S. (eds.) RoboCup 2001. LNCS (LNAI), vol. 2377, pp. 275–280. Springer, Heidelberg (2002)

Cooperative/Competitive Behavior Acquisition Based on State Value Estimation of Others

Kentaro Noma[1], Yasutake Takahashi[1], and Minoru Asada[1,2]

[1] Dept. of Adaptive Machine Systems, Graduate School of Engineering
Osaka University
[2] JST ERATO Asada Synergistic Intelligence Project
Yamadaoka 2-1, Suita, Osaka, 565-0871, Japan
{kentaro.noma,yasutake,asada}@ams.eng.osaka-u.ac.jp
http://www.er.ams.eng.osaka-u.ac.jp

Abstract. The existing reinforcement learning approaches have been suffering from the curse of dimension problem when they are applied to multiagent dynamic environments. One of the typical examples is a case of RoboCup competition since other agents and their behaviors easily cause state and action space explosion. This paper presents a method of hierarchical modular learning in a multiagent environment by which the learning agent can acquire cooperative behaviors with its teammates and competitive ones against its opponents. The key ideas to resolve the issue are as follows. First, a two-layer hierarchical system with multi learning modules is adopted to reduce the size of the state and action spaces. The state space of the top layer consists of the state values from the lower level, and the macro actions are used to reduce the size of the action space. Second, the state of the other to what extent it is close to its own goal is estimated by observation and used as a state value in the top layer state space to realize the cooperative/competitive behaviors. The method is applied to 4 (defence team) on 5 (offence team) game task, and the learning agent successfully acquired the teamwork plays (pass and shoot) within much shorter learning time (30 times quicker than the earlier work).

1   Introduction

Recently, there have been increasing number of studies on cooperative/competitive behavior acquisition in a multiagent environment by using reinforcement learning methods [3,4,6,8,10]. In such an environment, the state and action spaces for the learning can be easily exploded since not only objects but also other agents should be involved in the state and action spaces, and therefore the sensor and actuator level descriptions may cause information explosion that disables the learning methods to be applied within practical learning time. Kalyanakrishnan et al. [6] showed that the learning can be accelerated by sharing the learned information in the 4 on 5 game task. However, they need still long learning time since they directly use the sensory information as state variables to decide the situation. Stefan et al. [3]

U. Visser et al. (Eds.): RoboCup 2007, LNAI 5001, pp. 101–112, 2008.
© Springer-Verlag Berlin Heidelberg 2008

achieved the cooperative behavior learning task between two real robots by introducing the macro action that is abstracted action code predefined by the designer. However, only the macro actions do not seem sufficient to accelerate the learning time in a case that more agents are included in the environment. Therefore, the sensory information should be also abstracted to reduce the size of the state space.

A modular learning system is suitable for observing/learning/executing a number of behaviors in parallel, and various modular architectures have been proposed so far [5,7,12,2]. Each module is responsible for learning to achieve a single goal. One arbiter or a gate module is responsible for merging information from the individual modules in order to derive a single action performed by the robot. The prediction of other's behavior is important to realize the cooperative (competitive) behaviors with (against) others in general. Takahashi et al. [11] proposed a method to infer the other's intention by observation based on the idea that the increase of the state value (the larger the state value, the closer to the goal) means the other intends to achieve the corresponding goal regardless of the differences of viewpoint and/or action to achieve the goal. If this prediction capability is incorporated into the learning system, the learner can efficiently acquire the desired behaviors.

This paper presents a method of hierarchical modular learning in a multiagent environment by which the learning agent can acquire cooperative behaviors with its teammates and competitive ones against its opponents. The key ideas to resolve the issue are as follows. First, a two-layer hierarchical system with multi learning modules is adopted to reduce the size of the state and action spaces. The state space of the top layer consists of the state values from the lower level, and the macro actions are used to reduce the size of the physical action space. Second, the state of the other to what extent it is close to its own goal is estimated by observation and used as a state value in the top layer state space to realize the cooperative/competitive behaviors. The method is applied to 4 (defence team) on 5 (offence team) game task, and the learning agent successfully acquired the teamwork plays (pass and shoot) within much shorter learning time (30 times quicker than the earlier work).

2   Multi Module Learning System with Other's State Value Estimation Modules

2.1   Architecture

Fig.1 shows a basic architecture of the proposed system, i.e., a two-layered multi-module reinforcement learning system. The bottom layer (left side of this figure) consists of two kinds of modules: action modules and estimation ones that infer the other's state value. The top layer (right side of the figure) consists of a single gate module that learns which action module should be selected according to the current state that consists of state values sent from the modules at the bottom layer. The selected module then sends action commands based on its policy.

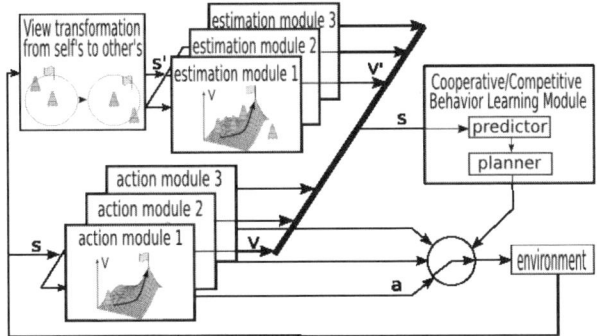

Fig. 1. A multi-module learning system

2.2   Action Module

An action module of the lower layer has a reinforcement learning module which estimates state values for the action.

Fig. 2. Agent-environment interaction   **Fig. 3.** Sketch of a state value function

Figure 2 shows a basic model of reinforcement learning. An agent can discriminate a set S of distinct world states. The world is modeled as a Markov process, making stochastic transitions based on its current state and the action taken by the agent based on a policy π. The agent receives reward r_t at each step t. State value V^π, the discounted sum of the reward received over time under execution of policy π, will be calculated as follows:

$$V^\pi = \sum_{t=0}^{\infty} \gamma^t r_t. \tag{1}$$

In case that the agent receives a positive reward if it reaches a specified goal and zero else, then, the state value increases if the agent follows a good policy π (see Figure 3). The agent updates its policy through the interaction with the environment in order to receive higher positive rewards in future. For further details, please refer to the textbook of Sutton and Barto[9] or a survey of robot learning[1]. Here, we suppose that the state values in each action module have been already acquired before the learning of the gate module.

2.3   Other's State Value Estimation Module

The role of the other's state value estimation module is to estimate the state value that indicate the degree of achievement of the other's task by observation, and to send this value to the state space of the gate module at the top layer. In order to estimate the degree of achievement, the following procedure is taken.

1. The learner acquires the various kinds of behaviors that the other agent may take as macro actions.
2. The learner estimates the sensory information observed by the other through the 3-D reconstruction of its own sensory information.
3. Based on the estimated sensory information of the other, each other's state value estimation module estimates the other's state value by assigning the state value of the corresponding action module of its own.

2.4   Cooperative/Competitive Behavior Learning Module

As shown in Figure 1, the gate module receives state values of lower modules, that is, the action modules and the other's state value estimation ones, and constructs a state space with them. The state space of the gate module is constructed as direct product of the variables of the state values. In order to adopt a discrete state transition model described above, the state space is quantized appropriately. The action set of the gate module is constructed with all action modules of the lower layer as macro actions.

3   Task and Assumptions

The game consists of the offence team (five players and one of them can be the passer) and the defence team (four players attempt to intercept the ball). The offence player nearest to the ball becomes a passer who passes the ball to one of its teammates (receivers) or shoot the ball to the goal if possible while the opposing team tries to intercept it (see Fig. 4).

Only the passer learns its behavior while the receivers and the defence team members take the fixed control policies. The receiver becomes the passer after receiving the ball and the passer becomes the receiver after passing the ball. After one episode, the learned information is circulated among team members through communication channel but no communication during one episode. The action and estimation modules are given a priori.

The offence (defence) team color is magenta (cyan), and the goal color is blue (yellow) in the following figures. The game restarts again if the offense team successfully scores a goal, kicks the ball outside of the field, or the defense team intercepts the ball from the opponent.

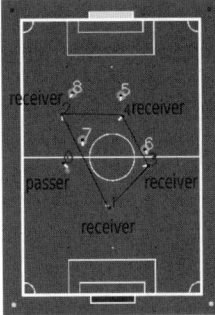
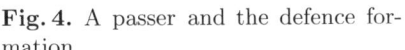

Fig. 4. A passer and the defence formation

Fig. 5. A real robot

3.1 Offence Team

The passer who is the nearest to the ball learns the team player behavior by passing the ball to one of four receivers or dribbling and shooting the ball to the goal by itself. After its passing, the passer shows a pass-and-go behavior that is a motion to the goal during the fixed period of time automatically. The receivers face to the ball and move to the positions so that they can form a rectangle by taking the distance to the nearest teammates (the passer or other receivers) (see Fig. 4). The initial positions of the team members are randomly arranged inside their territory.

3.2 Defence Team

The defence team member who is nearest to the passer attempts to intercept the ball, and each of other members attempts to "block" the nearest receiver. "Block" means to move to the position near the offence team member and between the offence and its own goal (see Fig. 4). The offence team member attempts to catch the ball if it is approaching. In order to avoid the disadvantage of the offence team, the defence team members are not allowed inside the penalty area during the fixed period of time. The initial positions of the team members are randomly arranged inside their territory but outside the center circle.

3.3 Robots and the Environment

Fig. 5 shows a mobile robot we have designed and built. Fig. 6 shows the viewer of our simulator for our robots and the environment. The robot has an omni-directional camera system. A simple color image processing is applied to detect the ball, the interceptor, and the receivers on the image in real-time (every 33ms.) The left of Fig. 6 shows a situation the agent can encounter while the right images show the simulated ones of the normal and omni vision systems. The mobile platform is an omni-directional vehicle (any translation and rotation on the plane).

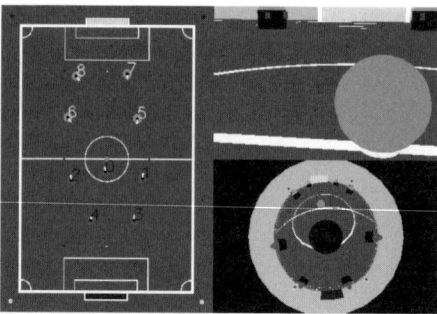

Fig. 6. Viewer of simulator

We suppose that the omni directional vision system provides the robot with 3-D construction of the scene. This assumption is needed for the other's state value estimation module since it is needed to estimate the sensory information observed by other robots.

4   Structure of the State and Action Spaces

4.1   State/Action Spaces for the Gate Module

The passer is only one learner, and the state and action spaces for the lower modules and the gate one are constructed as follows. The action modules are four passing ones for four individual receivers, and one dribble-shoot module. The other's state value estimation modules are the ones to estimate the degree of achievement of ball receiving for four individual receivers, that is how easily the receiver can receive the ball from the passer. These modules are give in advance before the learning of the gate module.

The action spaces of the lower modules adopt the macro actions that the designer specifies in advance to reduce the size of the exploration space without searching at the physical motor level.

The state space S for the gate module consists of the following state values from the lower modules:

- four state values of passing action modules corresponding to four receivers,
- one state value of dribble-shoot action module, and
- four state values of receiver's state value estimation modules corresponding to four receivers.

In order to reduce the size of the whole state space, these values are binarized, therefore its size is 2^4 x 2 x 2^4=512.

The rewards are given as follows:

- 10 when the ball is shot into the goal (one episode is over),
- -1 when the ball is intercepted (one episode is over),

 – 0.1 when the ball is successfully passed,
 – 0.3 when the ball is dribbled.

When the ball is out of the field or the pre-specified time period elapsed, the game is called "draw" and one episode is over.

4.2   State Space for the Passing Module

The state space of the passing module S is defined on the omni directional camera image as follows (see Fig. 7(a)):

 – the smallest angle among angles between the receiver and one the defence players who is nearer to the passer than the receiver (θ_1), and
 – the angle between the receiver and one of the defence players who is nearest to the passer (θ_2).

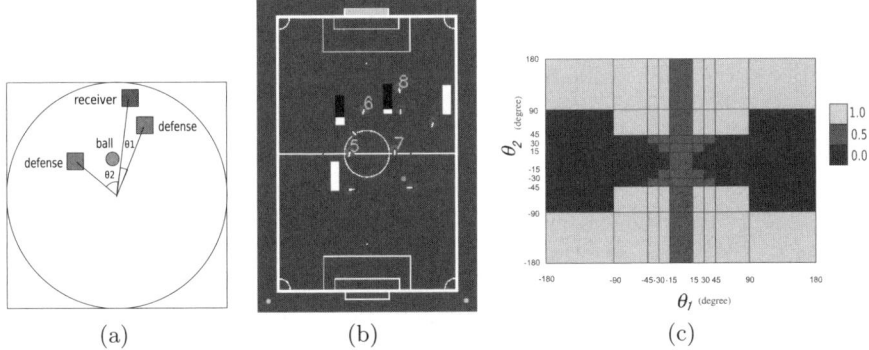

(a)   (b)   (c)

Fig. 7. State variable (a), examples of state values (b), and state value map of the pass module (c)

The both angles are quantized into ten levels including an invisible case, therefore the total number of states is 100. An example of the state values of four receivers is shown in Fig. 7(b) where the passer is the robot 3 (hereafter, r3 in short), and the color bars near four robots (r0, r1, r2, and r4) indicate the state values of the pass modules for four receivers, respectively. The higher the bar is, the higher the state value is. Since the pass courses for r1 and r2 are not intercepted by the defence players, their state values are high while the state values for r0 and r4 are low since their pass courses are intercepted by defence players.

The state value map is shown in Fig. 7(c) that indicates the smaller the angle between the receiver and the defence player is, the lower the state value is. The black region (one region is separated in the figure) is inexperience area.

4.3   State Space for the Dribble-Shoot Module

The state space of the dribble-shoot module S is defined on the omni directional camera image as follows (see Fig. 8(a)):

- the angle between the opponent goal and one of the defence players who is nearest to the passer (θ_1),
- the angle between the ball and one of the defence players who is nearest to the passer (θ_2),
- the distance to the nearest defence player (r), and
- the angle between the both edges of the opponent goal (θ_3) that represents the distance to the goal.

These state values are quantized into eight, five, eight, and seven, respectively. The total number of states is 8 x 8 x 5 x 7 = 2240.

The state value map of the dribble-shoot module in terms of θ_1 and r with fixed values of θ_2 and θ_3 is shown in Fig. 8(b) that indicates the nearer the defence player is, the smaller the state value is.

Two examples of the state values of the passer expected to take a role of a shooter is shown in Fig. 9 where the color bars near the passer indicate the state values of the dribble-shoot modules. The higher the bar is, the higher the state value is. Since the passer (r1) is near the goal and no defence players around in Fig. 9 (left), the state value is high while the state value of the passer (r3) in Fig. 9 (right) is low since it is located far from the goal and the defence players are around it.

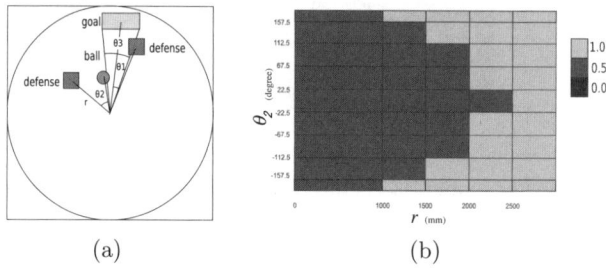

(a)   (b)

Fig. 8. State variables (a) and state values (b) for the dribble and shoot module

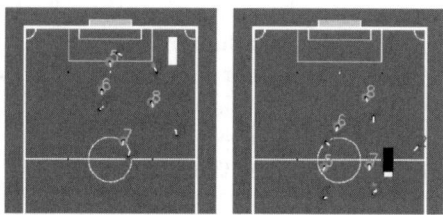

Fig. 9. Two examples of the state values: high (left) and low (right)

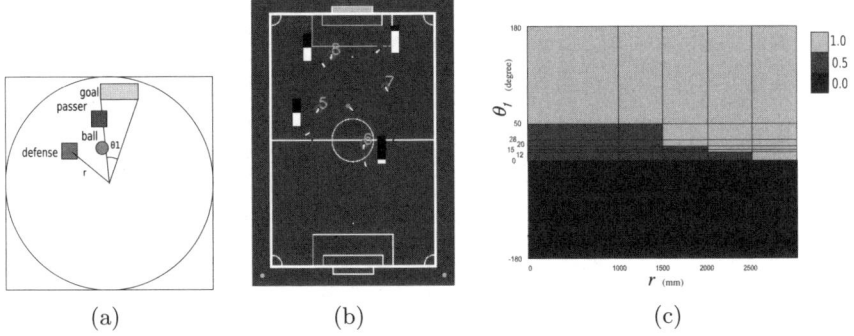

Fig. 10. State variables (a), examples of state values (b), and state value map (c) of the receiver module

4.4   State Space for the Receiver's State Value Estimation Module

The passer infers each receiver's state that indicates how easily the receiver can shoot the passed ball to the goal by reconstructing its TV camera view of the scene from the passer's omnidirectional view. Since we suppose that the passer has already learned the shooting behavior, the passer can estimate the receiver's state value by assigning its own experienced state of the shooting behavior.

The state space S for the receiver's state value estimation module consists of:

- The distance to the nearest defence player (r)
- The angle between the both side edged of the opponent goal (θ_1) that represents the distance to the goal (see Fig. (10(a)).).

The both are quantized into five and seven levels, therefore the number of states are 5 x 7 = 35.

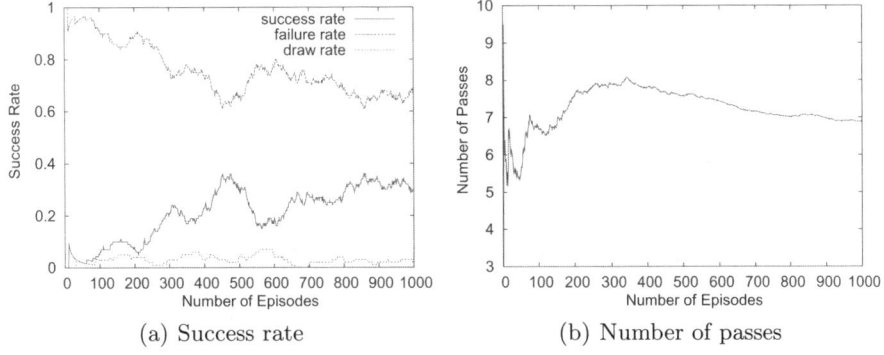

(a) Success rate

(b) Number of passes

Fig. 11. Success rate and the number of passes

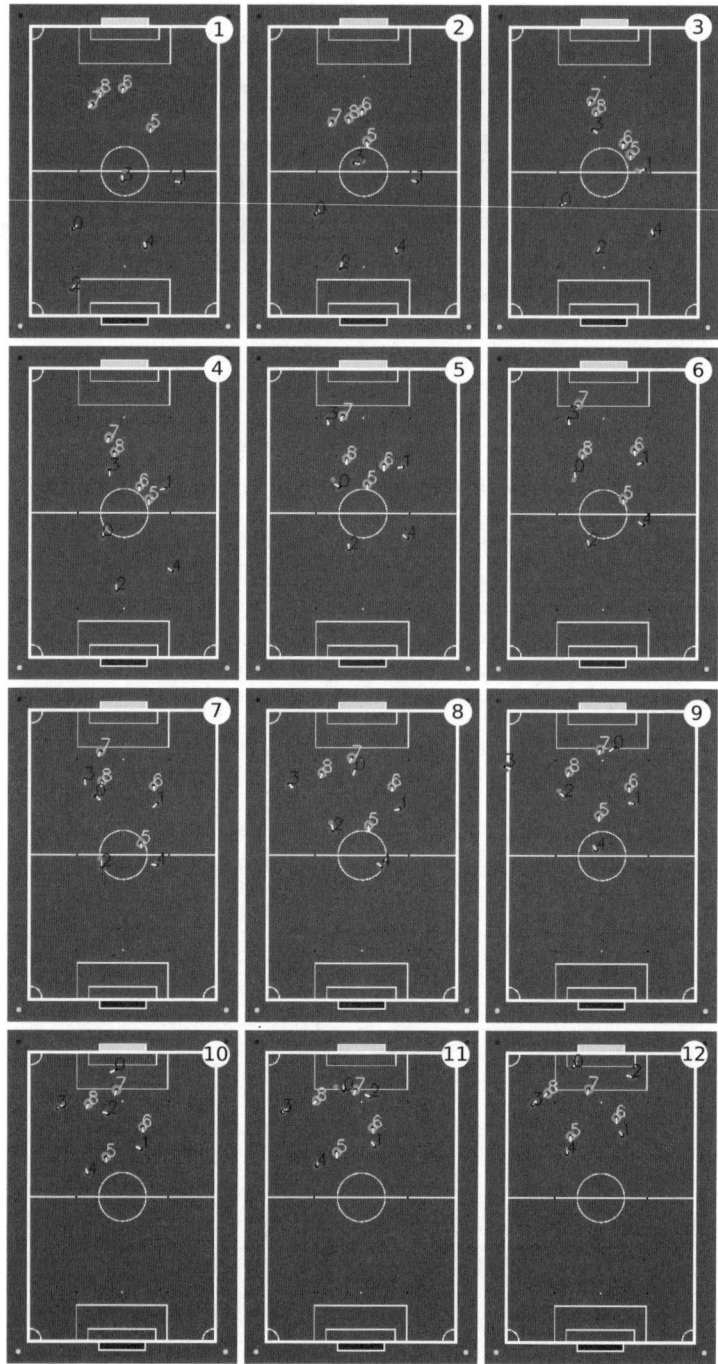

Fig. 12. An example of the acquired behavior

An examples of the state values of the receiver's state value estimation modules is shown in Fig. 10(b) where the color bars near the four receivers indicate their state values. The higher the bar is, the higher the state value is. Since the receiver (r0) is near the goal and no defence players around, the state value is high while the state values of other receivers (r1, r2, and r4) are low since it is located far from the goal and/or the defence players are around it.

The state value map of the receiver's state value estimation module in terms of θ_1 and r is shown in Fig. 10(c) that indicates the nearer (further) the defence player is and the further (nearer) the goal is, the smaller (larger) the state value is. The black region is inexperienced area.

5   Experimental Results

The success rate is shown in Fig. 11(a) where the action selection is 80% greedy and 20% random to cope with new situations. Around the 900th trial, the learning seems to have converged at 30% success, 70% failure, and 10% draw. Compared to the results of [6] that has around 30% success rate with 30,000 trials, the learning time is drastically improved (30 times quicker). Fig. 11(b) indicates the number of passes where it decreases after the 350 trials that means the number of useless passes decreased.

In cases of the success, failure, and draw rates when 100% greedy and 100% random are 55%, 35%, 10%, and 2%, 97%, 1%, respectively. The reason why the success rate in case of 100% greedy is better than in case of 80% greedy seems that the control policies of the receivers and the defence players are fixed, therefore not so new situations happened.

An example of acquired behavior is shown in Fig. 12 where a sequence of twelve top views indicates a successful pass and shoot scene.

6   Conclusion

We have used the state values instead of the physical sensor values and macro actions instead of the physical motor commands, and adopted the receiver's state value estimation modules that infer how easy for each receiver to receive the ball in order to accelerate the learning. As a result, we have much improved the learning time (30 times quicker!) compared to the result of the existing method [6] that has 32% success with communication and 23% without communication at around the 30,000th trial when the learning seems to have converged.

References

1. Connell, J.H., Mahadevan, S.: ROBOT LEARNING. Kluwer Academic Publishers, Dordrecht (1993)
2. Doya, K., Samejima, K., Katagiri, K.i., Kawato, M.: Multiple model-based reinforcement learning. Technical report, Kawato Dynamic Brain Project Technical Report, KDB-TR-08, Japan Science and Technology Corporation (June 2000)

3. Elfwing, S., Uchibe, E., Doya, K., Chirstensen, H.I.: Multi-agent reinforcement learning: Using macro actions to learn a mating task. In: Proceedings of 2004 IEEE/RSJ International Conference on Intelligent Robots and Systems, vol. 4, pp. 3164–3169 (2004)
4. Ikenoue, S., Asada, M., Hosoda, K.: Cooperative behavior acquisition by asynchronous policy renewal that enables simultaneous learning in multiagent environment. In: Proceedings of the 2002 IEEE/RSJ Intl. Conference on Intelligent Robots and Systems, pp. 2728–2734 (2002)
5. Jacobs, R., Jordan, M., Nowlan, S., Hinton, G.: Adaptive mixture of local experts. Neural Computation 3, 79–87 (1991)
6. Kalyanakrishnan, S., Liu, Y., Stone, P.: Half field offense in robocup soccer: A multiagent reinforcement learning case study. In: Proceedings CD RoboCup (2006)
7. Singh, S.P.: Transfer of learning by composing solutions of elemental sequential tasks. Machine Learning 8, 323–339 (1992)
8. Stone, P., Sutton, R.S., Kuhlmann, G.: Scaling reinforcement learning toward robocup soccer. Journal of Machine Learing Research 13, 2201–2220 (2003)
9. Sutton, R.S., Barto, A.G.: Reinforcement Learning: An Introduction. MIT Press, Cambridge (1998)
10. Takahashi, Y., Edazawa, K., Asada, M.: Multi-module learning system for behavior acquisition in multi-agent environment. In: Proceedings of 2002 IEEE/RSJ International Conference on Intelligent Robots and Systems, pp. CD–ROM 927–931 (October 2002)
11. Takahashi, Y., Kawamata, T., Asada, M.: Learning utility for behavior acquisition and intention inference of other agent. In: Proceedings of the 2006 IEEE/RSJ IROS 2006 Workshop on Multi-objective Robotics, pp. 25–31 (2006)
12. Whitehead, S., Karlsson, J., Tenenberg, J.: Learning multiple goal behavior via task decomposition and dynamic policy merging. In: Connell, J.H., Mahadevan, S. (eds.) ROBOT LEARNING, ch.3, pp. 45–78. Kluwer Academic Publishers (1993)

Beyond Frontier Exploration

Arnoud Visser, Xingrui-Ji, Merlijn van Ittersum,
Luis A. González Jaime, and Laurenţiu A. Stancu

Intelligent Systems Laboratory Amsterdam,
Universiteit van Amsterdam, The Netherlands
http://www.science.uva.nl/research/isla

Abstract. This article investigates the prerequisites for a global exploration strategy in an unknown environment on a virtual disaster site.
Assume that a robot equipped with a laser range scanner can build a detailed map of a previous unknown environment. The remaining question
is how to use this information on this map for further exploration.

   On a map several interesting locations can be present where the exploration can be continued, referred as exploration frontiers. Typically, a
greedy algorithm is used for the decision which frontier to explore next.
Such a greedy algorithm only considers interesting locations locally, focused to reduce the movement costs. More sophisticated algorithms also
take into account the information that can be gained along each frontier.
This shifts the problem to estimate the amount of unexplored area behind
the frontiers on the global map. Our algorithm exploits the long range
of current laser scanners. Typically, during the previous exploration a
small number of laser rays already passed the frontier, but this number
is too low to have major impact on the generated map. Yet, the few rays
through a frontier can be used to estimate the potential information gain
from unexplored area beyond the frontier.

1 Introduction

RoboCup Rescue is a competition in which (teams of) fully autonomous robots
visit a hypothetical disaster site. This situation is either simulated in the real
world [1] or a virtual world within the USARSim simulator [2]. The task for
the robots in the competition is to explore the site and locate victims. There
is a limited amount of time in which the robots can explore. Afterwards the
competing teams will be scored on a various criteria, among them are the size
of the explored area, the quality of the map and most importantly, the number
of located victims (for a more detailed list and scoring see [3]).

   An important problem in the competition is the *autonomous exploration* problem; to decide on the basis of the current map where to send the robot to improve
the future map [4]. A correct choice would improve the competition score, which
depends on the explored area and the quality of the map. Predictions about
what would be visible on the edges of the current map could help to make better
decisions for the robot.

U. Visser et al. (Eds.): RoboCup 2007, LNAI 5001, pp. 113–123, 2008.

In this work we build upon the contribution of the UvA Rescue team [5], which provided a fully autonomous agent system that controls up to eight virtual robots in the USARSim simulator [2]. The system already has a state of the art method [6] for simultaneously locating and mapping (SLAM) unknown environments, based on the Manifold approach [7] combined with the Weighted Scan Matching algorithm [8]. The exploration strategy of the robots is kept simple. The behavior is reactive and makes decision based on direct measurements, not on the current map. The goal of this work is to improve the exploration strategy by intelligently using global information that can be derived from the map. As demonstrated in [9], efficient allocation of the search effort can outperform simple exploration strategies. In the article the focus is on the exploration behavior of a single robot (or agent).

The outline of this paper is as follows. In Sect. 2 we will introduce some theoretical background behind this research. In Sect. 3 our algorithm will be worked out. The robustness of our method on maps from the RoboCup Rescue Competition will be demonstrated in Sect. 4. Finally, we draw our conclusions in Sect. 5.

2   Background

Exploration is the problem of directing a robot through the environment so that the knowledge about the external world is maximized [10]. Knowledge about the external world for a mobile robot is typically stored on a map m. Increasing the knowledge stored on a map can mean that the uncertainty about information on the map is reduced, or that new information is added to the map. The latter means that the map coverage is extended with parts of the external world that the robot has not seen before. Knowledge about the map m can be passively acquired, while the robot is wandering around busy with other tasks (for instance finding victims), as demonstrated by [11]. Here the focus is on autonomous exploration; the planning of the next exploration action a which will increase the knowledge about the world the most. Before this estimate is worked out in more detail, it should be noticed that such an exploration action can be quite complex from navigational point of view. Executing such an exploration action can mean that large parts of the current map are traversed, which can only be efficiently done with the availability of on-line path-planning functionality.

For a mobile robot it is important to remember were obstacles are located. This information can be represented with an occupancy grid map [12], where each grid cell indicates the probability $p(x)$ if that location x is occupied or free. Active exploration can been seen as minimizing the *information entropy $H(m)$* [13] of the probability distribution $p(x)$ for all x on the map m, which requires an integration over the complete occupancy grid map:

$$H(m) = -\int_{x \in m} p(x) log(p(x)) \tag{1}$$

When all grid cells are initialized as unknown by giving them a uniform value of $p(x) = 0.5$, the entropy of the map $H(m)$ is maximal. When the boundaries of the map m are not known, the limits of the integral are slowly extended when new areas are discovered. The interest is not in the absolute value of the entropy, but in the difference in entropy before $H(m)$ and after $H(m|a)$ an exploration action a; the *information gain* $\Delta I(a)$ [14,15,16]. Remember that the exploration action a could be a complex maneuver, consisting of a number of controls u_i and observations z_i for multiple timesteps i.

$$\Delta I(a) = H(m|a) - H(m) \tag{2}$$

Because the set of possible exploration actions can grow very fast when predictions are needed multiple timesteps in the future, this set is approximated. In existing exploration methods the number of exploration actions is reduced by considering only the path to a finite number of candidate observation points. Typically, those candidate observation points are chosen on the boundary of explored and unexplored areas; frontier-based exploration [17].

One of the most interesting approaches to generate and select those candidate observation points is the presented by González-Baños [18]. They model exploration frontier with free curves; polylines which indicate where the laser range scanner reported values larger than a threshold r_{max}. Near those free curves a number of candidate observation points are considered . This number of candidate points is generated randomly with a Monte-Carlo method, and for each point q they simulate a number of laser scans through the free curve. The amount of area $A(q)$ covered by those rays (with a maximum length r_{max}) is taken in account as a measure of the potential information gain $\Delta I(a_q)$ for the observation z_q at the observation point q.

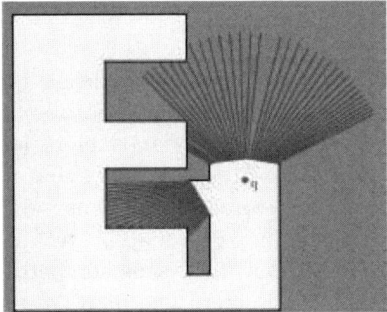

Fig. 1. The potential information gain of a candidate observation point q is the area $A(q)$ that may be visible through the two free edges; this area is estimated by casting rays from q. Courtesy from [18].

The area $A(q)$ is an estimation for the information gained from the observation z_q. This implicitly ignores the information that could be gained by the observations $z_1, \ldots, z_{q-1}$ along the path to the observation point q. When the robot traverses mainly well known regions on its path to point q this is a reasonable assumption. Yet, the exploration action a_q consists not only of a number of observations $z_1, \ldots, z_q$, but also of a number of controls $u_1, \ldots, u_q$ to drive the path. Because control is never perfect, confidence about the location of the robot is lost for every control step u_i. Probabilities spread out over the map, resulting in a loss of information. An optimal exploration action a^* can only be chosen when the cost of traveling along a path $u_1, \ldots, u_q$ is taken into account. Typical traveling cost functions are the distance traveled, the time taken or the energy expended. González-Baños has chosen to use as cost-function the length $L(q)$ of the path $u_1, \ldots, u_q$ to a candidate observation point q. They combine the cost of the path $u_1, \ldots, u_q$ and the estimated gain of the observation z_q for the evaluation action a_q into the following value function $V(q)$:

$$V(q) = A(q)e^{-\lambda L(q)} \tag{3}$$

The constant λ can be used balance the cost of motion $L(q)$ against the expected gain of information $A(q)$.

For the observation points found in this article an equivalent value function could be calculated. Note, however, that the area $A(q)$ of González-Baños is extrapolated from the current map by simulating a number of laser-rays through the frontier, while in our case the area $A(q)$ is directly estimated from the laser range measurements. Another difference is the generation of observation points. González-Baños generates multiple candidate observation points on a short random distance from the exploration frontiers. In our approach, per frontier a single candidate observation point is generated, in the center of the exploration frontier.

3   Estimation of Exploration Frontiers and Observation Points

A good autonomous exploration algorithm should navigate the robot to an optimal observation point. This point will be close to an exploration frontier. To find such an exploration frontier is not trivial. Exploration frontiers can be found based on an occupancy grid map. The probability that a point is an obstacle or not on a certain location can be stored in occupancy grid with arbitrary resolution. An example of such occupancy grid map is given in Fig. 2, an actual map produced during the Virtual RoboCup Rescue competition by the UvA Rescue team with a resolution of 1 centimeter. The map gives a top view of an office environment, where clearly three corridors are visible that are well explored, and a number of adjacent rooms that are not entered yet.

Grid points on a map can be combined to regions, when the edges of the regions can be found. As can be seen from the example (Fig. 2), the boundaries of the safe region are only sharp along walls. Inside the rooms and at the end of the corridors the boundaries of the safe region are fuzzy. Selecting an absolute threshold for this boundary is difficult [19]. Still, a human can clearly distinguished safe regions and indicate the regions that should be further explored (observation regions). What is difficult, also for a human, is the precise location of the boundary between those regions; the exploration frontier.

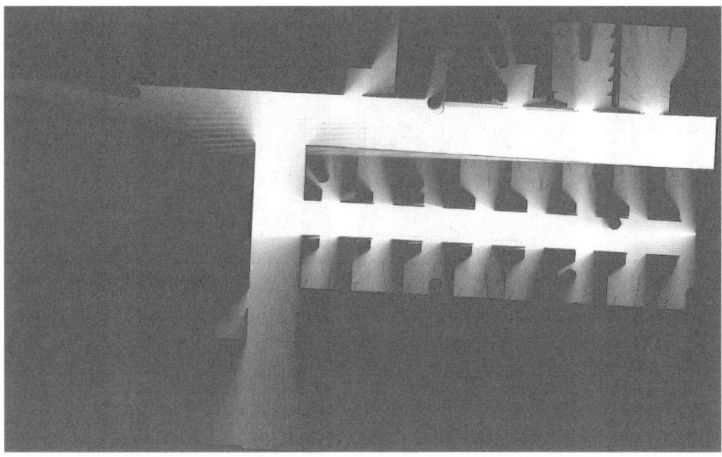

Fig. 2. The occupancy grid map produced during the semi-final by one the robots

The method used in this paper to distinguish safe regions from observation regions is based on a simple ray-casting technique. Ray-casting is used to generate an occupancy grid from the scan-data stored in the manifold [11]. The trick is to generate two occupancy grids at the same time; one with a short range constraint r_{safe} and one with a long range constraint r_{max} equal to maximum range of the laser scanner. A typical value for r_{safe} is 2 meters and for r_{max} 20 meters. The occupancy grid with a short range constraint r_{safe} generates a conservative estimate of the obstacle free space; the safe region.

An example of such safe region is given in Fig. 3.a. The three corridors of the office environment can be recognized in this picture. From the contour of the safe region the exploration frontier can be derived. The contour of the safe region is indicated in Fig. 3.b. The exploration frontier is only a part of the contour, the other part of the contour are walls (Fig. 3.c). The part of the safe region contour that is no wall can be identified as the exploration frontier (Fig. 3.d).

The same ray-tracing can be repeated with the long range constraint r_{max}. With the long range constraint a less conservative estimate of free space is generated. The areas are probably free of obstacles, but not guaranteed to be safe. The result is visualized in Fig. 4. Outside the corridors new contours are visible: the rooms along the corridor. These contours are the areas which are probably free, but not guaranteed to be safe: the observation regions (indicated in yellow). These observation regions are not equivalent with unknown areas; there also exist large parts of the map where the probability $p(x)$ is still on its initial value. The frontiers between the observation regions and the safe regions, as shown in Fig. 3.d, are also given (indicated in grey). The convex frontiers contours are extended with a small white point which is the center of the contour. These centers can be associated with potential observation points. From each point in the safe region the path to such a potential observation point q can be estimated with a breath-first algorithm. An example of such path is indicated in green, from the start position indicated with a large white point in the upper-right corner.

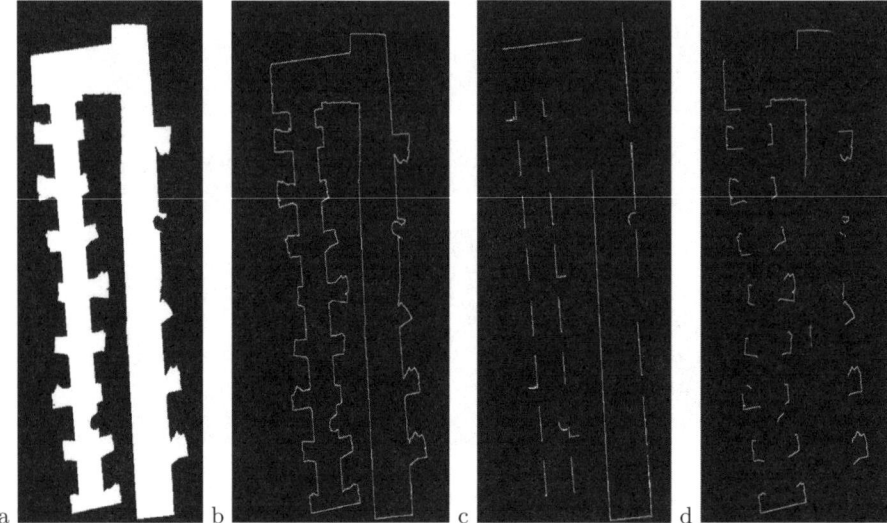

Fig. 3. The safe region of Fig. 2 (90° degrees rotated). From left to right respectively a) The surface. b) The contour. c) The part of the contour which is an obstacle (wall). d) The part of the contour which is free (frontier).

Fig. 4. The interpreted map of Fig. 2. The yellow contours indicate the observation regions. The grey contours indicate the exploration frontiers. The large white circle indicates the current position of the robot. The small white circles indicates potential observation points. The green line indicates a possible path to those observation points. Red lines indicate the walls.

In this example one can also see that the process of frontier estimation is not completely failsafe. Going from one corridor to another, the robot makes a sharp turn (top of Fig. 3). During such turn the confidence in the location estimate can drop. At that moment multiple laserscans of the same wall do not completely overlap, and the wall as edge is not sharp. In that case a part of the wall is seen as unknown, and such a contour could be incorrectly identified as frontier. The effect is visible near the upper right corner in Fig. 3.d. To prevent false positives like this, the following consistence check is designed. All frontiers are tested if they are concave or convex. Only convex frontiers generate candidate observation points q in Fig. 4. Many of the false frontiers can be removed because they are concave. The result is a slight increase of the number of false negatives, as indicated in the Sect. 4.

4   Results

In the previous sections we have illustrated our methods on the map given in Fig. 2. In this map several potential observation points were identified, as shown in Fig. 4. To test the reliability of our method the number of observation points is compared against the number of points that should have been found. After the 2006 competition the environment used during the RoboCup was made available for inspection[1]. A top view of this environment is visible in Fig. 5.a. The three corridors explored on the map can be found in the upper-right corner of Fig. 5.a.

With the provided environment as reference, the number of doorways and corridors that the robot has passed during its exploration can be counted. For the map given in Fig. 2 in total 2 corridors and 20 doorways to 20 rooms are passed. There is another doorway to a 21st room, but this doorway is blocked by a victim. The algorithm skipped this doorway correctly. The results are summarized in Table 1. Next to the expected and found number of doorways, the number of false positives and false negatives are given. This is done for both the exploration frontiers (both convex and concave) and the observation points (center of convex exploration frontier). One can see that the number of false positives is reduced by only selecting convex exploration frontiers.

To demonstrate the robustness of the algorithm, the procedure is repeated for two other maps that could have been encountered during the competition. The first map is tour that begins and ends in the lobby, the light-grey area at the bottom of Fig. 5.a. During this tour 6 corridors and 7 doorways to rooms should have been found. For the majority of the corridors and rooms an observation point is found, as can be seen from Table 1 and Fig. 5. The missed corridor and room are located in the left lower corner. The robot came through the narrow passage at the left and turned back towards the lobby. Due to this turn the robot did not get a clear view into the corner. The doorway to the room is visible; the 6th corridor stays mainly hidden behind the robot. The combined frontier of the corner, doorway and corridor was irregular of shape and not convex, which resulted in a false negative.

---

[1] http://sourceforge.net/projects/usarsim

Table 1. Experimental results

	expected frontiers & observation points	**exploration frontiers**			**observation points**		
		found	*false positives*	*false negatives*	**found**	*false positives*	*false negatives*
Three corridors (Fig. 2)	22	27	6	1	21	0	1
Lobby loop (Fig. 5)	13	17	5	1	11	0	2
Yellow arena (Fig. 6)	9	17	8	0	9	1	1

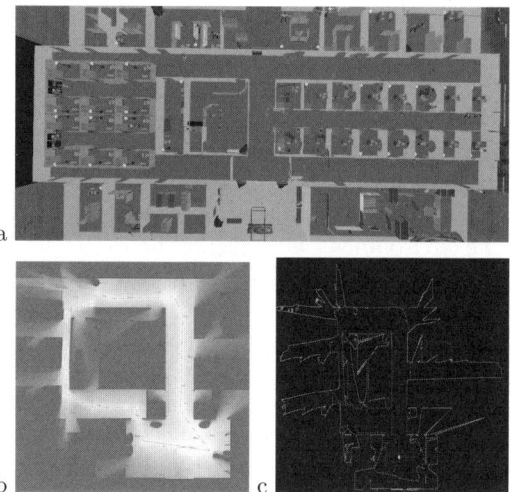

Fig. 5. Overview of the indoor area used for the 2006 RoboCup Virtual League Competition. Fig. b and c show the results of exploring a loop starting and ending in the large lobby at the bottom of Fig. a. On this map 11 of the 13 observation points are found.

Last, but not least, the algorithm was tested on the Yellow arena. This is classical benchmark in the RoboCup Rescue competition, where an irregular office-maze is build with flexible walls. The Yellow arena is also visible in Fig. 5.a, the large room to the right of the lobby. Fig. 6.a gives a closer look at this environment. In this office-maze it is less obvious to indicate what the frontiers are that should have been explored. For instance, central in the Yellow arena is a bed. The algorithm indicated with the three green paths that an observation should be made at the left, at the right and under the bed. This was classified as a correct decision. Another aspect is the open space in the rooms. The rooms were sometimes so large that frontiers appeared in the corners. These frontiers in the corner have a convex shape, and could be selected if the observation space behind the frontier was large enough. Checking the corners of a room is probably quite

Fig. 6. Results of the Yellow-Arena map. Fig. b and c show the results of exploring a loop starting and ending in the curved wall in the center of the figures. On this map 9 observation points are found (one false positive).

robust, but probably not highly efficient. Both the false positive and negative were related with a corner. At the bottom left of Fig. 6.c a doorway is missed, because that corner was not explored from close enough distance. On the other hand an observation point is generated to check the tiny space behind the 'W'-shaped obstacle. Fortunately many other observation points are generated which much more area behind the frontier, which make them far more attractive for exploration. This observation point was classified as a false positive.

Overall, these experiments, summarized in Table 1, demonstrate the robustness of the algorithm. The algorithm generated a limited number of potential observation points. The impact of the false negatives (4 of the 44 potential observation points were missed) on the exploration behavior will be minor. As long as there are enough candidate observation points, the robots can coordinate their actions and distribute the points over the team. They can optimize their effort by optimizing a joint value function equivalent with equation (3).

5   Conclusion

In this report an algorithm is proposed to generate a limited number of observation points, and to estimate the information that could be gained at each location. The method generates exploration frontiers on the contours of safe regions. One observation point is generated per convex exploration frontier. The potential information gain for each observation point is estimated based on the area of obstacle free space beyond the exploration frontier. This estimate of the area is based on measurements, and not on extrapolations from the current map. The algorithm shows good results in office-environments.

In our future research it will be demonstrated how much the exploration efficiency will increase by selecting the observation point with highest potential information gain and the lowest travel costs.

Acknowledgments

The authors thank Max Pfingsthorn and Bayu Slamet for providing us with the virtual robot control software, including the outstanding mapping capabilities, which was an excellent starting point for further research.

A part of the research reported here is performed in the context of the Interactive Collaborative Information Systems (ICIS) project, supported by the Dutch Min istry of Economic Affairs, grant nr: BSIK03024.

References

1. Jacoff, A., Messina, E., Weiss, B., Tadokoro, S., Nakagawa, Y.: Test arenas and performance metrics for urban search and rescue robots. In: Proceedings of the 2003 IEEE/RSJ International Conference on Intelligent Robots and Systems (2003)
2. Balakirsky, S., Scrapper, C., Carpin, S., Lewis, M.: Usarsim: providing a framework for multi-robot performance evaluation. In: Proceedings of PerMIS 2006 (2006)
3. Balakirsky, S., Carpin, S., Kleiner, A., Lewis, M., Visser, A., Wang, J., Ziparo, V.A.: Towards heterogeneous robot teams for disaster mitigation: Results and performance metrics from robocup rescue. Journal of Field Robotics (to appear, 2007)
4. Hasegawa, B.R.: Continues observation planning for autonomous exploration. Master's thesis, Massachusetts Institute of Technology (2004)
5. Pfingsthorn, M., Slamet, B., Visser, A., Vlassis, N.: Uva rescue team 2006 robocup rescue - simulation league. In: Lakemeyer, G., Sklar, E., Sorrenti, D.G., Takahashi, T. (eds.) RoboCup 2006: Robot Soccer World Cup X. LNCS (LNAI), vol. 4434, Springer, Heidelberg (2007)
6. Pfingsthorn, M., Slamet, B., Visser, A.: A scalable hybrid multi-robot slam method for highly detailed maps. In: Robocup 2007: Robot Soccer World Cup XI. Lecture Notes on Artificial Intelligence (to be published, 2007)
7. Howard, A., Sukhatme, G.S., Matarić, M.J.: Multi-robot mapping using manifold representations. In: Proceedings of the IEEE - Special Issue on Multi-robot Systems (2006)
8. Pfister, S.T., Kriechbaum, K.L., Roumeliotis, S.I., Burdick, J.W.: A weighted range sensor matching algorithm for mobile robot displacement estimation. IEEE Transactions on Robotics and Automation (to appear, 2007)
9. Bourgault, F., Göktoğan, A., Furukawa, T., Durrant-Whyte, H.F.: Coordinated search for a lost target in a bayesian world. Advanced Robotics 18(10), 979–1000 (2004)

10. Thrun, S., Burgard, W., Fox, D.: Probabilistic Robotics (Intelligent Robotics and Autonomous Agents). The MIT Press (2005)

11. Slamet, B., Pfingsthorn, M.: Manifoldslam: a multi-agent simultaneous localization and mapping system for the robocup rescue virtual robots competition. Master's thesis, Universiteit van Amsterdam (2006)

12. Moravec, H.: Sensor fusion in certainty grids for mobile robots. AI Magazine 9, 61–74 (1988)

13. Fox, D., Burgard, W., Thrun, S.: Active markov localization for mobile robots. Robotics and Autonomous Systems 25, 195–207 (1998)

14. Roy, N., Burgard, W., Fox, D., Thrun, S.: Coastal navigation - mobile robot navigation with uncertainty in dynamic environments. In: Proceedings of the IEEE International Conference on Robotics and Automation, pp. 34–40 (1999)

15. Simmons, R.G., Apfelbaum, D., Burgard, W., Fox, D., Moors, M., Thrun, S., Younes, H.: Coordination for multi-robot exploration and mapping. In: AAAI/IAAI, pp. 852–858 (2000)

16. Sim, R., Roy, N.: Global a-optimal robot exploration in slam. In: Proceedings of the IEEE International Conference on Robotics and Automation (ICRA), Barcelona, Spain (2005)

17. Yamauchi, B.: A frontier based approach for autonomous exploration. In: Proceedings of IEEE International Symposium on Computational Intelligence in Robotics and Automation, Monterey, July 10-11, 1997 (1997)

18. González-Baños, H.H., Latombe, J.C.: Navigation Strategies for Exploring Indoor Environments. The International Journal of Robotics Research 21(10-11), 829–848 (2002)

19. van Ittersum, M., Xingrui-Ji, Gonzalez, L., Stancu, L.: Natural boundaries. Report, Universiteit van Amsterdam (2007)

Robot Building for Preschoolers

Peta Wyeth[1] and Gordon Wyeth[2]

[1] University of Nottingham, Jubilee Campus, Nottingham NG8 1BB, UK
[2] University of Queensland, St Lucia, Queensland 4072, Australia
peta.wyeth@nottingham.ac.uk,
wyeth@itee.uq.edu.au

Abstract. This paper describes Electronic Blocks, a new robot construction element designed to allow children as young as age three to build and program robotic structures. The Electronic Blocks encapsulate input, output and logic concepts in tangible elements that young children can use to create a wide variety of physical agents. The children are able to determine the behavior of these agents by the choice of blocks and the manner in which they are connected. The Electronic Blocks allow children without any knowledge of mechanical design or computer programming to create and control physically embodied robots. They facilitate the development of technological capability by enabling children to design, construct, explore and evaluate dynamic robotics systems. A study of four and five year-old children using the Electronic Blocks has demonstrated that the interface is well suited to young children. The complexity of the implementation is hidden from the children, leaving the children free to autonomously explore the functionality of the blocks. As a consequence, children are free to move their focus beyond the technology. Instead they are free to focus on the construction process, and to work on goals related to the creation of robotic behaviors and interactions. As a resource for robot building, the blocks have proved to be effective in encouraging children to create robot structures, allowing children to design and program robot behaviors.

Keywords: Educational robotics, robot construction kit, robot programming environment.

1 Introduction

Robot building and programming allows children to become creators of technology. As designers and builders of technology, children become more deeply engaged with technology education than they might from more conventional classroom activities. However it is only in recent years that classroom robot building has become possible for children in middle and secondary school, with the advent of resources such as the LEGO® RCX™ brick. Now, with these type of resources available, children from around the world have become engaged in robot building and programming, as evidenced by the success of programs such as RoboCup Junior.

In the specialist area of early childhood education there remains the challenge for educators to develop educational programs which include technology that is suitable to the unique needs and abilities of this age group. There are concerns about the

U. Visser et al. (Eds.): RoboCup 2007, LNAI 5001, pp. 124–135, 2008.

young children's physical and cognitive readiness to use computers and other technological artifacts. Robot building and programming is a case in point. Given that young children, between the ages of three and six years, are only just acquiring the rudiments of notational systems and struggle with symbolization in language, pictures, three-dimensional objects and pretend play [1][2][3], it is apparent that existing technology is unsuitable for all but the most gifted in this age group.

This paper details the use of Electronic Blocks designed for young children aged between four and eight. Electronic Blocks are a new resource for technology education which have been designed and built to provide an appropriate means through which young children are able to create and program simple robots. The paper describes the blocks, and illustrates their effectiveness from observations of a two week study of four and five year old children as they used the Electronic Blocks.

2 Background

In identifying the way in which technology should be used in early childhood education, Yelland [4] looks towards environments that are stimulating and encourage active exploration of objects and ideas. Such environments facilitate quality technology education – the technology becomes a resource which allows young children to be involved in the design and production processes to produce various outcomes. Resnick [5] agrees and asserts that the

Best computational tools do not simply offer the same content in new clothing; rather, they aim to recast areas of knowledge, suggesting fundamentally new ways of thinking about the concepts in that domain, allowing learners to explore concepts that were previously inaccessible.

Resnick and his group at the MIT media lab based their research on this philosophy. They started with the development of LEGO/Logo [6] which combined the LEGO Technic product with the Logo programming language. It was the first robotic construction kit ever made widely available [7]. Unfortunately, each construction built in a LEGO/Logo environment by necessity was connected to a computer via wires. This led to a lack of mobility and was the greatest limitation of LEGO/Logo [7].

The Programmable Brick is a successor to this research. The Programmable Brick is a tiny computer embedded inside a LEGO brick that children use to build systems that behave and respond to their environment [8]. Children included the Programmable Brick into their regular LEGO constructions and then wrote Logo computer programs to make their creations react and behave. The second generation of the Programmable Brick – the Red Brick – was specifically designed for robustness and ease of manufacture and this Brick was widely used in classroom settings [7]. The success of the Red Brick is highlighted by the final version of the Programmable Brick – the LEGO™ RCX® Brick which is now a commercially available product, and is widely used by children in robot competitions such as RoboCup Junior.

A construction kit called Cricket was developed as a successor to the Programmable Brick. Crickets are small Programmable Bricks that, in addition to connecting to motors and sensors, can communicate with each other via infrared light [8]. The communication ability of Crickets allows children to think about systems of communicating entities and explore the behaviors that arise from Cricket interactions. Like the Programmable Brick, Crickets are fully programmable with children being able to write and download computer programs into the Crickets from a desktop computer [9].

The development of *curlybot*, under the direction of Hiroshi Ishii at MIT, has occurred in parallel to the development of Crickets. *curlybot* is aimed at children in their early stages of development - ages four and up [10]. It is an autonomous two-wheeled vehicle with embedded electronics that can record how it has been moved on any flat surface and then play back that motion accurately and repeatedly. Children can use *curlybot* to develop intuitions for advanced mathematical and computational concepts, like differential geometry, through play away from a traditional computer [10]. In preliminary studies conducted by the developers of *curlybot*, they found that children learned to use *curlybot* quickly.

3   Electronic Blocks

Electronic Blocks aim to provide the same rich robot building and programming experiences as the Programmable Brick, but with intuitive tangible interface of *curlybot*. The Electronic Blocks are physical building blocks of a size and shape familiar to the target age group (LEGO® Duplo™ Primo™ blocks). The programmability and intelligence of the blocks has been created by placing electronics inside. Some blocks have sensor inputs and others have action outputs. When connected together, the output of sensor blocks control the input of action blocks. Logic blocks can act as intermediary structures to change the effect of a sensor. Any number of blocks can be stacked together to create a huge variety of robotic vehicles and structures that interact with the environment and each other.

3.1   Functional Design

There are three kinds of Electronic Blocks: sensor blocks, action blocks and logic blocks. Sensor blocks are capable of detecting light, touch and sound. Each block has an input attached to its upper connector and an output attached to its lower connector (see Figure 2). The input is off unless it explicitly receives an on signal. The input and the sensor are logically ORed together to produce the output. As a result when two or more sensor blocks are stacked in any way on an action block, any sensor input will trigger the action block.

Action blocks produce some kind of output. The *light* block produces light, the *sound* block produces sound and the *movement* block is capable of motion. All action blocks have two connectors on top, each capable of triggering the action; both inputs are ORed together to produce the output. They are physically constrained by a base plate with no connectors so that they cannot be placed on top of another block and have to be positioned at the bottom of a block stack.

Fig. 1. The Electronic Blocks in action - a remote control car built from the blocks. The child has built a torch from a touch and light block (close-up on right), which is being used to trigger a light sensor on a motion block.

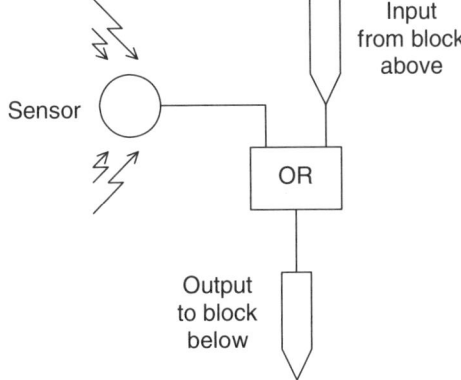

Fig. 2. The functional implementation of a sensor block. The sensor is ORed with any signals from blocks above.

Logic blocks have an intermediary role. Placed between a sensor block and an action block they have the ability to alter the expected action. Logic blocks provide users with the capability to:

- produce an action if a particular stimulus is not received (*not*),
- toggle the input so that in the first instance the stimulus from the environment will "turn the action on" and the second instance of the stimulus will "turn the action off" (*toggle*),
- stretch a short signal so that the action will stay on for two seconds after the stimulus stops (*delay*), and
- only produce an action if input signals are received simultaneously through both inputs (*and*).

With the exception of the *and* block, these blocks are single connector blocks with an input attached to the upper connector and an output attached to the lower connector. The *and* block has two inputs and two outputs. The *and* block has two upper connectors which may receive an input signal. The block works as a logical

AND – it must receive an input from both connectors to produce an output. The output signal produced is attached to both lower connectors.

3.2 Physical Block Design

All sensor blocks are yellow. Readily understandable icons identify the different functions of the sensing blocks: for example, an eye for a seeing block. The functionality of the action blocks is somewhat self-evident from the physical structure of the blocks. The sound and light blocks are also adorned with explanatory icons. Each different logic block type has distinctive icons and colors to assist their identification. It is difficult to choose meaningful icons for these blocks. What icon explains "and" to a preschooler? The icons were chosen to have readily understood adult meanings: for example, & for "and".

Fig. 3. The complete set of Electronic Blocks. The sensor blocks are to the left, the logic blocks are in the centre and the action blocks to the right.

3.3 Electronic Block Communication

Electronic Blocks are designed so that there is no need for children to attach wires or fix connectors to enable blocks to pass signals from one to the next. Each of the block's upper connector (or connectors) corresponds to a dome on the LEGO blocks. A block's lower connector (or connectors) is found in the hollow at the base of the block. Electronic Block communication is achieved optically, allowing for imprecise positioning of one block on the other.

4 Preschooler's Interactions with Electronic Blocks

One study of the Electronic Blocks was specifically designed for preschool children, aged between 4 and 5 years. This study was primarily focused on assessing the extent to which the Electronic Blocks allowed children to build and program simple robots. The study for this age-group is structured in such a way as to observe the children using Electronic Blocks in a natural, open-ended, free-play setting. It took place at a University Campus Preschool with twenty-eight children aged between four and six years. Fifteen of the participants were female, thirteen were male.

4.1 Study Procedure

The study spanned two weeks. Three sessions per week were conducted and each session lasted 90 minutes. For each session the Electronic Blocks were set up in an area within the indoor play area. A video camera and audio equipment were used to record children's interactions with the blocks. All children within the Preschool Room were free to participate in the study. However, due to the number of Electronic Blocks available, a limit of four children using the blocks at any time was imposed. The investigator actively participated in all evaluation sessions, providing children with ideas on how they might use the blocks, answering their questions, helping them to solve problems, and encouraged working in pairs or groups.

Before the first of the six sessions the Preschool teacher introduced the researcher to the children and the intention of the study was simply explained. The Electronic Blocks were then demonstrated to the entire group, with the functionality of each block briefly explained. Initial explanations of the Electronic Blocks primarily focused on the sensor blocks and action blocks. The idea was to introduce participants to the less complex Electronic Block concepts. Children were provided with the opportunity to become familiar with the functionality of these blocks before moving on to the more complex combinations involving logic blocks. By sessions 5 and 6, the involvement of the researcher was reduced. While available to help them if they ask for assistance, the researcher did not play an active role in stimulating the children's play experiences with the blocks.

4.2 Preschool Observations

The video of preschoolers using Electronic Blocks was examined to obtain usage analysis of the preschoolers' interactions with the blocks. Specifically, the video was analyzed to determine:

- the number of times each children interacted with the Electronic Blocks;
- the duration of interactions with Electronic Blocks; and
- the number of structures children built while using the blocks.

This data has enabled an evaluation of the Electronic Blocks to determine whether they were an effective resource for robot building and programming. Specifically, the data has been analyzed to determine the preschoolers'

- patterns of usage,
- interactions with the blocks,
- level of involvement in building a variety of constructions, and
- level of understanding of Electronic Block functionality.

Patterns of Usage
Of the 31 preschoolers who attended the preschool over the period of the evaluation, 28 chose to participate. Fifteen of the participants were female, thirteen were male. Of the preschoolers who used the Electronic Blocks, 20 used the blocks on more than one occasion. Children on average played with the blocks between two and three times

during the six days of evaluation, with females visiting the blocks slightly more frequently (an average of 2.5 visits for the females versus 2.3 visits for the males).

The average amount of time each child spent playing with the blocks in a single session was 15 minutes. Females spent an average of 12 minutes interacting with the Electronic Blocks in a single session, while males, on average, interacted with the blocks for 18 minutes in a single session. The longest time spent playing with the blocks in one session was 47 minutes while the shortest period of time was two minutes. Overall average length of visits remained reasonably consistent across visits, ranging between 11 and 16 minutes.

Interactions with Electronic Blocks

The video evidence shows that on average each child built a working block stack every two minutes. While one participating child failed to build anything during the evaluation period, other children built block constructions at an increased rate. Construction included adding a block or blocks to an existing stack or creating a stack from scratch. On average, girls created a different structure every two and a half minutes, and boys created one every one minute and forty seconds. It is interesting to note that while some children were avid builders others were content to build one particular structure and play with it for a long period of time. One example of note is where one child built a remote control car and then played with it for 15 minutes.

In general, boys were involved in building more structures than girls. On average, boys built 21 structures over the duration of the evaluation while girls built 10 structures. The girls built, on average, five structures per visit, while the boys build 10 structures per visit.

There were examples where children were observed using the Electronic Block structures to stimulate other play. The construction of Electronic Block structures did not appear to be their primary activity but rather an activity which complemented their pretend play adventures. Another noteworthy issue concerns construction activities which are primarily about process rather than outcome. There were some children who were not concerned with the output they produced and the act of construction was their motivation for taking part. In these cases the children tended to build elaborate stacks of blocks, the largest stack consisting of thirteen blocks. The children were primarily involved in building interesting structures with the blocks with no consideration for what the outcome would be.

Types of Construction

The children were involved in a wide variety of construction activities using the range of Electronic Blocks. Analysis of the video data indicates that the *movement* block was the action block of choice. While the *light* block was also popular, children were more likely to create structures with car bases than with the other two action blocks. All children who interacted with the blocks created a moving vehicle at some stage in their construction activity. The sensor blocks appeared equally popular. Children used the *seeing*, *hearing* and *touch* blocks to activate the *movement* blocks, and many were successful at creating a "remote control" car using a *seeing* block attached to the *movement* block and then activating a separate *light* block to make the car move. This became very popular and for over 70% of the evaluation period at least one child was playing with a remote control vehicle that they had built.

The *toggle* and *not* blocks were used on more occasions than the other logic blocks. The children would often use *not* blocks to activate their action blocks. Children would use the *toggle* block when they did not want to keep providing an environmental stimulus for some action block. Some children also enjoyed using *delay* blocks but the *and* block was used sparingly throughout the evaluation.

A large majority of the constructions undertaken by the children contained either two or three blocks. Very few structures were built which had five or more blocks. The addition of blocks to structures, particularly logic blocks, in some cases confused the children, and as the study progressed children built large stacks less frequently. The data captured shows that initially children were willing to build large stacks with four or five blocks, but this tended to drop off once children began to grasp the functionality of the blocks.

4.3 Case Study: Ben and Kathy

In addition to the usage analysis, a case study of two preschoolers using the Electronic Blocks has been included to highlight salient points. This case study is based on the preschool video footage.

Ben has put a touch block and a seeing block on a movement block. He touches it to make it move. Kathy has built a car with a touch block on it also. She pulls the touch block off and as she picks up a seeing blocks she says "and an eye one". "I need a torch" she says picking up a light block. Ben takes the touch block off his car and places it on a light block. Kathy places a touch block on top of her torch then touches it and checks that the light is working. She shines it at the seeing block to make her car move. Ben shines the torch he has made at his seeing block. His car moves.

Kathy takes the touch block off her torch and the seeing block off her movement block. She moves over to the box where all the Electronic Blocks are being stored. "Ben, wanna see these ones?" she asks as she picks up a not block. Ben takes the touch block off his light block. "I'll show you what these ones always do" says Kathy. She places the not block on a movement block and states, "They just make the car go." The car moves across the mat.

Ben leans over and takes the not block off the car. He tries puts it on his movement block (it still has the seeing block attached). "They're a non ... they're a non block ... they're a non stop block," says Kathy. The seeing block on Ben's car has skewed slightly making it difficult to slip the not block on to the spare hump. Ben gives up trying and places it on a sound block and then on a light block.

Kathy goes to the Electronic Blocks box and picks up a toggle block. She says to Ben "If you put it on it just goes and if you take it off, it stops!" illustrating her point by placing the toggle block on a movement block and then taking it off. The movement block moves when the toggle is attached. Ben picks up his torch (the not block with the light block) and takes it over to his car. He shines the light at the seeing block to make the car move.

Seven of the ten types of Electronic Blocks were used in the case study. The children didn't use the *hearing* block, the *and* block or the *delay* block. They used 13 blocks in total, a majority of which were sensor and action blocks. The constructions built during the case study include:

- A simple input/output stack created by Ben. This stack had a car base that was touch activated. The inclusion of a *seeing* block also meant it could have been light activated. Later in the case study, Ben built a touch activated torch, and consequently has transformed this input/output stack into an interacting block system.
- A simple touch activated car created by Kathy. This is an example of a simple input/output stack.
- An interacting block stack built by Kathy. The first stack was a touch activated light and the second structure was a *seeing* block stacked onto a car base.
- Two Electronic Block stacks that utilize logic created by Kathy. On both occasions Kathy uses a car base. In the first instance she stacks a *not* block on her *movement* block, on the second occasion she uses a *toggle* block.
- Output blocks activated by the *not* block. Two such stacks (on containing a *sound* block, the other a *light* block) were built by Ben.

The construction that took place in the case study is typical of the types of construction which took place during the Electronic Block study. Children were more likely to build complexity into their constructions by creating interacting block stacks or simple logic stacks. There were fewer examples of children building complexity into their constructions with the use of logic block combinations.

5 Discussion

Young children aged between three and six learn best while actively manipulating and transforming real materials. Therefore, educators argue, it is important that experiences with technology are empowered accordingly. Young children need to be able to play an active role in their encounters with technology, and in doing so develop images of machines and computers that they can control and program [11]. Electronic Blocks aim to address this issue. Unlike the computer and many other media used for technology education, the design of the Electronic Block allows both the input and the output to be physical. Of the interactive programming environments developed for use by young children, *curlybot* [10] is perhaps the only other resource in this category. However *curlybot* embodies a programming-by-demonstration while the Electronic Blocks allow children to create technological knowledge through constructive processes.

5.1 Understanding Block Functionality

The video footage provides clear evidence that the children were, in general, able to understand the functionality of the sensor and action blocks. The successful and repeated construction of working Electronic Block stacks reflects children's understanding of the resource. The case study demonstrates that both Kathy and Ben, for example, have a solid understanding of the ways in which sensor and action blocks work and the ways in which such stacks can be built to interact not only with the environment but also with each other. Of the twenty-eight children involved in the Electronic Block evaluation, only two failed to gain an understanding of the functionality of the sensor blocks and the action blocks. One of these children was content to watch others building with the blocks, while the other only built one Electronic Block structure.

The data presented provides evidence that the children felt most comfortable using sensor and action blocks to create simple structures, of both a stand-alone and interacting nature. Children mostly used logic blocks in simple stacks with an input, an output and the logic block in between. *Not* and *toggle* blocks were primarily used to support the construction of interacting block stacks.

Despite the successes children had in creating working sensor-action constructions, there are examples of misconceptions in this area. The most common error involved children trying to get an action block working without a sensor or logic block attached, or children trying to trigger a sensor block attached to an action block with an inappropriate signal. The investigator constantly stressed to the children the need to have a "yellow block" (a sensor block) in their stack. On numerous occasions during introductory sessions, the children would expect an output block to work without any, or with incorrect, input. The concept of inputs and outputs and the idea that the behavior of the action block is reliant on some signal from a sensor or a logic block caused children the greatest difficulty in their construction activities. Once the children understood this concept they were able to build any number of exciting creations and for those creations to exhibit the desired behaviors.

Fifteen of the twenty children who interacted with the blocks on more than one occasion became engaged in using the logic blocks, with the *not* and *toggle* blocks being used extensively during the evaluation. The *not* blocks were useful in that they created more action than they stopped, making more dynamic and interesting creations. Kathy introduced the *not* block to Ben in the case study. Her explanation and demonstration of its behavior indicated some understanding of the functionality of the *not* block. The *toggle* blocks were set up as effective on-off switches.

Many of the children struggled with the functionality of the logic blocks. Towards the end of the case study Kathy used a *toggle* block to make a car go without a sensor block. This worked in this instance because the *toggle* block was "switched on". This is not always the case. Kathy's corresponding comments indicated that she did not understand that sometimes the *toggle* block would be "off". The case study provides an insight into the difficulties that preschool children sometimes had when using the *toggle* block. The *toggle* block can be in one of two states: on or off. A child can't tell by just looking at the block which state the *toggle* is in. Only by observing the behavior of an action block that has a *toggle* block attached can the state of the toggle be determined. There was only a few examples of children showing clear understanding of the functionality of the *and* and *delay* blocks.

5.2 Play and Electronic Blocks

The Electronic Blocks study showed that children were interested in interacting with the blocks and enjoyed doing so. Time spent playing with the blocks reflected this interest and enjoyment. Significantly, the children remained interested in the blocks for the duration of the study. The study data highlighted the flexibility of the Electronic Blocks and their ability to appeal to children with different ability levels, interests and interaction styles. Many children appeared to become strongly engaged in Electronic Block activities. They were excited about the cars they were able to make move, the remote controls that they built to do so without direct contact with the vehicle, and the torches they were able to create with a light block and some kind of sensor input. It appeared that the children's enjoyment primarily stemmed from their ability to create their dynamic systems which interact with the physical world.

6 Conclusion

The Electronic Blocks were designed to address the challenge of developing a technology to allow preschool children the experience of constructing artificial agents, while addressing the unique needs and abilities for children of preschool age. The preschool study has shown that the Electronic Blocks interface is well suited to the needs of young children. Children are free to autonomously explore the functionality of the blocks as the complexity of the implementation is hidden from them.

Interaction with the Electronic Blocks primarily utilizes unstructured exploratory learning. Children's interactions with the blocks are best categorized as play. This play operates at several levels – the programming of a robot through a construction activity, the use of that agent in a variety of pretend play situations, and the ongoing creative revision of that agent. The Electronic Blocks are a resource which young children use to design, build and evaluate a large variety of robotic artifacts. They become creators of technology. In the process children become involved in meaningful technology education.

References

1. Santrock, J.W.: Child development, 8th edn. McGraw Hill, Boston (1998)
2. Shaffer, D.R.: Developmental Psychology: Childhood and adolescence. Wadsworth Group, Belmont (2002)
3. Sheingold, K.: The microcomputer as a symbolic medium. In: Pea, R.D., Sheingold, K. (eds.) Mirrors of minds: Patterns of experience in educational computing, pp. 198–208. Ablex Publishing Corporation, Norwood (1987)
4. Yelland, N.: Technology: changing the way we think and learn or maintaining the status quo. Australian Educational Computing 12(1), 3–8 (1997)
5. Resnick, M.: New Paradigms for Computing, New Paradigms for Thinking. In: di Sessa, A., Hoyles, C., Noss, R. (eds.) Computers and Exploratory Learning, pp. 31–43. Springer, Berlin (1995)
6. Resnick, M., Ocko, S.: LEGO/Logo: Learning Through and About Design. In: Harel, I., Papert, S. (eds.) Constructionism, Ablex Publishing, Norwood (1991)
7. Martin, F., Mikhak, B., Resnick, M., Silverman, B., Berg, R.: To Mindstorms and Beyond: Evolution of a Construction Kit for Magical Machines. In: Druin, A., Hendler, J. (eds.) Robots for Kids: Exploring New Technologies for Learning Experiences, pp. 9–33. Morgan Kaufman, San Francisco (2000)
8. Resnick, M., Martin, F., Berg, R., Borovoy, R., Colella, V., Kramer, K., Silverman, B.: Digital Manipulatives: New Toys to Think With. In: Proceedings of the CHI 1998, pp. 281–287. ACM Press, Los Angeles (1998)

9. Resnick, M., Berg, R., Eisenberg, M.: Beyond Black Boxes: Bringing Transparency and Aesthetics Back to Scientific Investigation. Journal of the Learning Sciences 9(1), 7–30 (2000)
10. Frei, P., Su, V., Mikhak, B., Ishii, H.: curlybot: Designing a New Class of Computational Toys. In: Proceedings of CHI 2000, pp. 129–136. ACM Press, The Hague (2000)
11. Resnick, M., Bruckman, A., Martin, F.: Pianos not stereos: Creating computational construction kits. Interactions 3(5), 41–50 (1996)

A Simulation Environment for Middle-Size Robots with Multi-level Abstraction[*]

Daniel Beck, Alexander Ferrein, and Gerhard Lakemeyer

Knowledge-based Systems Group
Computer Science Department
RWTH Aachen
Aachen, Germany
{dbeck,ferrein,gerhard}@cs.rwth-aachen.de

Abstract. Larger fields in the Middle-size league as well as the effort to build mixed teams from different universities require a simulation environment which is capable to physically correctly simulate the robots and the environment. A standardized simulation environment has not yet been proposed for this league. In this paper we present our simulation environment, which is based on the Gazebo system. We show how typical Middle-size robots with features like omni-drives and omni-directional cameras can be modeled with relative ease. In particular, the control software for the real robots can be used with few changes, thus facilitating the transfer of results obtained in simulation back to the robots. We address some technical issues such as adapting time-triggered events in the robot control software to the simulation, and we introduce the concept of multi-level abstractions. The latter allows switching between faithful but computionally expensive sensor models and abstract but cheap approximations. These abstractions are needed especially when simulating whole teams of robots.

1 Introduction

In several RoboCup leagues proposals for simulation environments have been made (e.g. [1,2,3]). In the Middle-size league (MSL) there exist a variety of different simulators. Nearly every team has implemented a simulation environment which is tailored to specific needs like the hardware platform in use, and the research focus of the particular team. Some teams are interested in high-level simulations, for instance to deploy reinforcement learning, others simulate low-level algorithms for localization, or vision, and others try to model prototypes for hardware developments in a simulation environment. Because of this diversity, the re-use of a simulator by another team is difficult, if not impossible.

[*] This work was partially supported by the German Science Foundation (DFG) in the Priority Program 1125, *Cooperating Teams of Mobile Robots in Dynamic Environments* and by the NRW Ministry of Education and Research (MSWF). Further support by the Bonn-Aachen International Center for Information Technology (B-IT) is gratefully acknowledged.

U. Visser et al. (Eds.): RoboCup 2007, LNAI 5001, pp. 136–147, 2008.

As the soccer fields in the MSL become larger (the size doubled compared to last year's competition) only very few teams can afford a full-size soccer field to test the robots and their behaviors. The number of players allowed per team was increased to six. To build and maintain a whole team of six robots will be hard for many teams. Further, the Technical Committee fosters the building of mixed teams and the team strategy of the mixed teams must be coordinated. A commonly accepted simulation environment which is able to satisfy the different needs of the teams and which is moreover able to simulate two teams of robots in a physical correct way will be of much more importance in the MSL in the coming years.

In this paper we propose a simulation environment for soccer robots in the Middle-size league. We envision a Middle-size simulation league where two teams can play simulated matches against each other with minimal changes to the original control software yielding realistic results. Our proposal founds on the 3D rigid physics simulation Gazebo. We briefly introduce the Gazebo framework and then show how on top of this a full simulation environment for the MSL can be developed, which allows for realistic simulations of low-level algorithms working on sensor data including image synthesis up to the strategy of a whole team of six robots. We discuss models for some of the most important sensors and actuators. The key to simulate whole games is the concept of multi-level abstraction, which we present at the end of the paper.

The rest of the paper is organized as follows. Section 2 introduces Gazebo and Player. Section 3 discusses other approaches to the simulator problem for RoboCup and argues why Gazebo is a good choice for our purposes. In Section 4 we sketch our models for realistic omni-drives, omni-directional cameras with realistic distorted images, and directed cameras on pan/tilt units. We also briefly address the accuracy of the simulation by comparing a differential drive in simulations with real data. In Section 5 we address the problem of how the robot control software must be adapted to fit to the simulation wrt. timing issues and introduce the concept of multi-level abstractions. Then we conclude.

2   Gazebo and Player

Our simulation environment relies on the simulator Gazebo [4] and the robot control server Player [5]. Gazebo is a 3D simulator and makes use of the freely available and constantly improving physics engine ODE [6]. Since Gazebo employs OpenGL for rendering of the simulated camera images sophisticated algorithms for rendering photo-realistic images might be integrated into Gazebo. Furthermore, Gazebo features a nice graphical user interface that allows to monitor the simulated world, inspect the current sensor readings and send commands to the actuators of the robots.

In Gazebo the simulation world, that is the 3D environment as well as the robots, is defined by means of a configuration file. Fig. 1 shows an example for the definition of one of our robots including a SICK laser range finder and a Sony camera on a pan-tilt unit. The positions of the devices are defined relative to the robot's coordinate system. Further parameters, e.g. the update frequencies of the laser scanner, can be specified.

The model file in Fig. 1 relates physical models to each other. The physical models are defined as C++ classes that describe the geometrical structure of the device, how sensor data are generated, and how data and commands are exchanged with external applications. A model can describe a physical object that possibly integrates a sensor or actuator. Since ODE is a rigid body simulator models have to be described in terms of those (which is sufficient for a robot simulator). Rigid bodies can be connected to each other by means of several different types of joints. Actuators are modeled by applying forces on those joints.

The Player network protocol [5] allows to exchange data and commands with a robot and is designed in a device-independent fashion. The communication between server and clients makes use of several task-specific interfaces (e.g. a camera interface to retrieve images). Device-dependent drivers are required for the communication with the hardware. Libraries, implementing the Player network protocol, exist for various programming languages. Since Player only defines the protocol and does not impose a certain architecture on the control software it can be easily integrated into an existing control software.

```
<model:RCBot>
  <id>bot1</id>
  <xyz>0 0 0</xyz>
  <updateRate> 25 </updateRate>
  <model:SickLD>
   <id>laser1</id>
   <xyz>-0.131 0 0.101 </xyz>
   <scanRate> 20 </scanRate>
  </model:SickLD>
  <model:SonyD100>
   <id>front1</id>
   <xyz>0.1 0 0.325</xyz>
  </model:SonyD100>
</model:RCBot>
```

Fig. 1. Excerpt from a world-file

3   Why Gazebo Is a Good Choice for the MSL

In the last decades a large number of robot simulators have been developed. Due to the considerable variation of application requirements of those simulators not all of these approaches are equally well-suited for providing a close-to-reality runtime environment for the control software. In the following we focus on a couple of selected characteristics of robot simulators and discuss related approaches under this aspect.

A (computer) simulation always constitutes an abstracted view of the system that is to be examined. Relating to robot simulators the level of abstraction ranges from representing a robot as a dot in a two-dimensional world (e.g. M-ROSE [7]) to simulators that attempt to physically correctly simulate the robot, its sensors, and all other objects in the simulation world. The latter kind of simulators are so-called *high-fidelity* simulators. Since we intend to use the simulator not only as a test-bed for the high-level decision making components we require a high-fidelity simulator.

Simulators that have an integrated physics engine compute the motion dynamics of the robot according to the physical properties of the models (e.g. its mass, the friction coefficients of surfaces, etc.). This means that the simulation model, the mathematical model on basis of which the progression of the simulation is computed, are the laws of physics. Other robot simulators rely on kinematic (e.g. SimRobot [8]) or probabilistic motion models (e.g. M-ROSE [7]). The probabilistic motion models are computed from observations that describe how the position of the robot changes as a reaction to certain movement commands. Most modern robot simulators are based on physics engines (e.g. USARSim [1], ÜberSim [2], Webots [3]) since they deliver very realistic results without the need for extensive evaluation of the real robots in order to get a proper kinematic or probabilistic motion model.

Another characteristic of a robot simulator is how the control software is coupled with the simulator. A very tight coupling is realized by SimRobot [8] which directly integrates the controller for the robot into one binary with the simulator itself. Most simulators provide interfaces that allow external applications to communicate with the simulated robots. Still, the control software might be integrated into the simulation loop as it is the case for the Simulation league RoboCup Soccer Simulator [9]. It is based on the Spark simulator framework [10] which integrate the Spades middle-ware [11]. Spades implements the so-called *software-in-the-loop* architecture. Its approach is to notify the control software when the simulation of a frame is finished, give it some time to do its computation, and then proceed with the simulation. This does not correspond to the situation in the real world where the environment changes while the control software is deciding what action to take next. Only loosely coupling the control software and the simulator corresponds to the realistic model since the simulator progresses without taking care of the control software—it is the task of the control software to keep up with the simulation speed.

Some robot simulators are specialized on certain types of robots and/or scenarios and cannot (easily) be extended (e.g. the UCHILSIM simulator [12] only simulates the Sony Aibo robots and the RoboCup Soccer Simulator is specialized on simulating soccer games). For the configurable, multi-purpose simulators it is interesting in which way robots and other objects in the simulation world can be defined. In Übersim, for instance, the objects in the simulated world are made up of basic primitives which can be combined and parameterized in a configuration file. Gazebo on the other hand requires to actually program most parts of the models. The advantage of the first approach is that the simulated world can be changed without the need to recompile, but the latter approach admits the user the chance to bail out the full expressiveness of a programming language for the description of the models.

Currently, there are several robot simulator that meet our requirements. In particular, these are USARsim, Webots, Übersim, and Gazebo. USARsim is the simulator for the RoboCupRescue simulation league and is based on the Unreal Tournament game engine. Consequently, it features a high-quality rendering engine, a high overall stability, an integrated physics engine, and several tools for comfortably editing the simulated worlds. USARsim also supports Player. Übersim was primarily developed as a simulator for the small-size league and was later enhanced to serve as a simulator for vision-centric robots. It has been successfully used to simulate a self-stabilizing two-wheeled robot but it seems not to be actively worked on at the moment. Webots is a commercial product and contains besides the actual simulator a large library of hardware components for robots that can be integrated into the simulation and allows for certain types of robots to directly transfer software developed within the simulation to the robot.

In the end we opted for Gazebo because it is available under an open source license, comes with a viable documentation, is actively maintained, and supports the open network protocol Player. Also, Gazebo proved to be capable of simulating complex robots in realistic environments (cf. [13]).

4   Physical Modeling

Gazebo comes with a set of standard robot and sensor models. Robots like the RWI B21, the Pioneer 2AT, or sensors like the SICK LMS 200 laser range finder are already modeled. Gazebo follows an approach on the middle ground between fully scripted and fully implemented physical models. The geometry, mass distribution, friction, etc. of parts have to be programmed. From the basic models the robot's appearance and the position of devices are defined using an XML script as presented in Fig. 1.

Important for the MSL is to have basic models of commonly used actuators and sensors. Most importantly for the MSL is a model for an omni-drive and for an omni-directional camera. Unfortunately, such models are not built-in models of Gazebo, and especially these models are problematic in many other simulation environments. However, these models are not too difficult to model in Gazebo and we developed models for an omni-directional and a differential drive as well as for omni-directional cameras. These models can be used as a prototype implementation and easily adopted to other robot platforms. Fig. 2(a) shows the models of different drives. The left image shows two of our differential drive robots on an MSL soccer field; the right image shows our model of a prototypical omni-directional drive. The model of the omni-wheel constitutes a realistic description of the actual mechanics of the wheel, i.e. the small rollers are connected to the center wheel by means of rotational joints. Tricks like omitting the definition of the rollers and setting the friction coefficient of the center wheel to zero along the direction of the axle are not needed.

Fig. 2(b) shows images from our directed camera (left side) and our omni-directional camera (right side). The implementation of this model is based on the cube-mapping technique. Images of the environment are mapped to the faces of a cube; this texture can then be applied to the surface of a three-dimensional object—in our case a mesh object resembling the surface of a hyperbolic mirror. The realism of the synthesized images was further improved by integrating focal depth and shadow mapping (which can be observed in Fig. 2(a)).

 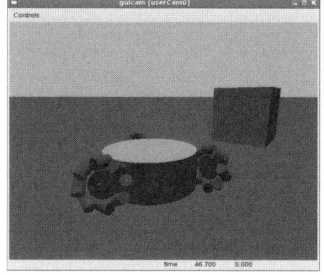

(a) Models of a soccer-field and robots with differential drive (left) and a prototype for an omni-drive robot platform (right).

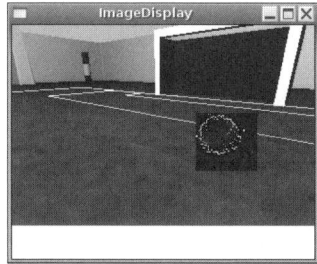 

(b) Simulated camera images of a directed and an omni-directional camera processed by the vision modules.

Fig. 2. Our Gazebo models for MSL robots

Since the simulation is based on a physics engine (ODE in our case) information about the dimensions and mass of the robot and its components have to be defined in the model. The more precise this information is the more realistic the results of the simulation will be.

In general, it is quite challenging to exhaustively measure the degree of realism of a simulation of a robot because it requires the knowledge of ground truth in the real world in order to compare virtual and real robots. To acquire some means about accuracy and quality of our simulation framework, we conducted several experiments comparing the simulation models with the real behavior of the robot components. An example for one of those experiments can be seen in Fig. 3. It compares the acceleration behavior of the the real robot (Fig. 3, left) and the simulated robot (Fig. 3, right) given a certain target velocity (the red line). Since, in our case, the parameters of the PID controller implemented in the real motor controller DSP were unknown we conducted the tests for several different parameter sets.

Regarding the quality of the simulated camera images (Fig. 2(b)) one first has to note that the markings of the ball stem from the vision algorithms which we use on our real robots. The vision software was used without any changes. The simulated camera images obviously differ from real camera images but the vision recognition modules yielded satisfactory result.

Finally, we tested the model of the SICK LD scanner. Gaussian noise is added to the exact distances such that the variance of the real device is met. The results

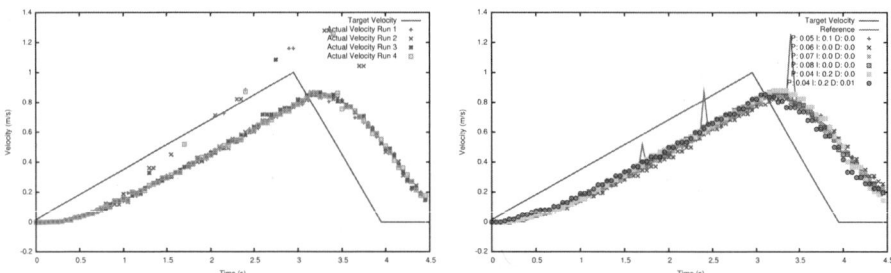

Fig. 3. These graphs compare how the real and the simulated robot react to a linear increase in the target velocity (the red lines). On the left, the actual velocities of the real robot are shown for several runs. The actual velocities of the simulated robot are shown for different parameter sets for the PID controller.

of our localization and collision avoidance algorithm which rely on the range data showed a good approximation of reality.

5   Connecting the Control Software

In this section we describe which changes are necessary to connect a control software to the simulator and show that the adaptations are only limited to the low-level hardware drivers and timing components. In other words, a control software designed for a real robot can be refitted quite easily (Sect. 5.1). Further, in Sect. 5.2, we introduce a concept that allows to simulate the robots on more abstract levels which is desirable for multi-robot scenarios, e.g. a robot soccer game. For reasons of space we can only roughly outline the concept and not give any technical details of how this concept manifests in the implementation of the robot's model.

The architecture of our simulation environment follows the client/server concept. The *simulation server* runs the simulator and a separate instance of the Player server for each simulated robot. Each robot is controlled by an instance of the control software which can be spread over several machines; these are called the *simulation clients*.

The general approach to integrate a Player client into an existing control software is to replace those parts of the system that are directly communicating with the robot's hardware by the appropriate functions provided by the client library to access the respective component in the simulation, i.e. instead of requesting a camera image from the frame-grabber an image is obtained from the simulated camera. Before data can be exchanged with the simulation server a connection to the Player server has to be established which is usually done once during the initialization of the control software.

Our control software consists of numerous, asynchronously running modules communicating by means of a central blackboard. Those modules are embedded into a hierarchy that defines which modules directly exchange data and/or commands with one another. The lowest level of this hierarchy comprises the modules that interact with the hardware components of the robot. According to the approach proposed above we implemented new modules that communicate with the simulation and these are started instead of the real drivers. Replacing the low-level modules was sufficient to provide the control software with the necessary input data and the capability to send commands to the actuators (the images in Fig. 2(b) for instance are the output of the unchanged image processing routines). In our experiments it showed that opening one connection to the Player server for each module separately produced too much overhead and slowed down the simulation server noticeably. The solution is to open only one single connection to the simulation server which is then shared by all modules. We remark that connecting a monolithic control software to the Player server would have been even easier.

5.1   Synchronization

Low-level as well as high-level components of a control software often use time and duration as part of their decision-making. Thus, time may have a major influence on the behavior of the robot. As a consequence, all timing related issues have to be handled w.r.t. the simulation-time instead of the real-time clock. This guarantees that temporal intervals are computed correctly.

In a robot control software it is quite usual that certain events are triggered at regular intervals like grabbing a camera image and processing it. When a simulated robot is controlled, the problem arises that the simulation may progress at a non-uniform speed. This means that the interval timer which triggers the event cannot be set to a constant interval time. Instead the interval time has to be adapted to the current simulation speed. Reasons for the changing simulation speed are that in certain situations more computations are necessary to determine the successor state. For example, collisions of objects lead to such situations.

The step-time is the time by which the simulation is progressed in each step. The ratio of the step time and the time it took to compute the last step is called the instantaneous simulation speed. As it can be seen in Fig. 4 the instantaneous simulation speed varies considerably and, consequently, it is impossible to obtain an accurate estimate of when the simulation-time will have increased by a certain amount of time. As a remedy we compute the estimate for the current simulation speed as the average over a history of fixed length of past instantaneous simulation speed values. In Fig. 4 the results of averaging over histories of different length are depicted. The estimated simulation speed is used to predict how long it takes to progress the simulation by a certain amount of time. Interval timers triggering certain events are then set accordingly.

As expected the accuracy of the prediction decreases with an increasing length of the interval. We tested intervals of lengths between 10 msec and 500 msec. For intervals between 10 msec and 100 msec (which are the most commonly used intervals in our control software) the average prediction error lies within a range of less than two percent of the interval length; for intervals between 150 msec and 500 msec the average prediction error is still less than four percent.

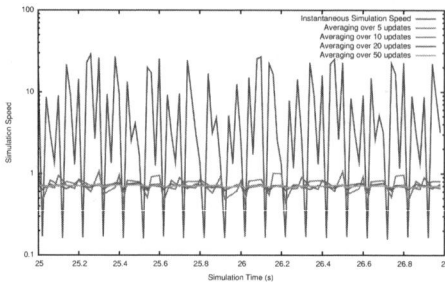

Fig. 4. The graph shows the erratic characteristics of the instantaneous simulation speed (the red line) and the results of smoothing over 5, 10, 20, and 50 consecutive instantaneous simulation speed values

In order to gain a resolution of the clock higher than the step-time of the simulation we extrapolate the simulation time between two consecutive updates taking into account the current estimate of the simulation speed.

Since in our control software there exists a single component that handles all timing related issues it was sufficient to exchange the low-level modules and to extend this time component such that it can provide the current simulation time and estimate the current simulation speed—the rest of the control software runs unchanged in the simulation environment.

5.2   Multi-level Abstraction

Simulating the robots on a device level, i.e. simulating the output of the real sensors, allows the complete control software (except those components that communicate with the hardware) to be tested with a simulated robot. Although this definitely is one of the objectives for the simulation environment, it would be helpful, in certain situations, to also have a more abstract simulation of the robots, for example, if a high-level concept should to be evaluated but lower level components providing appropriate input data are not (yet) available.

Usually, the way how the processing of sensor data is managed in a robot control software is the following. In a first step the sensor readings are aggregated, then relayed to one or several other components which take this data as input. The output of those modules might be forwarded to other components that do some further processing, and so on.

This is exactly where the concept of multi-level abstraction applies. Instead of simulating sensor readings the input data of higher-level components is directly provided by the simulation (cf. Fig. 5(a)). What kind of information that is depends on the component and on the kind of input data provided by lower-level components, respectively.

By selecting a higher abstraction level for a sensor it becomes unnecessary to run those components that normally generate the input data for a certain high-level component on basis of the sensor readings obtained from sensors since it is now simulated directly. This implies that high-level components can be tested

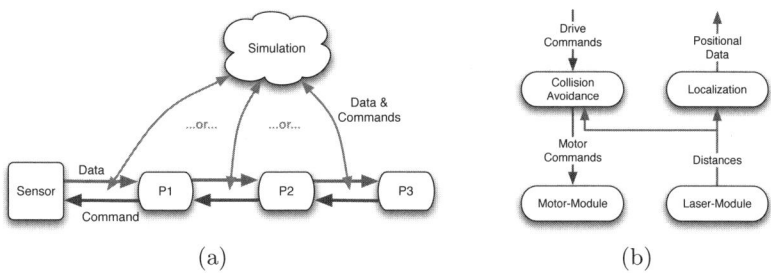

(a)   (b)

Fig. 5. (a) General idea of the concept of Multi-level Abstraction. Example: Suppose the sensor is a pan-tilt camera. Then P1 aggregates images from the camera and translates pan-tilt commands into the camera's protocol. P2 is a vision recognition module and provides information about the detected objects to P3. P2 might receive commands like "Search for object X in the images!" or "Look at world point (x, y, z)!". (b) In this example it is not reasonable to shift the functionality of the collision avoidance module to the server side in a higher level of abstraction since the computational complexity is not reduced despite of complete world knowledge.

and evaluated completely independent of the lower-level components. More precisely, the error-rate of the input data for a high-level component can be controlled directly. Usually, the input data of a high-level module is quite easy to compute given the fact that ground truth is known in the simulated world. The error-rate is then adjusted by adding noise to the absolutely correct data. A side-effect of not simulating raw sensor readings is that the computational complexity can be reduced because simulating realistic sensor readings and processing those data on the client-side, both rather expensive, can be omitted in a high-level simulation. High-level sensors not only have to generate high-level information. Also, they have to accept commands issued by the high-level component which they directly provide input data to. This is because those commands possibly lead to a change in the simulated world that affects the provided information, e.g. the visibility of an object depends on the viewing direction of the camera and thus a high-level pan-tilt camera has to accept pan-tilt-commands. A complete example is given in Fig. 5(a).

It has to be noted that it is not reasonable for every component to shift its functionality to the simulator in a high-level simulation. For instance, suppose the laser distance readings are processed by the localization component as well as by the collision-avoidance component as it is depicted in Fig. 5(b). A high-level simulation of the laser may provide the output of the localization component directly but it is of no advantage to also simulate the workings of the collision-avoidance component. This is because the knowledge about ground truth in the simulated world does not reduce the computational complexity for the problem of collision-avoidance—the same algorithms as in the control software would have to be implemented for a simulation of the collision-avoidance component.

The concept of multi-level abstraction as we implemented it opens up new possibilities for testing high-level components in a realistic environment. It has to be noted that even in a high-level simulation all the advantages of the integrated physics engine are still desirable (e.g. correct motion dynamics, collision detection, etc.). The reduction of the computational complexity comes in handy if the simulator or the control software have to be started on a slower machine or a greater number of robots is to be simulated simultaneously. Especially the rendering of simulated camera images is very expensive and consequently doing a high-level simulation of the cameras leads to a considerable increase of the average simulation speed.

Our implementation allows to select the abstraction level for every instance of a model separately and thus simulations with mixed levels of abstraction are possible.

6   Conclusion

The simulation environment we presented allows to realistically simulate robots on a device level. We successfully modeled omni-directional drives and integrated new rendering techniques into Gazebo that allow to simulate images of omni-directional cameras on the one hand, and on the other hand increase the realism of the rendered images by adding shadows and focal depth to the images. Thus, all components typically built into robots of the middle-size league can be simulated.

We presented the concept of multi-level abstraction. The idea is to simulate the robotic system on different abstraction levels. If the task is to develop low-level modules like navigation or localization algorithms, a low level of abstraction is needed. In that case one simulates the robotic system in great detail with the down-side of a more complex simulation (for example, when images have to be rendered). Here, the simulation environment gives precious information how the implemented algorithm work in a near-to-reality environment. Since only minimal and especially no structural changes are necessary to make an existing control software control a simulated instead of a real robot, the transfer of software developed with the simulation to a real robot is facilitated. Due to the standardized network protocol Gazebo can be exchanged by the 2D simulator Stage [5] effortlessly, where, for instance, learning tasks with more than real-time can be performed.

Following our vision that the simulation framework presented in this paper could serve as a standard simulation environment for the MSL where whole matches are simulated, one clearly needs a higher level of abstraction. Simulating on a behavior level, raw images do not have to be synthesized. In a real game, usually, this information is not logged either.

In the future we want to extend our framework with an automatic referee similar to the Simulation league. This sets the prerequisite to simulate whole games in a realistic way and with this establishing a Middle-size Simulation league, so to say. This should not be seen as a supplement for real games. It should give a means to increase the quality of the games at competitions. On the other hand, with such a standardized simulator the quality of research in the field of RoboCup is increased as the simulator can provide (simulated) ground truth data. Regarding Gazebo as the underlying engine we remark that we do not claim that this is the only possible choice for an MSL simulator. Gazebo just turned out to be very well suited to tackle this problem.

As another, more technical issue for future work, one weakness of the simulator must be addressed. For very large multi-robot simulations the (single) server/(multiple) clients architecture of the simulation environment does not scale too well. The problem lies in the fact that the simulation at present cannot be distributed. Therefore we intend to enhance the simulator such that it allows distributed simulations. In a first step it is planned to integrate distributed rendering into the simulator since image synthesis is one of the most expensive parts of the simulation.

While we focused on the MSL in this paper, it was shown that soccer and service robotics as well as service robotics and Gazebo can be married successfully as the example of [14,13] shows. Thus, the scope of our framework ranges much beyond Middle-size league soccer playing.

References

1. Scrapper, C., Balakirsky, S., Messina, E.: MOAST and USARSim: a combined framework for the development and testing of autonomous systems. In: Unmanned Systems Technology VIII; Proc. SPIE 2006 (2006)
2. Browning, B., Tryzelaar, E.: Übersim: a multi-robot simulator for robot soccer. In: AAMAS 2003, pp. 948–949. ACM Press (2003)
3. Michel, O.: Cyberbotics ltd. webots: Professional mobile robot simulation (2006), http://www.cyberbotics.com/publications/ars.pdf
4. Koenig, N., Howard, A.: Design and use paradigms for gazebo, an open-source multi-robot simulator. In: Proc. ICRA 2004, pp. 2149–2154 (2004)
5. Gerkey, B., Vaughan, R., Howard, A.: The player/stage project: Tools for multi-robot and distributed sensor systems. In: Proc. ICAR 2003, pp. 317–323 (2003)
6. last visited February, http://www.ode.org (2007)
7. Buck, S., Beetz, M., Schmitt, T.: M-ROSE: A multi robot simulation environment for learning cooperative behavior. In: Distributed Autonomous Robotic Systems 5 (2002)
8. Siems, U., Herwig, C., Röfer, T.: Simrobot, ein system zur simulation sensorbestückter agenten in einer dreidimensionalen umwelt, Number 1/94 in ZKW Bericht. Zentrum für Kognitionswissenschaften. Universität Bremen (1994)
9. last visited February, sserver.sourceforge.net (2007)
10. Obst, O., Rollmann, M.: Spark - a generic simulator for physical multi-agent simulations. In: Lindemann, G., Denzinger, J., Timm, I.J., Unland, R. (eds.) MATES 2004. LNCS (LNAI), vol. 3187, pp. 243–257. Springer, Heidelberg (2004)
11. Riley, P.F., Riley, G.F.: Next generation modeling iii - agents: Spades — a distributed agent simulation environment with software-in-the-loop execution. In: Proc. WSC 2003, pp. 817–825 (2003)
12. Zagal, J.C., del Solar, J.R.: Uchilsim: A dynamically and visually realistic simulator for the robocup four legged league. In: Nardi, D., Riedmiller, M., Sammut, C., Santos-Victor, J. (eds.) RoboCup 2004. LNCS (LNAI), vol. 3276, pp. 34–45. Springer, Heidelberg (2005)
13. Müller, A., Beetz, M.: Designing and implementing a plan library for a simulated household robot. In: Proc. CogRob 2006, AAAI 2006, pp. 119–128 (2006)
14. Schiffer, S., Ferrein, A., Lakemeyer, G.: Football is coming home. In: Proceedings of the International Symposium on Practical Congnitive Robots and Agents (2006)

Improving Robot Self-localization Using Landmarks' Poses Tracking and Odometry Error Estimation*

Pablo Guerrero and Javier Ruiz-del-Solar

Department of Electrical Engineering, Universidad de Chile
{pguerrer,jruizd}@ing.uchile.cl

Abstract. In this article the classical self-localization approach is improved by estimating, independently from the robot's pose, the robot's odometric error and the landmarks' poses. This allows using, in addition to fixed landmarks, dynamic landmarks such as temporally local objects (mobile objects) and spatially local objects (view-dependent objects or textures), for estimating the odometric error, and therefore improving the robot's localization. Moreover, the estimation or tracking of the fixed-landmarks' poses allows the robot to accomplish successfully certain tasks, even when having high uncertainty in its localization estimation (e.g. determining the goal position in a soccer environment without directly seeing the goal and with high localization uncertainty). Furthermore, the estimation of the fixed-landmarks' pose allows having global measures of the robot's localization accuracy, by comparing the real map, given by the real (a priori known) position of the fixed-landmarks, with the estimated map, given by the estimated position of these landmarks. Based on this new approach we propose an improved self-localization system for AIBO robots playing in a RoboCup soccer environment, where the odometric error estimation is implemented using Particle Filters, and the robot's and landmarks' poses are estimated using Extended Kalman Filters. Preliminary results of the system's operation are presented.

1 Introduction

Localization is a key feature of a mobile robotic system, which has been deeply investigated over the last years. Commonly a localization module is expected to filter two sources of error: (i) observational errors that are produced by the imperfections of the sensors and their models, and (ii) odometric errors that are produced by flaws in the modeling of the actuators and by events that are very difficult to model as slipping and collisions. It is not the aim of this paper to compare or to analyze different localization methods -as it is a very largely discussed matter- but to discuss how to improve the localization process.

Existent localization approaches filter simultaneously both sources of error, observational and odometric, making use of what we call *global information*, perceptions of objects with fixed and known global pose. However, we believe that, in addition to the global information, additional sources of information, what we call *local information*, can be exploited by localization methods. We use the word "local" in its temporal and spatial meanings. Spatially local information corresponds to information

---

* This research was partially supported by FONDECYT (Chile) under Project Number 1061158.

U. Visser et al. (Eds.): RoboCup 2007, LNAI 5001, pp. 148–158, 2008.

that is only useful in a reduced region of the space, while temporally local information corresponds to information only valid in a short period of time. Temporally local information corresponds mainly to mobile objects, whose poses' estimates are valid for a short period of time. Spatially local information corresponds mainly to view-dependent objects or textures, whose perceptions are valid in a reduced neighborhood. The spatially restricted utility may be at least due to three reasons: (i) the object is only observable from a restricted region of the space, for example a design in the floor or a visual feature or detail only perceptible from close positions, (ii) its appearance changes from different points of view (most of the natural objects do not have a radial symmetry and also non isotropic light may make them appear different from different points of view), and/or (iii) several identical -or difficult to distinguish- objects are present in different places in the space, which could easily lead to confusion, for example, a tree in a forest, a tile or texture in the floor, or a chair in a classroom. SLAM approaches can deal with objects locally observable or with non-symmetric appearance by creating and maintaining a pose estimate for each of them, or of their different appearances treated as different objects. However, this could lead a system to maintain millions of estimates, which is computationally infeasible and practically senseless. Nevertheless, it is possible to think in a SLAM-like approach that maintains locally relevant information with the purpose of estimating the odometric error. We believe such an approach is more biologically inspired. For instance, humans are able to correct their odometry even when they have no knowledge of the environment.

In this context we propose improving the classical self-localization approach by estimating, independently from the robot's pose, the robot's odometric error and the landmarks' poses. This allows using, in addition to fixed landmarks, dynamic landmarks such as temporally local objects (mobile objects) and spatially local objects (view-dependent objects or textures), for estimating the odometric error, and therefore improving the robot's localization. Moreover, the estimation or tracking of the fixed-landmarks' poses allows the robot to carry out certain tasks, even when having high uncertainty in the localization estimation. This is especially valuable when performing attention demanding tasks, like pursuing a ball, which forbid the use of active vision in order to get more (standard) landmarks' perceptions. Another nice feature of the proposed system is that the robot is able to correct its odometry even when it is totally lost. The latter ability may be useful in several situations as for example, when shooting the ball to a recently seen goal, by correcting the relative robot's pose estimation with only observations of the ball. In this sense, we believe our approach also goes in the direction towards performing tasks with much less use of global localization, as certainly humans do.

Furthermore, the estimation of the fixed-landmarks' pose allows having global measures of the robot's localization accuracy, by comparing the real map, given by the real (a priori known) position of the fixed-landmarks, with the estimated map, given by the estimated position of these landmarks.

Based on the described new self-localization approach, we propose an improved self-localization system for AIBO robots playing in a RoboCup soccer environment, that implements odometric error estimation using Particle Filters, and robot's and landmarks' poses estimation using Extended Kalman Filters.

How and when to select spatially local observations as valid landmarks is a topic not addressed in this article. In the current implementation we consider temporally local observations, mobile objects, such as the ball and robot players in a soccer environment.

This paper is organized as follows. In section 2 is presented some related work. The improved self-localization approach is described in section 3. In section 4 some features of the proposed approach are discussed. In section 5, preliminary results are presented. Finally, in section 6 some conclusions and future work are given.

2   Related Work

Standard Bayesian-based robot self-localization fuses odometric information with perceptual information coming from different sensors. Thus, odometry is employed for predicting the next robot pose state using a cinematic model of the robot, while perceptual information from landmarks is employed for correcting this prediction using an observational model. For implementing these two steps, the most employed Bayesian filters are Kalman [6] and Particle Filters [3][5]. Kalman Filtering is a very well-known technique for parameter estimation, whose main drawback is the linearity and Gaussianity assumptions. EKF extends the Kalman Filtering idea by linearizing the measurement and plant (in this case the robot) models, but still has the assumption of Gaussianity [4]. On the other hand, particle filters overcome the drawbacks of assuming linearity and Gaussianity, by implementing a "factored sampling" of the processes' conditional densities [5]. Particle Filtering is very popular in the Computer Vision community where the most employed implementation is called *Condensation* [5], while in the mobile robotics community the most used implementation is the *Monte Carlo Localization – MCL* algorithm. However, particle filters have an important drawback: their performance depends strongly on the number of particles. In specific applications as robot soccer, in which the computational resources are limited, the number of employed particles may not be very high (normally between 50 and 200), and therefore particle filters do not clearly outperforms EKF. As a fact, in the RoboCup soccer leagues successful teams use either EKF, MCL, or mixtures of both (see for example [8] or [9]).

Nevertheless, it is not our intention to analyze or to compare different Bayesian filters and their application to the robot self-localization problem, but to improve the standard Bayesian-based robot self-localization by including new independent stages for estimating the robot's odometric error and the landmarks' poses. To the best of our knowledge this idea is novel, and has not being implemented before in robot localization systems. Although decoupling the odometry estimation from the landmark position estimation has been proposed in the SLAM literature [10], there is a strong implicit assumption in these works, which is that all the landmarks will remain static forever. In some SLAM approaches detected objects are tracked and characterized as mobile or static [11]. However the mobile object's information is not used for estimating the odometric error.

In visual odometry approaches (see for example [12][13]) local visual features (e.g. Harris or SIFT features) are employed for estimating the robot relative movements (the odometry) by detecting and matching the features between consecutive frames. The main differences with our approach are: (1) In our approach the estimated odometry is used in the robot localization process, and also for updating the high-level tracking of landmarks. Traditional visual odometry approaches do not include high-level tracking of landmarks, therefore the use of the estimated odometry is much simpler; (2) Visual odometry approaches use local features, while in our case high-level fixed or moving landmarks are employed (e.g. a ball or a goal in a soccer environment). We believe that using high-level landmarks is more robust because: (i) local features cannot be detected in any environment or moment (e.g. in a robot soccer carpet no local features can be detected), and (ii) when analyzing images corresponding to real-world environments the process of matching local features can produce a larger number of false matches (due to shadows, highlights, symmetry problems, too many detected features, etc.) than the one of matching high-level features.

3   Proposed Self-localization Using Landmarks' Tracking and Odometry Error Estimation

As already mentioned, the basic idea of the proposed approach is to estimate, independently from the robot's pose, the robot's odometric error and the landmarks' poses. For achieving this two new processes are included: High-Level Tracking (*HL-Tracking* module) and Odometry Error Estimation (*OEE* module). As can be observed in figure 1, the operation of all modules is tightly interconnected. First, in the *HL-Tracking* module the pose of the observed landmark ($\mathbf{x}_{l_*,k}^-$), either static or mobile, is *early* predicted using the odometry information ($\mathbf{u}_{k-1}$). Then, the odometric error ($\mathbf{e}_k$) is estimated in the *OEE* module using the information of current observations (coming from *Vision*) ($\mathbf{z}_k$), and the corresponding landmark's early estimated pose. Afterwards, in the *HL-Tracking* module the estimated odometric error is used for estimating the new landmarks' poses ($\{\mathbf{x}_{l_i,k}\}$). Finally, the corrected odometry and the

landmarks' poses are employed for estimating the robot's localization ($\mathbf{x}_{R,k}$).

In the next sections the operation of all modules is described in detail for the case of a RoboCup four-legged environment using AIBO robots. The pseudo code and equations are detailed in tables 1-3.

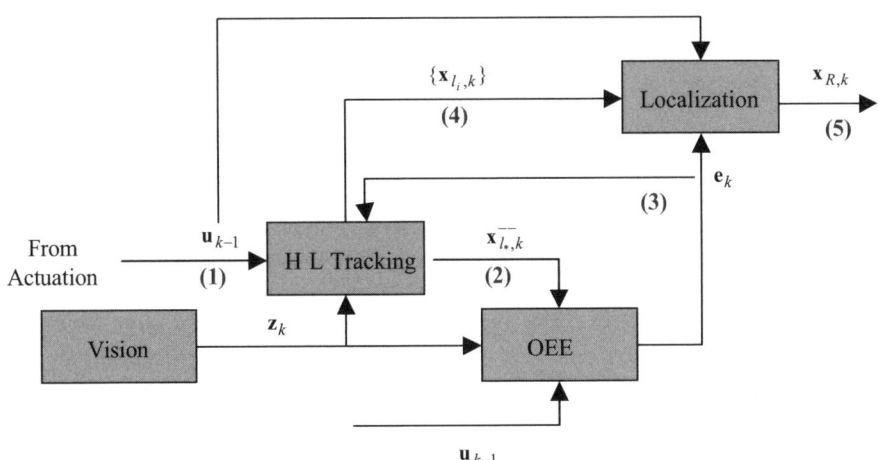

Fig. 1. Block diagram of the system. Two stages are added between vision and localization: HL-Tracking and Odometry Error Estimation.

3.1 Vision

Our vision system (RoboCup four-legged league scenario) is based on color segmentation of the images, and rule based perceptors for the relevant objects (ball, robots, beacons and goals) (see detailed description in [1]). The vision system also includes a recently proposed context filter, which takes into account the coherence between current and past detections, as well as scene and situation contexts, to filter incoherent detections [2].

3.2 HL-Tracking

In this module the state of every detected and coherent object/landmark ($\mathbf{x}_{l_i,k}$) is tracked (estimated). For the fixed objects, the state corresponds to their 2D pose, relative to the robot, for the mobile ones, a relative velocity is also added (see figure 2). The pose of any object includes a 2D position, and may include a relative orientation, if this is distinguishable (for objects with radial symmetry as ball and beacons it is not possible to notice their orientation).

The current implementation of the HL-Tracking stage consists of one independent EKF for each object. The prediction of each EKF has two stages: (i) early prediction (step (1) in pseudo code shown in table1), where the pose of the l_* landmark associated with the current observation $\mathbf{z}_k$, is predicted using the last executed odometry $\mathbf{u}_{k-1}$, and (ii) a standard prediction stage (step (6) in pseudo code), where the poses of all landmarks ($\{\mathbf{x}_{l_i,k}\}$) is predicted using the corrected odometry ($\mathbf{u}_{k-1} + \mathbf{e}_k$). The correction stage is standard and considers only the observed landmark l_* (step (7) in pseudo code). In the case of mobile objects, the correction that the filter takes is standard and very straightforward, since the predicted observation may be extracted directly from the state -the observation model Jacobian H is equal to the identity or some submatrix-.

For the system to be able to quickly detect and recover from kidnaps, all tracked estimates of objects' poses in HL-Tracking has a smoothed object coherence indicator (see [2] for details).

3.3 Odometry Error Estimation

The odometry error estimation (*OEE*) stage is implemented using a particle filter, as in MCL, but in this case each particle represents a hypothesis for the accumulated odometric error (instead of the global pose of the robot, as in MCL). Consequently, the state of each particle is a pose (x, y, θ) relative to a coordinate system centered in an odometry error-free pose. Then, the particles are drawn over the same coordinate system shown in figure 2.

In the sampling stage of the OEE (step (3) in pseudo code), it is considered that the expected odometric error is zero, thus, we only add noise covariance to the particles, coming from the standard odometry. Given any odometry $\mathbf{u}_{k-1}$, an a priori odometry error covariance $\mathbf{Q}_{k-1}$ is used to scatter the particles. The weighting stage consists in the calculation of weights for each particle (step (4) in pseudo code). A particle will have a higher weight when it better explains the difference between the observed and estimated poses of the observed objects. Finally, in the Resampling Stage (step (2) in

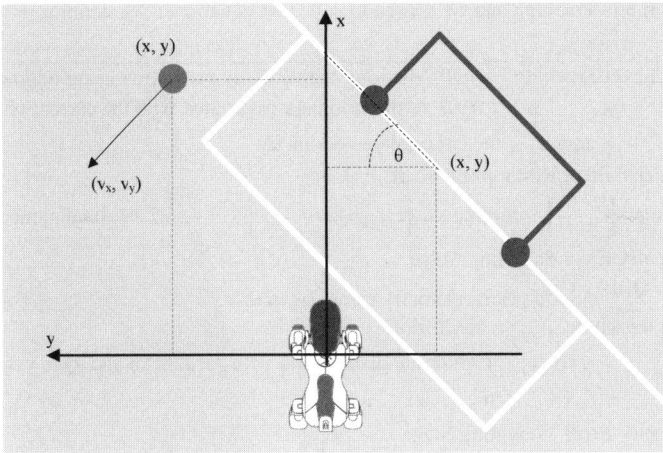

Fig. 2. Tracked objects and their state in the HL-Tracking stage. Static objects are represented by their relative pose, which includes a position and may include a relative orientation (blue goal). For mobile objects, a relative velocity is also added (orange ball).

pseudo code), particles are resampled according to their weights. A particle is copied a number of times that has an expected value proportional to its weight. As Resampling is the first stage in our implementation, it considers the weights calculated in the previous iteration of the system.

Every time an observation arrives, the system executes these three steps, and finally the estimated odometric error (e_k) and its covariance ($Q_{e,k}$) are statistically calculated over the particles (step (5) in pseudo code), and used as additional predictive inputs for the HL-Tracking and Localization stages. Finally, the odometric error estimate is subtracted from each particle, to set the new odometry error estimate to zero.

3.4 Localization

We have implementations of standard robot's localization modules based on EKF, MCL, and mixtures of them [1]. However, for the proposed approach, we have implemented the robot's localization using a standard EKF filter (see pseudo code in table 2). The main new features are: (i) the corrected odometry ($u_{k-1} + e_k$) is used for predicting the new robot's pose (step (8) in the pseudo code), and (ii) the filtered relative poses of all landmarks, fixed and mobile, are employed as the filter's observations in the correction stage (step (9) and (10) in pseudo code).

4 Discussion

4.1 Is It Good for a Robot to Be Egocentric?

A clear difference between the proposed approach and most of the existent approaches for localization is that, in this one, all the analysis is made in reference to the robot

Table 1. HL-Tracking and OEE pseudo code and equations. See definitions in table 3.

(1) High-Level Tracking Early Prediction Stage // Early prediction using odometry

$\mathbf{x}^-_{l_*,k} = f_l(\mathbf{x}_{l_*,k-1},\mathbf{u}_{k-1},0)$ // l_* the landmark associated with the current observation $\mathbf{z}_k$,

and $\mathbf{u}_{k-1}$ the last executed odometry

(2) Odometry Error Resampling Stage

Resample $\left\{\mathbf{x}'_{p_j,k}\right\}$ according to $\left\{\mathbf{x}_{p_j,k-1},\omega_{p_j,k-1}\right\}$, $j=1,...,T$ //T: total number of particles

(3) Odometry Error Sampling Stage

$\mathbf{Q}_{k-1} = \mathbf{Q}(\mathbf{u}_{k-1})$ // a priori odometry error covariance

For each particle $p_j, j=1,...,T$ do:

$\quad\mathbf{u}_{p_j,k} \sim N(0,\mathbf{Q}_{k-1})$ //Normal distribution with zero mean and $\mathbf{Q}_{k-1}$ covariance

$\quad\mathbf{x}_{p_j,k} = f_p(\mathbf{x}'_{p_j,k},\mathbf{u}_{p_j,k},0)$

(4) Odometry Error Weighting Stage

For each particle $p_j, j=1,...,T$ do:

$\quad\mathbf{v}_{p_j,k} = \mathbf{z}_k - \mathbf{h}_{l_*}(f_l(\mathbf{x}^-_{l_*,k},\mathbf{u}(\mathbf{x}_{p_j,k}),0),0)$ // $\mathbf{z}_k$ the current observation, and l_* the

corresponding landmark

$\quad\tilde{\omega}_{p_j,k} = e^{-\frac{\mathbf{v}^T_{p_j,k}\mathbf{R}^{-1}_{l_j,k}\mathbf{v}_{p_j,k}}{2}}$

$\quad\omega_{p_j,k} = \dfrac{\tilde{\omega}_{p_j,k}}{\sum_n \tilde{\omega}_{p_n,k}}$ // weights normalization

(5) Odometry Error Statistics Calculation

$\mathbf{e}_k = \sum_j \omega_{p_j,k}\mathbf{x}_{p_j,k}$ // estimated odometry error

$\mathbf{Q}_{\mathbf{e},k} = \sum_j \omega_{p_j,k}\mathbf{x}_{p_j,k}\mathbf{x}^T_{p_j,k}$ // estimated odometry error covariance

For each particle $p_j, j=1,...,T$ do:

$\quad\mathbf{x}_{p_j,k} = \mathbf{x}_{p_j,k} - \mathbf{e}_k$

(6) High-Level Tracking Odometry Prediction Stage

For each landmark $l_i, i=1,...,L$ do: // The new poses of all landmarks are predicted

$\quad\mathbf{x}^-_{l_i,k} = f_l(\mathbf{x}_{l_i,k-1},(\mathbf{u}_{k-1}+\mathbf{e}_k),0)$ // Prediction using the corrected odometry

$\quad\mathbf{P}^-_{l_i,k} = \mathbf{A}_{l,k}\mathbf{P}_{l_i,k-1}\mathbf{A}^T_{l,k} + \mathbf{W}_{l_i,k}\mathbf{Q}_{\mathbf{e},k}\mathbf{W}_{l_i,k}$

(7) High-Level Tracking Correction Stage // Only the observed-landmark's pose is corrected

$\mathbf{v}_{v,k} = \mathbf{z}_k - \mathbf{h}_{l_*}(x^-_{l_*,k},0)$ // $\mathbf{z}_k$ the current observation, and l_* the corresponding landmark

$\mathbf{M}_{l_*,k} = \mathbf{H}_{l_*,k}\mathbf{P}^-_{l_*,k}\mathbf{H}^T_{l_*,k} + \mathbf{V}_{l_*,k}\mathbf{R}_{l_*,k}\mathbf{V}^T_{l_*,k}$

$\mathbf{K}_{l_*,k} = \mathbf{P}^-_{l_*,k}\mathbf{H}^T_{l_*,k}\mathbf{M}^{-1}_{l_*,k}$

$\mathbf{x}_{l_*,k} = \mathbf{x}^-_{l_*,k} + \mathbf{K}_{l_*,k}\mathbf{v}_{v,k}$

$\mathbf{P}_{l_*,k} = (\mathbf{I} - \mathbf{K}_{l_*,k}\mathbf{H}_{l_*,k})\mathbf{P}^-_{l_*,k}$

Table 2. EKF Localization pseudo code and equations. See definition in table 3.

(8)   Localization Prediction Stage // Robot pose prediction using the corrected odometry

$$\mathbf{x}^-_{R,k} = f_R\,(\mathbf{x}_{R,k-1},(\mathbf{u}_{k-1}+\mathbf{e}_k),0)$$

$$\mathbf{P}^-_{R,k} = \mathbf{A}_{R,k}\mathbf{P}_{R,k-1}\mathbf{A}^T_{R,k} + \mathbf{W}_{R,k}\mathbf{Q}_{e,k}\mathbf{W}^T_{R,k}$$

(9)   Localization Data Association Stage

$$\mathbf{v}_{l,k}=\varnothing$$

For each landmark $l_i, i=1,...,L$ do: //Landmarks are filtered out using an innovation threshold g

$$\mathbf{v}_{l_i,k} = \mathbf{x}_{l_i,k} - h_{R,l_i}\,(\mathbf{x}^-_R,0)$$

$$\mathbf{M}_{R,l_i,k} = \mathbf{H}_{R,l_i,k}\mathbf{P}^-_R\mathbf{H}^T_{R,l_i,k} + \mathbf{V}_{R,l_i,k}\mathbf{P}_{l_i,k}\mathbf{V}^T_{R,l_i,k}$$

$$if\,(\mathbf{v}_{l_i,k}\mathbf{M}_{R,l_i,k}\mathbf{v}^T_{l_i,k}\le g^2)\ then\ \mathbf{v}_{l,k} = \mathbf{v}_{l,k}\cup\mathbf{v}_{l_i,k}$$

(10) Localization Correction Stage // Robot pose correction

$$\mathbf{K}_{R,k} = \mathbf{P}^-_{R,k}\mathbf{H}^T_{R,k}\mathbf{M}^{-1}_{R,k}$$

$$\mathbf{x}_{R,k} = \mathbf{x}^-_{R,k} + \mathbf{K}_{R,k}\mathbf{v}_{l,k}$$

$$\mathbf{P}_{R,k} = (\mathbf{I}-\mathbf{K}_{R,k}\mathbf{H}_{R,k})\mathbf{P}^-_{R,k}$$

Table 3. Variables, matrices and functions definitions

Variable/Matrix	Definition
$\mathbf{x}_{l_i,k}$, $\mathbf{P}_{l_i,k}$	Landmark l_i state vector and covariance matrix.
$\mathbf{x}_{R,k}$, $\mathbf{P}_{R,k}$	Robot state vector and covariance matrix.
$\mathbf{Q}_k$	A priori odometry error covariance
f_l, f_R	Landmarks and robot process models.
$\mathbf{u}_{k-1}$	Robot odometry.
$\mathbf{e}_k$ and $\mathbf{Q}_{e,k}$	Estimated odometric error and its covariance.
$\mathbf{x}_{p_j,k}$ and $\omega_{p_j,k}$	Position and weigh of particle p_j (odometry error estimation)
f_p	Cinematic model of the particles (odometry error estimation)
$\mathbf{A}_{R,k}/\mathbf{W}_{R,k}$	The Jacobian matrix of the partial derivatives of f_R with respect to the state vector and process noise, respectively.
$\mathbf{A}_{l,k}/\mathbf{W}_{l,k}$	The Jacobian matrix of the partial derivatives of f_l with respect to the state vector and process noise, respectively.
$\mathbf{H}_{R,k}/\mathbf{V}_{R,k}$	The Jacobian matrix of the partial derivatives of h_R with respect to the state vector and observational noise, respectively.
$\mathbf{H}_{R,l_i,k}/\mathbf{V}_{R,l_i,k}$	The Jacobian submatrix, corresponding to the landmark l_i, of the partial derivatives of h_R with respect to the state vector and observational noise, respectively.
$\mathbf{H}_{l_i,k}/\mathbf{V}_{l_i,k}$	The Jacobian matrix of the partial derivatives of h_{l_i} with respect to the state vector and observational noise, respectively.
$\mathbf{R}_{l_i,k}$	Landmarks observational noise covariance.

instead of making it from a global point of view. One could argue that all the previous formulation could be transported to a global approach by representing all the local information in a global coordinate system. We believe that being egocentric (taking a self-centered coordinate system for most of the calculations) is a good decision because: (i) many (may be most) of the tasks a robot must perform can be executed with only local information, for example, a robot does not need to know its global pose neither the global pose of the ball to approach to it, and (ii) even high level tasks that need global information normally result in low-level tasks that can be performed locally.

4.2 Towards Playing Soccer with Much Less Use of Localization

Human soccer players can effectively perform most of their tasks with a very poor estimation of their global pose in the field. They make extensive use of local information to: go to the ball, shoot to the goal, pass, keep close to the goal (in the case of the goalie), keep the ball inside of the field, mark opponents, etc. Even strategically positioning, which could be argued to be a localization-dependent task is performed with extensive use of local information; players do not only tend to be close to one static part of the field, they also (and may be more important) tend to maintain certain positions relative to their teammates, opponents and the ball. We believe our work is a step towards that direction, because it allows a robot to correct its odometry, and thus its relative estimates of non-seen objects, with local information.

5 Results

Preliminary results that illustrate the operation of the system are presented. Figure 3 shows a sequence of egocentric local maps, relative to the robot (coordinate system shown in figure 2), in selected moments of a real movement's sequence (data are collected from the robot and displayed in a visualization software). The sequence corresponds to the following situation: the robot walks from the yellow goal area to the center of the field, while panning its camera. In colors are shown the blue goal and the beacons (with exaggerated radiuses) estimations carried out by HL-Tracking. In the center of each map, the OEE particles appear, where lighter ones corresponds to those having a higher score. In order to make the functioning of the OEE visible, the particles' positions, relative to the center of the egocentric map, are zoomed ~4x with respect to the HL-Tracking estimation. In the tested sequence, the odometry was specially poorly calibrated, with a high bias (the accumulated odometry was of ~600cm, while the actual movement of the robot was of ~240cm). However OEE combined with HL-Tracking was able to keep tracking of the observed objects and correct the non-seen ones (yellow landmarks after they are left behind).

We have performed several experiments as the one already illustrated. In these experiments we have seen that when the robot perceives the fixed landmarks (the ones defining the map) regularly, the accuracy of the proposed robot's localization approach is similar that the one obtained when using a standard EKF, and no OEE or HL-Tracking stages (variations are less than 1% in accuracy). However, when the robot executes attention demanding tasks as approaching the ball without active vision behaviors for looking for the fixed landmarks, or turning with the ball while

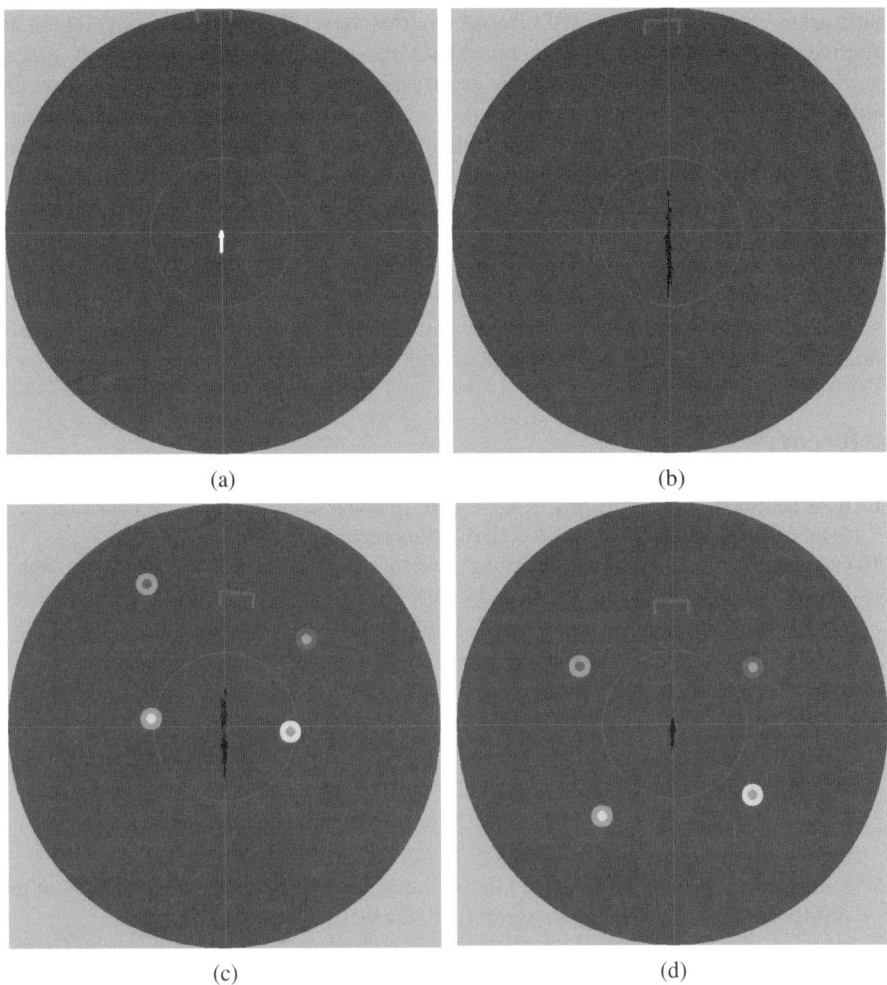

(a)   (b)

(c)   (d)

Fig. 3. Egocentric local maps in selected moments of a movement sequence: the robot walks from the yellow goal area to the center of the field, while panning its camera. a) The robot starts with a perception of the blue goal, all particles are together, with a high score. b) As odometry arrives, the particles start scattering, which allows the odometry correction. c) The robot is walking, in an intermediate point, d) The robot arrives to the center of the field.

preparing a goal-kick, the accuracy of the robot's odometry estimation is 14% better than in the case when the OEE and HL-Tracking stages are not used, while the robot's localization is 6% more accurate.

6   Conclusions and Future Work

In this article, an improvement over the classical robot's localization approach was proposed, in which, in addition to the robot's pose, the robot's odometric error and the landmarks' poses are estimated. Based on this new approach, we developed an improved self-localization system for AIBO robots playing in a RoboCup soccer environment. In this system odometric error estimation is implemented using Particle Filters, while robot's and landmarks' poses are estimated using Extended Kalman Filters. Preliminary results show that, when the robot executes attention-demanding tasks, the accuracy of the robot's odometry estimation is 14% better than in the case when the new estimation modules are not used, while the robot's localization is 6% more accurate.

Currently we are carrying out a better characterization of the proposed system. In addition, we are developing an extension to the presented system, which consists in using the robot's odometric error for modeling and correcting the permanent odometric error using an on-line trained neural network.

References

1. Ruiz-del-Solar, J., et al.: UChile Kiltros 2007 Team Description Paper. In: RoboCup 2007 Symposium, Atlanta, USA, July 9 – 10 (CD Proceedings) (2007)
2. Guerrero, P., Ruiz-del-Solar, J., Palma-Amestoy, R.: Spatiotemporal Context in Robot Vision: Detection of Static Objects in the RoboCup Four Legged League. In: Proc. 1st Int. Workshop on Robot Vision(in 2nd Int. Conf. on Computer Vision Theory and Appl. – VISAPP 2007), Barcelona, Spain, March 8 – 11, pp. 136–148 (2007)
3. Doucet, A., Andrieu, C., Godsill, S.: On Sequential Monte Carlo Sampling Methods for Bayesian Filtering. Statistics and Computing 10(3), 197–208 (2000)
4. Welch, G., Bishop, G.: An introduction to the Kalman Filter. Tech. Report TR 95-041 (Update May 23, 2003), Department of Computer Science, Univ. of North Carolina at Chapel Hill (2003)
5. Isard, M., Blake, A.: Condensation – Conditional Density Propagation for Visual Tracking. Int. Journal of Computer Vision 29(1), 5–28 (1998)
6. Kalman, R.E.: A New Approach to Linear Filtering and Prediction Problems. Trans. of the ASME - Journal of Basic Engineering 82, 35–45 (1960)
7. Siegwart, R., Nourbakhsh, I.: Introduction to Autonomous Mobile Robots. MIT Press (2004)
8. Polani, D., Browning, B., Bonarini, A., Yoshida, K. (eds.): Proc. of the RoboCup 2003 Symposium, Padova, Italy, July 9 - 11 (2003) (CD Proceedings)
9. Nardi, D., Riedmiller, M., Sammut, C., Santos Victor, J. (eds.): Proc. of the RoboCup 2004 Symposium, Lisbon, Portugal, July 4 – 5 (2004) (CD Proceedings)
10. Montiel, J., Davison, A.: A Visual Compass based on SLAM. In: Proc. of the ICRA 2006, Orlando, USA, May 15-19, pp. 1917–1922 (2006)
11. Wang, C., Thorpe, C., Thrun, S.: Online Simultaneous Localization and Mapping with Detection and Tracking of Moving Objects: Theory and Results from a Ground Vehicle in Crowded Urban Areas. In: Proc. of the ICRA 2003, Taipei, Taiwan, May 12-17, pp. 842–849 (2003)
12. Cheng, Y., Maimone, M.W., Matthies, L.: Visual Odometry on the Mars Exploration Rovers. IEEE Robotics and Automation Magazine 13(2), 54–62 (2006)
13. Konolige, K., Agrawal, M.: Frame-frame Matching for Realtime Consistent Visual Mapping. In: Proc. 1st Int. Workshop on Robot Vision (in 2nd Int. Conf. on Computer Vision Theory and Appl. – VISAPP 2007), Barcelona, Spain, March 8 – 11, pp. 13–26 (2007)

Generating Dynamic Formation Strategies Based on Human Experience and Game Conditions*

John Atkinson and Dario Rojas

Department of Computer Sciences
Universidad de Concepcion, Concepcion, Chile
atkinson@inf.udec.cl

Abstract. In this paper, a new approach to automatically generating game strategies based on the game conditions is presented. A game policy is defined and applied by a human coach who establishes the attitude of the team for defending or attacking. A simple neural net model is applied using current and previous game experience to classify the game's parameters so that the new game conditions can be determined so that a robotic team can modify its strategy on the fly. Results of the implemented model for a robotic soccer team are discussed.

Keywords: Robotics Game Strategies, Team Formation, Multi-Agent Systems.

1 Introduction

A team's playing strategy is a human football team's main asset. For human players, the strategy is fixed by a coach who defines the players' positions and roles on the football field based on his/her perception on the game conditions and the players' abilities. Accordingly, providing an adaptive playing strategy should involve defining and obtaining the game's current conditions. In order to decide which actions and formation must be taken (and therefore, which low-level behaviors must be accomplished) a team must gather information to determine whether this is doing well or not.

For human football teams, it is relatively easy to determine the game conditions. Several criteria are taken into account including the chances to score, the position on the field, the number of catchings, the score, etc. However, processing this perception information on autonomous robots is not that easy as there are diverse constraints such as processing capabilities, available time, errors with sensors and those of Multi-Agent Systems (MAS) running on a dynamic environment [1].

From a MAS perspective, the playing strategy for the 4-legged robotic competition becomes a significant component due to recent advances on robust vision

---

* This research is partially sponsored by the National Council for Scientific and Technological Research (FONDECYT, Chile) under grant number 1070714 *"An Interactive Natural-Language Dialogue Model for Intelligent Filtering based on Patterns Discovered from Text Documents"*.

U. Visser et al. (Eds.): RoboCup 2007, LNAI 5001, pp. 159–170, 2008.

and localization techniques which currently provide a more accurate and precise perception from the environment. Because of the dynamic and underlying uncertain nature of the league (i.e., there is distributed autonomous intelligence rather than centralized control), new methods are required to generate effective playing strategies which should not be resource-demanding. To this end, a new adaptive approach to team formation using simple and efficient connectionist techniques is proposed to enable a robotic team to effectively adapt its strategy and positions as the game goes on, depending on diverse parameters obtained from the environment's sensorial information.

This paper is organized as follows: in section 2, related robotics and simulation approaches to team formation are discussed, section 3 proposes a new neural net model for team formation and role assignment based on the conditions of the game, in section 4 the main experiments using our model and different role selection strategies are discussed. Finally, section 5 highlights the main conclusions, drawbacks and issues of this research.

2   Related Work

Designing perception systems for autonomous robots participating in robotic soccer competitions is one of the most important challenges. Recent advances on hardware and software for these robotic applications such as perception and locomotion have been promising in terms of providing powerful and robust robotic control systems.

Nevertheless, no significant progress has been reported in team adaptation and cooperation (4-legged) for behavior-based systems. Most of the research on MAS for the 4-legged robotic competition focuses on solving individual problems for each agent (i.e., decision making, navigation, etc). Hence the domain knowledge is indirectly being considered in a nearly reactive way, that is, decisions are purely made based on explicit triggering rules specified by the programmers into the agents' code.

Since that there is no deep analysis of the game conditions, current playing strategies are only based on the ball's current position, and no cumulated experience or human feedback is considered to improve the agents' performance. Furthermore, most of the state-of-the-art research on four-legged robotic teams use machine learning approaches so as to provide agents with some individual basic skills (i.e., reactive tasks) or cooperative capabilities but no efforts are put into getting information regarding the conditions of the game.

A novel approach to decision making based on roles assignment from auctions is proposed by [2]. Agents bid for their roles on the field, and a captain agent (the *bidder*) determines which agent offers the most for some role. Each bid includes the agent's position and the distance to the ball. Offers are then optimized by using Genetic Algorithms which allow the agents to bid for the best possible offer as the system evolves. However, the agents' actions are not taken into consideration and no estimates regarding the conditions of the game are provided.

A slightly different role assignment strategy is proposed by [8] in which decisions are distributedly made, this is, no coordinator or captain is designed. For this, each agent reacts to the team mates' positions and the ball's position by minimizing a distance function so that the closest agent will be given the attacker role whereas the others take defense and support roles. The distributed nature of the model provides a fair approximation of the overall view of the world among agents which allows them to determine the ball's position more precisely.

In order to deal with some of these issues, [9] developed a simulated training agent capable of classifying the opponent's model using a Bayesian method which is previously trained with manually defined rules. Based on the real-time opponent's model and the environment's status, the training agent creates centralized plans using adaptive *Simple Temporal Networks* to be distributedly executed by the players in which best plans are selected by using a hill-climbing search strategy. Overall, the approach does not provide a clear notion of the conditions of the game hence a team can not determine how good/bad the executed actions are. This problem is partially overcome by using a multi-agent system in which agents cooperate to generate different models of the world: a local model and a shared model. However, the approach focuses on error reduction of self-localization rather than cooperation to determine the conditions of the game.

In the context of the *RoboCupSoccer Simulation* league, some approaches determine the strategy of the opponent so to dynamically adapt to the environment. Game conditions are so computed from statistical parameters such as the ball's average position, number of corners, number of goals, etc., which are difficult to obtain accurately. Other approaches explore the generation of agent-agent advises. Here, a training agent is an *adviser* for the other players and produces a set of Markov rules obtained from a set of abstract states contained in the previous games' logs (environment's status, agents' actions, etc). One of the drawbacks is that obtaining these logs involves a global view of the environment which may not be easily available for the four-legged league [5].

Role and task assignment problems have also been formally studied by [3] in the context of MRTA (*Multi-Robot Task Allocation*). One of the drawbacks of this analysis is that the dependence between tasks/roles assigned to robotic agents is not considered. Note that this is a key issue as most of the tasks allocated to one agents has a strong relation with tasks assigned to the other agents of the robotic team. Although the approach can be applied to roles distribution, this always gets the same task distribution to a given scenario as Greedy search is used [6].

3   An Adaptive Model for Computing Game Conditions

One of the most important tasks of a game strategy is to determine whether the executed actions are correct. We need to compute the environments' characteristics which establish when the game conditions are favorable or unfavorable.

There are several approaches to determine the game conditions, however, most of them focus on simulations.

We propose a new model to automatically generate dynamic formation strategies based on the game conditions. The approach computes parameters that determine a good or bad game with no need to manually define input data such as difference of score and number of agents. The relation between these data and the game conditions are automatically obtained by using a learning approach which takes into account human experience, other agents' perceptions and tasks, etc. The agents' experience (i.e., actions performed in a time period) provides criteria on the game conditions with no need to deal with complex perception data.

The *Game Conditions* (GC) of a team establish how favorable/unfavorable the conditions of the game are. The value of the GC can not be obtained instantaneously so the condition can be seen as a sequence of events occurring in a period of time. Thus a GC represents a cumulative form of the current game condition. Unfavorable conditions for the team occur whenever the current game condition shows better choices of losing the game. On the contrary, when favorable conditions occur, the choices of winning the game are better than rather losing it. Near-zero values indicate that the conditions are either uncertainty or balanced.

Computing the GC based on the players' activities provides us with rich information to make further decisions (i.e., difference of score is not a determining factor by itself as a goal can happen by a chance). For instance, if a team is doing very well and even so there is a tie (i.e., there is no useful feedback information on 0-0 scores), then obtaining favorable GC for the team allows it to put more effort into offensive tasks so that this can increase the chances of scoring.

In the proposed model, experiences are seen as a set of actions performed by the agents in a period of time T and include:

- **Player:** Represents any player but the keeper. The player's experience is computed by counting the following actions: *Actions* (AC) is the total number of actions the player intents to perform on the ball, *Blockades of the ball* (BL), *Shots* (SH) is the number of attempts to score on the opponent's defense zone, *Changes of Positions* (CP) is the total number of changes of positions, *Defense to Attacker* (DA) is the number of changes of position from defense to attack, *Attacker to Defense* (AD) is the number of changes of position from attack to defense, *Total Time* (T) is the elapsed time since the beginning of the sampling, *Defense Time* (DT) is the time spent to perform defense tasks on the defense zone, *Support Time* (ST) is the time spent to perform support tasks on the center of the football field, *Attacker Time* (AT) is the time spent to perform attack tasks on the opponent's defense zone, and *Idle Time* (IT) is the time spent in which no tasks are executed.
- **Goal Keeper:** The keeper avoids the opponent team to score and owns an exclusive field's area. Since the goal keeper does not change its position nor perform roles exchanging, only the action parameters on the ball are required. Note that the only enabled actions for the robotic team are catching the ball and shooting the ball, both of which can be considered a kind

of "catching". Hence both actions can be thought as one parameter called *Warning time* (WT) which is the time the goal keeper is in risk.

– **Captain:** The team's captain gathers the players' game data and produces its own parameters. In addition, this provides us with information required to calculate the difference of scores (DS) such as the *Scored Goals* (SG) and the *Received Goals* (RG).

Note that we are interested to quantify the intent of acting rather than determining the effect these produce. This outcome will be learned from a Neural Network by putting all the game's parameters together.

The parameters required to obtain the players' experience are sampled every T seconds. Every time a sampling process gets started, the corresponding experience values are set to 0 and data are normalized to values between 0 and 1. Each normalized experience's parameter can be seen in table 1.

Table 1. Experiences Computed for Each Agent's Role

Experience	Agent Type	Meaning
$E_{BL} = \frac{BL}{AC}$	Player	Blockades
$E_{SH} = \frac{SH}{AC}$	Player	Shots
$E_{DA} = \frac{DA}{CP}$	Player	Defense to Attack change
$E_{AD} = \frac{AD}{CP}$	Player	Attack to Defense change
$E_{DT} = \frac{DT}{T}$	Player	Defense Time
$E_{ST} = \frac{ST}{T}$	Player	Support Time
$E_{AT} = \frac{AT}{T}$	Player	Attacker Time
$E_{IT} = \frac{IT}{T}$	Player	Idle Time
$E_{DS} = \frac{1}{1+e^{SG-RG}}$	Captain	Difference of Score
$E_{WT} = \frac{WT}{T}$	Goal Keeper	Risk Time

The difference of score $(SG(t) - RG(t))$ is then represented as a sigmoid function for differences between -10 and $+10$. Note that the four-legged league does not allow (absolute) differences higher than 10.

In order to compute the GC, a Neural Network (NN) based model is proposed to map players' actions into game conditions. The approach is capable of finding optimum relationships between the players' actions so that game conditions can be computed and then transferred to the team's strategy.

Training of the NN is carried out by performing nine ten-minute robotic soccer simulations. This aimed to obtain initial parameters so to investigate the feasibility of the proposed model. Since that the competition has time and resources constraints, simple NN models have been used. In particular, a Back-Propagation Neural Net was implemented based on [4]. A usual sigmoid function was used as an activation function in the hidden and output layers [4] and this provided fair results on the different configurations. Initially, a full-connected neural network is fed with random values of weights with learning rate values between 0.2 and 0.3. Experiments suggested that learning rates of $\eta = 0.3$ produced the best results as for convergence rates and error drops.

Since that every agent can provide its own set of experiences, an individual vector was generated to combine the whole set of the team's experiences. To this end, every agent's experiences (figure 1) were averaged by computing the values E_{DS} and E_{WT} so to generate the following input vector to the NN:

$$\mathbf{X_{xp}} = (\sum_{i=1}^{N} \frac{E_{BL}^i}{N}, \sum_{i=1}^{N} \frac{E_{TI}^i}{N}, \sum_{i=1}^{N} \frac{E_{DA}^i}{N}, \sum_{i=1}^{N} \frac{E_{AD}^i}{N}, \sum_{i=1}^{N} \frac{E_{TD}^i}{N},$$

$$\sum_{i=1}^{N} \frac{E_{TS}^i}{N}, \sum_{i=1}^{N} \frac{E_{TA}^i}{N}, \sum_{i=1}^{N} \frac{E_{TO}^i}{N}, E_{DG}, E_{TR}) \tag{1}$$

where i represents the $i - th$ agent, and N is the total number of agent of a team (no keeper is considered).

Determining the game conditions is performed by getting information from a human expert. Real data were obtained by simulating a football game. A total of 9 games of 10-minutes each were performed with no side changes. Every 30 seconds, the game is stopped so to ask the expert to assess the game with a fitness ranging from 0 to 1 concerning the team's performance, which is regarded to as $GC_{hum} \in [0,1]$. The expert must provide a value close to zero whenever he/she thinks the GC is unfavorable, whereas a value close to 1 is provided whenever the conditions are seen as favorable. For each time interval, the vector generated from equation 1 is obtained so that the corresponding input vector (X_{xp}) and expected output (GC_{hum}) are produced to train the NN.

Using this method, 171 training data were generated from which 40% was used for training purposes and 60% was used for testing the net. Results of the training tasks suggest that the model is well correlated with the expert's score.

Graphics in figure 1 shows the assessment of the trainer GC_{hum} versus the difference of score DS. For GC_{hum} values close to zero, a correlation with the negative difference of score can be observed. In addition, whenever the conditions are favorable, DS gets closer to the maximum (1). However, whenever the difference is minimum, the trainer provides a broad number of assessments. For example, for a low difference of score $(DS = 0.5)$, the game conditions have been assessed from 0.1 to 0.9 by the trainer which is far more significant.

A metric for assessing the team's attitude was designed. This represents the attacking and defending efforts of a team for each of the two kinds of games (defense and attack). Attitude provides us with information regarding the defensive or attacking attitude of a team depending on which tasks the team is willing to perform most.

Accordingly, the attitude indicates the proportion of resources the team is willing to spend for defending and attacking during attack/defense games. Thus, the lower the value of attitude, the better the willingness to defend is. Attitude values close to 0 indicate that the team is willing to defend. On the contrary, if the value is close to 1, the team is more willing to perform attacking tasks.

Furthermore, a policy is defined as a team's most preferred attitude based on the GC. A human trainer is responsible to define and set the policy before the game gets started. Thus, a policy is a function on the game conditions and

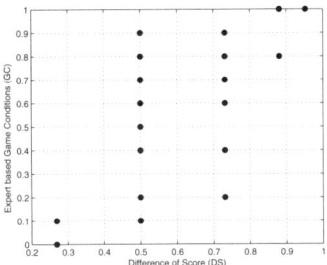

Fig. 1. (a) Difference of Score (DS) versus Game Conditions according to the human expert (GC_{hum})

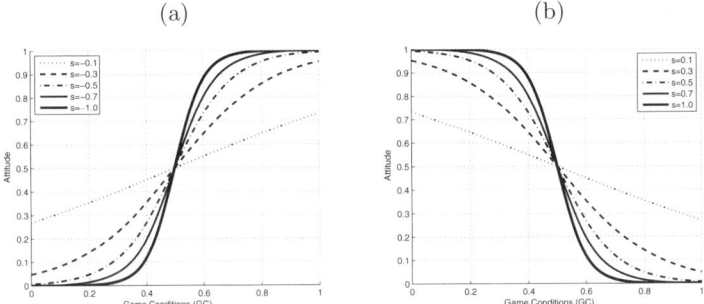

Fig. 2. Proposed Functions for policy. a) Defense Policy b) Offensive Policy.

provides an attitude to be adopted by the team, this is: $attitude = policy(GC) \in [0,1]$. Graphics in figure 2 show the different policies a team may adopt. For example, figure 2(a) represents a policy generating a defensive attitude providing that GC are unfavorable for the team. A policy that produces an offensive attitude, assuming unfavorable conditions, can be seen at figure 2(b). Here, the human coach aims to set the game policy for the team, so the policy function which fits the expectation is provided to the model.

To select the most suitable team formation according to its *policy*, a selection strategy based on the *Roulette Wheel*, commonly used for some implementation of Genetic Algorithms, was applied [7].

Team formation involves assigning a defined area for the agents when they are not acting on the ball. This of area assignment (aka. *home*) is a specific position in axis X of the field divided into three areas: defense, central and offense. From here, the selection algorithm will pick a home distribution for each agent based on the team's attitude and current formation, keeping in mind that sudden changes on the team formation are not desirable as the agents would need to move a lot in order to reach the home position. Accordingly, a team formation is defined as:

$$\mathbf{F} = (F_x, F_y, F_z) \quad with \quad F_x + F_y + F_z = N$$

where F_x, F_y, F_z represent the number of players having a *defense home*, a *central home* and a *attack home*, respectively, and N is the number of agents in the formation. The fitness of a formation F can be computed based on the team's attitude and the closeness to the previous formation:

$$fitness_{att}(\mathbf{F}) = 1 - |att - (F_x * (\frac{0}{N}) + F_y * (\frac{0.5}{N}) + F_z(\frac{1}{N}))| \qquad (2)$$

where att is the attitude value obtained from the game conditions and the policy. Next, the closeness-based fitness is determined by calculating the similarity between the current ($\mathbf{F'}$) and the new formation ($\mathbf{F}$). This is computed from the Euclidean distance between both formations and normalized to the maximum distance (i.e., the distance between formations $(3, 0, 0)$, $(0, 0, 3)$ and $(0, 3, 0)$, that is, $\sqrt{18}$):

$$fitness_{sim}(\mathbf{F}, \mathbf{F'}) = \frac{\sqrt{(\mathbf{F} - \mathbf{F'})^2}}{\sqrt{18}} \qquad (3)$$

Both fitness functions are weighted according to an expert's defined parameter $\alpha \in [0, 1]$. Thus, the fitness of a formation is evaluated from the current formation and current attitude as follows:

$$fitness(\mathbf{F}, \mathbf{F'}) = \alpha fitness_{sim}(\mathbf{F}, \mathbf{F'}) + (1 - \alpha) fitness_{att}(\mathbf{F}) \qquad (4)$$

Our selection algorithm computes the *fitness* for all the possible formations $\mathbf{F_i}$ with $i \in [1, 10]$. An elitist criterion is used to pick the M formations having the highest *fitness*. Next, the fitness proportional to $fitness_{pi}$ is calculated for each selected formation as: $fitness_{pi} = \frac{fitness_i}{\sum_{i=1}^{10} fitness_i}$

Obtained fitnesses ($fitness_{pi}$) are then sorted and a random number β with uniform distribution is chosen. The algorithm cumulates the $fitness_{pi}$ in descending order until the value is greater or equal to β. Afterwards, the best formation having the last fitness $\mathbf{F_i}$ is selected. The outcome of the algorithm is the new formation $\mathbf{F_{new}}$ which represents the best fitness based on the team's attitude and the expert's criteria.

4   Evaluation and Results

The benefits of using the model for determining the game conditions on the fly were investigated by carrying out a series of experiments. This aimed to assess the robustness of the approach in terms of different time intervals, bandwidth efficiency, and the improvement of the team's performance compared to a different team in which no cooperation strategy is provided.

The model is capable of operating on different time intervals with no dependency on the sampling period T used for the training phase. The tolerance of the neural net model to different run-time intervals was assessed by performing a series of games under two approaches. The first approach considers sampling with continuous cumulation of experience, and the second one involves sampling with cumulation at independent intervals of experience. For these, nine games

(a)

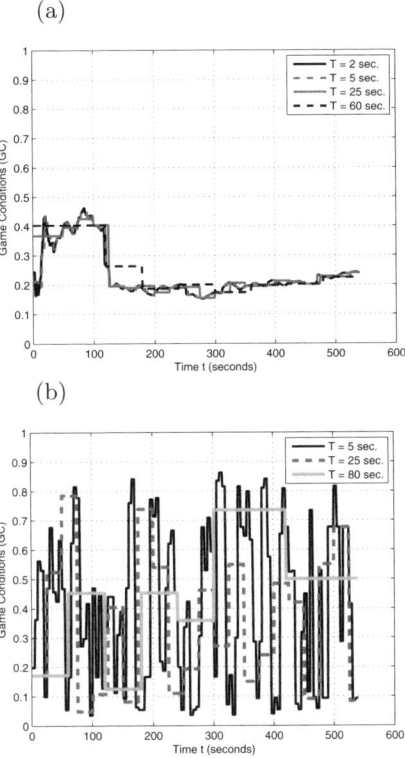

(b)

Fig. 3. (a) GC for continuous cumulated experience. (b) GC for independent-interval cumulated experiences.

were performed for training purposes. Changes to the conditions during these games having the worst performance can be seen at figure 3. The games do not use the dynamic formation strategy and so a fixed formation involving two defenders and one attacker is provided only. Runs in figure 3(a) show that despite having different sampling time intervals, the model is still capable of computing a correct value for GC, meaning that for the same game, the sampling time is not significant whenever the approach no.1 is applied.

The effects of the net having experience removed between samplings can be seen in figure 3(b). Based on the approach no. 2, results suggest that agents represent only game conditions from time $t - T$ to t, and accordingly GC is perceived as an *evaluation* of the last sampled time interval.

The efficiency of the model for dynamic team formation was assessed by performing 24 testing games in which 4 unseen opponent agents teams were used. These games used different approaches for cumulating experience, each of which was tested using three time intervals (10, 25, 60). The opponent teams were *Team 1* (potential fields based navigation), *Team 2* (dynamic role assignment with agent always gets the ball), *Team 3* (fixed formation involving two defenders and one central), *Team 4* (stands in the way of the opponent agents).

Table 2. Results of the Testing Games for the Model Team

Team	+ Score	- Score	Difference	Total Score
Baseline	34	36	-2	18
Model	54	38	16	37

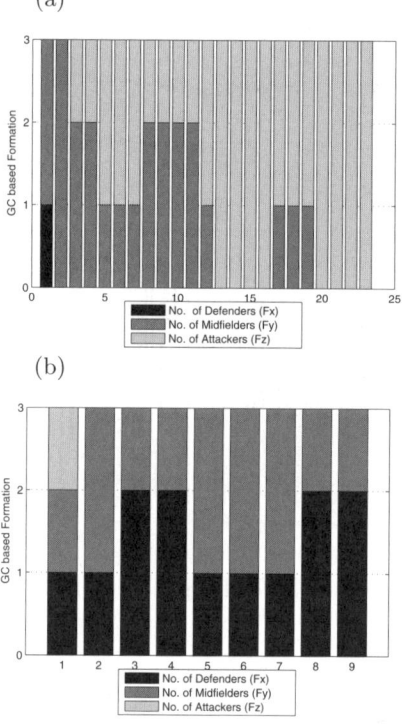

(a)

(b)

Fig. 4. Two Experimental Formations: (a) Formation with T=25. (b) Formation with T=60.

The model was implemented and based on the simple individual skills of team no. 3 which allowed us to assess the dynamic formation strategy. This team is referred to as the *Model Team* or **mTeam** whose testing parameters were as follows: defensive policy with a stepness of $s = 0.2$, sampling time every T seconds with $T \in \{10, 25, 60\}$, 4 agents. The same 24 games were performed by using the team no. 3 as baseline and having a fixed formation (a central, a defender, an attacker). This team does not use any adaptive formation strategy and is referred to as the *Base Team* or **bTeam**.

Performance was then assessed by assigning scores to the *bTeam* and *mTeam*. The obtained score for each team is counted as for human football: the winner team gets 3 points and the loser gets 0 points. If there is a tie, each team gets 1 point. Some key observation can be made from results in table 2.

Changes on the game conditions for both teams can be observed from graphics in figure 1. Dark line represents the game conditions for the *mTeam* which uses the dynamic formation strategy whereas the dotted line represents the conditions for the *bTeam*. In both scenarios, the teams' defensive policy allows them to keep the conditions favorable, which can be seen by looking at the scores, the obtained differences, and the game conditions (GC).

Graphics in figure 4 show the changes to the formations during the games. Performance of formations in figure 4(a) suggests that the initial formation is *centralized* with most of the players standing at the center of the field. As the game goes on, the formation becomes more offensive as the game conditions advice them to do so (GC close to 0.8). The fixed and rigid structure of the *bTeam* refrains it from scoring a single point due to the lack of supporters. Furthermore, the *bTeam* frequently tended to lose the ball at the middle of the field.

The performance of the formation of figure 4(b) can be seen in graphics of figure 1(b), in which the initial formation is *uniform* (i.e., there is a defender, an attacker and a supporter).

Our team takes a defensive formation involving two defenders and one central as the opponent team performs most of the offensive and risky actions. As the time goes on, game conditions become favorable for the model team as this keeps the ball most of the time. Overall, the strategy produces a more centralized formation using two centrals and one defender, and as a consequence favorable conditions are kept by scoring 3 goals.

5   Conclusions

A new approach to dynamic team formation using a simple Neural Net based model is described. The model is a mixture of simple neural net approaches, heuristics-based evaluation, multi-agent systems techniques and the human expert's experience so as to provide a robust method to compute game conditions which in turn allows the team to dynamically modify its positions and roles on the fly.

The experiments and real testing show that the neural net's prediction level as being trained by a human expert is well correlated with the automatically trained model. This suggests that the function applied to automatically generate samples is both robust and resource-efficient. The final design and implementation for our team also provides some interesting insights. Neural nets being trained at periods of time in which previous experience was removed before starting a new period of sampling proved to be useful to measure the change of conditions in a specific period of time. Hence this model may be applied to check whether some decision on a performed strategy had an instant effect on the game or not.

References

1. d'nverno, M., Luck, M.: Understanding Agent Systems, 2nd edn. Springer, Heidelberg (2004)
2. Frias, V., Sklar, E., Parsons, S.: Exploring auction mechanisms for role assignment in teams of autonomous robots. In: Nardi, D., Riedmiller, M., Sammut, C., Santos-Victor, J. (eds.) RoboCup 2004. LNCS (LNAI), vol. 3276, pp. 532–539. Springer, Heidelberg (2005)
3. Gerkey, B., Mataric, J.: A formal analysis and taxonomy of task allocation in multi-robot systems. The International Journal of Robotics Research 23(9), 939–954 (2004)
4. Hagan, M., Demuth, H., Beale, M.: Neural Network Design. Martin Hagan (2002)
5. Kuhlmann, G., Knox, W., Stone, P.: Know thine enemy: A champion robocup coach agent. In: Twenty-First National Conference on Artifical Inteligence (AAAI 2006), Boston, MA, July 2006, AAAI Press (2006)
6. Lerman, K., Jones, C., Galstyan, A., Mataric, M.: Analysis of dynamic task allocation in multi-robot systems. The International Journal of Robotics Research 25(3), 225–241 (2006)
7. Mitchell, M.: An Introduction to Genetic Algorithms. The MIT Press (1996)
8. Quinlan, M.J., Nicklin, S.P., Hong, K., Henderson, N., King, R.: The 2005 nubots team report. Technical report, School of Electrical Engineering and Computer Science, The University of Newcastle, Australia (2006)
9. Riley, P., Veloso, M.: Planning for distributed execution through use of probabilistic opponent models. In: Proceedings of the Sixth International Conference on AI Planning and Scheduling, pp. 72–81 (2002)

Model-Based Reinforcement Learning in a Complex Domain

Shivaram Kalyanakrishnan, Peter Stone, and Yaxin Liu

Department of Computer Sciences
The University of Texas at Austin
Austin, TX 78712-0233
{shivaram,pstone,yxliu}@cs.utexas.edu

Abstract. Reinforcement learning is a paradigm under which an agent seeks to improve its policy by making learning updates based on the experiences it gathers through interaction with the environment. *Model-free* algorithms perform updates solely bas ed on observed experiences. By contrast, *model-based* algorithms learn a model of the environment that effectively simulates its dynamics. The model may be used to simulate experiences or to plan into the future, potentially expediting the learning process. This paper presents a model-based reinforcement learning approach for Keepaway, a complex, continuous, stochastic, multiagent subtask of RoboCup simulated soccer. First, we propose the design of an environmental model that is partly learned based on the agent's experiences. This model is then coupled with the reinforcement learning algorithm to learn an action selection policy. We evaluate our method through empirical comparisons with model-free approaches that have been previously applied successfully to this task. Results demonstrate significant gains in the learning speed and asymptotic performance of our method. We also show that the learned model can be used effectively as part of a planning-based approach with a hand-coded policy.

1 Introduction

The reinforcement learning (RL) [12] problem is usually modeled as a Markov Decision Process (MDP) [10], which is of the form (S, A, R, T, γ). S is the set of states in the environment, and A the set of actions available to the agent. $R : S \times A \to \mathbb{R}$ is the reward function for the task: it returns the real number reward provided to the agent for taking an action from a given state. The dynamics of the environment are encapsulated in the transition function $T : S \times A \times S \to [0, 1]$; given a state and action, T returns a probability distribution over next states to which the agent may be transported. A (deterministic) policy $\pi : S \to A$ specifies the action to be taken by the agent from any given state. Every policy π can be associated with an action value function $Q : S \times A \to \mathbb{R}$ that computes the expected long-term discounted reward the agent will accrue by following π after taking some action a from some state S. $\gamma \in [0, 1]$ is a discount factor in

U. Visser et al. (Eds.): RoboCup 2007, LNAI 5001, pp. 171–183, 2008.
© Springer-Verlag Berlin Heidelberg 2008

the expected long-term reward. The problem is to solve for an *optimal* policy π, i.e., one that maximizes $\max_a Q^\pi(s, a)$ for every state s, defined by:

$$Q^\pi(s, a) = R(s, a) + \gamma \sum_{s' \in S} T(s, a, s') Q^\pi(s', \pi(s')). \tag{1}$$

In most practical settings, the agent must act in the environment to gather *experiences*, using which it can improve its policy. An experience (or *transition*) is of the form (s, a, r, s'), where s is the agent's state, a an action taken from s, r the reward received, and s' the state to which the agent moves. Theoretical guarantees establish that under some conditions, the optimal policy can indeed be learned by making temporal difference updates based on the observed experiences, for instance, through methods like Q-learning [16]. Nonetheless, it is seldom possible in real world tasks to meet the conditions necessary for convergence. Solutions to complex tasks invariably have to adopt an engineering approach and exploit their underlying structure to the extent possible.

In this paper, we explore the potential of model-based methods in scaling RL to complex tasks. Whereas model-free methods like Q-learning interpret the policy directly through the action value function Q, model-based methods seek to decouple Q into its "components" T and R, termed the transition and reward models of the task respectively. By doing so, it becomes possible to use the *model* (T and R together) to *simulate* experiences that can be used to update Q, instead of solely relying on ones gathered from the environment. More specifically, the model can be used to explore parts of the state space that are possibly underrepresented in the observed experiences. Hence, simulating experiences using the model can potentially improve the quality of the solution, while achieving economy in sample complexity. A further benefit gained from learning T and R individually is the advantage of separating the dynamics of the environment from the objective of the task at hand, offering the flexibility to share parts of the solution with different tasks in similar environments.

Model-based methods have been applied successfully in the past to several challenging problems. In domains such as game-playing, a partial or complete model of the environment is sometimes available, but determining the action selection policy can still be challenging owing to factors like the intractability of searching through the state space [14,15]. On the other hand, for many real-world domains, learning the environmental model is itself a substantial undertaking. In past efforts involving learning the model [2,9], the environment is typically a physical system that is sampled at some regular frequency, and the actions are control signals perturbing the state of the system. By contrast, in Keepaway, the domain we consider for our experiments, the actions are abstract, high-level skills, which last for extended, variable durations of time. Keepaway is a large-scale, complex, multiagent task involving both teammates and adversaries, which are part of the environment being modeled. The approach we follow is to partially learn the model for this task, and partially describe it using simple rules. This necessarily approximate model is then used in our Model-based Policy Improvement (MBPI) algorithm to examine if it can still help expedite learning.

The remainder of the paper is organized as follows. Section 2 describes the Keepaway task, and Section 3 presents our design of a model for this task. Section 4 provides details of the model-based RL algorithm. In Section 5 we present experimental results evaluating our method, providing comparisons with other algorithms that have been applied to Keepaway. Section 6 discusses related work, and Section 7 concludes.

2   Keepaway Task Description

Keepaway [11] is a subtask of simulated RoboCup soccer [8] played between a team of m *keepers* and a team of n *takers* inside a rectangular region. The objective of the keepers is to maintain possession of the ball (have it close enough to be kicked), while the takers try to steal it. The task is episodic – each episode starts with the ball in possession of one of the keepers, and ends when some taker gets the ball or it goes outside the region of play. The version of Keepaway we consider for our experiments involves 3 keepers and 2 takers ($3v2$) inside a $20m \times 20m$ region, as depicted in Figure 1. We proceed to describe how Keepaway is framed as a reinforcement learning problem, outlining the challenges it poses.

A complete state description in Keepaway would include the positions and velocities of the players and the ball, the players' body and neck angles, their stamina levels, and so on. However, we find that their positions alone convey most of the information required for the purpose of learning. Since the players and ball may occupy any position inside the region of play, the state space is continuous. Furthermore, the players are provided noisy sensations of state.

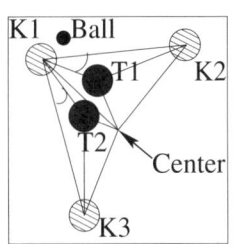

Fig. 1. $3v2$ Keepaway. K_1, K_2, and K_3 are keepers. T_1 and T_2 are takers.

The keepers are the learning agents: the task each keeper has to learn is which action to take when it gets possession of the ball. This being the case, it becomes necessary to define the concept of a state only when some keeper has possession. In each state, the keeper closest to the ball is denoted K_1; the other keepers are denoted $K_i, i = 2, 3, 4, \ldots, m$, K_i being the i-th *closest* keeper to K_1. Similarly, the takers are denoted $T_i, i = 1, 2, 3, \ldots, n$, T_i being the i-th *closest* taker to K_1. K_1 is the keeper that must choose an action to execute. The behaviors of the takers and keepers who are without possession are fixed: the takers try to intercept the ball, while $K_2, \ldots, K_m$ attempt to move to positions to which a pass from K_1 is likely to succeed.

Figure 1 illustrates the indexing of keepers and takers, also marking out distances and angles among the players and the center of the field. These serve as *abstract* features derived from the players' positions, which are used as inputs to the function approximator representing the action value function. We refer the reader to Stone, Sutton and Kuhlmann [11] for a detailed description of these abstract state features. Notice that there are 13 for $3v2$ Keepaway.

The actions that are available to K_1, when it has possession of the ball, are **HoldBall**, by which it keeps the ball with itself, and **PassBall**(i), $i = 2, 3, 4, \ldots, m$, which is a direct pass to the K_i. While it is convenient to treat **HoldBall** and **PassBall**(i) as actions, they are really high-level skills or *options* [13] implemented through a series of low-level actions like **Turn** and **Kick**. Passes can last a variable number of simulator cycles; so the task is effectively a Semi-Markov Decision Process [4]. The transition dynamics of the extended high-level actions, which are necessarily stochastic because of the keepers' noisy actuators, thus become susceptible to even greater irregularity. Also, the dynamics are not smooth, as some actions can lead to terminal states.

The reward provided for taking an action from a state is simply the number of cycles elapsed until the next state is reached. Since the task is episodic, no discounting is required. Maximizing expected long-term reward corresponds to maximizing the expected overall duration of the episode, also called the *hold time*. **HoldBall**() typically lasts 1-2 cycles; **PassBall**(i) can last between 4 and 12 cycles, depending on the distance the pass has to travel. A cycle of simulation lasts 100 milliseconds in real time. In 3v2 Keepaway, a random policy that chooses uniformly among the actions (**HoldBall**, **PassBall(2)**, **PassBall(3)**) registers a hold time of about 4.7 seconds.

In our experiments, we use the same version of 3v2 Keepaway as used by Kalyanakrishnan and Stone [6], but with one minor change. In their version, K_1 executes **HoldBall** through a series of kicks close to its body that take it away from the direction of the takers. In our implementation, K_1 simply stops the ball once it is kick-able, and subsequently leaves it untouched. We find that this helps our model-based approach by simplifying the transition dynamics. Interestingly, informal testing reveals that it also leads to better performance with the model-free methods successfully applied earlier [6,11]. We compare all these algorithms using our version of **HoldBall**.

3   Learning the Model

In this section, we describe our design of a model for Keepaway. The precise requirements of the model are that given state s and action a, it predict a distribution over next states s', as well as the reward r for the transition. Since the actions are disparate, high-level skills, we maintain separate models for each action. Figure 2 lays out the schematic design. Though Keepaway is indeed a stochastic domain, we adopt the simple approach of approximating its dynamics using a *deterministic* model, i.e., the model returns a unique next state s' instead of a distribution over next states. Since some transitions can lead to terminal states, we employ a separate predictor to compute t, a boolean value indicating whether a given transition is terminal. Likewise, a separate predictor computes the real-valued transition reward r.

Our main objective is not building an accurate model in itself, but rather to evaluate the advantages of using a model in conjunction with the RL algorithm. We find it sufficient for this purpose to specify parts of the model using intuitive, hand-coded rules, but nonetheless, necessary to derive other parts of it by applying machine learning. As shown in Figure 2, the next state s' is computed by applying a simple rule to the current state s. The rule simply assumes the players do not change their positions between s and s'. In case of the **HoldBall** action, the ball's position in s' is predicted to be the same as K_1's, and if the action is **PassBall(i)**, the ball is predicted to occupy the same position as K_i. Figure 3 illustrates through an example from $3v2$ Keepaway how the next state prediction is made for a given state and action.

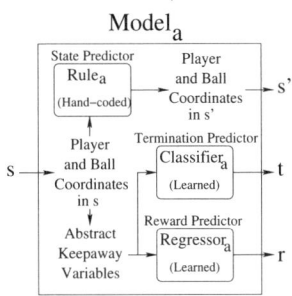

Fig. 2. Schematic Diagram of Keepaway Model

In our model, the termination and reward predictors are trained through supervised learning using the observed experiences. The reward predictor for each action is a single-layer neural net with 10 hidden nodes. Its inputs are the abstract state features derived from the Keepaway state (see Section 2), over which effective generalization is possible. The output is a real-valued prediction of the reward. The termination predictor is a single-layer neural net with 5 hidden nodes. It takes the same inputs as the reward predictor, but computes a boolean-valued output instead. In $3v2$ Keepaway, only roughly 10% of all transitions are terminal; nonetheless, we increase the weight of terminal transitions in the training distribution to present each termination predictor an equal number of terminal and non-terminal transitions.

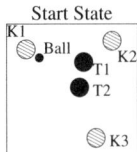 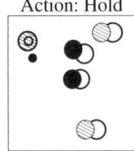 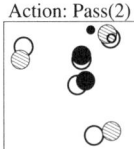 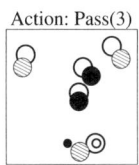

Fig. 3. At left is a start state. The subsequent figures show true (shaded) and predicted (outlined) next states reached after each action is taken from the start state.

The operation of the model is summarized as follows. Given a set of training experiences $D = \{(s, a, r, s')\}$, we fit a model as $M = learnModel(D)$, the learning restricted to the termination and reward predictors for each of the actions. Subsequently, M can be used to make predictions; given state s and action a, the predictions are of the form $s' = M.predictNextState(s, a)$, $t = M.predictTermination(s, a)$, and $r = M.predictReward(s, a)$. In Section 4, we explain how the model is employed under the Model-based Policy Iteration (MBPI) algorithm.

Table 1 lists the prediction errors of the models for the three actions in $3v2$ Keepaway. The entries are averages from 5 independent runs – in each run a model is learned based on transitions from 50 episodes (an episode typically comprises 10-20 transitions) during which the keepers follow a random policy. This model is then tested for 1000 episodes, again following random action selection. For the purpose of computing prediction errors in the positions of players, we use the same ordering $(K_1, K_2, K_3, T_1, T_2)$ in s' as seen in s. Thus, if keeper K_A is K_1 in s, and has passed the ball to K_B, K_A is still considered K_1 in s' while computing the error. Of course, the real ordering of s' is used while computing abstract features for s'.

We carried out $3v2$ Keepaway inside a $20m \times 20m$ region; distances among players are typically 5-$15m$. Notice that for the **HoldBall** action, the prediction errors for all the players' positions are less than $1.0m$; this is because the action itself typically lasts only 1-2 cycles, during which the players do not move very far. The errors are much higher for the pass actions, and indeed higher for **PassBall(3)** than **PassBall(2)** because of the longer distance the pass has to travel. The reward predictions errors are quite small for **HoldBall**, and within about 2 cycles for the pass actions. For the **PassBall(i)** actions, the misclassification probabilities of terminal and non-terminal transitions are comparable. The high error in classifying terminal **Hold** actions arises because of insufficient training data: only a very small fraction of **HoldBall** actions terminate while following a random policy. We recognize that there is scope to improve the accuracy of the model; in particular, the accuracy of the state predictor (see Section 5). Nonetheless, the measure we seek to evaluate in this paper is not the accuracy of the model itself, but the performance achieved by the RL algorithm employing the model. The algorithm is described in the next section.

Table 1. Errors in the positions are root mean squared values of the distance (in meters) between true and predicted positions. Terminal and Non-terminal errors are the fractions of terminal and non-terminal actions misclassified. Reward errors are root mean squared values of the difference (number of cycles) between true and predicted values.

Action	Position						Terminal	Non-terminal	Reward
	K_1	K_2	K_3	T_1	T_2	Ball			
HoldBall	0.63	0.89	0.91	0.81	0.96	0.64	0.93	0.004	0.33
PassBall(2)	3.62	3.88	4.03	2.85	2.89	3.74	0.16	0.13	2.07
PassBall(3)	4.03	3.78	4.78	2.85	2.92	4.98	0.17	0.12	1.96

4   Using the Model

The central idea underlying our Model-based Policy Improvement (MBPI) algorithm is to use the gathered experiences to learn a model of the environment, and then use this model extensively to simulate transitions based on which the action value function is updated. The model and the learned policy are improved iteratively, as we describe in Algorithm 1.

We begin with some initial Q function (line 1). A policy is interpreted from Q through the *selectAction()* function (line 9), which can implement, for instance, ϵ-greedy action selection. A batch of experiences D is collected by following this policy for some fixed number e of episodes (lines 6-14). Once the experiences are

Algorithm 1. Model-based Policy Improvement

1: $Q \leftarrow Q_0$. //Initialize action value function.
2: $D \leftarrow \emptyset$. //Initialize memory of experiences.

3: //Improve Q iteratively.
4: **repeat**

5:     // **Experience Generation**
6:     **for** e episodes **do**
7:       $s \leftarrow startStateFromEnvironment()$.
8:       **repeat**
9:         $a \leftarrow selectAction(Q)$.
10:         $r \leftarrow rewardFromEnvironment()$.
11:         $s' \leftarrow nextStateFromEnvironment()$.
12:         $D \leftarrow D \cup (s, a, r, s')$.
13:       **until** s' is terminal.
14:     **end for**

15:     // **Model Learning**
16:     $M \leftarrow learnModel(D)$.

17:     // **Policy Improvement**
18:     **for** n iterations **do**
19:       $s \leftarrow randomStartStateFromSimulator()$.
20:       $d \leftarrow 0$.
21:       //Simulate trajectories of depth $depth$.
22:       **repeat**
23:         $a \leftarrow selectActionSimulate(Q, s)$.
24:         $r \leftarrow M.predictReward(s, a)$.
25:         $t \leftarrow M.predictTermination(s, a)$.
26:         // Update Q based on simulated transitions.
27:         **if** t **then**
28:           $Q(s, a) \leftarrow Q(s, a) + \alpha(r - Q(s, a))$.
29:         **else**
30:           $s' \leftarrow M.predictNextState(s, a)$.
31:           $Q(s, a) \leftarrow Q(s, a) + \alpha(r + \gamma \max_{a'} Q(s', a') - Q(s, a))$.
32:           $s \leftarrow s'$.
33:         **end if**
34:         $d \leftarrow d + 1$.
35:       **until** $t = true$ or $d = depth$.
36:     **end for**

37: **until** Q has converged.

collected, they are used to learn a model M of the environment (line 16). Q is now updated using transitions that are simulated using M (lines 18-36). This is accomplished by generating trajectories of depth $depth$ using M, beginning with some random start state (line 19) and following an action selection policy specified by the function $selectActionSimulate()$ (line 23). Once Q is updated, it is used to generate the next batch of experiences; a new model is learned and the process continues until Q converges. MBPI is similar to Lin's experience replay algorithm [7], applied to Keepaway by Kalyanakrishnan and Stone [6], which differs from it in the following manner: in experience replay, no explicit model is learned, and the policy improvement occurs through (depth 1) updates solely involving the experiences stored in D.

In all our experiments, we have fixed the values of parameters and choices for subroutines through informal experimentation. The function approximation scheme we use for representing Q is the same used by Stone *et al.* [11] and Kalyanakrishnan and Stone [6] – a separate CMAC [1] for each action, taking as input the 13 abstract state features computed from the state. Each CMAC employs 32 one-dimensional tilings along each feature, the tile widths being $3.0m$ for features corresponding to distances, and $10°$ for those corresponding to angles. The *selectAction*() function implements ϵ-greedy action selection, with $\epsilon = 0.01$. We fix the number of experiences in each batch, e, to 50. By setting all CMAC weights to zero in Q_0, the initial action value function, the policy followed for generating the first batch of experiences is random.

The set of experiences D used in every iteration to learn the model comprises *all* the past experiences collected thus far; we find that this yields better performance than obtained by only keeping the most recent batch (or some recent window) of experiences in D. The termination and reward neural networks for each action are trained using supervised learning. For the termination predictors, $200,000$ backprop updates are made with a learning rate of 0.0001, picking terminal and non-terminal transitions in D with equal likelihood. $20,000$ backprop updates using randomly chosen experiences from D are made in the case of the reward predictors, with a learning rate of 0.0005. While making learning updates to the function approximator representing Q using experiences simulated by the learned model M, we fix the number of (Q-learning) *updates* to $30,000$, each made with a learning rate of 0.025. We find that doing so offers more stability than fixing the number of iterations n, under which the actual number of updates would depend on the size of D. The start states of the trajectories are randomly chosen start states from the transitions in D, and *selectActionSimulate*() implements random action selection.

5   Experimental Results and Discussion

In this section, we present the results of our experiments on $3v2$ Keepaway. Figure 4(a) shows the performance of our model-based policy iteration algorithm, using $depth = 1$ while simulating trajectories (MBPI-1). It is compared with experience replay (ER), which achieves the best asymptotic performance on $3v2$ Keepaway among the batch methods considered by Kalyanakrishnan and Stone [6], and simple on-line learning (OL) [11], where a single Q-learning update is made after every transition. We find ER to achieve its best performance by making $30,000$ Q-learning updates during the policy improvement phase, with learning rate 0.025. Interestingly, the same values were found the best for MBPI-1. For OL, we used a learning rate of 0.125, the same used by Stone *et al.* [11] in their Sarsa-based OL implementation. Figure 4(a) shows that MBPI outperforms ER and OL right from the beginning, and also betters their asymptotic performance (Figure 4(b) shows OL continuing until $20,000$ episodes). At 200 episodes, MBPI-1 registers a higher hold time than ER and OL with p-value at most $p < 5 \times 10^{-9}$, under a single-tailed t-test. The best performance achieved by MBPI-1 (11.62 seconds, 450 episodes) exceeds those of ER (9.24 seconds, 250 episodes) and OL (9.48 seconds, $11,000$ episodes) with p-value at most $p < 10^{-13}$.

The main reason MBPI-1 and ER achieve an order of magnitude gain in sample complexity over OL is that they make more efficient use of the collected

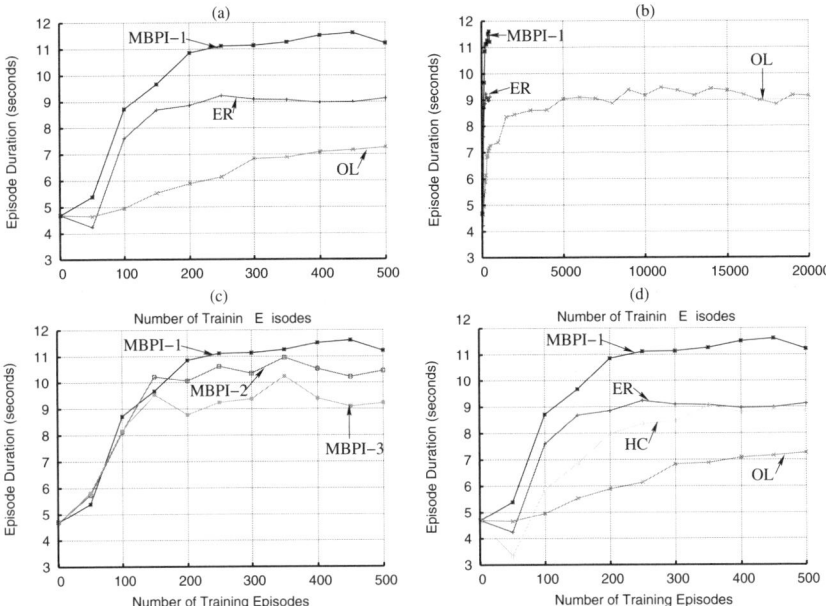

Fig. 4. The graphs show on the x axis the number of training episodes; on the y axis the hold time achieved by some policy. The reported hold time is the average over 200 episodes in which the policy is frozen and executed. For the first 500 episodes of training, we evaluate the policies at intervals of 50 episodes. The algorithms making batch updates do so every 50 episodes; they are evaluated immediately *after* the update. Each curve is an average of at least 25 independent runs.

experiences through batch updates. The updates made by ER are strictly based on observed experiences, which reflect the true dynamics of the environment. Further, only states that are reachable by following the policy used while generating the experiences get backed up. In contrast, MBPI explores more parts of the state space by following trajectories randomly generated using the model. Experiences generated by the (approximate) model are likely to be somewhat inaccurate, and the states visited along the simulated trajectories may not be reached in practice. But despite the inaccuracy, the exploration can potentially result in the discovery of desirable states and thus improve the policy.

Lin [7] compares ER with *relaxation planning*, a model-based approach in a discrete, grid-world domain. In case the agents have to learn the model, then the model-based approach performs *worse* than ER; however, when the agents are provided a perfect model to begin with, the algorithms have comparable performances. In our experiments with 3v2 Keepaway, MBPI-1, under which the model both has to be learned and used, consistently outperforms ER. We conjecture that for domains with small state spaces, the observed transitions may themselves be sufficiently representative of the dynamics of the domain, but as the size of the problem increases, this may cease to be the case, and the extrapolation afforded by the model may prove beneficial. In this paper, it is our intention to compare MBPI and ER on Keepaway by studying them in isolation, but in principle, it is possible to combine them by making policy updates from both observed and simulated transitions. More specifically, it may be possible to offset the noise introduced by an incorrect model by making sufficient updates based on true experiences. Section 6 discusses Sutton and Barto's Dyna-Q algorithm [12], which takes a related approach.

Figure 4(c) shows the effect of increasing the depth of the simulated trajectories in our model-based algorithm. It seems plausible that deeper trajectories will enhance the exploration of the state space, boosting performance. On the other hand, due to the noise in the model predictions, simulated transitions are likely to deviate more from the true transitions deeper in the trajectory. In our experiments, we notice that with increasing depth (MBPI-2 and MBPI-3), the performance of the model-based approach *degrades* progressively. To diminish the adverse effect of noise deep in the trajectories, we decay the learning rates for updates made deeper down, still keeping the sum of the learning rates along each trajectory constant (at 0.025) so that the comparisons among the experiments remain fair. Despite using a sharp decay factor (0.01), the performance of MBPI-2 and MBPI-3, as seen in Figure 4(c), fall significantly short of MBPI-1's. Nonetheless, MBPI-2 (10.96 seconds, 350 episodes) still achieves higher performance than ER (9.24 seconds, 250 episodes) and OL (9.48 seconds, 11,000 episodes), with p-values at most $p < 2 \times 10^{-3}$.

Surely, a major reason for the loss in performance when exploring deeper is the approximation in our model. Since the dynamics of Keepaway are stochastic, a deterministic model is bound to be inaccurate. Further, the function approximation scheme used in the model may not be sufficiently expressive. Past efforts in modeling physical systems have focused on learning precise models, and indeed modeling environmental noise as well [9]. It is a promising avenue for future research to develop a more accurate model for Keepaway, and examine if it can be used to plan deeper into the future. Nevertheless, the performance gain offered by MBPI-1 is evidence that model-based approaches can be viable even with an approximate model, on a task that is itself continuous and stochastic.

Algorithm 2. Hand-coded Policy (Model M, State s)

1: $A_{non-terminal} \leftarrow \{a | M.predictTermination(s, a) = false\}$.
2: **if** $A_{non-terminal} = \emptyset$ **then**
3:     Return $random(\textbf{HoldBall}, \textbf{PassBall(2)}, \textbf{PassBall(3)})$.
4: **else**
5:     **if** $\textbf{HoldBall} \in A_{non-terminal}$ **then**
6:         Return **HoldBall**.
7:     **else if** $\textbf{PassBall(2)} \in A_{non-terminal}$ **then**
8:         Return **PassBall(2)**.
9:     **else**
10:        Return **PassBall(3)**.
11:    **end if**
12: **end if**

In our MBPI algorithm, the learned model is used to update the Q function through which the action selection policy is interpreted. While this conforms with the traditional RL approach of learning the Q function, it is not necessary for putting the model to use. Algorithm 2 lays out a hand-coded policy that uses an available model to select the action to take. In fact, this policy only makes use of the termination predictor of the model, implementing the following intuitive strategy: from any state, choose **HoldBall** if the model predicts it will not terminate the episode. If it is predicted to terminate, try **PassBall(2)** in a similar manner, and then **PassBall(3)**. If all actions are predicted to terminate, simply choose a random one. Note that the hand-coded policy is myopic: it doesn't perform lookahead to take actions that will avoid future bad states. In this way it is handicapped when compared to the learning algorithms.

Figure 4(d) plots the performance of this hand-coded policy. As with MBPI, it begins with a random model that is updated every 50 episodes; however, the policy followed in between the updates is the hand-coded policy. After 350 episodes, this policy registers 9.04 seconds of hold time, which is within 0.5 seconds of the best reached by OL and ER. The purpose of this experiment is not to highlight the performance of the hand-coded policy in itself, but to illustrate that a model can be useful even independent of the action value function. Here, it is necessary to improve the model iteratively, but one can imagine scenarios where a model is available from past experiences or adapted from related tasks. A model-based approach provides the flexibility to re-use parts of the solution in a natural way. It would be promising as part of future research to adapt the model-based approach followed here to interact with similar tasks in the RoboCup soccer domain, for instance, 4v3 Keepaway [11] and Half Field Offense [5]. Another possible avenue for research is to employ the environmental model as part of a planning algorithm for solving the task.

6   Related Work

In their expository textbook, Sutton and Barto [12] investigate the relationship between model-based RL and planning. They present the Dyna-Q algorithm, in which an environmental model is learned and used to simulate experiences for updating the Q function along with direct updates based on real experiences. Dyna-Q is enhanced by using Prioritized Sweeping, a technique whereby the model-based updates are concentrated around the regions where the Q function is changing rapidly. The main motivation for our work is indeed to extend the qualitative results of model-based approaches like those seen in the simple, relatively small, discrete domains considered by Sutton and Barto to a realistic, high-dimensional, continuous task. Our MBPI algorithm is similar to their Trajectory Sampling method, where model-based updates are based on the *on-policy* distribution of experiences. In our case, an ϵ-greedy policy is used while interacting with the environment, but a random policy is used to generate trajectories for the model-based updates. The complexity of Keepaway and the real-time constraints of the RoboCup simulator force us to make model-based updates *off-line*, whereas it is possible to make such updates on-line in the example domains used to illustrate Dyna-Q and Prioritized Sweeping.

In several past efforts of model-based approaches, learning the model is itself the key issue. Ng *et al.*[9] successfully learn a model for helicopter control. The state space is described by 8 body coordinates, and 4 continuous actions serve as control signals to maneuver the helicopter every 50th of a second. A stochastic model is learned using locally weighted regression; an action policy is derived from this model using the PEGASUS algorithm. 3v2 Keepaway has a state space of higher dimension (13), with states being more temporally distant. Also, actions are abstract, high-level skills, unlike control signals to the helicopter that perturb its state smoothly. Additionally, in Keepaway, it is actually necessary to iteratively gather experiences based on updated versions of the policy (about 5-6 times using MBPI-1) in order to achieve high performance. The helicopter control policy, on the other hand, is learned based on a single batch of experiences obtained by a human pilot flying the helicopter.

Other approaches involving modeling physical systems include, among others, those of Atkeson and Santamaría [2], and Boone [3]. The former investigate a pendulum swing-up problem with 2 state variables and 1 continuous action; the latter considers the Acrobot problem, having 4 state variables and 3 discrete actions. Apart from having fewer state variables and actions, these physical world tasks have smoother transition dynamics than Keepaway: a key component of our Keepaway model is the termination predictor, which is not required for the pendulum and acrobot tasks. Nonetheless, the main results from these tasks concur with ours – that model-based RL can greatly reduce sample complexity, while improving the quality of the learned solution.

Experience Replay is a model-free batch learning method due to Lin [7], which has been applied to Keepaway by Kalyanakrishnan and Stone [6]. The results in this paper show that our model-based approach registers faster learning and better asymptotic performance than experience replay on Keepaway. We compare and contrast the approaches in Section 5.

7   Conclusion

We examine the viability of using model-based RL for Keepaway, a complex, stochastic, continuous, high-dimensional, multiagent task. The actions in this task are abstract, high-level skills that can last variable periods of time, making it novel from a model-learning perspective. Our model is partially specified through simple rules, partially learned, and then used as a subroutine in the RL algorithm to learn the action selection policy. Empirical results demonstrate that such a model-based RL algorithm can yield significant gains in sample complexity and asymptotic performance when compared to model-free approaches that have been applied to Keepaway successfully in the past. Also, we show that a model can be used effectively with other static policies, lending flexibility to the learned solution. Problems for future work include improving upon our design of the Keepaway model, using it for knowledge transfer among related tasks, and applying it with planning-based approaches.

Acknowledgements

This research was supported in part by NSF CISE Research Infrastructure Grant EIA-0303609, NSF CAREER award IIS-0237699, and DARPA grant HR0011-04-1-0035.

References

1. Albus, J.S.: Brains, Behavior, and Robotics. BYTE Books, Peterborough (1981)
2. Atkeson, C., Santamaría, J.: A comparison of direct and model-based reinforcement learning. In: IEEE International Conference on Robotics and Automation, vol. 4, pp. 3557–3564 (April 1997)
3. Boone, G.: Efficient reinforcement learning: model-based acrobot control. In: IEEE International Conference on Robotics and Automation, vol. 1, pp. 229–234 (April 1997)
4. Bradtke, S.J., Duff, M.O.: Reinforcement learning methods for continuous-time Markov decision problems. In: Tesauro, G., Touretzky, D., Leen, T. (eds.) Advances in Neural Information Processing Systems, vol. 7, pp. 393–400. The MIT Press (1995)
5. Kalyanakrishnan, S., Liu, Y., Stone, P.: Half field offense in RoboCup soccer: A multiagent reinforcement learning case study. In: Proceedings of the RoboCup International Symposium 2006 (June 2006)
6. Kalyanakrishnan, S., Stone, P.: Batch reinforcement learning in a complex domain. In: The Sixth International Joint Conference on Autonomous Agents and Multiagent Systems (May 2007)
7. Lin, L.-J.: Self-improving reactive agents based on reinforcement learning, planning and teaching. Machine Learning 8, 293–321 (1992)
8. Chen, M., Foroughi, E., Heintz, F., Huang, Z., Kapetanakis, S., Kostiadis, K., Kummeneje, J., Noda, I., Obst, O., Riley, P., Steffens, T., Wang, Y., Yin, X.: Users manual: RoboCup soccer server — for soccer server version 7.07 and later. In: The RoboCup Federation (August 2002)
9. Ng, A.Y., Kim, H.J., Jordan, M.I., Sastry, S.: Autonomous helicopter flight via reinforcement learning. In: Thrun, S., Saul, L., Schölkopf, B. (eds.) Advances in Neural Information Processing Systems 16, MIT Press, Cambridge (2004)
10. Puterman, M.L.: Markov Decision Processes: Discrete Stochastic Dynamic Programming. John Wiley and Sons, New York (1994)
11. Stone, P., Sutton, R.S., Kuhlmann, G.: Reinforcement learning for RoboCup-soccer keepaway. Adaptive Behavior 13(3), 165–188 (2005)
12. Sutton, R.S., Barto, A.G.: Reinforcement Learning: An Introduction. MIT Press, Cambridge (1998)
13. Sutton, R.S., Precup, D., Singh, S.P.: Between MDPs and semi-MDPs: A framework for temporal abstraction in reinforcement learning. Artificial Intelligence 112(1-2), 181–211 (1999)
14. Tesauro, G.: Practical issues in temporal difference learning. In: Moody, J.E., Hanson, S.J., Lippmann, R.P. (eds.) Advances in Neural Information Processing Systems, vol. 4, pp. 259–266. Morgan Kaufmann Publishers, Inc. (1992)
15. Tsitsiklis, J.N., Roy, B.V.: Feature-based methods for large scale dynamic programming. Machine Learning 22(1-3), 59–94 (1996)
16. Watkins, C.J.C.H., Dayan, P.: Q-learning. Machine Learning 8(3-4), 279–292 (1992)

HMDP: A New Protocol for Motion Pattern Generation Towards Behavior Abstraction

Norbert Michael Mayer, Joschka Boedecker, Kazuhiro Masui, Masaki Ogino, and Minoru Asada

Dept. of Adaptive Machine Systems,
Graduate School of Engineering, Osaka University, Osaka, Japan and
Asada S.I. Project, ERATO JST, Osaka, Japan
{michael,joschka,masui,ogino,asada}@jeap.org

Abstract. The control of more than 20 degrees of freedom in real-time is one challenge of humanoid robotics. The control architecture of an autonomous humanoid robot often consists of two parts, namely a real-time part that has direct access to the motors or RC servos, and a non-real-time part, that controls the higher-level behaviors and sensory information processing such as vision and touch. As a result motion patterns are developed separately from the other parts of the robots behavior. In research, particularly when including developmental processes, it is often necessary that the design or the evolution of motion patterns is integrated in the overall development of the robot's behavior. This is indeed one of the main principles of the embodied intelligence paradigm. The main aim of this work is to define a flexible way of describing motion patterns that can be passed to the motion controller which in turn executes them in real-time. As a result, the Harmonic Motion Description Protocol (HMDP) is presented. It allows the motions to be described as vectors of coefficients of harmonic motion splines. The motion splines are expressed as human-readable ASCII strings that can be passed as a motion stream. Flexibility is achieved by implementing the principle of superposition of several motion patterns. In this way also closed loop control is achievable in principle. Moreover, the HMDP can be implemented into the (deleted for blind review) project of the 3D soccer simulation league as a standard way to communicate motion patterns between the agent and the simulation interface and/or real humanoid robots.

1 Introduction

Many developers of autonomous robot systems experience difficulties when designing a control system that is at the same time capable of high level sensor processing, in particular vision sensors, and motor control. The solution is in most cases a hybrid design using 2 CPUs, one for motor control and one for sensor processing. The sensor data processing is often done by a PC like system with a broadband multitasking operating system (Windows, Linux) that usually does not have real-time capabilities. The motor control is done by a micro-controller

U. Visser et al. (Eds.): RoboCup 2007, LNAI 5001, pp. 184–195, 2008.
© Springer-Verlag Berlin Heidelberg 2008

that performs predefined motion patterns. The demanded motion pattern is communicated between both CPUs in some way, e.g., by serial bus. In particular in humanoid robots the motion control part has to be a real-time system in order to avoid jerky motions.

During the development process of the robot's behavior usually problems arise from this hybrid design. Whereas a PC-like system is always accessible, ready for changes, the motor controller can only be accessed via specialized editors and development tools, debuggers etc. Changes of motor controller programs can only be realized by flashing the limited memory that is available on the controller board. The programming of the motor-controller is mostly in done in C by using many custom definitions that depend on the design features of the motor controller which vary in dependence of the product line and the manufacturer. Moreover, the real-time behavior is managed by a series of interrupts that are again dependent on the type of the controller.

The problems usually result in a development process in which the motion patterns are developed separately from the design of the overall behavior. This seems acceptable in robot systems that do not require a big set of motions and do not have many degrees of freedom.

In humanoid robots, however, this design principle is not really satisfying. This is particularly true for soccer playing humanoid robots. Whereas humans have an infinite number of motion patterns available, the typical motion number of patterns of robots that participate a the RoboCup is normally less then 10, e.g. strong kick, soft kick, walking, turn, several goal keeper behaviors. A first step would be to allow for the activation of several motion patterns at the same time. This can be used for looking for the ball and walking forward independently in parallel. Furthermore, it can be used to balance out perturbations from the walking process. Thus, in addition to the normal walking process a weak pattern can be added that can stabilize the motion pattern. For this purpose it is necessary that the exact phase relation between both motion patterns is controllable. This is one requirement for the protocol.

Splines and harmonic functions have been used in various projects in different fields so far. Greszczuk and Terzopoulos [1] describe learning of muscle-actuated locomotion through control abstraction in order to generate realistic animations for computer graphics applications. They employ artificial animals with many degrees of freedom and abstract learned controller functions using Fourier analysis. These compact controller representations are then synthesized in learning of higher-level behaviors which benefits from the dimensionality reduced form of these controllers.

Another example from the field of computer graphics is given in [2]. Here, Fourier expansions of experimental data of actual human behaviors are taken as a basis to interpolate and extrapolate locomotion for an animated human figure. The authors describe how rich variations of human locomotion can be achieved by superposition of different Fourier expansions. Furthermore, the abstraction of the movements allows different parameters of the animation (like

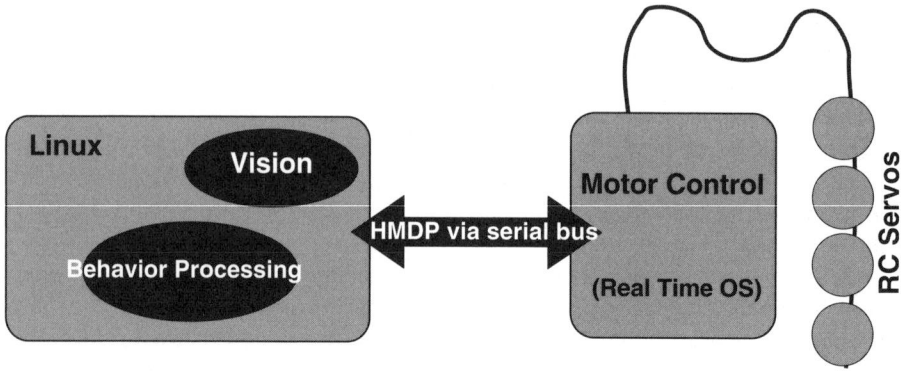

Fig. 1. Possible implementation of HMDP in a robot environment: Higher level behaviors are processed in a Linux micro PC (e.g. Geode). The PC sends motion patterns over the serial bus to micro controller. They are executed in real time.

e.g. step-length, speed, or hip position) to be controlled interactively to tweak the resulting animation.

In robotics, we find application of splines e.g. for trajectory generation of mobile robots (see [3] for the case of controlling an all-wheel steering robot). The idea of control abstraction for a more compact representation of movements is realized with different methods, for instance Fourier analysis for cyclic motion patterns as in [4] or using hierarchical nonlinear principal component analysis to generate high-dimensional motions from low-dimensional input [5].

In the following section we outline the requirement specifications that follow from the above mentioned motivations. We then outlined the principle of superposition of motion, its advantages and potential problems in section 3. The syntax of the HMDP as implemented in our software is described in section 4. To illustrate the work with the protocol, we provide an example using an experimental graphical user interface in section 5. Finally, we present a possible role of the HMDP in the 3D2Real project [6], and close with a discussion.

2   HMDP Requirement Specifications

To define the specifications of the HMDP more precisely:

– The HMDP includes messages that are submitted from the PC to the micro-controller and response messages from the micro-controller to the PC.
– The protocol allows for the PC side to set the current time as an integer and also to set the maximal time value after which the current time on the micro-controller is set to zero again.
– The protocol defines motion patterns in terms of splines. In order to allow for periodic motion patterns that can be repeated an arbitrary number of times the set of base functions is defined as a set of sines and cosines.
– The protocol activates motion patterns including the information at what time the motion pattern is activated, and its amplitude. It also defines which

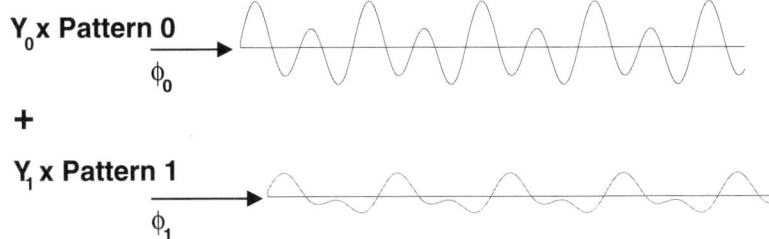

Resulting motion pattern (schematic)

Fig. 2. Motion superposition: By using HMDP two or more motions can be superposed by defining the amplitudes Y_i and the phase shift ϕ_i. The resulting motion pattern is the sum of both initial patterns.

step of the motion pattern is assigned to what time step of the motion controller (motion phase assignment).

- The design of the HMDP includes the management of the motion patterns on the micro-controller side. It is possible to activate several motion patterns at the same time. The resulting motion pattern is the superposition of all activated motion patterns (motion superposition principle).
- The protocol allows to read out values of sensors that are connected to the micro-controller. In particular, it allows to read out the the the angle of the servo positions at a particular time step. The message for a sensor request consists of a the information of the time at which the sensor value should be read out and the name of the particular sensor. As soon as the time for read out is reached the time value the sensor name and the sensor value is sent from the micro-controller to the PC.

3   Harmonic Motion Splines and Motion Superposition Principle

In this section we outline the principle with which motion patterns can be expressed in terms of motion splines; how they can be superimposed and under which circumstances the superposition of motions is useful.

In the following we discuss the harmonic motion splines for a robot with A actuators. Currently a mere position control is considered. We have spline functions that describe motion patterns $f_{p,a}(t)$. These are expressed in terms of discrete finite series of sine and cosine functions:

$$f_{p,a}(t) = c_0 +$$

$$\sum_{0 \leq n \leq max} c_{2n+1,a} \sin(\rho \times \omega_{n,p} \times t) + c_{2n+2,a} \cos(\rho \times \omega_{n,p} \times), \quad (1)$$

where $\rho = \frac{\pi}{2N}$, $0 \leq p < P$ is the index of the pattern and $0 \leq a < A$ the index of the actuator. The set wave numbers $\omega_{n,p}$ is specified when the pattern is initialized. As a convention of the current HMDP for a specific pattern p it is identical for all actuators a.

The vector $\mathbf{f}_p(t) = \{f_{1,p} \ldots f_{a,p}\}$ expresses then the state vector of the robot, i.e., the positions of all servos, given only the pattern p is active with an amplitude of 1. The final position that is sent to the servo is then:

$$\mathbf{F}(t) = \sum_{p<P} R_p(t)\mathbf{f}_p(t - \phi_p) \quad (2)$$

Where the amplitude $R_p(t)$ and the offset ϕ_p are transmitted when the pattern is activated. Before the onset or change of the amplitude of a motion pattern the value ϕ_p, $Y_{new,p}$, T_{start0}, T_{start1} have to be transmitted, $R_p(t)$ is then determined by

$$\begin{array}{ll} t < T_{start0} & : R_p(t) = Y_{old,p} \\ t \in [T_{start0}, T_{start1}] & : R_p(t) = (Y_{new,p} - Y_{old,p})/(T_{start1} - T_{start0}) \times t \quad (3) \\ t > T_{start1} & : R_p(t) = Y_{new,p} \end{array}$$

In other words the amplitude is changed in a linear way from the previous value to the current value. However, the value of ϕ_p changes at the time T_{start0}.

The design makes several types of messages necessary:

- **Pattern initialization message:** This message determines the ID of the pattern p. It also sends information about the used wave numbers $\omega_{n,p}$ for all n.
- **Coefficient transmission:** Since this can be a large set of information and since the buffer of of the receiving device is limited it seems useful to separate this type of message from the first message. So in this second type of message all coefficients $c_{i,a}$ with $0 \leq i \leq 2n+1$ and $0 < a < A$ have to be submitted in "digestible" message sizes.
- **Use Pattern messages:** For using the patterns the on set times T_{start0} and T_{start1} as well as the the new amplitude Y_{new} and ϕ_p have to be used.
- **Sensor reading commands:** Those commands should contain a time value that specify for the motion controller at what phase of the motion the sensor value i.e. that potentiometer value of the joint should be read.
- **Time managing commands:** The time management has to be covered, in some way. The higher level controller should roughly know the current time value of the motor controller. The overflow of the time counter needs to be managed. The increment of the time counter should be changeable.

In the current approach the structures that manage the patterns are organized in a static manner. Also dynamic ways to storage the patterns seem possible.

The total motion of the control output according to equation 2 is the super-position of all active motion patterns in their current amplitude. The virtues of this superposition might not be directly obvious in the general case. In the following we go into three different examples where the superposition of motion patterns is useful.

First, examples for periodic movements: Independent movements concern non overlapping sets of actuators and are applied by simply running both patterns at the same time. With respect to humanoid robotics this can be done by looking for the ball – that is: moving the head and walking at the same time. Both patterns can have different wave numbers and can be applied completely independent from each other. Here it is necessary that both movements do not interfere with each other, e.g., that the limbs collide under certain circumstances. In addition it is only possible to have one pattern with dynamic effects on the whole body of the robot active at any time. In the case of walking and looking, only the walking would have an effect on the dynamic of the whole body of the robot. This kind of combination is only possible if both movements concern completely non overlapping sets of joints and one movement pattern leaves the joint that concerns the other movement in a default position.

The second example would be two movements at the same frequency. Where the first movement is the default behavior and second movement is a response of the control to some perturbation. As an example, take vibrations during walking; these can be damped by adding a regulatory movement on top of the standard movement.

The third example would be parametric non-periodic movements, like kicking. Here, the kicking direction can be superposed to a standard kicking behavior.

Limitations of the HMDP approach are closed loop control tasks that require inevitably control reactions below the limit of the reaction time of the combined system motor controller higher behavior controller.

4 HMDP Syntax

In the following we describe the set of messages and the way parameters and numbers are defined.

4.1 Characters Used in HMDP Syntax

It is a subset of the ASCII characters. At the current stage the HMDP uses: numbers ([0..9,a..f]), capital letters ([A..Z]) and the symbols $*$, $<$, $>$, $+$,$-$, @ ,! and &. Carriage return (decimal code 10,13) defines the end of a command line, after which the line is parsed in the processor.

4.2 Check Sum Feature

All commands can be sent in a check sum mode. The check sum is calculated as: $A = (\sum_i X_i) \; modulo \; 16$, where A is itself expressed as a hex-value([0-9,a-f]).

4.3   Long Command Feature

In some microprocessors command lines over 10 digits may become unsafe since the hardware serial-bus chips provide only a buffer of around 10 characters. Therefore it seems useful to break long messages down into shorter parts that are written into the program buffer. An & character at the end indicates that the current command line is continued after the carriage return. In combination with the check sum feature messages up to a defined length limit can be sent safely.

4.4   At-Time Execute Command Feature

With this feature a command - usually a command to read out sensor values - is executed at a specified time. In the current implementation the slave assumes that the at-time commands are sent in sequence, i.e., the command that should be executed first is sent first etc. The slave does not sort the commands and would wait until the time for the next command in its batch is reached and then look for the start time of the following at-time command. Thus, if one sends an at-time command for time 100 and then an at-time command for time 50, both commands are executed at 100 and 101, respectively. If the time counter is already behind the time given in the at-time command (like in the previous example) the command is executed in the next time cycle.

4.5   Types of Numbers

Numbers are transmitted as hex numbers that include the numbers [0..9] and the letters [A..E]. Currently, three types of numbers are used: integers, rational numbers, and real values.

- Integers are transmitted as conventional hex numbers.
- Rational numbers are used to express wave numbers. Since most motor-controllers can only emulate floating arithmetic by software, rationals seem to be faster than real values. They are expressed by a set of two subsequent integers.
- Floats are described in a standard semi logarithmic way by transmitting the exponent and the mantissa as integer values including their signs.

4.6   Internal States

The robot can be set into 3 distinct internal states. Depending on the activated state different groups of commands have different effects. It is important to note that if a command is used in the wrong state the real-time property may be disrupted, the command may have an undesired effect, or no effect at all.

- State 0: The motion machine is deactivated and commands can be directed towards the motors. Command groups 0,I can be used.

– State 1: HMDP state. The motion machine is on and overwrites motor commands (group I). Instead the HMDP commands have to be used that control the motion machine. At-time commands are possible but may interfere with the timer and therefore the motion may not be precise (lost ticks may appear).
– State 2: "Plastic" state of the robot. The robot can be set manually into a state, and keeps the current posture, while changes in the current posture are possible by applying force to the servos.

4.7   Group 0 Administrative Commands

This group communicates state variables and other information of the current state of the system. The syntax is usually $> XX$ for a command to the slave and $< XX$ for requested information. In addition, some commands for copying motion patterns into the flash memory are provided. Servos can be turned on and off. For the protection of the servos allowed ranges of position values of the servos are defined. Zero positions[1] can be re-defined, and the list of available servos can be requested. These commands are typically used when the robot is in state 0, but can also be used in the other two states. However, in state 2 they affect the real-time property, and ticks may be lost.

4.8   Group I Static Posture Commands

The commands affect the posture of the robot directly. In state 0 these may be used to control postures, which is useful in the development phase of the the the motions and in order to calibrate the zero positions. In addition, positions can be read from the controller. There, the controller can distinguish between the actual current position and the position that is targeted by the controller.

4.9   Group II HMDP Commands

An HMDP message starts with a systematic set of key characters which simplifies the parsing of the protocol. An additional initial character in front of every message can be custom defined in order to make it possible to add HMDP to already existing messaging systems. Apart from the custom defined header the first character of each HMDP messages can be either a P, a T, or S, indicating a time-related, pattern-related or sensor-related command.

5   Example for a Visual Motion Design Tool

We programmed a graphical user interface (GUI) (see Fig. 3) in which the coefficients can be found be defining manually frequencies and support points. The

---

[1] Zero positions are the calibrated values of the servos of a robot in an upright position and the arms in a certain diagonal angle.

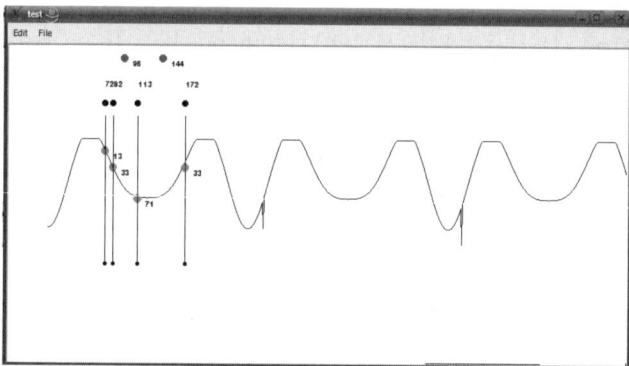

Fig. 3. Experimental graphical user interface for motion design MotionDesigner: Currently the wave numbers have to chosen by hand (pink dots). Then a function (black curve) is interpolated between support points (green dots).

program calculates the coefficients and produces HMDP messages, that define this particular motion pattern. Coefficients c_i are calculated by solving the system of equations that are defined by

$$x(t) = c_0 +$$
$$\sum_{0 \leq n \leq max} c_{2n+1,a} \sin(\rho \times \omega_{n,p} \times t) + c_{2n+2,a} \cos(\rho \times \omega_{n,p} \times t). \quad (4)$$

Since a set $x_i(T_i)$ for a certain T_i and all $\omega_{n,p}$ are defined by the user, we have a set of linear equations that can be solved by deriving the pseudo inverse. In dependence of the ratio between the number of $x_i(T_i)$ and the number of coefficients c_j we get an under- or over-defined system of linear equations.

The program currently controls a virtual motion controller. The motion control part is C-code and can readily be implemented into a standard motion controller ICs down to the level of PIC chips or similar.

6   Possible Role in the 3D2Real Project

One problem for the RoboCup project is that throughout the leagues a lot of work is duplicated, and collaboration is rather sparse between the different leagues. This is not a desirable situation as know-how is not transferred effectively, and progress is slower than it could be since resources are bound to solve the same problems over and over again. To address this situation, the 3D2Real project was initiated in 2006.

The main idea of this project is to try and use synergy effects from a collaboration between researchers in the Humanoid and the Soccer Simulation League (SSL). This collaboration includes a joint road map for the near future of both leagues, as well as the specification of standards and the development of tools that can be used in both leagues.

Traditionally, the SSL and the HL in RoboCup have had rather different research topics. While researchers in the HL mainly worked on the design and the low-level control of their robots, participants in the SSL were concerned with high-level strategies and collaboration. In recent years, however, there have been developments which might bring both leagues closer to each other. In the SSL, there have been continuing efforts to introduce more realism into the rather abstract simulation in order to ensure that the developed strategies can be transferred more easily to real robots. Humanoid robot simulation is the preferred choice for many participants of the SSL in order to achieve this. In the HL, on the other hand, the first multi-robot games have been held, and the great progress in controlling the robots allows researchers to approach issues of collaboration and coordination which have been extensively studied in the SSL. In short, both leagues are beginning to come closer to each other, and joint efforts in the development of tools and architectures that allow easier transfer of knowledge and technologies could speed up the mutual progress towards the 2050 goal of RoboCup.

The goal we envision for the 3D2Real project is to have the finals of the soccer simulation league using real robots by the year 2009 or 2010. For this ambitious goal several steps are necessary in the next years to create the necessary infrastructure and tools. First, the 3D simulator of the SSL [7] has to be completed, and a real robot prototype has to be implemented as a simulation model, the XML-based format *RoSiML* as used in the *SimRobot* simulator [8] seems promising. According to the proposed road map, a technical challenge would be held at RoboCup 2008 to test the ability to use the agent code of SSL participants on a predetermined real robot. From 2008 until 2009, we propose the development of a *central parts repository* (CPR). This would be a collection of real robot designs, sensor and actuator models, complete robots, as well as controllers for certain architectures. Participants of both HL and SSL contribute to this repository according to their expertise and interest. The format would again be the *RoSiML* mentioned above. These contributions become a mandatory part for the HL qualification from 2009, and should be continued (at least) until 2010, even after the 3D SSL final has taken place using real robots.

The HMDP introduced in this work could be used as a standard for the motion description of the simulated and the real robots. In the simulation, the agents are connected to the simulator over the network. This means that they have to send messages (currently in ASCII strings) back to the simulator in order to specify, e.g., desired positions for motors. Since many agents connect to the simulator at once during a game this can lead to a high volume of network traffic that can cause severe problems for the server. If the HMDP were used for the description of motion patterns, longer messages describing the motions would only have to be sent sporadically when new patterns have to be set. Thus, the HMDP would provide a good solution for very related problems in the 3D soccer simulation league, and the humanoid league (as described in the introduction), and might facilitate running the same code on simulated and real humanoids eventually.

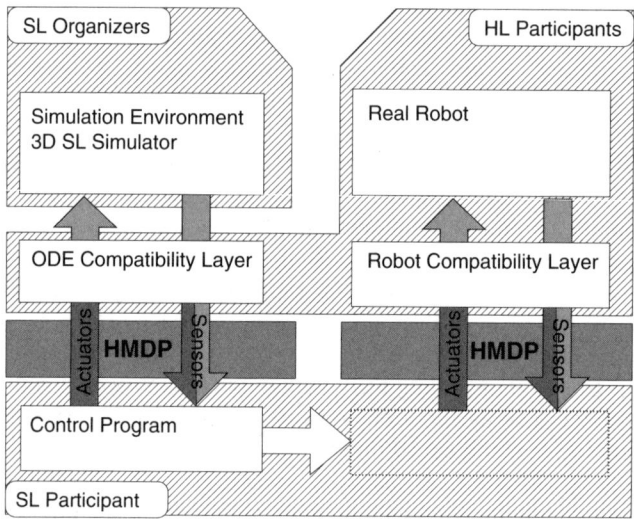

Fig. 4. 3D2Real project: Layout of the control architecture. The hatched boxes show how the different leagues contribute to the complete system architecture of the 3D2Real project. The control program for simulation system and real robot system are identical.

7 Discussion

The intention of this work is to provide a standard for the description of motion patterns for several purposes. It can be seen as an abstract but flexible motion description protocol in a server and client architecture. Moreover, it can be used for simulation purposes and at the same time for connection between a higher-level behavior control unit and a real-time motion controller. Its merits are biggest in situations where motions patterns should be changed "on-the-fly", like in a scenario that uses developmental or evolutionary methods to change the motion generation, but it also allows for a great flexibility in motion execution in general.

The representation of motions by Fourier coefficients in the protocol and the superposition principle are valuable properties for behavior abstraction, i.e., the synthesis and blending of new, higher-level behaviors from compact representations of lower-level ones. This is one aspect we want to explore further in the future, as well as the implementation of currently missing features like variable time increments, sensor readings, and overflow of time values that can happen at least theoretically.

Furthermore, the 3D2Real project is one example inside RoboCup where the HMDP seems to be an appropriate tool. The plan is now to test and improve the HMDP in dependence on results within the 3D2Real project.

Acknowledgements

We greatfully acknowledge the support of this work by the Japan Science and Technology Agency (JST), and a fellowship for young scientists from the Japan Society for the Promotion of Science (JSPS).

References

1. Greszczuk, R., Terzopoulos, D.: Automated learning of muscle-actuated locomotion through control abstraction. In: Proceedings of SIGGRAPH 1995 (1995)
2. Unuma, M., Anjyo, K., Takeuchi, R.: Fourier principles for emotion-based human figure animation. In: Proceedings of the 22nd annual conference on Computer graphics and interactive techniques, pp. 91–96. ACM Press (1995)
3. Howard, T., Kelly, A.: Trajectory and spline generation for all-wheel steering mobile robots. In: Proceedings of the 2006 IEEE/RSJ International Conference on Intelligent Robots and Systems (IROS 2006), October 2006, pp. 4827–4832 (2006)
4. Schmidt, H., Sorowka, D., Piorko, F., Marhoul, N., Bernhardt, R.: Control system for a robotic walking simulator. In: Proceedings of the 2004 IEEE International Conference on Robotics and Automation (2004)
5. Tatani, K., Nakamura, Y.: Reductive mapping for sequential patterns of humanoid body motion. In: Proceedings of the 2nd International Symposium on Adaptive Motion of Animals and Machines (2003)
6. Mayer, N.M., Boedecker, J., da Silva Guerra, R., Asada, M.: 3d2real: Simulation league finals in real robots. In: Lakemeyer, G., Sklar, E., Sorrenti, D.G., Takahashi, T. (eds.) RoboCup 2006: Robot Soccer World Cup X. LNCS (LNAI), vol. 4434. Springer, Heidelberg (2007)
7. Obst, O., Rollmann, M.: SPARK – A Generic Simulator for Physical Multiagent Simulations. Computer Systems Science and Engineering 20(5) (September 2005)
8. Laue, T., Spiess, K., Röefer, T.: SimRobot - a general physical robot simulator and its application in robocup. In: Bredenfeld, A., Jacoff, A., Noda, I., Takahashi, Y. (eds.) RoboCup 2005. LNCS (LNAI), vol. 4020. Springer, Heidelberg (2006)

A Fuzzy Controller for Autonomous Negotiation of Stairs by a Mobile Robot with Adjustable Tracks

Winai Chonnaparamutt and Andreas Birk*

School of Engineering and Science
Jacobs University Bremen
Campus Ring 1, D-28759 Bremen, Germany
a.birk@iu-bremen.de

Abstract. Tracked mobile robots with adjustable support tracks or flippers are popular promising solutions for negotiating rough terrain and 3D obstacles. Though many according robot bases are in principle physically capable of climbing stairs, it is a non-trivial control-task for a remote tele-operator, especially when the user can not directly see the robot like in search and rescue scenarios. To limit training requirements and to ease the cognitive load on operators, respectively to enable fully autonomous rescue robots, we developed a fuzzy controller for this task, which adjusts the drive forces and the posture of the flipper. The design of the controller is guided by observing the strategies of a trained user when tele-operating a robot with unlimited visual information. In doing so, an Open Dynamics Engine (ODE) simulation of our robot is used where the full set of all physical parameters is accessible for analysis. Based on this data, it is shown in several experiments that the controller is not only capable of climbing stairs but that it does so in a more efficient manner than the human user who served as training model.

1 Introduction

There are many different options for locomotion systems, each with its particular pros and cons. Also within search and rescue robotics, many different approaches are used (figure 1), ranging from wheeled over legged to serpentine systems [1][2][3][4][5][6][7]. But tracked locomotion is often considered to be the most versatile locomotion system in difficult environments as it can handle large obstacles and loose soil, hinders and small holes and ditches. Compared to e.g. wheeled system, a tracked system can also develop higher thrust or gross traction force particularly for the operation over weak terrain [8]. This type of locomotion is the most suitable to surmount obstacles, negotiate stairways, and is able to adapt to terrain variations [9][10]. For unstructured environments, e.g., a collapsed building, a construction site, tracked locomotion might hence be the most suitable choice. Accordingly, there are various tracked robots have been employed for hazard missions [11][12][13][14][15].

---

* Previously International University Bremen (IUB).

U. Visser et al. (Eds.): RoboCup 2007, LNAI 5001, pp. 196–207, 2008.

Fig. 1. Examples of the different locomotion approaches for rescue robots: Gryphon-I, Genbu, CUL robot system, Quadruped Jumping Robot, Scout robot, MOIRA, and Marsupial robots [1][2][3][4][5][6][7](from left to right, top to bottom)

But it is almost impossible to select the right parameters for a single pair of tracks. Sometime the footprint of the robot and hence the length of the tracks should be small, for example when negotiating narrow passages. The footprint should on the other hand be large for climbing large obstacles like slopes or stairs. The common solution to this problem is to use additional tracks that can change their posture relative to the main robot body. Examples of variable configuration robots are shown in figure 2. They are successful in many application domains related to rescue robotics [16][17][18] including the RoboCup rescue league [19,20,21][22,23,24].

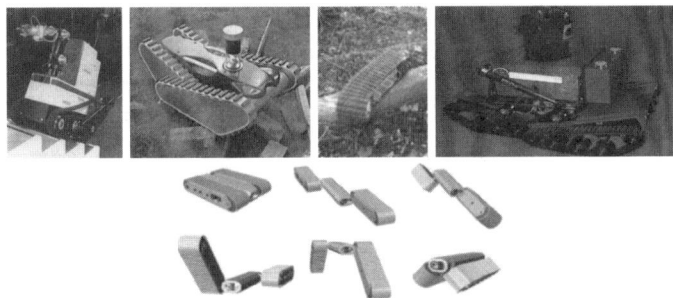

Fig. 2. Variable tracked configuration robots [16]: Pandora [17], AURORA [18], link-type tracked vehicle [25], and NUGV [26]

All aforementioned systems are designed and used for teleoperation. But though these systems are in principle very capable, it is sometimes tremendously hard to negotiate obstacles when the operator can not directly observe the robot. This even holds for commercial systems used in the military domain as indicated by the results of the European Land Robotics Trial 2006 [27]. Intelligent high level motion control is hence of interest. An according functionality can be used to relief the operator, respectively to allow for autonomous operations.

2   The Locomotion System of Jacobs Robotics

The work presented here is based on the "rugged robot" or short rugbot platform [28]. It is a variable configuration robot with a special support track or flipper. The flipper mechatronics are based on a special design, which make it particularly robust against shocks and requires smaller joint forces than other state of the art designs. The underlying design and its implementation are described in detail in [29]. Rugbots are capable of climbing stairs (figure 3) and various other obstacles including random step fields. But we also experienced that there is a tremendous difference between "the robot is in principle capable of climbing stairs" and "a remote operator can climb stairs with the robot". The first statement refers to situations where the operator can see the robot from a global perspective much like the view in figure 3. Then, stair climbing is relatively easy. It is on the other hand much more difficult if this is to be done by a remote operator who is only provided with the sensor views of the robot itself. Hence, autonomous negotiation of complex obstacles is not only of interest for pure fully autonomous operations but also to assist teleoperation.

There are two main disadvantages when doing experiments with autonomous motion control over complex obstacles in the real world. First, crucial properties of the interaction between the system and its environment are very difficult to measure. Especially, contact forces, energy efficiency, etc. are difficult to measure with decent accuracy and meaningful spatial and temporal resolution. Second, experiments in the early testing phase can easily go wrong and pose high risks for the robot as well as the experimenter, e.g., when the robot falls from the stairs. Third, it is very tedious to test the system with various environment parameters, e.g., step widths of stairs, let alone to do this in an exhaustive and controlled manner. A commonly used strategy for the design process under such circumstances is hence in general to start with a high fidelity physical simulation.

Here, the Open Dynamics Engine (ODE) [30] is used. Open Dynamics Engine (ODE) is an open source, high performance library for simulating articulated rigid body dynamics, e.g. ground vehicles, legged creatures, and moving objects in virtual environments. ODE is platform independent with an easy to use

Fig. 3. Rugbot climbing stairs in its teleoperation mode (from left to right, top to bottom)

Fig. 4. Rugbot's model in ODE

Fig. 5. Examples of obstacles: stairs (left) and a random step field (right)

C/C++ API. It is designed to be used in interactive or real-time simulation. Its major features are:

- ODE uses a highly stable integrator, so the simulation errors should not grow out of control. ODE emphasizes speed and stability over physical accuracy.
- ODE has hard contacts. This means that a special non-penetration constraint is used whenever two bodies collide.
- ODE has a built-in collision detection system that provides fast identification of potentially intersecting objects, through the concept of spaces.

There are two main components in ODE: a dynamics simulation engine, i.e., world and rigid body, and a collision detection engine, i.e., space and geom (geometry object). The first has the information about the position, velocity, and mass of the rigid body. The latter is given information about the shape of each body. For the first engine, a body is an object which is affected by forces, while a geom of the second engine is an object that can collide with other geoms. A body and a geom together represent all properties of a simulated object.

The composition of rugbot's body in the simulation are a main body, a flipper ball screw, and four locomotion belts. There are six bodies in the simulation: rugbot main body, four wheel bodies, and a flipper body. Four hinge-2 joints are used for connecting four wheels to the rugbot main body, while the flipper is connected to rugbot via a hinge joint. The overall configuration can be seen in figure 4. To drive a force is applied to the body and torque is applied to the four hinge-2 joints with a fixed speed. The driving force of comes hence from the force on rugbot's body and the torque of the joints. Autonomous driving requires the adjustment of the driving force and of the posture of the flipper. The environment features several obstacles including stairs, ramps and a random step fields (figure 5).

3   Mimicking Human Operators

The intelligent motion controller is separated in two levels: high-level control and low-level control systems. The high-level controller is based on fuzzy logic [31][32], which is a popular choice for this kind of task [33]. The low-level controller is in charge of handling the motors, especially for the flipper. It is based on PID controllers and its behavior has also been intensively studied. Note that the modeling of the robot was done down to a detailed physical simulation of the motor properties. As the low-level controller operates on a different, much faster time scale than the high level controller, it is neglected in the rest of this paper.

The main idea for the design of the high level controller is to observe the control patterns used by a human operator who has "perfect" information about the situation of the robot. These control patterns are then turned into fuzzy control laws, which are autonomously carried out by the robot.

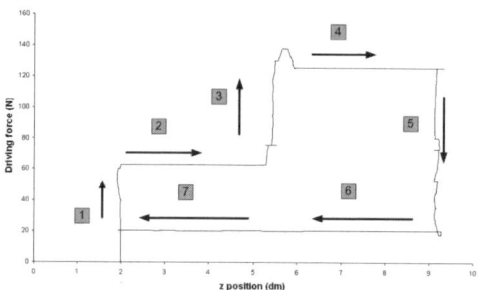

Fig. 6. The relation between the z position and the driving force as measured in a teleoperation run when climbing stairs; the arrows and numbers indicated the sequence of changes by the operator over time

Figure 6 shows for example the relation between the z position and the driving force in a typical teleoperation run when climbing a stair. This type of data was collected over multiple runs. The operator had each time a best possible view, i.e., optimal information about the situation of the robot. Similarly, figure 4 shows the relation between the pitch angle and the flipper angle. This data was then used for designing the high level controller.

The high level controller consists of two fuzzy logic modules, one for the driving force and one for the posture of the flipper. Both run in parallel in the

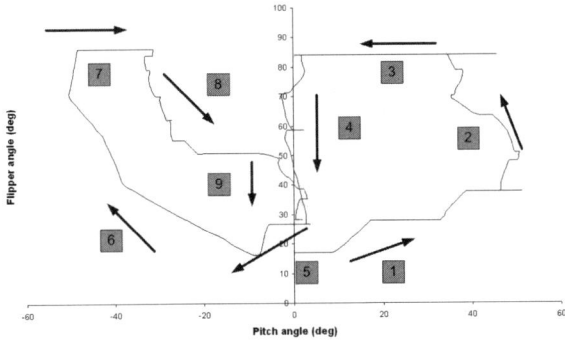

Fig. 7. The relation between the pitch angle and the flipper angle measured in teleoperation run when climbing stairs

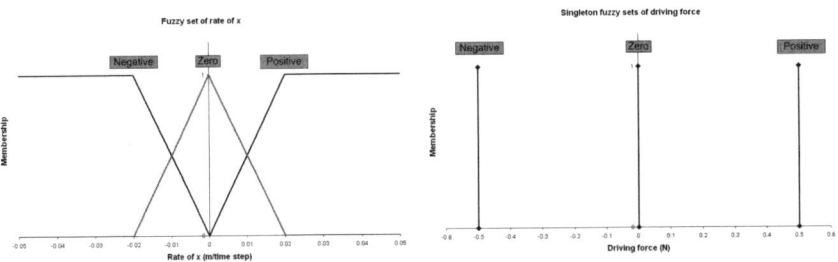

Fig. 8. The input fuzzy set of the driving force controller in x direction (left). The output fuzzy set of driving force controller: the rate of driving force (right).

spirit of behavior-oriented control. The fuzzy controller for the driving force was implemented first as it is a prerequisite for designing the fuzzy controller for the flipper.

The inputs of the driving force controller are the changing rate of locomotion in x and z directions, i.e., Δx and Δz. Based on the analyzed data, the input fuzzy sets were defined as shown in figure 8. The output of the controller is the rate of driving force, ΔF (figure 8). Both Δx and Δz have three membership sets each. Each membership also has the same values (unit: m per sampling time): zero, plus at 0.02, and negative at -0.02. The rate of driving force also has three membership sets (unit: N): zero, plus at 5, and minus at -5. The fuzzy rules, based on Mamdani fuzzy rules type, are shown in the table 1 The fuzzy inference method to find the output for each rule is based on the Mamdani minimum reference method. An output fuzzy set is converted to be a real number by using center-of-gravity method for singletons [34][35].

As mentioned before, the second fuzzy logic module of the high level controller adjust the posture of the flipper. It runs in parallel to the drive force controller. The inputs for the flipper fuzzy controller are the pitch angle of Rugbot, α, and its rate, $\Delta\alpha$. The output is the rate of the moving angle of the small track

Table 1. Fuzzy rules for the driving force fuzzy controller

Δx	Δy	ΔF
negative	negative	positive
negative	zero	positive
negative	positive	positive
zero	negative	negative
zero	zero	positive
zero	positive	positive
positive	negative	zero
positive	zero	negative
positive	positive	zero

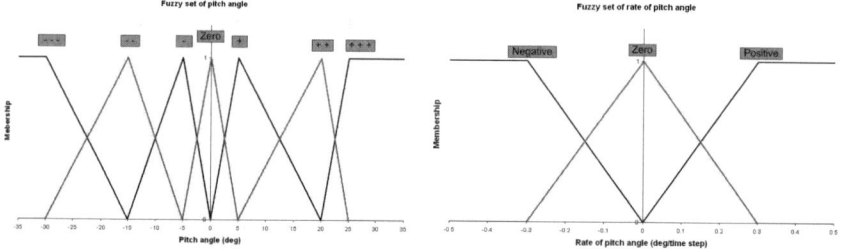

Fig. 9. The input fuzzy sets of the flipper fuzzy controller: the pitch angle of the robot and its rate

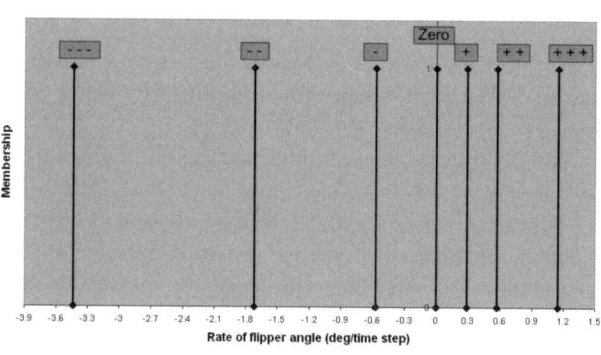

Fig. 10. The output fuzzy set of the flipper fuzzy controller: the rate of flipper angle

relative to the ball screw that drives the mechanism, $\Delta\theta_F$. The fuzzy sets of inputs and output are shown in figures 9 and 10 respectively.

The pitch angle has seven membership sets (unit: degree): zero, most plus at 30 (+ + +), more plus at 15 (+ +), plus with 5 (+), most negative at -30 (- - -), more negative at - 15 (- -), and negative at -5 (-). The rate of pitch angle has

Table 2. Fuzzy rules for the flipper fuzzy controller

	$\Delta\alpha$		
	negative	**zero**	**positive**
α	$\Delta\theta_F$	$\Delta\theta_F$	$\Delta\theta_F$
- - -	+	+	-
- -	+ +	+ + +	- - -
-	zero	zero	zero
zero	zero	zero	zero
+	- - -	zero	zero
+ +	zero	+	+ +
+ + +	zero	+	+ + +

three membership sets (unit: degree per sampling time): zero, plus at 0.3, and minus -0.3. The rate of flipper angle has seven membership sets (unit: degree per sampling time): zero, most plus at 0.4 (+ + +), more plus at 0.2 (+ +), plus with 0.1 (+), most negative at -1.2 (- - -), more negative at - 0.6 (- -), and negative at -0.2 (-). Based on Mamdani fuzzy rules type, the fuzzy rules of the controller are defined as in the table 2:

By using the Mamdani minimum reference method for inference and the center-of-gravity method for singletons for deffuzification, we got the results of the controller as shown in figure ??.

4   The General Performance

The autonomous controller was intensively tested in various experiments. First and foremost, it is indeed successful in reliably moving the robot stairs up and down without any human intervention. The robot adjusts its driving force and flipper posture such that its center of gravity is supported, it gets good traction, and that it moves in the desired direction.

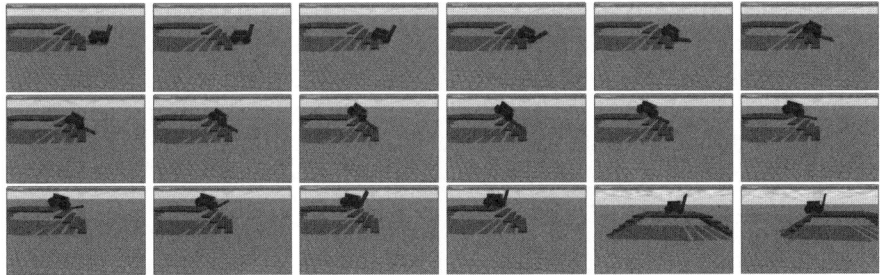

Fig. 11. The autonomous controller moves the robot reliably up on stairs (from left to right, top to bottom)

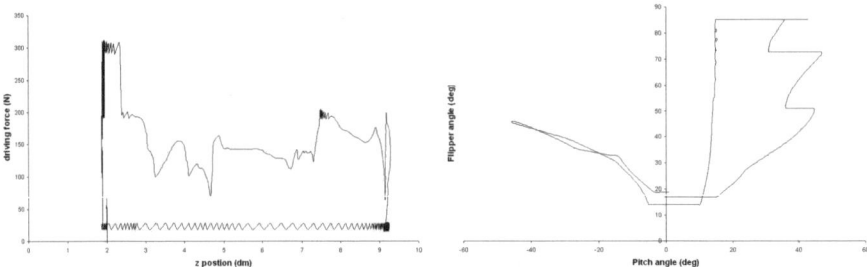

Fig. 12. The relation between the z position and the driving force when climbing a stair in an autonomous run (left). The relation between the pitch angle of the robot and the flipper angle while driving in autonomous mode (right).

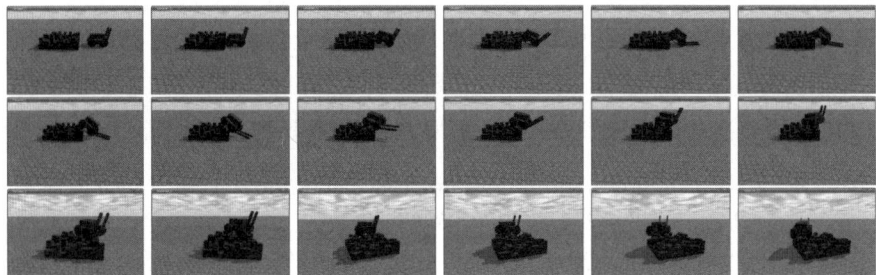

Fig. 13. The controller was designed based on observing the inputs of human operators during stair climbing. But it generalizes to other obstacles. Here, the robot autonomously negotiates a Random Step Field with the controller (from left to right, top to bottom).

Second, it does so more effectively than a human operator does. Concretely, the overall amount of forces, i.e., the energy put into the system, is smaller when the autonomous controller is driving the robot compared to cases when a human operator does exactly the same task. Note that this already holds for the case where the human operator has "perfect" visual information about the robot, i.e., the operator can freely place his viewpoint in the environment. The quantitative amount of this increased effectiveness is of course user and mission dependent. Butthe qualitative properties of this effect can be seen when comparing human control (figures 6 and) with the autonomous one (figure 12). It can be noticed that the autonomous controller causes much smoother changes. This benefit is also well-known from other applications of fuzzy logic to control.

Finally, there are important results with respect to the general usefulness of the controller. It has been designed based on observations of human operator input during tele-operated stair climbing. This is an important, but nevertheless very specialized type of locomotion task. On the other hand, the nature of the task comprises of all basic challenges, namely adjusting the center of gravity, making contact with support points for sufficient traction, and adjusting the driving force, which can be found for any other obstacles. And it indeed turned out that the controller does well on other obstacles as well, including ramps and as difficult terrain as modeled by Random Step Fields (figure 13).

5   Conclusion

A fuzzy controller for autonomous negotiation of stairs was presented. The main idea for the design of the controller is to observe the control patterns of human operators during tele-operation under optimal conditions, i.e., with "perfect" visual information. The controller is tested in an ODE simulation with a detailed physical model of the robot and various environment parameters. The autonomous controller is first of all capable of driving the robot up and down stairs. Second, it is not only very reliable in doing so, but it outperforms human operators in terms of efficiency. Finally, the autonomous controller seems to generalize very well as it manages to also move the robot autonomously over other obstacles including random step fields.

Acknowledgments

The authors gratefully acknowledge the financial support of *Deutsche Forschungs-gemeinschaft* (DFG).

Please note the name-change of our institution. The Swiss Jacobs Foundation invests 200 Million Euro in **International University Bremen (IUB)** over a five-year period starting from 2007. To date this is the largest donation ever given in Europe by a private foundation to a science institution. In appreciation of the benefactors and to further promote the university's unique profile in higher education and research, the boards of IUB have decided to change the university's name to **Jacobs University Bremen**. Hence the two different names and abbreviations for the same institution may be found in this article, especially in the references to previously published material.

References

1. Debenest, P., Fukushima, E.F., Hirose, S.: Development and control of a buggy robot for operation on unstructured terrain. In: International Conference on Intelligent Robots and System, vol. 1, pp. 763–768 (2002)
2. Kimura, H., Hirose, S.: Development of genbu: Active wheel passive joint articulated mobile robot. In: International Conference on Intelligent Robots and System, vol. 1, pp. 823–828 (2002)
3. Tokuda, K., Osuka, K., Yano, S., Ono, T.: Concept and development of general rescue robot cul. In: International Conference on Intelligent Robots and Systems, vol. 3, pp. 1902–1907 (1999)
4. Kikuchi, F., Ota, Y., Hirose, S.: Basic performance experiments for jumping quadruped. In: International Conference on Intelligent Robots and Systems, vol. 3, pp. 3378–3383 (2003)

5. Stoeter, S.A., Rybski, P.E., Gini, M., Papanikolopoulos, N.: Autonomous stair-hopping with scout robots. In: International Conference on Intelligent Robots and System, vol. 1, pp. 721–726 (2002)
6. Osuka, K., Kitajima, H.: Development of mobile inspection robot for rescue activities: Moira. In: International Conference on Intelligent Robots and Systems, vol. 3, pp. 3373–3377 (2003)
7. Murphy, R.: Marsupial and shape-shifting robots for urban search and rescue. IEEE Intelligent Systems 15, 14–19 (2000)
8. Wong, J., Huang, W.: "wheels vs. tracks" - a fundamental evaluation from the traction perspective. Journal of Terramechanics 43, 27–42 (2006)
9. Hardarsson, F.: Locomotion for difficult terrain. Technical Report TRITA-MMK-1998:3, Mechatronics Lab, Dept. of Machine Design (1997)
10. Wong, J.Y.: Theory of Ground Vehicles, 3rd edn. John Wiley and Sons, Inc. (2001)
11. Houdini: Reconfigurable in-tank robot. Technology Development Data Sheet, Red Zone Robotics, Inc.
12. Isozaki, Y., Nakai, K.: Development of a work robot with a manipulator and a transport robot for neuclear facility emergency preparedness. Advanced Robotic 16(6), 489–492 (2002)
13. Rison, B., Wedeward, K.: Locomotion lecture slide of introduction to design course, http://www.ee.nmt.edu/~wedeward/EE382/SP02/locomotion.pdf
14. Dan, Z., Tianmiao, W., Jianhong, L., Guang, H.: Modularization of miniature tracked reconnaissance robot. In: Proceedings of the 2004 IEEE International Conference on Robotics and Biomimetics, pp. 490–494 (2004)
15. Li, B., Ma, S., Liu, J., Wang, Y.: Development of a shape shifting robot for search and rescue. In: Proceedings of the 2005 IEEE International Workshop on Safety, Security and Rescue Robotics, pp. 25–31 (2005)
16. Iwamoto, T., Yamamoto, H.: Mechanical design of variable configuration tracked vehicle. Journal of Mechanical Design 112, 289–294 (1990)
17. Schempf, H., Mutschler, E., Piepgras, C., Warwick, J., Chemel, B., Boehmke, S., Crowley, W., Fuchs, R., Guyot, J.: Pandora: autonomous urban robotic reconnaissance system. In: International Conference on Robotics and Automation, 1999, vol. 3, pp. 2315–2321 (1999)
18. Schempf, H.: Aurora - minimalist design for tracked locomotion. Springer Tracts in Advanced Robotics 20, 453–466 (2003)
19. Lee, W., Kang, S., Kim, M., Shin, K.: Rough terrain negotiable mobile platform with passively adaptive double-tracks and its application to rescue missions. In: Proceedings of the 2005 IEEE International Conference on Robotics and Automation, 2005. ICRA 2005, pp. 1591–1596 (2005)
20. Kang, S., Lee, W., Kim, M., Shin, K.: Robhaz-rescue: rough-terrain negotiable teleoperated mobile robot for rescue mission. In: IEEE International Workshop on Safety, Security and Rescue Robotics, SSRR, pp. 105–110 (2005)
21. Lee, W., Kang, S., Kim, M., Park, M.: Robhaz-dt3: teleoperated mobile platform with passively adaptive double-track for hazardous environment applications. In: IEEE/RSJ International Conference on Intelligent Robots and Systems (IROS), vol. 1, pp. 33–38 (2004)
22. Lee, W., Kang, S., Lee, S., Park, C.: Robocuprescue - robot league team ROBHAZ-DT3 (south korea). In: Bredenfeld, A., Jacoff, A., Noda, I., Takahashi, Y. (eds.) RoboCup 2005. LNCS (LNAI), vol. 4020. Springer, Heidelberg (2006)

23. Kadous, M.W., Kodagoda, S., Paxman, J., Ryan, M., Sammut, C., Sheh, R., Miro, J.V., Zaitseff, J.: Robocuprescue - robot league team CASualty (australia). In: Bredenfeld, A., Jacoff, A., Noda, I., Takahashi, Y. (eds.) RoboCup 2005. LNCS (LNAI), vol. 4020. Springer, Heidelberg (2006)
24. Tsubouchi, T., Tanaka, A.: Robocuprescue - robot league team Intelligent Robot Laboratory (japan). In: Bredenfeld, A., Jacoff, A., Noda, I., Takahashi, Y. (eds.) RoboCup 2005. LNCS (LNAI), vol. 4020. Springer, Heidelberg (2006)
25. Lee, C.H., Kim, S.H., Kang, S.C., Kim, M.S., Kwak, Y.K.: Double-track mobile robot for hazardous environment applications. Advanced Robotics 17, 447–459 (2003)
26. Blackburn, M., Bailey, R., Lytle, B.: Improved mobility in a multidegree-of-freedom unmanned ground vehicle. In: SPIE Proceeding on Unmanned Ground Vehicle Technology VI. IEEE (2004)
27. ELROB: European land-robot trial (elrob) (2006)
28. Birk, A., Pathak, K., Schwertfeger, S., Chonnaparamutt, W.: The iub rugbot: an intelligent, rugged mobile robot for search and rescue operations. In: IEEE International Workshop on Safety, Security, and Rescue Robotics (SSRR). IEEE Press (2006)
29. Chonnaparamutt, W., Birk, A.: A new mechatronic component for adjusting the footprint of tracked rescue robots. In: Lakemeyer, G., Sklar, E., Sorrenti, D.G., Takahashi, T. (eds.) RoboCup 2006: Robot Soccer World Cup X. LNCS (LNAI), vol. 4434. Springer, Heidelberg (2007)
30. Smith, R.: Open dynamics engine, http://www.ode.org/
31. Zadeh, L.A.: Fuzzy sets. Information and Control, 338–353 (1965)
32. Zadeh, L.A.: Fuzzy logic. IEEE Computer, 83–93 (1988)
33. Ying, H.: Fuzzy Control and Modeling. IEEE Press (2000)
34. Jantzen, J.: Design of fuzzy controllers. Technical report, Technical University of Denmark, Department of Automation (1998)
35. Niku, S.B.: Introduction to Robotics: Analysis, Systems, Applications, vol. 9. Prentice Hall (2001)

Solving Large-Scale and Sparse-Reward DEC-POMDPs with Correlation-MDPs*

Feng Wu and Xiaoping Chen

Multi-Agent Systems Lab,Department of Computer Science,
University of Science and Technology of China,
Hefei, 230026, China
wufeng@mail.ustc.edu.cn, xpchen@ustc.edu.cn

Abstract. Within a group of cooperating agents the decision making of an individual agent depends on the actions of the other agents. A lot of effort has been made to solve this problem with additional assumptions on the communication abilities of agents. However, in some real-world applications, communication is limited and the assumptions are rarely satisfied. An alternative approach newly developed is to employ a correlation device to correlate the agents' behavior without exchanging information during execution. In this paper, we apply correlation device to large-scale and spare-reward domains. As a basis we use the framework of infinite-horizon DEC-POMDPs which represent policies as joint stochastic finite-state controllers. To solve any problem of this kind, a correlation device is firstly calculated by solving Correlation Markov Decision Processes (Correlation-MDPs) and then used to improve the local controller for each agent. By using this method, we are able to achieve a tradeoff between computational complexity and the quality of the approximation. In addition, we demonstrate that, adversarial problems can be solved by encoding the information of opponents' behavior in the correlation device. We have successfully implemented the proposed method into our 2D simulated robot soccer team and the performance in RoboCup-2006 was encouraging.

1   Introduction

Multi-Agent systems often require coordination to ensure that a multitude of agents will work together in a globally coherent manner under uncertainty. For some problems, each self-organizing agent has to cooperate to optimize a joint reward function, while having different local observations and limited communication [Kaelbling et al., 1998]. RoboCup [Kitano et al., 1997] is a good example of a cooperative multi-agent system in which the soccer-playing robots like human soccer players have to coordinate their actions by different limited messages.

The infinite-horizon Decentralized Partially Observable Markov Decision Process (DEC-POMDP) framework is one way to model these problems. And

---

* This work is supported by the NSFC 60275024 and the 973 programme 2003CB317000.

U. Visser et al. (Eds.): RoboCup 2007, LNAI 5001, pp. 208–219, 2008.

Bounded Policy Iteration (BPI) [Bernstein et al., 2005] is currently the leading approximate algorithm which guarantees both bounded memory usage and monotonic value improvement for all initial state distributions. It defines a joint controller to be a set of local controllers along with a correlation device. On each iteration, a node is chosen from one of the local controllers or the correlation device, and its parameters are updated through the solution of a linear program. Namely, an iteration is guaranteed to produce a new controller with value at least as high as the old for every possible initial state distribution. A major drawback of this approach is that it scales exponentially in the number of agents. When we apply it to our soccer simulated team, it can be very slow for the major characteristic of sparse-reward structures which means the joint reward functions are zero everywhere, except for a few states.

In this paper, we just go one step further by developing an alternative approach to handle large-scale DEC-POMDPs with sparse-reward structures.Our new method which aims to reduce, as efficiently as possible, the runtime, solves these problems as follow: a correlation device is firstly calculated by solving Correlation Markov Decision Processes (Correlation-MDPs) and then used to improve the local controller for each agent. Our experimental results show its efficiency and it runs substantially faster, which achieves a tradeoff between computational complexity and the quality of the approximation.

The rest of the paper is organized as follows. The next section gives some comparison of the related works. After that, the DEC-POMDP model, and the basic idea of BPI are introduced. Then, we present our new algorithm and some experimental results in the RoboCup domain.

2   Related Work

Over the last six years, researchers have proposed a wide range of optimal and approximate algorithms for decentralized multi-agent planning. In this paper we focus on cooperation aspect. One important class of algorithms is called MAA*: A Heuristic Search Algorithm for Solving Decentralized POMDPs [Zser et al., 2005], where multi-agent A* (MAA*), the first complete and optimal heuristic search algorithm for solving decentralized POMDPs with finite horizon was presented. But the algorithm runs out of time very quickly, because the search space grows double exponentially.

The previous approach that is closest in spirit to ours is called Team coordination among robotic soccer players [Matthijs et al., 2002]. It is based on the idea of dynamically distributing roles among the team members and adds the notion of a global team strategy (attack, defend and intercept). Utility functions are used for estimating how well suited a robot is for a certain role. But inconsistencies sometimes occur.

A DEC-POMDP can also be seen as a partially observable stochastic game (POSG) with common payoffs [Emery-Montemerlo et al., 2004]. In this approach, the POSG is approximated as a series of smaller Bayesian games. Interleaving planning and execution, this algorithm finds good solutions for short horizons, but it still runs out memory after horizon 10.

3   The DEC-POMDP Model and BPI Algorithm

The family of Markov decision processes describes discrete stochastic systems that evolve under the influence of one or multiple controllers. With each transition of the system is associated a reward value, the objective of the controller is to select precisely a sequence of actions that maximizes the collection of rewards in the long run. For the case of several distributed but cooperative controllers, their objective is to act selfishly as to maximize the reward collected by the team.

3.1   The DEC-POMDP Model

We base our work on the DEC-POMDP framework introduced by Bernstein [Bernstein et al., 2002], although alternative definitions are equally allowed.

Definition 1 (DEC-POMDP). An n-agent DEC-POMDP is given as a tuple $\langle I, S, \{A_i\}, \{O_i\}, P, R \rangle$, where

- I is a finite set of agents indexed $1, ..., n$
- S is a finite set of states
- A_i is a finite set of actions available to agent i and $\overrightarrow{A} = \times_{i \in I} A_i$ is the set of joint actions, where $\overrightarrow{a} = \langle a_i, ..., a_n \rangle$ denotes a joint action
- O_i is a finite set of observations for agent i and $\overrightarrow{O} = \times_{i \in I} O_i$ is the set of joint observations, where $\overrightarrow{o} = \langle o_1, ..., o_n \rangle$ denotes a joint observation
- P is a set of Markovian state transition and observation probabilities, where $P(s', \overrightarrow{o} | s, \overrightarrow{a})$ denotes the probability that taking joint action $\overrightarrow{a}$ in state s results in a transition to state s' and joint observation $\overrightarrow{o}$
- $R : S \times \overrightarrow{A} \to \mathcal{R}$ is a reward function

In this paper, we consider the case in which the process unfolds over an infinite sequence of stages, At each stage, all agents simultaneously select an action, and each receives the global reward and a local observation. The objective of the agents is to maximize the expected discounted sum of rewards received. We denote the discount factor γ and require that $0 \leq \gamma < 1$. In order to be optimal, the Markov assumption requires a policy to depend on the whole information available to the agent at time t, namely its complete history of past observations and actions. For infinite horizon problems however, this would require a controller to have infinite memory, which is not always possible. Therefore, our algorithm uses stochastic finite-state controllers (FSCs) to represent policies.

Definition 2 (FSC). A stochastic finite-state controller (FSC) is a policy graph, defined as a tuple $\langle Q_i, \psi_i, \eta_i \rangle$, where

- Q_i is a finite set of controller nodes
- $\psi_i : Q_i \to \Delta A_i$ is an action selection function
- $\eta_i : Q_i \times A_i \times O_i \to \Delta Q_i$ is a transition function

The functions ψ_i and η_i parameterize the conditional distribution $P(a_i, q'_i | q_i, o_i)$. For the case of decentralized problems with multiple controllers, the goal is it to find a set of FSCs, one for each agent, such that their concurrent execution maximizes the expected discounted sum of rewards received. The agents' controllers determine the conditional distribution $P(\overrightarrow{a}, \overrightarrow{q}', | \overrightarrow{q}, \overrightarrow{o})$.

Recently, a memory-bounded dynamic programming algorithm was proposed for infinite-horizon DEC-POMDPs [Bernstein et al., 2005] . It extends a joint controller to allow for correlation among the agents. To do this, an additional finite-state machine, called a correlation device is introduced, which provides extra signals to the agents at each time step. The device operates independently of the DEC-POMDP process, and thus does not provide the agents with information about the other agents observations. By using correlated joint controllers, higher value can be achieved than with independent joint controllers of the same size.

Definition 3 (Correlation Device). A correlation device is a tuple $\langle C, \psi \rangle$, where

- C is a set of states
- $\psi : C \rightarrow \Delta C$ is a state transition function

To improve a correlated joint controller, either the correlation device or one of the local controllers can be changed. Both improvements can be done via a bounded backup, which involves solving a linear program.

Following an improvement, the controller can be reevaluate through the solution of a set of linear equations. It has been proofed that performing either of two updates cannot lead to a decrease in value for any initial state distribution. The runtime is polynomial in the sizes of the DEC-POMDP and the joint controller, but exponential in the number of agents.

However, in large-scale and sparse-reward domains, improving the correlation device is very difficult because of the characteristic of sparse reward structures. It can take a very long time for rewards to propagate to distinct states. Thus, it is often possible to get no improvement for just a few of steps. Long steps of search is rather inefficient for large scale problems such as RoboCup. It is obviously worse if multiple choices exist at each state. For example, in the decision making of the RoboCup 2D Simulation League, agents gain non-zero reward for their joint defense actions only when some agent of the team kicks or tackles the ball. Currently, most of the opponents process ball with high quality. Thus, it usually takes thousands of time steps to steal the ball for opponents.

4   Dynamic Programming for Correlation-MDPs

In this section, we describe the Dynamic Programming (DP) algorithm to calculate the correlation device as an approximate alternative to BPI. This method, analogous to the belief propagation, operates by solving a MDP, which can be regarded as a rewards propagation process. For sparse reward structures, each distinct state will have non-zero reward after the DP algorithm.

We can translate the above method to our multi-agent decision making problem by giving the correlation device a concrete meaning.

Definition 4 (Correlation Device State). A correlation device state $c \in C$ is a set of joint actions[1], where $\forall \overrightarrow{a_1}, \overrightarrow{a_2} \in c, |\bar{R}(\overrightarrow{a_1}) - \bar{R}(\overrightarrow{a_2})| \leq \varepsilon$ and $\bar{R}(\overrightarrow{a})$ is the reward function. The reward function for c is $R(c) = max_{\overrightarrow{a} \in c} \bar{R}(\overrightarrow{a})$.

Thus, how to compute the correlation device states is the main job of our method. In RoboCup 2D defensive decisions, the majority of the joint actions has no immediate rewards, or in other wards, it is very difficult in the immediate direct rewards given when the proceeds are sparse reward structures. Only when all opponents can not process the ball any longer, our agents make that defensive effectiveness and have non-zero reward. Running for a better opponent team, it may need to spend tens, hundreds or even more of the cycles to reach this ultimate goal. So far, solving this type of DEC-POMDP problems with the existing methods is not very satisfied.

In the DEC-POMDP model, the reward function $R(s, \overrightarrow{a})$ indicates that joint actions should link to a particular state for the need to obtain rewards. Thus, the pairs of specific state and joint actions which have the maximum rewards can be certainly established. In RoboCup 2D defensive decisions, the structure of rewards is sparse. In order to assess the states of those with zero reward, first is to be done with the goal of state to propagate the reward to distinct states. The propagation process can be described by the following definition of Correlation-MDPs.

Definition 5 (Correlation-MDP). A Correlation-MDP is given as a tuple $\langle \bar{S}, \bar{A}, \bar{P}, \bar{R}, \gamma \rangle$, where

- $\bar{S}$ is the set of states of the DEC-POMDP model
- $\bar{A} = \times_{i \in I} A_i$ is a set of joint actions
- $\bar{R} : S \rightarrow \Re$ and $\bar{R} = max_{a \in \bar{A}} \{R(s, \overrightarrow{a})\}$, where $R(s, \overrightarrow{a})$ is the reward function of the DEC-POMDP model
- $\bar{P} : S \times S \rightarrow [0, 1]$ can be defined:

$$\bar{P}(s'|s) = \frac{P(s') \cdot P(s|s')}{\sum_{s_i \in S} P(s_i) \cdot P(s|s_i)} \quad {}^{2} \tag{1}$$

where $P(s|s') = max_{\overrightarrow{a} \in \bar{A}} \{P(s|\overrightarrow{a}, s')\}$, $P(s|s_i) = max_{\overrightarrow{a} \in \bar{A}} \{P(s|\overrightarrow{a}, s_i)\}$, $P(s')$ and $P(s_i)$ are the probability of s' and s_i
- γ is the discount factor

A Correlation-MDP compared to the original DEC-POMDP can be viewed as a reverse model. What Correlation-MDP considers is the shift from the target

---

[1] Note that a correlation device is a finite-state machine, any reasonable definition of the states is allowed.

[2] Known as Bayes formula.

state to the initial one. However, in light of a DEC-POMDP model, it can only provide such a probability $P(s', \overrightarrow{o}|s, \overrightarrow{a})$, which is from the initial state to the target after the execution of joint actions. And the equation (1) is one of the possible solutions (from $P(s|s')$ to $\bar{P}(s'|s)$). In our model, $P(s')$ and $P(s_i)$ are the probability of s' and s_i, which can be used to control the emergence of a particular state (for example, in the area of inside and outside penalty, the strategy for both is different in the RoboCup 2D simulation league). It is easy to encode the information of opponents behavior into the correlation device in this way. Solving a Correlation-MDP means finding a policy $\pi : \bar{S} \to \bar{A}$ that maximizes the expected reward for each state $s \in \bar{S}$. In fact, the whole process is to build a tree from a root which is the target of the strategy. For the sake of better understanding this process, let us consider a simplified example of a soccer defense situation in Fig. 1. and the tree built by the DP algorithm is in Fig. 2.

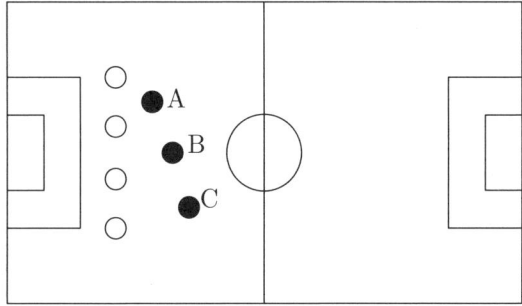

Fig. 1. White circles represent our members, and black ones represent opponents A (the top one), B (the middle one), C (the bolttom one). Each agent only has four type of actions: (formation), (mark, A), (mark, B), (mark, C). The target is to make all the opponents marked. Our side is left in the pitch.

In order to control the time complexity and precision we define $maxChildren$ to limit the maximum number of children for each parent. **Algorithm 1** shows the pseudo-code of the DP algorithm.

Proposition 1 (Algorithm 1). *The **Algorithm 1** has a linear space complexity with respect to the maxChildren, and the worst case time complexity is $O(maxChildren \times |\bar{S}|^2)$.*

Proof. The main loop of the algorithm (line 5-14) depends linearly on the size of $\bar{S}$. Inside this loop, the remaining critical operation is in line 9, where the children of each node is updated. Once the variable $maxChildren$ is fixed, the number of the children per node is not more than it. In every iteration of the algorithm, no more than $maxChildren$ subtrees are constructed. In the worst case, each node has to search all of the states to update its children, therefor, the upper limit

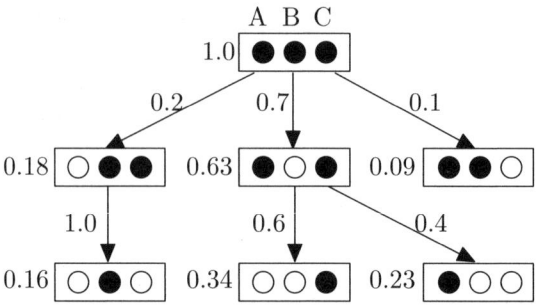

Fig. 2. The value of nodes is $\bar{R}(s)$, and the value of edges is $\bar{P}(s'|s)$. The discount factor is 0.9. The black circle means the opponent is marked, while the white one means not. Note that, in Fig. 1, the easiest action is to mark B, then is to mark A, and to mark C is the most difficult. Obviously the best defense strategy (Marking all the opponent successfully at the same time is usually impossible) should be marking C first, then A, and B finally. And the worst one may be marking A first, then B, and C finally, because the final step is much difficult by following this strategy: Opponent C may control the ball at that time and pass it to A or B, which is very dangerous in some case.

Algorithm 1. Compute the policy tree (correlation tree) Q_{t+1}

1: $Q_0 \leftarrow$ initialize all states in $\bar{S}$ as a tree
2: for each $s \in \bar{S}, V_0(s) \leftarrow \bar{R}(s)$
3: $maxChildren \leftarrow$ max number of children for the tree
4: $t \leftarrow 0$
5: loop:
6: $t \leftarrow t + 1$
7: for each $s \in \bar{S}$, do {
8:     $Q_{t+1} \leftarrow$ fullBackup(Q_t)
9:     find a set S' from $\bar{S}$ to satisfy:

   $- \forall s'' \in \bar{S} - \bar{S}', s' \in \bar{S}' \; P(s|s'') \leq P(s|s')$
   $- \forall s' \in \bar{S}' R(s') + V_t(s) \cdot \bar{P}(s'|s) \leq V_t(s')$
   $- |\bar{S}'| \leq maxChildren$

10:    $V_{t+1}(s') \leftarrow R(s') + V_t(s) \cdot \bar{P}(s'|s)$ for each $s' \in \bar{S}'$
11:    $V_{t+1}(s'') \leftarrow V_t(s'')$ for each $s'' \in \bar{S} - \bar{S}'$
12:    $Q_{t+1} \leftarrow$ set s as the parent of each $s' \in \bar{S}'$, set each $s' \in \bar{S}'$ as the children of s.
13: }
14: until $max_s|V_{t+1}(s) - V_t(s)| < \varepsilon$
15: return Q_{t+1}

---

for the construction is equal to $maxChildren|\bar{S}|$. By choosing $maxChildren$ appropriately, the desired upper limit of the tree length can be per-set, no more than $|\bar{S}|$. Thus, the amount of space grows linearly with $maxChildren$. For the worst case, this is $O(maxChildren \times |\bar{S}|^2)$.

Although the worst case time complexity is $O(maxChildren \times |\bar{S}|^2)$, the average time complexity is much smaller. Increasing the value of $maxChildren$ generally leads to both higher accuracy and time complexity on average. In practice, $maxChildren$ is usually necessary to achieve a tradeoff between computational complexity and the quality of the approximation. When $maxChildren$ is equal to $|\bar{S}|$, every iteration of **Algorithm 1** needs to consider all the possible states, which is the same as a linear program of BPI. The size of these states is exponential in the number of agents. A more detailed analysis of the DEC-POMDPs shows that most of the states are useless, especially with sparse reward structures.

According to the tree computed by **Algorithm 1**, the correlation device can be calculated easily. In general, the assumption below holds: Only when an agent finds a better, or when it finds higher reward cooperation strategy, the current one is changed. It means a rational agent will not choose the worse forms of cooperation from its own local observation. In the RoboCup 2D defensive decision making, an agent in the next step would be impossible to choose a strategy, although in accordance with their own local observation such a strategy might be possible, from the perspective of cooperation it is not likely to exist such a big leap. An upper bound estimate for the reward can be established by the tree calculated above.

Proposition 2 (Algorithm 2). *The **Algorithm** 2 returns the near-optimal value for $P(c'|c)$, which is proportional to $\frac{maxChildren}{|S|}$.*

Proof. The usage of $P(c'|c)$ is to correlate the joint controllers in BPI. According to BPI, the procedure for improving the correlation device works by looking for the best parameters satisfying the following inequality:

$$V(s, \overrightarrow{q}, c) \leq \sum_{\overrightarrow{a}} P(\overrightarrow{a}|c, \overrightarrow{q})[R(s, a) + \gamma \sum_{s', \overrightarrow{o}, \overrightarrow{q}', c} P(\overrightarrow{q}'|c, \overrightarrow{q}, \overrightarrow{a}, \overrightarrow{o}) \cdot$$

$$P(s', \overrightarrow{o}|s, \overrightarrow{a}) \cdot P(c'|c)V(s', \overrightarrow{q}', c')] \quad (2)$$

for all $s \in S$ and $\overrightarrow{q} \in \overrightarrow{Q}$.

Note that if $R(c') > R(c)$, the value of $P(c'|c)$ is definitely high by the concrete meaning of c' and c, since the inequality (2) implies that $V(s, \overrightarrow{q}, c) > V(s', \overrightarrow{q}', c')$ is not allowed. The basis of the computational process of **Algorithm 2** is an ideal tree constructed by **Algorithm 1** in which each state has the largest reward propagated from the target. For each node of the tree, the value of its parent is the upper bound of the possible rewards for the next step. Therefore, the rewards which could be calculated for the next step during execution time will range between current rewards and the upper bound. And the process of **Algorithm 2** is just on the premise that the information about agents local observations is unavailable and gives $P(c'|c)$ a reasonable approximation of the distribution while ensuring the inequality (1) non-reducing. This guarantees the near-optimality of the solution. In the process of **Algorithm 2**, if $maxChildren = |\bar{S}|$, all possible states need to be examined, which is the same

Algorithm 2. Compute the state transition function $p(c'|c)$

1: $Q \leftarrow$ a pre-computed correlation tree by **Algorithm 1**;
2: $S_0 \leftarrow$ all of the nodes in Q with the value $V(s) : |V(s) - R(c)| \leq \varepsilon$
3: $S_1 \leftarrow \emptyset, S_2 \leftarrow \emptyset$
4: $\delta \leftarrow 0, \gamma \leftarrow 0$
5: for each $s \in S_0$ {
6:   $\delta \leftarrow \delta + V(s) \cdot 1$
7:   if$(|V(s) - R(c')| \leq \varepsilon)$ $\gamma \leftarrow \gamma + V(s) \cdot 1$
8:   $s' \leftarrow$ the parent of s
9:   $S_1 \leftarrow S_1 \cup \{s'\}$
10:   $\delta \leftarrow \delta + V(s') \cdot \bar{P}(s|s')$
11:   if$(|V(s') - R(c')| \leq \varepsilon)$ $\gamma \leftarrow \gamma + V(s') \cdot \bar{P}(s|s')$
12:   $S' \leftarrow$ all the children of s' except s
13:   for each $s'' \in S'$ {
14:     if$(R(c) - V(s'') \leq \varepsilon)$ {
15:       $S_1 \leftarrow S_1 \cup \{s''\}$
16:       $\delta \leftarrow \delta + V(s'') \cdot \frac{\bar{P}(s''|s)}{\bar{P}(s|s')}$
17:       if$(|V(s'') - R(c')| \leq \varepsilon)$ $\gamma \leftarrow \gamma + V(s'') \cdot \frac{\bar{P}(s''|s')}{\bar{P}(s|s')}$
18:     }
19:   }
20: }
21: return $\frac{\gamma}{\delta}$

as a linear program for BPI, and if $maxChildren < |\bar{S}|$, a lot of states which are useless for the target will be eliminated. Thus, the algorithm works efficiently and the precision is proportional to $\frac{maxChildren}{|\bar{S}|}$.

In **Algorithm 2**, by choosing the value of $maxChildren$, the states which make small contribution to the goal have not been carried out in order to reduce the amount of calculation while guarantee high accuracy. This technique is effective, especially for the sparse reward domain. Further analysis of the RoboCup 2D defense problem shows that many joint actions for a special state are useless. BPI algorithm gives these useless joint actions the same needs of assessment, while our algorithm takes full account of this characteristic, thereby maintaining the high accuracy with a substantial amount of the reduction of the runtime. The parameters $maxChildren$ present an important trade-off: its increase generally increases both precision and runtime. Thus, we can examine this trade-off and identify the best parameters for concrete problems. Though we are not able to give theoretically strict proof on this issue and the exact solution for the accuracy of the two algorithm in this paper, the following experiments proved that our approach is more effective.

Although Correlation-MDPs are very usefull for our soccer robot team, it has two limitations when extended to other applications. Firstly, it requires initial state distribution as input. Secondly, the joint actions of agents and their effort should be easily modeled.

5   Experiments

We performed an experimental feasibility study in RoboCup domain that compares our algorithm and BPI [Bernstein et al., 2005], currently the leading algorithm for solving infinite-horizon DEC-POMDPs with quality guarantees. Below, we describe our experimental methodology, the specifies of the problems, and our results.

As noted above, the correlation device operates independently of the DEC-POMDP process. Thus, the experiments took place in two phases. First, either our algorithm or BPI was run to calculate the correlation device. Secondly, the correlation devices were used to improve local controllers with the some improving procedure.We applied both the BPI and our algorithm to compute the correlation device offline. In BPI we first chose a device node c, and considered changing its parameters for just the first step. New parameters must yield value at least as high for all states and nodes of the other local controllers. For our algorithm, we applied a Correlation-MDP model for all states, and calculated it with some high reward states fixed[3] . Then the correlation device was born by **Algorithm 2**. The computation is performed offline in a centralized way and the final solution is a correlation device which can then be executed by multiple agents in a decentralized way.

In order to encode the information of opponents behavior, we use some learning methods [Ubbo Visser et al., 2003] to determine the value of P(s) in Equation (1). For example, some opponents prefer attacking from the midway, then the probability of midway defending states will be increased; while some rivals like attacking from the sideway, then the probability of sideway defending states will be increased. But a discussion of the learning algorithm is beyond the scope of this paper and thus the following performance comparison does not include results for it.

Experiment 1: We determined how our algorithm and BPI trade off between the number of agents and runtime for the RoboCup Simulation 2D League with the fixed threshold reward (0.8). Our results show that our algorithm is faster than BPI when the number of agents is bigger than 3. For example, our algorithm needed 458.2ms and BPI needed 799.5ms to compute a solution that is only 8 agents under consider (there are 11 agents for each team in the RoboCup Simulation 2D League). Fig. 3 presents the performance comparison.

Experiment 2: We then determined how our algorithm and BPI trade off between runtime and the reward for the RoboCup Simulation 2D League when the number of agents is fixed (7 agents). Our results show that our algorithm is still faster than BPI with the same reward value, by fixing the number of agents. For example, our algorithm needed 533.2ms and BPI needed 1111.9ms to compute a solution that the reward is 0.9. Fig. 4 presents the performance comparison.[4]

---

[3] The key parameter is the maximum number of children for each node, $maxChildren$, which is related to the runtime and precision. In our experimental domain, $maxChildren = 3$ is sufficient to produce the best solution.

[4] All results are generated on a 2.2GHz/1GB machine using a C++ implementation.

RoboCup-2006 has provided an ideal test bed for our algorithm which implemented in our 2D simulation robot team. This team used the framework and algorithm described in the previous section to improve its highlevel strategy. The main motivation was to improve upon the coordination during defense. Since many different factors contribute to the overall performance of the team, it is difficult to measure the actual effect of the coordination with our new algorithm clearly. However, using this approach, we won all the matches except one ended in a draw 0-0 in the RoboCup-2006 in Bremen, German.

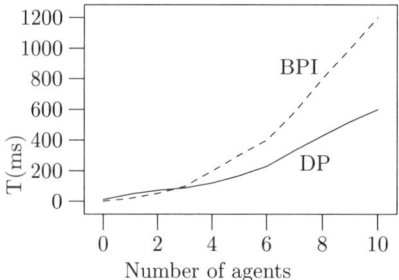 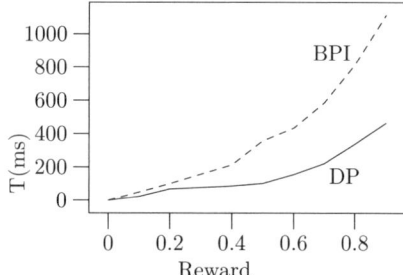

Fig. 3. The result of Experiment 1 with 0.8 as the threshold reward and the discount rate is $\gamma = 0.9$

Fig. 4. The result of Experiment 2 with 7 agents and he discount rate is $\gamma = 0.9$

6   Conclusion and Future Work

The decision making in the RoboCup 2D Simulation League can be modeled with DEC-POMDPs. Despite recent advances in solving DEC-POMDPs, state-of-the art solution methods are still either inefficient [Bernstein et al., 2005] or cannot provide guarantees on the quality of the resulting policy. In this paper, we presented a solution method, that avoids both of these shortcomings. Our experimental results show that the algorithm performs very well. The aim of this paper was to provide a first experimental feasibility study to demonstrate its potential. It is future work to study the theoretical properties of this method in more depth (for example, analyze its complexity or extend its error analysis), and extend it (for example, to handle more adversarial problems). The algorithm and representations used in this work open up multiple research avenues for developing effective approximation algorithms for the DEC-POMDP model in the RoboCup domain.

Acknowledgments

We thank Changjie Fan and Benjamin Johnston for helpful discussions of this work and other members of our team (Jiliang Wang, Jingnan Cai) for their contribution to WE2006 on which the experiments of this paper based. We also thank the anonymous reviewers for their valuable comments on the early version of this paper.

References

[Kitano et al., 1997] Kitano, H., Asada, M., Kuniyoshi, Y., Noda, I., Osawa, E., Matsubara, H.: RoboCup: A Challenge problem for AI. AI Magazine

[Kaelbling et al., 1998] Kaelbling, L.P., Littman, M.L., Cassandra, A.R.: Planning and acting in partially observable stochastic domains. Artificial Intelligence 101, 101–134 (1998)

[Bernstein et al., 2002] Bernstein, D.S., Givan, R., Immerman, N., Zilberstein, S.: The complexity of decentralized control of Markov decision processes. Mathematics of Operations Research 27(4), 819–840 (2002)

[Bernstein et al., 2005] Bernstein, D.S., Hansen, E.A., Zilberstein, S.: Bounded Policy Iteration for Decentralized POMDPs. In: Proceedings of the Nineteenth International Joint Conference on Artificial Intelligence(IJCAI), Edinburgh, Scotland, pp. 1287–1292 (July 2005)

[Ubbo Visser et al., 2003] Visser, U., Weland, H.-G.: Using online learning to analyze the opponents behavior. In: Kaminka, G.A., Lima, P.U., Rojas, R. (eds.) RoboCup 2002. LNCS (LNAI), vol. 2752, pp. 78–93. Springer, Heidelberg (2003)

[Rabinovich et al., 2003] Rabinovich, Z., Goldman, C.V., Rosenschein, J.S.: The Complexity of Multiagent Systems: The Price of Silence. In: Proceedings of the Second International Joint Conference on Autonomous Agents and Multi-Agent Systems(AAMAS), Melbourne, Australia, pp. 1102–1103 (2003)

[Matthijs et al., 2002] Spaan, M.T.j., Groen, F.C.A.: Team coordination among robotic soccer players. In: Kaminka, G.A., Lima, P.U., Rojas, R. (eds.) RoboCup 2002. LNCS (LNAI), vol. 2752. Springer, Heidelberg (2003)

[Emery-Montemerlo et al., 2004] Emery-Montemerlo, R., Gordon, G., Schneider, J., Thrun, S.: Approximate solutions for partially observable stochastic games with common payoffs. In: Kudenko, D., Kazakov, D., Alonso, E. (eds.) AAMAS 2004. LNCS (LNAI), vol. 3394, pp. 136–143. Springer, Heidelberg (2005)

[Zser et al., 2005] Szer, D., Charpillet, F., Zilberstein, S.: MAA*: A Heuristic Search Algorithm for Solving Decentralized POMDPs. In: Proceedings of the 21st Conference on UAI (2005)

Heuristic Reinforcement Learning Applied to RoboCup Simulation Agents

Luiz A. Celiberto Jr.[1,2], Carlos H.C. Ribeiro[2], Anna H.R. Costa[3], and Reinaldo A.C. Bianchi[1]

[1] Centro Universitário da FEI
Av. Humberto de Alencar Castelo Branco, 3972.
09850-901 – São Bernardo do Campo – SP, Brazil
celibertojr@uol.com.br, rbianchi@fei.edu.br
[2] Instituto Tecnológico de Aeronáutica
Praça Mal. Eduardo Gomes, 50.
12228-900 – São José dos Campos – SP, Brazil
carlos@ita.br
[3] Laboratório de Técnicas Inteligentes
Escola Politécnica da Universidade de São Paulo
Av. Prof. Luciano Gualberto, trav. 3, 158. 05508-900 – São Paulo – SP, Brazil
anna.reali@poli.usp.br

Abstract. This paper describes the design and implementation of robotic agents for the RoboCup Simulation 2D category that learns using a recently proposed Heuristic Reinforcement Learning algorithm, the Heuristically Accelerated Q–Learning (HAQL). This algorithm allows the use of heuristics to speed up the well-known Reinforcement Learning algorithm Q–Learning. A heuristic function that influences the choice of the actions characterizes the HAQL algorithm. A set of empirical evaluations was conducted in the RoboCup 2D Simulator, and experimental results show that even very simple heuristics enhances significantly the performance of the agents.

Keywords: Reinforcement Learning, Cognitive Robotics, RoboCup Simulation 2D.

1 Introduction

Reinforcement Learning (RL) techniques have been attracting a great deal of attention in the context of multiagent robotic systems. The reasons frequently cited for such attractiveness are: the existence of strong theoretical guarantees on convergence [9], they are easy to use, and they provide model-free learning of adequate control strategies. Besides that, they also have been successfully applied to solve a wide variety of control and planning problems.

However, one of the main problems with RL algorithms is that they typically suffer from very slow learning rates, requiring a huge number of iterations to converge to a good solution. This problem becomes worse in tasks with high dimensional or continuous state spaces and when the system is given sparse

U. Visser et al. (Eds.): RoboCup 2007, LNAI 5001, pp. 220–227, 2008.

rewards. One of the reasons for the slow learning rates is that most RL algorithms assumes that neither an analytical model nor a sampling model of the problem is available *a priori*. However, in some cases, there is domain knowledge that could be used to speed up the learning process.

As a way to add domain knowledge to help in the solution of the RL problem, a recently proposed Heuristic Reinforcement Learning algorithm – the Heuristically Accelerated Q–Learning (HAQL) [1] – uses a heuristic function that influences the choice of the action to speed up the well-known RL algorithm Q–Learning. This paper investigates the use of HAQL to speed up the learning process of teams of mobile autonomous robotic agents acting in a concurrent multiagent environment, the RoboCup 2D Simulator. It is organized as follows: section 2 describes the Q–learning algorithm. Section 3 describes the HAQL and its formalization using a heuristic function. Section 4 describes the robotic soccer domain used in the experiments, presents the experiments performed, and shows the results obtained. Finally, Section 5 summarizes some important points learned from this research and outlines future work.

2   Reinforcement Learning and the Q–Learning Algorithm

Consider an autonomous agent interacting with its environment via perception and action. On each interaction step the agent senses the current state s of the environment, and chooses an action a to perform. The action a alters the state s of the environment, and a scalar reinforcement signal r (a reward or penalty) is provided to the agent to indicate the desirability of the resulting state.

The goal of the agent in a RL problem is to learn an action policy that maximizes the expected long term sum of values of the reinforcement signal, from any starting state. A policy $\pi : \mathcal{S} \to \mathcal{A}$ is some function that tells the agent which actions should be chosen, under which circumstances [5]. This problem can be formulated as a discrete time, finite state, finite action Markov Decision Process (MDP). The learner's environment can be modeled [6] by a 4-tuple $\langle \mathcal{S}, \mathcal{A}, \mathcal{T}, \mathcal{R} \rangle$, where:

- $\mathcal{S}$: is a finite set of states.
- $\mathcal{A}$: is a finite set of actions that the agent can perform.
- $\mathcal{T} : \mathcal{S} \times \mathcal{A} \to \Pi(\mathcal{S})$: is a state transition function, where $\Pi(\mathcal{S})$ is a probability distribution over $\mathcal{S}$. $T(s, a, s')$ represents the probability of moving from state s to s' by performing action a.
- $\mathcal{R} : \mathcal{S} \times \mathcal{A} \to \Re$: is a scalar reward function.

The task of a RL agent is to learn an optimal policy $\pi^* : \mathcal{S} \to \mathcal{A}$ that maps the current state s into an optimal action(s) a to be performed in s. In RL, the policy π should be learned through trial-and-error interactions of the agent with its environment, that is, the RL learner must explicitly explore its environment.

The Q–learning algorithm was proposed by Watkins [10] as a strategy to learn an optimal policy π^* when the model ($\mathcal{T}$ and $\mathcal{R}$) is not known in advance. Let

$Q^*(s, a)$ be the reward received upon performing action a in state s, plus the discounted value of following the optimal policy thereafter:

$$Q^*(s, a) \equiv R(s, a) + \gamma \sum_{s' \in S} T(s, a, s') V^*(s'). \tag{1}$$

The optimal policy π^* is $\pi^* \equiv \arg\max_a Q^*(s, a)$. Rewriting $Q^*(s, a)$ in a recursive form:

$$Q^*(s, a) \equiv R(s, a) + \gamma \sum_{s' \in S} T(s, a, s') \max_{a'} Q^*(s', a'). \tag{2}$$

Let $\hat{Q}$ be the learner's estimate of $Q^*(s, a)$. The Q–learning algorithm iteratively approximates $\hat{Q}$, i.e., the $\hat{Q}$ values will converge with probability 1 to Q^*, provided the system can be modeled as a MDP, the reward function is bounded ($\exists c \in \mathcal{R}; (\forall s, a), |R(s, a)| < c$), and actions are chosen so that every state-action pair is visited an infinite number of times. The Q learning update rule is:

$$\hat{Q}(s, a) \leftarrow \hat{Q}(s, a) + \alpha \left[r + \gamma \max_{a'} \hat{Q}(s', a') - \hat{Q}(s, a) \right], \tag{3}$$

where s is the current state; a is the action performed in s; r is the reward received; s' is the new state; γ is the discount factor ($0 \leq \gamma < 1$); α is the learning rate.

An interesting property of Q–learning is that, although the exploration-exploitation tradeoff must be addressed, the $\hat{Q}$ values will converge to Q^*, independently of the exploration strategy employed (provided all state-action pairs are visited often enough) [6].

3   The Heuristically Accelerated Q–Learning Algorithm

The Heuristically Accelerated Q–Learning algorithm [1] was proposed as a way of solving the RL problem which makes explicit use of a heuristic function $\mathcal{H} : \mathcal{S} \times \mathcal{A} \rightarrow \mathfrak{R}$ to influence the choice of actions during the learning process. $H_t(s_t, a_t)$ defines the heuristic, which indicates the importance of performing the action a_t when in state s_t.

The heuristic function is strongly associated with the policy: every heuristic indicates that an action must be taken regardless of others. This way, it can be said that the heuristic function defines a "Heuristic Policy", that is, a tentative policy used to accelerate the learning process. It appears in the context of this paper as a way to use the knowledge about the policy of an agent to accelerate the learning process. This knowledge can be derived directly from the domain (prior knowledge) or from existing clues in the learning process itself.

The heuristic function is used only in the action choice rule, which defines which action a_t must be executed when the agent is in state s_t. The action choice rule used in the HAQL is a modification of the standard $\epsilon - Greedy$ rule used in Q–learning, but with the heuristic function included:

$$\pi(s_t) = \begin{cases} \arg\max_{a_t} \left[\hat{Q}(s_t, a_t) + \xi H_t(s_t, a_t)\right] & \text{if } q \leq p, \\ a_{random} & \text{otherwise,} \end{cases} \quad (4)$$

where:

- $\mathcal{H} : \mathcal{S} \times \mathcal{A} \to \Re$: is the heuristic function, which influences the action choice. The subscript t indicates that it can be non-stationary.
- ξ: is a real variable used to weight the influence of the heuristic function.
- q is a random value with uniform probability in [0,1] and p $(0 \leq p \leq 1)$ is the parameter which defines the exploration/exploitation trade-off: the greater the value of p, the smaller is the probability of a random choice.
- a_{random} is a random action selected among the possible actions in state s_t.

As a general rule, the value of the heuristic $H_t(s_t, a_t)$ used in the HAQL must be higher than the variation among the $\hat{Q}(s_t, a_t)$ for a similar $s_t \in S$, so it can influence the choice of actions, and it must be as low as possible in order to minimize the error. It can be defined as:

$$H(s_t, a_t) = \begin{cases} \max_a \hat{Q}(s_t, a) - \hat{Q}(s_t, a_t) + \eta & \text{if } a_t = \pi^H(s_t), \\ 0 & \text{otherwise.} \end{cases} \quad (5)$$

where η is a small real value and $\pi^H(s_t)$ is the action suggested by the heuristic.

As the heuristic is used only in the choice of the action to be taken, the proposed algorithm is different from the original Q–learning only in the way exploration is carried out. The RL algorithm operation is not modified (i.e., updates of the function Q are as in Q–learning), this proposal allows that many of the conclusions obtained for Q–learning to remain valid for HAQL [1].

The use of a heuristic function made by HAQL explores an important characteristic of some RL algorithms: the free choice of training actions. The consequence of this is that a suitable heuristic speeds up the learning process, and if the heuristic is not suitable, the result is a delay which does not stop the system from converging to a optimal value.

4   Experiment in the RoboCup 2D Simulation Domain

One experiment was carried out using the RoboCup 2D Soccer Server [7]: the implementation of a defense team, with a goalkeeper and a first defender (fullback) that have to learn how to minimize the number of goals scored by the opponent. In this experiment, the implemented team have to learn while playing against a team composed of two striker agents from the UvA Trilearn 2001 Team [2].

The space state of the learning agents is composed by its position in a discrete grid with N x M positions the agent can occupy, the position of the ball in the same grid and the direction the agent is facing. This grid is different for the goalkeeper and the defender: each agent has a different area where it can move, which they cannot leave. These grids, shown in figure 1, are partially overlapping, allowing both agents to work together in some situations. The direction that the

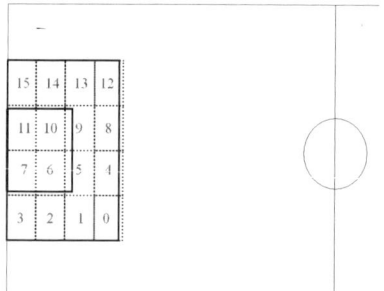 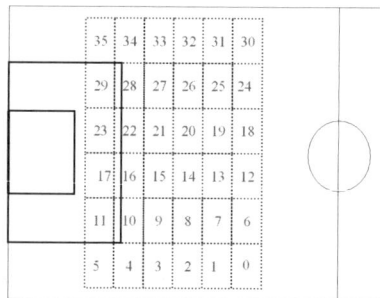

Fig. 1. Discrete grids that compose the space state of the goalkeeper (left) and the defender (right)

agent can be facing is also discrete, and reduced to four: north, south, east or west.

The defender can execute six actions: turnBodyToObject, that keeps the agent at the same position, but always facing the ball; interceptBall, that moves the agent in the direction of the ball; driveBallFoward, that allows the agent to move with the ball; directPass, that execute a pass to the goalkeeper; kickBall, that kick the ball away from the goal and; markOpponent, that moves the defender close to one of the opponents.

The goalkeeper can also perform six actions: turnBodyToObject, interceptBall, driveBallForward, kickBall, which are the same actions that the defender can execute, and two specific actions: holdBall, that holds the ball and moveToDefensePosition, that moves the agent to a position between the ball and the goal.

All these actions are implemented using pre-defined C++ methods defined in the BasicPlayer class of the UvA Trilearn 2001 Team. "The BasicPlayer class contains all the necessary information for performing the agents individual skills such as intercepting the ball or kicking the ball to a desired position on the field" [2, p. 50].

The reinforcement given to the agents were inspired on the definitions of rewards presented in [3], and are different for the agents. For the goalkeeper, the rewards consists of: ball caught, kicked or driven by goalie = 25; ball with any opponent player = -25; goal scored by the opponent = -100. For the defender, the rewards are: ball kicked or passed to the goalie = 15; ball with any opponent player = -10; goal scored by the opponent = -15.

The heuristic policy used for the goalkeeper and the defender is described by two rules: if the agent is not near the ball, run in the direction of the ball, and; if the agent is close to the ball, do something with it. Note that the heuristic policy is very simple, leaving the task of learning what to do with the ball and how to deviate from the opponent to the learning process. The values associated with the heuristic function are defined using equation 5, with the value of η set to 200. This value is computed only once, at the beginning of the game. In all

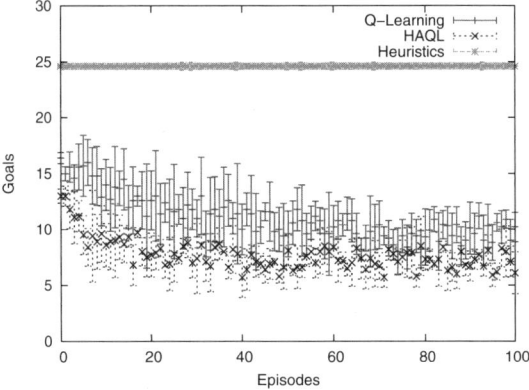

Fig. 2. Average goals scored against the defense agents using the Q–Learning and the HAQL algorithms, for training sessions against two UvA Trilearn attack agents

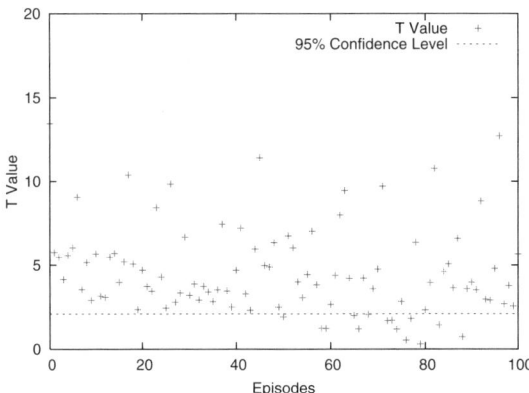

Fig. 3. Results from Student's t test between Q–learning and HAQL algorithms, for defense agents training against two UvA Trilearn attack agents

the following episodes, the value of the heuristic is maintained fixed, allowing the learning process to overcome bad indications.

In order to evaluate the performance of the HAQL algorithm, this experiment was performed with teams of agents that learns using the Q–learning algorithm, the HAQL algorithm and using agents that acts based only on a heuristic rule (without learning capabilities). The results presented are based on the average of 10 training sessions for each algorithm. Each session is composed of 100 episodes consisting of matches taking 3000 cycles each. During the simulation, when a teams scores a goal all agents are transferred back to a pre-defined start position.

The parameters used in the experiments were the same for the two algorithms, Q–learning and HAQL: the learning rate is $\alpha = 1.25$, the exploration/ exploitation rate $p = 0.05$ and the discount factor $\gamma = 0.9$. Values in the Q table were randomly initiated, with $0 \leq Q(s, a, o) \leq 1$. The experiments were programmed in C++ and executed in a Pentium IV 2.8GHz, with 1GB of RAM on a Linux platform.

Figure 2 shows the learning curves for both algorithms when the agents learn how to play against a team composed of two strikers from the UvA Trilearn Team 2001 [2]. It presents the average goals scored by the opponent team in each episode. It is possible to verify that Q–learning has worse performance than HAQL at the initial learning phase, and that as the matches proceed, the performance of both algorithms become more similar, as expected.

Another important information contained in this figure is the number of goals scored against a defense team that uses only the heuristic policy to select which action must be done, at a given time. As it can be seen, this team will receive an average of 24 goals in each episode, performing worst than any of the other two algorithms. This shows that the heuristic policy by itself is not a complete solution to the problem, but only an indication of some actions that should be taken, at certain times.

Student's t–test [8] was used to verify the hypothesis that the use of heuristics speeds up the learning process. For the experiments described in this section, the value of the module of T was computed for each episode using the same data presented in figure 2. The result, presented in figure 3, shows that HAQL performs clearly better than Q–learning until the 60th episode, with a level of confidence greater than 95%. After the 60th episode, the results became closer. But it can be seen that HAQL still performs better than Q–learning.

Finally, table 1 shows the cumulative number of goals made by the strikers at the end of 100 episodes (averaged for 10 training sessions). What stands out in this tables is that, due to a lower number of goals scored against the HAQL at the beginning of the learning process, this algorithm receives significanly less goals than the Q–learning algoritm (with a statistical confidence $> 99.9\%$).

Table 1. Cumulative goal the end of 100 episodes (average for all training sessions)

Algorithm	Cumulative goal score
Q–leaning	(1177 ± 51)
HAQL	(836 ± 10)

5   Conclusion and Future Works

This paper presented the use of the Heuristically Accelerated Q–Learning (HAQL) algorithm to speed up the learning process of teams of mobile autonomous robotic agents acting in the RoboCup 2D Simulator.

The experimental results obtained in this domain showed that agents using the HAQL algorithm learned faster than ones using the Q–learning, when they were trained against the same opponent. These results are strong indications that the performance of the learning algorithm can be improved using very simple heuristic functions.

Due to the fact that the reinforcement learning requires a large amount of training episodes, the HAQL algorithm has been evaluated, so far, only in simulated domains. Among the actions that need to be taken for a better evaluation of this algorithm, the more important ones include:

- The development of teams composed of agents with more complex space state representation and with a larger number of players.
- Working on obtaining results in more complex domains, such as RoboCup 3D Simulation and Small Size League robots [4].
- Comparing the use of more convenient heuristics in these domains.
- Validate the HAQL by applying it to other the domains, such as the "car on the hill" and the "cart-pole".

References

[1] Bianchi, R.A.C., Ribeiro, C.H.C., Costa, A.H.R.: Heuristically Accelerated Q-Learning: a new approach to speed up reinforcement learning. In: Bazzan, A.L.C., Labidi, S. (eds.) SBIA 2004. LNCS (LNAI), vol. 3171, pp. 245–254. Springer, Heidelberg (2004)

[2] de Boer, R., Kok, J.: The Incremental Development of a Synthetic Multi-Agent System: The UvA Trilearn 2001 Robotic Soccer Simulation Team. Master's Thesis, University of Amsterdam (2002)

[3] Kalyanakrishnan, S., Liu, Y., Stone, P.: Half field offense in RoboCup soccer: A multiagent reinforcement learning case study. In: Lakemeyer, G., Sklar, E., Sorenti, D., Takahashi, T. (eds.) RoboCup-2006: Robot Soccer World Cup X, Springer, Berlin (2007)

[4] Kitano, H., Minoro, A., Kuniyoshi, Y., Noda, I., Osawa, E.: Robocup: A challenge problem for ai. AI Magazine 18(1), 73–85 (1997)

[5] Littman, M.L., Szepesvári, C.: A generalized reinforcement learning model: Convergence and applications. In: Procs. of the Thirteenth International Conf. on Machine Learning (ICML 1996), pp. 310–318 (1996)

[6] Mitchell, T.: Machine Learning. McGraw Hill, New York (1997)

[7] Noda, I.: Soccer server: a simulator of robocup. In: Proceedings of AI symposium of the Japanese Society for Artificial Intelligence, pp. 29–34 (1995)

[8] Spiegel, M.R.: Statistics. McGraw-Hill (1998)

[9] Szepesvári, C., Littman, M.L.: Generalized markov decision processes: Dynamic-programming and reinforcement-learning algorithms. Technical report, Brown University, Department of Computer Science, Brown University, Providence, Rhode Island 0, 1996. CS-96-11 (2912)

[10] Watkins, C.J.C.H.: Learning from Delayed Rewards. PhD thesis, University of Cambridge (1989)

Pareto-Optimal Offensive Player Positioning in Simulated Soccer

Vadim Kyrylov[1] and Serguei Razykov[2]

[1] Rogers State University; Claremore, OK 74017 USA
vkyrylov@rsu.edu
[2] Simon Fraser University – Surrey; Surrey, British Columbia V3T 0A3 Canada

Abstract. The ability by the simulated soccer player to make rational decisions about moving without ball is a critical factor of success. Here we limit our scope to the offensive situation, i.e. when the ball is controlled by own team, and propose a systematic method for determining the optimal player position. Existing methods for accomplishing this task do not systematically balance risks and rewards, as they are not Pareto optimal by design. This may result in overlooking good opportunities. One more shortcoming of these methods is over simplifications in predicting the situation on the field, which may lead to performance loss. We propose two new ideas to address these issues. Experiments demonstrate that this results in a substantial increase in the team performance.

1 Introduction

The ability by the soccer player making rational decisions about where to move without the ball is the critical factor of success both in a real-life and simulated game. With the total of 22 soccer players in two teams, an average player must be indeed purposefully moving to some place on the field more than 90 per cent of the time. Thus one should be expecting about 90 per cent impact on the whole game from any improvement in player positioning.

In this paper, we propose the algorithm for determining a good position on the soccer field for the artificial player when the ball is beyond its control. We limit here the consideration to the *offensive* positioning, i.e. when the ball is possessed by own team. In the offensive situation, the major purpose of co-coordinated moving to some individual positions on the field by the players without the ball is creating opportunities for receiving passes and scoring the opponent goal. Therefore, player positioning is not a standalone task; rather, it is part of a coordination mechanism. In this study, we deliberately isolate positioning from ball passing because in our recent paper we have proposed a Pareto-optimal method for ball passing [1]. Now we want the decisions about positioning to be optimized in the similar sense.

During the last 10 years, RoboCup scholars have addressed the positioning problem in different ways [2-8]. In order to demonstrate the unique features of our approach, here we propose a generic two-level model that includes the existing methods for player positioning.

The upper level determines so-called *reference* position for the player where it should be moving unless it is involved in intercepting or kicking the ball. Player coordination on

U. Visser et al. (Eds.): RoboCup 2007, LNAI 5001, pp. 228–237, 2008.
© Springer-Verlag Berlin Heidelberg 2008

this level is addressed only implicitly. The lower level is responsible for fine tuning this reference position by transforming it into the *target* position. The decision-making rules and/or optimality criteria used in such methods are normally reflecting the algorithm designer's vision of the soccer game. Alternatively, the decision making rules are derived by learning algorithms.

The first ever comprehensive study of the player positioning problem was presumably made in [2]. In this method, the higher-level reference position can be regarded as a fixed point in the field assigned to each player with respect to its role in the team formation. The lower control level allowed the player to deviate from this default position within some area in order to move to the calculated target position with respect to current situation. Although this method implemented in *CMUnited* proved to be a good start, later on it was criticized in [3] for the limited flexibility on the upper control level. A method based on a set of logical rules proposed in [3] has addressed these shortcomings in the *FCPortugal* who outperformed *CMUnited*.

The development that followed, did not demonstrate much improvement on the lower control level, though. The next very successful team, *UvA Trilearn* [4] that had outplayed *FCPortugal* in several competitions, implemented somewhat simpler player positioning method. In our classification, the early version of this method dated 2001-2002 is completely located on the higher control layer. In particular, this method does not take into account fine details such as the opportunity to receive a pass. This shortcoming was later addressed using so-called coordination graphs [5]. This lower-level model combines decision making about passing the ball and receiving the pass in an elegant way; implemented in *UvA Trilearn*, it helped to become the World RoboCup winner in 2003. However, we believe that this model could be further improved, as it was using heuristics rather than rigor multi-criteria optimization.

One more group of the improvements to player positioning is a set of methods based on learning algorithms [6, 7, 8]. Like the coordination graph, some of these methods do not treat positioning as a standalone player skill, which makes theoretical comparisons difficult. More difficulties arise while trying to elicit meaningful decision making rules and especially address the convergence issue of learned rules to optimal decision making algorithms based on explicitly formulated decision criteria. Thus we are leaving methods based on learning beyond the scope of this study.

The main objective of this paper is to provide an improved solution for low-level decision making about player positioning based on the known tactical principles of the real-life soccer [9, 10]. Because the recent studies addressed the lower control level rather superficially, it looks like our closest prototype is described in [2]. That was indeed the first method that was using both control levels. On the lower level, it maximizes the following utility function:

$$U(P) = \sum_{i=1}^{n} dist(P, O_i) + \sum_{j=1}^{n-1} dist(P, T_j) - (dist(P, G))^2, \qquad (1)$$

where

P – the desired position for the player in anticipation of a pass;
n – the number of agents on each team;
O_i – the current position of each opponent, $i = 1,\ldots, n$;
T_j – the current position of each teammate, $j = 1,\ldots, (n\text{-}1)$;
G – the position of the opponent goal;
$dist(A, C)$ – distance between positions A and C.

One advantage of this utility function is robustness; the two sums suppress the sensor noise. It repulses the player without ball from other players and encourages advancing to the opponent goal while seeking its best position. This indeed is reflecting the soccer tactics.

We would also mention three shortcomings. First, from the tactical standpoint, criterion (1) does not encourage players to deliberately get open for receiving a pass; this can only happen by chance. Second, the contribution of the remotely located players is exaggerated; increasing distance from the closest opponent by say 5 meters has same effect as increasing distance by 5 meters for the opponent located on the other end of the field. Third, from the mathematical standpoint, the authors clearly indicate that this is a vector optimization problem; indeed, each addendum in (1) could stand for a criterion. However, the reduction to single-criteria optimization is questionable. Aggregating several criteria in a linear combination is indeed theoretically acceptable if all criteria functions and constraints are convex [11]. However, the first two addendums in (1) are multi-modal functions of P; hence they are non-convex. So this single-criterion optimization does not guarantee that all potentially acceptable player positions would be considered. This issue could be resolved by using the Pareto optimality principle.

Dynamics is one more difficulty with optimizing player position on low level. Reaching an optimal position takes some time, and the player must be persistent in doing so. In our experiments we have found that random factors in the simulated soccer tend to make the computation of the optimal target very unstable from cycle to cycle. In what follows, we discuss this issue.

Therefore, the unique contribution of this paper is in that we (1) resolve the time horizon issue; (2) propose a new set of decision making criteria based on real-life soccer tactics; and (3) implement a Pareto-optimal solution to the optimization problem with these criteria.

Section 2 addresses the time horizon issue. Section 3 explains how feasible alternative player positions could be generated. Section 4 introduces five optimality criteria and the algorithm for finding the Pareto-optimal solution. Section 5 provides experimental results and conclusions.

2   Predicting the Situation for Player Positioning

The issue is what the planning time horizon should be once the player decides to go to some *target* position. A straightforward approach is just setting some fixed time horizon τ, say 1 to 3 seconds, and trying to extrapolate the movements of the players and the ball for this time interval. However, our experiments have shown that this method does not work well. This is because of that, once the ball is kicked by some player, the rest players revise their previous intentions and change their directions of movement. Neglecting these abrupt changes if the next ball kick occurs before the prediction time expires would result in poor decisions. On the other hand, forecasting new movements by the players when the ball gets under close control is difficult.

While trying to resolve this issue, it makes sense to see how human players are acting in this situation [9, 10]. In the offensive positioning, the teammate with the ball may be in two different states: (a) chasing the freely rolling ball while staying the fastest player to it or (b) keeping the ball close to itself while being able to pass it at any time. In case (a) the human player without ball can easily predict the situation and estimate time remaining until the ball will be intercepted. So nothing abrupt will likely occur while the ball is rolling freely. Thus a human player predicts the situation and plans his actions accordingly until the ball is intercepted. This gives the idea of the time horizon that we should be using in our model. During this time the player without ball must concentrate on reaching good position before the teammate has gained close control of it. No communication is really necessary, and it does not make much sense substantially changing the originally optimized target position while the ball is rolling. Only minor corrections to it may be necessary as the world model is updated.

In case (b), however, the situation is hard to predict, as the player with the ball can kick it at almost any time in a wide range of directions. In this case human players, if they have chance to become pass receivers, are watching carefully the teammate with the ball in order to be prepared to intercept the ball. The time horizon can be communicated by the active player to the partners; so communication is highly important in this situation. With or without the communication, time horizon is very limited, anyway. So players in this case only can do short-time planning to adjust their positions; the major objective is implementing the team strategy by moving to what we call here the *reference* position determined by current situation. In many cases during the game player may be indeed too far away from this position; so the major concern is reaching it as fast as stamina permits.

Thus we come to the idea of the variable time horizon during that player behavior should be staying persistent. While the ball is rolling freely, its movement follows the laws of physics and is easily predictable. All players (if they are rational) are acting in the ways that are also rather easy to predict if their decision making model is known. Predicting teammate movements thus is possible with high precision, Figuring out the upper-level positioning algorithm of the opponent team is possible using opponent modeling implemented in the online coach. This problem could be simplified by concentrating only on the situations when the ball is rolling freely.

Therefore, for these situations we are using the prediction horizon τ that is equal to the time remaining until the ball will be intercepted by a teammate. (If the fastest to the ball is the opponent player, the situation changes to defensive; this lies outside the scope of this study.) With this approach, once the ball has been kicked, the player estimates the interception point and time τ. This is the time remaining to plan its actions and move to the best possible position on the field before the ball will be kicked in the new direction. Because τ is known, the player selects only such optimal target position that could be indeed reached during this time. While the ball is rolling, the player is persistently moving to this position.

The model for predicting the situation comprises three components: the ball, the friendly and the opponent player movements. The ball movement can be predicted with high precision, as the underlying physics is simple. The movements by teammates can be also predicted with precision, as their control algorithms and to some extent their perceived states of the world are available to the player in question. The fastest teammate to the ball will be intercepting it by moving with the maximal speed;

so its position can be predicted even more precisely. The rest teammates will be moving towards the best positions determined with yet another precisely known algorithm which could be used for predicting their positions. However, in our method we do not use such a sophisticated approach; in regards of each teammate without the ball, we assume that it is just moving with a constant velocity. This velocity is estimated by the player using own world model. Of the opponent players, the fastest one will be presumably also chasing the ball by following the trajectory that can be predicted fairly precisely. For the rest opponents possible options include assuming same positioning algorithm as for the teammates.

If the ball is kickable by the teammate, we would suggest that τ should be set to a small constant value τ_{min} which should be determined experimentally. So while the ball is closely controlled by the teammate, this would allow the player to continue adjusting its position within this limited time horizon.

3 Identifying the Feasible Options

Decision making is always a choice from a set of alternatives. In the discussed problem, the player first generates a set of feasible options and evaluates those using different criteria. Then the multi-criteria algorithm is applied to find the single best option by balancing the anticipated risks and rewards.

Once the ball interception point and the remaining time τ have been determined by the player without ball, it generates a set of alternative positions in the vicinity of the reference position. Because the decision space (xy-plane) is continuous, it contains infinite number of such alternatives. With highly nonlinear and non-convex decision criteria, searching such space systematically would be hardly possible. Therefore, we use a discrete approximation, with the alternative positions forming on the xy-plane a grid about the default position. To reduce computations, we would like to keep the number of points in this grid minimal. The grid step determines the total number of the alternatives to be searched. The rule of thumb is setting the step equal to the radius of the player reach for kicking the ball. Increasing it might result in lost opportunities. Using a smaller step makes sense only if we have enough computational time to further fine tune the balance of different optimality criteria (see more about it in the next section).

Of these alternative positions, the player is only interested in those that could be reached in time τ. This allows eliminating part of the grid that is lying beyond the player reach. One more constraint that helps eliminating poor alternatives is the maximal distance from the reference position. The alternatives lying outside the field or creating risk of offside are also eliminated.

Figure 1 displays the field when the ball is kicked by red player #5. Arrows show the predicted positions of all objects at the moment when the ball is intercepted by red player #7. Highlighted are the alternative positions for red player #8. The area of responsibility is filled with small gray points; the reference position being the center of this area. The bigger blue points show the reachable positions of which this player must choose the best.

Fig. 1. The area of responsibility for red player #8 (gray dots) and reachable positions (blue dots) before the ball is intercepted by red #7

Fig. 2. The Pareto set for red player #8 (bigger dots) and the optimal solution

4 Criteria for Decision Making and the Optimization Algorithm

Each feasible alternative position has its pros and cons that an intelligent player is taking into account while choosing the best option. These decision criteria should be reflecting the soccer tactics; in particular they should be measuring anticipated rewards and risks. We propose slightly different criteria sets for attackers, midfielders, and defenders because their tactical roles differ indeed [9, 10].

For the attackers the criteria set is, as follows.

1. All players must maintain the formation thus implementing the team strategy. So the distance between the point in the feasible set and the reference position should be minimized.

2. All attackers must be open for a direct pass. Thus the angle between the direction to the ball interception point and the direction to the opponent located between the evaluated position and the interception point must be maximized.

3. All players must maintain open space. This means that the distance from the evaluated point to the closest opponent should be maximized.

4. The attackers must keep an open path to the opponent goal to create the scoring opportunity. So the distance from the line connecting the evaluated point and the goal center to the closest opponent (except the goalie) should be maximized. This criterion is only used in the vicinity of the opponent goal.

5. The player must keep as close as possible to the opponent offside line to be able to penetrate the defence. So, the player should minimize the x-coordinate distance between the point in the feasible set and the offside line (yet not crossing this line).

Note that each criterion appears to have equal tactical importance; this observation will be used while discussing the optimization procedure below.

Criteria for midfielders and defenders differ in that they do not contain criteria 4 and 5 that encourage the opponent defense penetration. Instead, these players should be creating opportunities for launching the attack. This is achieved by minimizing the opponent player presence between the evaluated position and the direction to the opponent side of the field.

These criteria are conflicting, as it is hardly possible to optimize all them simultaneously. This situation is well known in the literature on systems analysis and economics; a special paradigm called the *Pareto optimality principle* allows to eliminate wittingly inappropriate so-called dominated alternatives [11]. These are the points that could be outperformed by at least some other point in the feasible set by at least one criterion. So only the non-dominated alternatives making so-called Pareto set should be searched for the 'best' balance of all criteria. Balancing requires additional information about the relative importance of these criteria, or their weights. If the criteria functions and the feasible set are all convex, then the optimal point could be found by minimizing the weighed sum of the criteria (assuming that they all must be minimized) [11]. However, because in our case the criteria functions may have several extremes, there is no hope for such a simple solution.

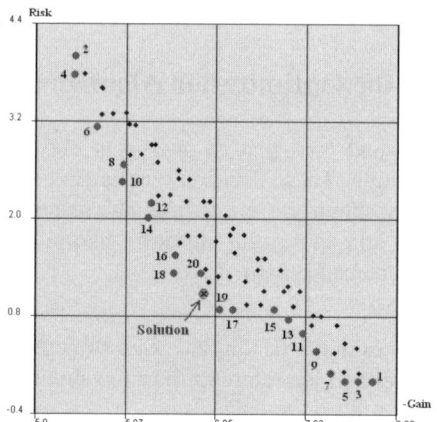

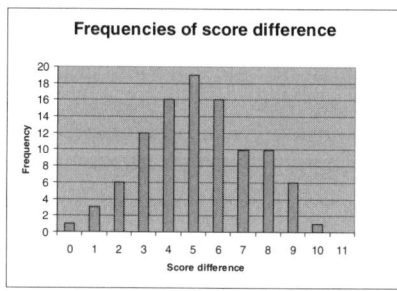

Fig. 3. The criteria space. Numbers at the points in the Pareto set show the elimination order.

Fig. 4. A histogram of the score difference in 100 games

The way out has been proposed in our recent work [1], where a method for searching the balanced optimal point in the finite Pareto set was presented. This method is based on the sequential elimination of the poorest alternatives using just one criterion at a time. With N alternatives in the Pareto set, it requires $N-1$ steps. The criterion for the elimination on each step is selected randomly with the probability proportional to the weight of this criterion. Hence more important criteria are being applied more frequently. The sole remaining option after $N-1$ steps is the result of this optimization. This method works for any non-convex and even disconnected Pareto set. Its computational complexity is $O(N^2)$.

In this application, we have further simplified the decision making procedure by assuming that all criteria have equal importance. Thus instead of randomly selecting the criteria on each step of elimination, our procedure is looping through the criteria in the deterministic order.

If the total number of the alternatives is too small, this would result in only near-optimal decision. Better balancing of the conflicting criteria is possible with increased N. So we propose to estimate the available computational time in current simulation cycle and select larger N if time permits. This optimization algorithm is scalable indeed. It is also robust, because even with small N the decisions returned by it are still fairly good.

Although we actually have five optimality criteria, for the purpose of illustration we have aggregated them all in just two: the *Risk*, which is combination of criteria 2 and 3, and *Gain* which aggregates criteria 1, 4, and 5. The signs of the individual criteria in these aggregates were chosen so that both Risk and -Gain must be minimized.

Figures 2 and 3 illustrate the configuration of the Pareto set in the decision and criteria space, respectively. Of the total of 21 points in the Pareto set 20 are eliminated as shown in Figure 3; the remaining point is the sought solution. Note that the Pareto frontier is non-convex.

The optimal point is reachable and is located at less than the maximal distance of the reference position. It is lying on the way towards the opponent goal and far away from the predicted positions of the two threatening opponents, yellow #10 and #6. This point is open for receiving the pass from the anticipated interception point. This is indeed a well-balanced solution to the positioning problem for the red player #8. With non-aggregated five criteria we can only expect even better decisions.

5   Experimental Results and Conclusion

We have conducted experiments with the purpose to estimate the sole contribution of the proposed method for the lower-level optimized player positioning compared with only strategic, higher-level positioning.

Measuring the player performance using existing RoboCup teams is difficult because new features always require careful fine tuning with the existing ones. For this reason, we decided to compare two very basic simulated soccer teams. The only difference was that the experimental team had player positioning on two levels and the control team just on one level. Players in both teams had rather good direct ball passing and goal scoring skills and no dribbling or holding the ball at all. Thus any player, once gaining the control of the ball, was forced to immediately pass it to some teammate. In this setting, the ball was rolling freely more than 95 per cent of the time, thus providing ideal conditions for evaluating the proposed method.

To further isolate the effects of imperfect sensors, we decided to use *Tao of Soccer*, the simplified soccer simulator with complete information about the world; it is available as the open source project [12]. Using the RoboCup simulator would require prohibitively long running time to sort out the effects of improved player positioning among many ambiguous factors.

The higher-level player positioning was implemented similar to used in *UvA Trilearn* [4]; this method proved to be reasonably good indeed. Assuming that both goals are lying on x-coordinate axis, the coordinates of the reference position for *i*-th player are calculated as follows:

$$x_i = w*xhome_i + (1-w)*xball + \Delta x_i,$$
$$y_i = w*xhome_i + (1-w)*yball, \tag{2}$$

where w is the weight ($0<w<1$), ($xhome_i$, $yhome_i$) and ($xball$, $yball$) are the fixed home and the current ball positions respectively, Δx_i is the fixed individual adjustment of x-coordinate whose sign differs for the offensive and defensive situations and the player role.

Because players in the control team were moving to the reference positions without any fine tuning, ball passing opportunities were occurring as a matter of chance. In the experimental team, rather, players were creating these opportunities on purpose.

The team performance was measured by the game score difference. Figure 4 shows the histogram based on 100 games each 10 minutes long.

Only one game has ended in a tie; in all the rest 99 games the experimental team won. The mean and the standard deviation of the score difference are 5.20 and 2.14, respectively. By approximating with Gaussian distribution, we get 0.9925 probability of not loosing the game. The probability to have the score difference is greater than 1 is 0.975 and greater than 2 is 0.933. This gives the idea of the potential contribution of the low-level position optimization. With the smaller proportion of the time when the ball is rolling freely, this contribution will decrease. So teams favoring ball passing would likely benefit from our method more than teams that prefer dribbling.

The experimental results demonstrate that, by locally adjusting their positions using the proposed method, players substantially contribute to the simulated soccer team performance by scoring on the average about five extra goals than the opponent team that does not have this feature. This confirms that optimized player positioning in the simulated soccer is the critical factor of success. Although this method has been developed for simulated soccer, we did not rely much on the specifics of the simulation league. Therefore, we believe that the main ideas presented in this work could be reused with minor modifications in other RoboCup leagues.

Acknowledgements

The authors would like to thank Peter Stone and Marian Lekavy for their help with identifying the early studies of the player positioning methods.

References

1. Kyrylov, V.: Balancing Rewards, Risks, Costs, and Real-Time Constraints in the Ball Passing Algorithm for the Robotic Soccer. In: Lakemeyer, G., Sklar, E., Sorrenti, D.G., Takahashi, T. (eds.) RoboCup 2006: Robot Soccer World Cup X. LNCS (LNAI), vol. 4434. Springer, Heidelberg (2007)

2. Stone, P., Veloso, M., Riley, P.: The CMUnited-98 Champion Simulator Team. In: Asada, M., Kitano, H. (eds.) RoboCup 1998. LNCS (LNAI), vol. 1604, pp. 61–76. Springer, Heidelberg (1999)

3. Reis, L.P., Lau, N., Oliveira, E.C.: Situation Based Strategic Positioning for Coordinating a Team of Homogeneous Agents. In: Hannebauer, M., Wendler, J., Pagello, E. (eds.) ECAI-WS 2000. LNCS (LNAI), vol. 2103. Springer, Heidelberg (2001)

4. De Boer, R., Kok, J.: The Incremental Development of a Synthetic Multi-Agent System: The UvA Trilearn 2001 Robotic Soccer Simulation Team. Master's Thesis. University of Amsterdam (2002)

5. Kok, J., Spaan, M., Vlassis, N.: Multi-Robot Decision Making Using Coordination Graphs. In: Proceedings of the International Conference on Advanced Robotics (ICAR), Coimbra, Portugal, pp. 1124–1129 (June 2003)

6. Andou, T.: Refinement of Soccer Agents' Positions Using Reinforcement Learning. In: Kitano, H. (ed.) RoboCup 1997. LNCS, vol. 1395, pp. 373–388. Springer, Heidelberg (1998)

7. Nakashima, T., Udo, M., Ishibuchi, H.: Acquiring the positioning skill in a soccer game using a fuzzy Q-learning. In: Proceedings of IEEE International Symposium on Computational Intelligence in Robotics and Automation, July 16-20, 2003, vol. 3, pp. 1488–1491 (2003)

8. Kalyanakrishnan, S., Liu, Y., Stone, P.: Half Field Offense in RoboCup Soccer: A Multi-agent Reinforcement Learning Case Study. In: Lakemeyer, G., Sklar, E., Sorrenti, D.G., Takahashi, T. (eds.) RoboCup 2006: Robot Soccer World Cup X. LNCS (LNAI), vol. 4434. Springer, Heidelberg (to appear, 2007)

9. Beim, G.: Principles of Modern Soccer. Houghton Mifflin Company, Boston (1977)

10. Vogelsinger, H.: The Challenge of Soccer. Allyn and Bacon, Inc., Boston (1973)

11. Miettinen, K.: Nonlinear Multiobjective Optimization. Kluwer Acaemic Publishers, Berlin (1998)

12. Zhan, Y.: Tao of Soccer: An Open Source project (2006), https://sourceforge.net/projects/soccer/

13. Fisher, R.A., Bennett, J.H.: Statistical Methods, Experimental Design, and Scientific Inference. Oxford University Press, Oxford (1990)

Rational Passing Decision Based on Region for the Robotic Soccer

Xu Yuan and Tan Yingzi

School of Automation, Southeast University, Nanjing, China
xuyuan.cn@gmail.com, tanyz@seu.edu.cn

Abstract. For an agent to behave appropriately in an uncertain environment, efficient representation of knowledge and reliable reasoning mechanisms are at the core of design. This paper proposes a novel region based passing scheme for the robotic soccer. The scheme captures qualitative knowledge of soccer in a natural and efficient way. We implemented the *rational passing decision based on region*(RPDR) in our RoboCup simulation 3D team. Experiments show that our method outperforms the base line method, i.e. position searching approach.

1 Introduction

For an agent to behave appropriately in an uncertain environment, efficient representation of knowledge and reliable reasoning mechanisms are at the core of design. To date, researchers have proposed a top-down approach to model soccer knowledge[1], which was the basic motivation of this paper.

An important decision for a soccer robot to make is which action performs when it has control of the ball. For a given situation, the soccer robot may pass, dribble, shoot or clear. The decision on how to perform a pass can be an especially crucial one, since passing is the only cooperating action in a soccer game. In fact, dribbling, shooting and clearing can be thought as passing the ball to itself, to the goal, and to a safe area respectively. To properly devise a passing decision, previous research focuses on the use of searching[3,4,9,11], learning[7] and coordination[2] algorithms, but are not entirely satisfactory.

In this paper, we present passing decision theory based on regions in the soccer field. We propose a particular passing decision algorithm, that sets passing to regions instead of passing to position. This improves the passing quality given the constraints of its passing skill. Experiments show that our method outperforms position searching approach. The advantages of our method are: 1) it has a low computational complexity; 2) it does not need precise information; 3) it balances the success and reward of passing.

The paper is organized as follows. In section 2, we briefly describe the basic concepts of passing problem in RoboCup and related work of passing decision. The proposed algorithm is described in section 3. Section 4 discusses how the functionality and efficiency of our algorithm is experimentally validate, followed by conclusion and future work in section 5.

U. Visser et al. (Eds.): RoboCup 2007, LNAI 5001, pp. 238–245, 2008.

2   Ball Passing Decision in RoboCup

If the robot has already decided to pass the ball, two problems need to be addressed: 1) passing decision; 2) passing skill. The latter is an intrinsic design of a soccer robot. Using the kicking device, a passing instruction defines how the robot kicks the ball from its current position to a given *destination position* such that it shoots over a blocking player but is still suitable for receiving when it reaches the destination. We represent a pass as $Pass(p)$ where p stands for the *destination position*. Before the passing is implemented, the passing decision is essential.

2.1   Passing Decision

In soccer theory, good passing makes it difficult for the opponent's team to defend, and is essential for good tactics[5]. To achieve this, one will need to solve a problem which will be referred to as the **optimal passing decision** problem and can be stated as follows: *find the position in the field where the outcome is the highest when the ball is passed to this position in a given situation.*

Unfortunately, the problem depends on many uncertain factors, making it impractical to solve. For example, the opponent's movement is an essential factor, but can never be exactly predicted. The presence of uncertainty changes radially the way in which an agent makes decisions[12]. The right thing to do, the **rational passing decision**, therefore, *depends on both the likelihood of a successful pass and the rewards of passing.*

2.2   Related Works of Passing Decision

However, making a rational passing decision is not a straightforward task. The principle difficulty arises from the fact that such a problem consists of many different elements, which have to be operated together in an appropriate way. Previous research focuses on the use of searching[3,4,9,11], learning[7] and coordination[2] algorithms to address the different elements. Although these methods are useful, the results are not entirely satisfactory. Reasons are as follows.

- The total number of possible passing positions is extremely large. There is too much work to list the complete set of evaluation functions, and too hard to use the numerous rules to produce result. For example, to compute probabilities of passes it is necessary to do thousands of interception time calculations[9].
- The performances of these methods are extremely dependent on the quality of their evaluation functions. But some evaluation functions need precise information, which is not possible in the RoboCup domain.
- It is not possible to design a single decision rule suitable for all situations, since different variables can be dominant for different situations. For example, a player making the same decisions while playing with teams of different styles will result in a bad outcome.

3   Rational Passing Decision Based on Region

The word "region" often appears in soccer theory books[5], such as *forward region* and *scoring region*. A **region** is defined as a set of positions with the same attribute. We simplify the passing decision problem by replacing the position with a region, in which the positions' outcome are all high. However, the passing skill only accept the position as its argument as mentioned above. We define passing the ball to a region as passing the ball to the centroid of the region. Passing the ball to the centroid has the lowest probability of having the ball fall out of the designated region. As a result, replacing position with the region is more likely to overcome the influence of uncertainty and represent the situation on the field well.

In the ball passing domain, regions can be further classified into different categories with different attributes: *tactical region, dominate region, passable region,* and *falling region.*

Tactical Region: The field is divided into different tactical regions, which indicate different tactical rewards and strategy purposes. We can set the priorities of tactical regions with tactical rewards according to soccer theory. Additionally, the on-line coach can also change tactical rewards according to the situation on the field. Tactical regions derived from [5] can be seen on Figure 1.

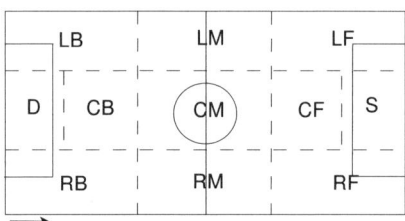

Fig. 1. Tactics regions on the field. The field is divided into eleven regions. Three sides: left(L), center(C), right(R); three rows: back(B), midfield(M), forward(F); and special regions: scoring(S), danger(D).

Dominant Region: A dominant region is defined as a region such that the player can arrive earlier than any other player[13]. The domain region of player n can be defined as follows:

$$\boldsymbol{Rd_n} = \{\boldsymbol{p} \mid \boldsymbol{p} \in \boldsymbol{R}, \mathrm{T}(n, \boldsymbol{p}) < \min_{i \in \boldsymbol{N} \setminus n} (\mathrm{T}(i, \boldsymbol{p}))\} \qquad (1)$$

where $\mathrm{T}(i, \boldsymbol{p})$ returns the shortest time for player i to move to position $\boldsymbol{p}$, $\boldsymbol{R}$ is the whole soccer field, $\boldsymbol{N}$ corresponds the set of every player's number. If the ball is in the player's dominant region, he is able to reach the ball before any other player can since he is fastest.

Passable Region: The player with the ball is limited by only passing the ball to the passable region. The reason for this is that the player can only kick the ball with limited power and direction. Furthermore, the opponent defenders may block part of the passing angle. The passing region is defined as follows:

$$Rp = \{p \mid p \in R, |p_b - p| < d_{max}, \Phi(p_b - p) \in \Phi_p \cap \overline{\Phi_b}\} \qquad (2)$$

where p_b is the position of ball, d_{max} is for maximum passing distance, $\Phi(p)$ returns the direction of vector p, Φ_p stands for the directions the ball can be kick to, and Φ_b stands for the directions which are blocked by other players.

Falling Region: Because of noise of the kick device and uncertainty of the ball movement, the balls which are kicked in the same initial situation and the same kick force may fall in different positions. In *RoboCup Simulation 3D Server*[8], the virtual kick device has three kinds normally distributed noise: horizontal angle error, latitudinal angle error and power error. The falling region is defined as follows:

$$Rf = \{p \mid |\,|p_b - p| - |p_b - p_g|\,| < D_e(|p_b - p_g|), |\Phi(p_b - p) - \Phi(p_b - p_g)| < \phi_e)\} \qquad (3)$$

where p_g is the position of destination, $D_e(d)$ denotes the distance error while passing distance is d, and ϕ_e stands for the direction error of passing.

The intersection of the *falling region*, *dominate region* and *passable region* denotes the likelihood of a successful pass. Within these regions, we propose the **Rational Passing Decision based on Region** as follows:

1. Figure out different regions;
2. Calculate the intersection of regions;
3. Select the destination region.

The pseudo code is given in Algorithm 1. We use a polygon as an approximate alternative to a region since the polygon holds the same attribute within the region. All required algorithms can be found in computational geometry books[6] such as computing the intersection of polygons.

The region captures qualitative knowledge of passing in a natural and efficient way. First, it has a low computational complexity. The time complexity of our algorithm is only in reference to the number of players, because we use regions instead of positions. Second, the regions are rarely influenced by the noise of perception and passing to a region allows larger kicking errors than passing to a position allowing the agent to reach rational decisions without precise information given unexpected noise in perception and action. Lastly, the selected region's area indicates the likelihood of success and its location denotes the tactical reward of passing. The large region with a high reward is favored thus it balancing the success and reward of passing.

4   Experiments

In this section we report our empirical results. In the early stage, we applied the RPDR to the keepaway[10] problem. Based on these experiments, we implemented the RPDR in our RoboCup soccer simulation 3D team—SEU-3D.

Algorithm 1. Pseudo-code of Rational Passing Decision based on Region

---

Define: $A(R)$ return the area of region R
Define: $C(R)$ return the centroid of region R
Define: n the passer number
Define: Rt_i the ith tactical region on the field
Define: $reward_i$ the tactical value of ith tactical region on the field, it is set according to tactics in soccer theory and situation on the field
Define: $\mathcal{RT} = \{Rt_1, \ldots, Rt_{11}\}$ the set of tactical regions order by its tactical value
$Rp \leftarrow$ the passable region of the passer
$v_m \leftarrow 0$
$R_m \leftarrow NULL$
for all Rt_i in $\mathcal{RT}$ **do**
  $Rp_i \leftarrow Rp \cap Rt_j$
  for all j such that $1 \leq j \leq 11$ and $j \neq n$ **do**
    $Rd_j \leftarrow$ the dominate region of the teammate j
    $Rp_{ij} \leftarrow Rp_i \cap Rd_j$
    if $Rp_{ij} \neq NULL$ **then**
      $Pp_{ij} \leftarrow C(Rp_{ij})$
      $Rf_{ij} \leftarrow$ the falling region when passing the ball to Pp_{ij}
      $R \leftarrow Rf_{ij} \cap Rp_{ij}$
      $v \leftarrow A(R)/A(Rf_{ij}) * reward_j$
      if $v > v_m$ **then**
        $v_m \leftarrow v$
        $R_m \leftarrow R$
      end if
    end if
  end for
end for
return $C(R_m)$

4.1   Keepaway

Keepaway is a subproblem of soccer in which one team: the keepers, tries to maintain possession of the ball within a limited region while the opposing team: the takers, attempts to gain possession[10]. The advantage of keepaway is that it is more suitable for comparing different methods directly than the full robot soccer task. We have developed a customized monitor (See Figure 2), which can set up keepaway scene in the *RoboCup Soccer Simulation 3D Server*[8].

In our experiment, the takers' behavior is specified with a fixed policy: the taker who fastest to the ball tries to clear the ball out while others run to fixed positions. In order to compare different passing decision methods, keepers have the same low-level skills and policy except the passing decision. Two different passing decision making methods are tested in the keepaway task:

Position Searching Passing Decision: Given discreet positions in the keepaway region as candidates, evaluate each position by the difference of minimal distance to takers and minimal distance to other keepers. The position with

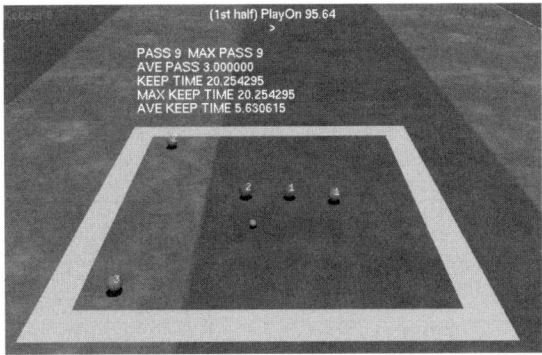

Fig. 2. A snapshot of a 3 vs. 2 keepaway episode in a 10m × 10m region. The keepers are red, takers are blue, and the ball is yellow.

the highest evaluation value is chosen. In our implementation, the sampling interval is 1m in both coordinates and the fringe positions are eliminated to avoid passing the ball out.

Rational Passing Decision based on Region: A simple version of RPDR, there is only one tactical region in the keepaway task.

Table 1. The average costing time(in milliseconds) and keeping time(in seconds) with their standard deviations for different passing decisions

Keepers	3 vs. 2		4 vs. 3	
	Costing Time	Keeping Time	Costing Time	Keeping Time
PSPD	0.0926	14.6 ± 1.1	0.2426	25.0 ± 2.9
RPDR	0.0780	17.2 ± 1.0	0.1365	36.2 ± 2.8

The keepers with different passing decisions played on the following two different keepaway tasks: 3 vs. 2 run on a 10m × 10m field, and 4 vs. 3 run on a 15m × 15m field. In each cycle, the agent makes passing decision 1000 times and records the total costing time: therefore we can get an average cost time of making a passing decision[1]. The results are summarized in Table 1. The results show that our method outperforms the other method. First, the average costing time of RPDR is less than PSPD, especially in a larger field. It is important to note that there is only one evaluation function in our test. Therefore, the costing time of PSPD with many evaluation functions is much longer than the time we tested. Second, the keeping time of RPDR is longer than PSPD meaning the players with RPDR played better.

---

[1] The server is configured to allow agents thinking so long time.

Table 2. Results and several statistics of all the games played by our team with different passing decisions. The 'Poss' column denotes the percentage of the total time during a match in which our team was in ball possession. The 'Def', 'Mid' and 'Att' respectively denote the percentage of the time in which the ball was located in our defensive, middle and attacking region of the field. The 'Pass' column denotes the total number of passing, and 'Suc.P' means the percentage of successful passing.

Opponent	Passing Decision	Score	Poss	Def	Mid	Att	Pass	Suc.P
FC Portgual 2006	PSPD	7 - 1	44%	42%	49%	9%	363	62%
	RPDR	**12 - 1**	**48%**	**21%**	**54%**	**25%**	**418**	**73%**
Wright Eagle 2006	PSPD	0 - 0	53%	11%	70%	19%	474	63%
	RPDR	**11 - 0**	**55%**	**3%**	**64%**	**33%**	**528**	**70%**
ZJU Base 2006	PSPD	2 - 0	51%	24%	50%	26%	380	65%
	RPDR	**14 - 0**	**54%**	**17%**	**50%**	**33%**	**437**	**74%**

4.2 Overall Team Results

To obtain an indication of the performance of the passing decision, we held a trial competition with our teams with different passing decisions on one side and the top three teams at *RoboCup-2006* on the other side. Our teams have the same low-level skills and policy: the only exception being the passing decision. One of our team uses RPDR while another uses the PSPD which is implemented as [3].

In this competition our two teams with different passing decisions played 10 games against each other team. The results are summarized in Table 2. Look at the results of the trial competition, it is clear that the team with the RPDR is better than the other one. First, the percentage of successful passes of RPDR is about 10% higher than the PSPD. Furthermore, the passing of RPDR is more aggressive since the ball stayed on the opponent's field longer. As a result, the RPDR team achieved more goals. Additionally, the RPDR team performs well against every team while the PSPD can not even win against a defensive team. This shows that RPDR is flexible enough to balance the success and reward of passing.

5 Conclusion and Future Work

In this paper, we have described and investigated the use of the region model for passing decision as qualitative soccer theory. As mentioned above, previous passing decision making methods are useful in RoboCup simulation teams, however, their performances are not entirely satisfactory. The region captures qualitative knowledge of passing in a natural and efficient way. After figuring out regions on the field, RPDR then chooses the final passing regions based on known regions and tactical purpose that can be found in soccer theory books. We have provided empirical evidence to show: 1) it has a low computational complexity; 2) it allows the agent to arrive at rational decisions without precise information even when there is unexpected noise in perception and action; 3) it balances the success and reward of passing. As a result of the above reasons, we believe that RPDR is a feasible approach for a passing decision in RoboCup.

For future research, we plan on implementing an on-line tactics adjusting mechanism based on the tactical regions, and applying the RPDR to the middle-size robots of our university. We would like to extend the region to other decision problems, such as dribbling and clearing. Furthermore, we wish to investigate whether other qualitative soccer theory can be applied to RoboCup.

References

1. Dylla, F., Ferrein, A., Lakemeyer, G., Murray, J., Obst, O., Rofer, T., Stolzenburg, F., Visser, U., Wagner, T.: Towards a League-Independent Qualitative Soccer Theory for RoboCup. In: Nardi, D., Riedmiller, M., Sammut, C., Santos-Victor, J. (eds.) RoboCup 2004. LNCS (LNAI), vol. 3276, pp. 29–40. Springer, Heidelberg (2005)
2. Dawei, J., Shiyuan, W.: Using the Simulated Annealing Algorithm for Multiagent Decision Making. In: Lakemeyer, G., Sklar, E., Sorrenti, D.G., Takahashi, T. (eds.) RoboCup 2006: Robot Soccer World Cup X. LNCS (LNAI), vol. 4434. Springer, Heidelberg (2007)
3. Jiang, H., Du, X.F., Luo, D.J., Zhang, Y.F., Zhou, Y.: ZJUBase3D - Team Description 2006. In: Lakemeyer, G., Sklar, E., Sorrenti, D.G., Takahashi, T. (eds.) RoboCup 2006: Robot Soccer World Cup X. LNCS (LNAI), vol. 4434. Springer, Heidelberg (2007)
4. Kyrylov, V.: Balancing Gains, Risks, Costs, and Real-Time Constraints in the Ball Passing Algorithm for the Robotic Soccer. In: Lakemeyer, G., Sklar, E., Sorrenti, D.G., Takahashi, T. (eds.) RoboCup 2006: Robot Soccer World Cup X. LNCS (LNAI), vol. 4434. Springer, Heidelberg (2007)
5. Li, M.X., Wang, G.B.: Soccer Offensive Skill in Graphics. Beijing Sport University Press (June 2004)
6. de Berg, M., van Kreveld, M., Overmars, M., Schwarzkopf, O.: Computational Geometry: Algorithms and Applications, 2nd edn. Springer, New York (2000)
7. Riedmiller, M., Gabel, T.: Brainstormers 2D Team Description 2006. In: Lakemeyer, G., Sklar, E., Sorrenti, D.G., Takahashi, T. (eds.) RoboCup 2006: Robot Soccer World Cup X. LNCS (LNAI), vol. 4434. Springer, Heidelberg (2007)
8. Koler, M., Obst, O.: Simulation League: The Next Generation. In: Polani, D., Browning, B., Bonarini, A., Yoshida, K. (eds.) RoboCup 2003. LNCS (LNAI), vol. 3020. Springer, Heidelberg (2004)
9. Stone, P., McAllester, D.: An Architecture for Action Selection in Robotic Soccer. In: Proceedings AGENTS 2001, 5th International Conference on Autonomous Agents, Montreal, Quebec, Canada, May 28-June 1, 2001, pp. 316–323 (2001)
10. Stone, P., Sutton, R.S.: Keepaway Soccer: a machine learning testbed. In: Birk, A., Coradeschi, S., Tadokoro, S. (eds.) RoboCup 2001. LNCS (LNAI), vol. 2377, pp. 214–223. Springer, Heidelberg (2002)
11. Reis, L.P., Lau, N.: FC Portugal Team Description: RoboCup 2000 Simulation League Champion. In: Stone, P., Balch, T., Kraetzschmar, G.K. (eds.) RoboCup 2000. LNCS (LNAI), vol. 2019, pp. 29–40. Springer, Heidelberg (2001)
12. Russell, S.J., Norvig, P.: Artiticial intelligence: A modern approach, 2nd edn. Prentice Hall (2003)
13. Taki, T., Hasegawa, J.-i.: Visualization of Dominant Region in Team Games and Its Application to Teamwork Analysis. cgi. In: Computer Graphics International 2000 (CGI 2000), p. 227 (2000)

Automatic On-Line Color Calibration Using Class-Relative Color Spaces*

Pablo Guerrero, Javier Ruiz-del-Solar, Josué Fredes, and Rodrigo Palma-Amestoy

Department of Electrical Engineering, Universidad de Chile
{pguerrer,jruizd,jfredes,ropalma}@ing.uchile.cl

Abstract. In this article we present an automatic on-line color calibration system that makes extensive use of the spatial relationships between color classes in the color space. First, we introduce the definition of class-relative color spaces, where classes are represented in terms of their spatial relation to a base color class. Then, using class-relative color spaces, the system is able to remap classes from the already trained ones, which gives a starting point for training the remaining classes. The color-calibrating system also uses a feedback from the detected objects using the remapped (or partially trained) classes. As a result, the system is able to generate a complete color look-up table from scratch, and to adapt quickly to severe lighting condition changes. A particularity of our system is that it does not need to solve the natural ambiguity in color classes' intersections, but it is able to keep and use it during color segmentation using the concept of soft-colors.

1 Introduction

In the RoboCup Four Legged League, as in most of the RoboCup soccer leagues, objects are specifically colored to allow robots to recognize them easily. Most of the employed vision systems use color segmentation to take advantage of the color information. There several approaches for segmenting colors in real time [5], a common one is based in the use of a 3D look-up table (LUT). When the robot operates in a controlled environment having fixed lighting conditions, a fixed LUT performs well, and thus it can be trained off-line. However, this approach has two main flaws: a lot of time is needed for a human to calibrate the LUT, and the operation of the robot is strictly limited to artificial environments with highly controlled illumination. In this context, the development of automatic or adaptive color calibration systems has been intensely treated by the RoboCup community in the last years. The motivation is very clear: RoboCup is supposed to increasingly move to more realistic game conditions, which of course include natural lighting.

The presented work proposes an automatic on-line color calibration system, which allows the robot to build a LUT on-line, and to adapt it quickly to severe lighting condition changes. To our knowledge, the paper is innovative in three aspects: (i) spatial relationships between color classes are used in a very general fashion, which

---

* This research was partially supported by FONDECYT (Chile) under Project Number 1061158.

U. Visser et al. (Eds.): RoboCup 2007, LNAI 5001, pp. 246–253, 2008.

allows the system to be easily adapted to any other application with different objects and/or color classes, (ii) the intersections of the color classes (also called *soft-colors* in the literature) are also automatically trained and stored in the LUT, and (iii) non isotropic illumination is considered, and an automatic training procedure is proposed.

2 Related Work

A main stage in any automatic color calibration system is the extraction of pixels of colored objects to train the LUT or the classes' statistics. To obtain these pixels, some of the published approaches relay on the knowledge of the objects' shape and on the pose of the robot, and thus, on the relative positions of the fixed objects with respect to the robot [6][2]. Some other systems use scan lines and predefined transition rules based on simple spatial relations between color classes in the color space (for example: "cyan has a higher U component than green") [1]. Other systems use a priori membership distribution to track the classes' statistics by means of the EM algorithm [3]. Some of the proposed approaches make use of incremental layers or estimations of the color classes [1], where coarse layers are used to extract pixels that are used to train more precise layers. Regarding color information representation, there are several ways to represent color classes. The following are examples of proposed class representations, sorted by complexity and flexibility: cuboids [1], non-rotated ellipsoids (mean and uncorrelated variances of color components) [6], union of rotated ellipsoids (Gaussian Mixture) [3], and hybrids that bounds in different ways the different color coordinates [4]. Most of them do not allow intersection between classes so ambiguity must be solved before filling the LUT, attempting to minimize the expected classification error. Additionally, color constancy approaches (e.g. [8]) propose the existence of transformations or mappings, in a determined color space, that describe what happen to colors of an image when the lighting changes. If one applies such an approach literally to the color segmentation problem, one could preprocess each image, and get a transformed one that should be easily segmented with a previous LUT. But, this transformation should be applied to each pixel, which is a prohibitive task in real-time robotics. However, we are using the color space mapping idea to propose the remapping applied to color classes instead of pixels, which is a task that can be performed in real time.

3 Proposed Approach

The proposed automatic color calibration system starts its operation with the extraction of pixels corresponding to color classes that can be trained with a total lack of a priori knowledge. These color classes are green and white in our application (in the RoboCup soccer environment the field's lines and carpet can be detected without using color information), but from a more general point of view, they can be colors of any objects that can be detected without use of color information. The extracted classes' statistics are used in combination with a priori knowledge of the spatial relationships among the color classes to remap the rest of the classes. This remapped color classes are then used for the on-line detection of objects. Then, the system takes feedback from the detected objects to extract pixels of the respective classes, and makes a smooth transition from the *remapped estimations* of the color classes to *trained estimations* of the classes.

3.1 Basic Definitions: Colors, Color Classes and Color Classes Representation

The system works in the YUV color space because the AIBO camera takes images in this format. A point in the YUV color space will be named a *color*. A *color class* is a set of colors that can be observed in pixels corresponding to an object or an object part having a given human-defined color. For example the class "yellow" is the one that contains pixels belonging to a yellow goal or a yellow part of a beacon. The set of color classes Ω is defined by the application. As discussed in [6] (even when they chose a simpler representation), we have found that a correlated 3D Gaussian is enough to represent a color class. Thus, we have chosen to represent each class $\mathbf{K} \in \Omega$ by a mean and a covariance in the YUV space, $(\boldsymbol{\mu}_\mathbf{K}, \boldsymbol{\Sigma}_\mathbf{K})$. An *innovation threshold* $\lambda_\mathbf{K}$ is used to determine when a color belongs to any class $\mathbf{K}$:

$$\mathbf{K} = \left\{ \mathbf{c} \in [0,255]^3 \big/ (\mathbf{c} - \boldsymbol{\mu}_\mathbf{K})^T \boldsymbol{\Sigma}_\mathbf{K}^{-1} (\mathbf{c} - \boldsymbol{\mu}_\mathbf{K}) < \lambda_\mathbf{K} \right\} \tag{1}$$

Note that the size of a class is determined by its covariance matrix $\boldsymbol{\Sigma}_\mathbf{K}$ and its innovation threshold $\lambda_\mathbf{K}$. This class representation corresponds to an ellipsoid in the YUV space with possibly rotated axes and different radiuses. A value of $\lambda_\mathbf{K} = 10$ is found to be optimal when the class statistics are reasonably well estimated.

When lighting conditions change, color classes change their position and size in the color space. This makes a fixed color's LUT inapplicable in those situations. However, even when the lighting condition change drastically, and the color classes suffer severe modifications, some spatial relationships between them in the color space remain unaltered. Thus, given a color class $\mathbf{K}$, we can define a *K-relative color space* as one centered in $\boldsymbol{\mu}_\mathbf{K}$ and with a metric linearly transformed by a function of $\boldsymbol{\Sigma}_\mathbf{K}$. Any color $\mathbf{c}$ can be transformed to the $\mathbf{K}$-relative color space:

$$\mathbf{c}^\mathbf{K} = \sqrt{\boldsymbol{\Sigma}_\mathbf{K}^{-1}} (\mathbf{c} - \boldsymbol{\mu}_\mathbf{K}) \tag{2}$$

Where the square root of a matrix $\mathbf{A}$ is defined as a lower triangular matrix that satisfies: $\mathbf{A} = \sqrt{\mathbf{A}} \sqrt{\mathbf{A}}^T$. The square root is implemented using the Cholesky factorization [9]. We call $\mathbf{c}^\mathbf{K}$ the *K-relative representation* of $\mathbf{c}$. Analogously, given any color class $\mathbf{D}$, with mean and covariance $(\boldsymbol{\mu}_\mathbf{D}, \boldsymbol{\Sigma}_\mathbf{D})$, it can have its K-relative representation defined as $(\boldsymbol{\mu}_\mathbf{D}^\mathbf{K}, \boldsymbol{\Sigma}_\mathbf{D}^\mathbf{K})$, where,

$$\boldsymbol{\mu}_\mathbf{D}^\mathbf{K} = \sqrt{\boldsymbol{\Sigma}_\mathbf{K}^{-1}} (\boldsymbol{\mu}_\mathbf{D} - \boldsymbol{\mu}_\mathbf{K}) \; ; \; \boldsymbol{\Sigma}_\mathbf{D}^\mathbf{K} = \sqrt{\boldsymbol{\Sigma}_\mathbf{K}^{-1}} \boldsymbol{\Sigma}_\mathbf{D} \sqrt{\boldsymbol{\Sigma}_\mathbf{K}^{-1}}^T \tag{3}$$

Note that in particular, $(\boldsymbol{\mu}_\mathbf{K}^\mathbf{K}, \boldsymbol{\Sigma}_\mathbf{K}^\mathbf{K}) = (0, \mathbf{I})$, with $\mathbf{I}$ the 3x3 identity matrix.

3.2 Off-Line Training

In the proposed system color classes are trained manually using a procedure as the one described in [7]. The statistics $\left(\mu_D^K, \Sigma_D^K\right)$ are calculated and stored for every pair of classes $(K, D) \in \Omega^2$. This procedure is intended so that the system learns the spatial relationships between classes, and it is needed to be carried out it only once for a determined set of color classes. This is why we associate this procedure to the one when a human learns the colors for the first time. We have found that the system is robust enough to small changes in the actual colors of the objects (for example, the color of the carpet).

3.3 On-Line Operation

Our system maintains two estimations of the color classes, a *remapped estimation* and *trained estimation* (see explanation in the next sections). These two estimations are combined to obtain the resulting estimation that is used to fill a LUT. This resulting LUT is the output of the system. Fig. 1 shows the system's components and the information flow. In the next sections the different modules will be explained.

Pixels Extraction and Statistics Calculation: In this stage, acquired images are used to extract pixels from detected objects. A fixed maximum number of pixels colors are stored for each color class. The number of stored colors is selected to ensure that enough images are considered (approximately 10 images, with a mean number of extracted pixels per image of ~200 green pixels and ~40 for the rest of the classes). When necessary, oldest colors are rewritten by the newest ones. Green and white are extracted using scan lines. Scan lines are perpendicular to the horizon line and the scan is performed similarly as described in [1], but following upwards direction. For the sake of brevity, we will not describe in detail this procedure since it is not the focus of the paper. When using this procedure, it is not necessary to have a priori knowledge about the lighting conditions to extract green and white pixels because the visual sonar is based on Y channel transitions, thus we call green and white *self-sufficient classes*. This is why the visual sonar is the starting point of the system. Yellow, cyan, pink, orange, red and blue (or other colors in the case of applications different than RoboCup soccer) are extracted from detected objects having the corresponding color class. Of course, to detect these objects it is necessary to have a previous estimation of those color classes, which is not possible when the system starts or when the lighting conditions change. That is why we call these classes *dependent*.

Pixels selected to train a class are filtered (using (1), but considering a higher innovation threshold) according to their innovation with respect to the resulting estimation, to prevent outliers from damaging classes' statistics. Trained classes' statistics are recalculated after each image is processed using the stored colors. This recalculation is implemented in an efficient incremental fashion.

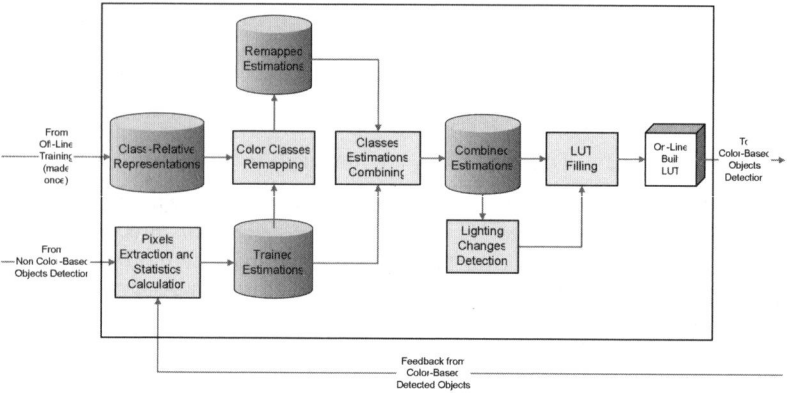

Fig. 1. Block diagram of the system. Trained and remapped color classes estimations are combined to get a resulting estimation which is used to fill the resulting LUT. The system is able to completely train and adapt the resulting LUT having prior knowledge of the spatial relationships between colors.

Color Classes Remapping: As discussed in [8], a linear mapping is not enough to cope with the possible color space transformations that may appear when lighting changes, but, one could locally approximate such a mapping with a linear transformation. We present a statistic method for remapping the non-trained color classes. To overcome the lack of extracted pixels for the dependent classes, we make use of the class-relative color spaces to create a first approximation of them. If any class $\mathbf{K}$ is already trained, a remapped estimation of any other class $\mathbf{D}$ can be obtained from its $\mathbf{K}$-relative representation and the $\mathbf{K}$ trained estimation:

$$\boldsymbol{\mu}_{\mathbf{D},r}^{\mathbf{K}} = \boldsymbol{\mu}_{\mathbf{K},t} + \sqrt{\Sigma_{\mathbf{K},t}}\,\boldsymbol{\mu}_{\mathbf{D}}^{\mathbf{K}} ; \; \Sigma_{\mathbf{D},r}^{\mathbf{K}} = \sqrt{\Sigma_{\mathbf{K},t}}\,\Sigma_{\mathbf{D}}^{\mathbf{K}}\sqrt{\Sigma_{\mathbf{K},t}}^{T} \tag{4}$$

When there is more than one trained class, the system uses *remapping weights* $\beta_{\mathbf{D}}^{\mathbf{K}}$ to determine how relatively important is the $\mathbf{K}$-relative color space to remap the class $\mathbf{D}$. Remapping weights are calculated as:

$$\beta_{\mathbf{D}}^{\mathbf{K}} = \frac{\alpha_{\mathbf{K}}\left\|\boldsymbol{\mu}_{\mathbf{D}}^{\mathbf{K}}\right\|^{-1}}{\displaystyle\sum_{\mathbf{K}'\in\Omega/\{\mathbf{D}\}}\alpha_{\mathbf{K}'}\left\|\boldsymbol{\mu}_{\mathbf{D}}^{\mathbf{K}'}\right\|^{-1}} ; \; \alpha_{\mathbf{K}} = \frac{N_{\mathbf{K}}}{T_{\mathbf{K}}} \tag{5}$$

Where $T_{\mathbf{K}}$ is the maximum number of extracted pixels for the class $\mathbf{K}$. Then, the remapped estimation of $\mathbf{D}$ is calculated as:

$$\boldsymbol{\mu}_{\mathbf{D},r} = \sum_{\mathbf{K}'\in\Omega/\{\mathbf{D}\}}\beta_{\mathbf{D}}^{\mathbf{K}'}\boldsymbol{\mu}_{\mathbf{D},r}^{\mathbf{K}'} ; \; \Sigma_{\mathbf{D},r} = \sum_{\mathbf{K}'\in\Omega/\{\mathbf{D}\}}\beta_{\mathbf{D}}^{\mathbf{K}'}\Sigma_{\mathbf{D},r}^{\mathbf{K}'} \tag{6}$$

Every class is remapped in function of the existing trained colors. This remapped estimation is stored to be combined with the trained estimation, if it already exists.

Classes Estimations Combining: A linear combination of remapped and trained estimations is used to get the resulting color class estimation:

$$\mathbf{\mu_D} = \alpha_D \mathbf{\mu}_{D,t} + (1-\alpha_D)\mathbf{\mu}_{D,r}; \ \mathbf{\Sigma_D} = \alpha \mathbf{\Sigma}_{D,t} + (1-\alpha)\mathbf{\Sigma}_{D,r} \tag{7}$$

The use of α_D allows a smooth transition from the use of the remapped estimation of the class (when no pixel has yet been trained) to the use of the trained estimation (when the maximum number of pixels has already been trained). This smooth transition has the objective of avoiding mistakes in the association of training pixels to partially trained classes.

LUT Filling: The LUT is filled when any of the classes' resulting estimation moves enough, from the used to build the current LUT, to make it obsolete. The LUT filling is efficiently implemented: For each pair (Y,U), the two solutions V_1 and V_2 of the quadratic equation $(\mathbf{c}-\mathbf{\mu_K})^T \mathbf{\Sigma_K^{-1}} (\mathbf{c}-\mathbf{\mu_K}) = \lambda_K$ are calculated, with $\mathbf{c} = (Y,U,V)$. If $V_1, V_2 \in \mathbb{R}$, the LUT is filled in the (Y,U) row, from V_1 to V_2 with class $\mathbf{K}$.

4   Results

We have tested our autonomous calibration system in real AIBO image sequences, with both the robot and its camera moving and partially controlled lighting conditions. To illustrate how the system creates a new LUT from scratch, figure 2 shows important events in the color calibration process, and how the segmentation improves as new images are processed. The whole sequence corresponds to a half turn of the robot around itself (~2 sec). From testing the system in several image sequences as the shown in fig. 2, we have concluded that the system is able to completely train a LUT from scratch.

   We compare the performance of the proposed method with Adaptive Color Distribution Transformation (ACDT) [10]. Fig. 3 shows the evolution of the *correctly classified pixel rate* (CCPR) over an image sequence[1]. The CCPR corresponds to the rate of pixels correctly classified inside the regions of the image occupied by actual objects. As can be seen from the curves, the system performs very similar to ACDT (CCPR≈40%) when the off line stage was trained in a different environment (UChile Lab). When the offline stage is performed in the same environment, the performance of the system is noticeably superior (CCPR≈55%).

   Processing time is a very relevant issue in mobile robotics systems, and even more when having limited processing power. Thus, we limit the frequency in which each of the operations is executed. This limitation is flexible and it is possible to balance the reactivity of the system versus the demanded processing time.

---

[1] The image sequence and its correspondent ground truth information was downloaded from http://www.dis.uniroma1.it/~spqr/cms/

Fig. 2. Example pictures from a video sequence obtained while the robot is making calibration from scratch (above), and the correspondent segmented images using the LUT obtained up to that moment (bellow). Some relevant events are the first detections of: the blue goal (left), the ball (center), and the pink yellow beacon (right).

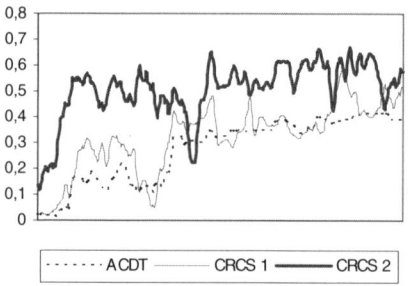

Fig. 3. Correctly Classified Pixel Rate evolution over a 394 image sequence for: Adaptive Color Distribution Transformation (ACDT), Class-Relative Color Spaces with an off-line stage in a different environment, with different illumination (CRCS1), and Class-Relative Color Spaces with an off-line stage in the same environment, with the same illumination (CRCS2)

Table 1. Processing times for each stage of the system

Stage	Frequency	Mean processing time (AIBO)
Training Classes	10Hz	6ms
Remapping Classes	0.5Hz	1.6ms
Combining Estimations	0.5Hz	0.5ms
Filling the LUT	~0.1Hz	26ms
Total time	1Hz	70ms

To convince the reader that the system is able to work on-line, table 1 shows the processing time consumed by each of the stages of the process and the frequency in which each of them is performed. The total time is presented with a frequency of 1Hz because the operations are not performed at the same time, so the presented total time is a mean over 1 second period. The presented processing times are measured in an AIBO CPU (64bit RISC, 576 MHz, Aperios). It is important to note that the system can be executed in real time over an AIBO CPU because, if necessary, some frequencies could be further reduced without a noticeable impact on the performance of the system, assuming that the lighting conditions will not change too often. With no frequency limitations, the entire process takes approximately 35ms, which is not good enough to play soccer but allows the robot to get a good LUT as quick as possible.

5 Conclusions

We have presented a novel approach for automatic calibration of a color segmentation system. Although the system is applied for a specific RoboCup soccer league, the presented framework is general enough to be used in other soccer leagues, and in other applications having any reasonable set of color labels. As is shown in the results section, the system is able to work online and to completely train a LUT from scratch. However, there are several efficiency improvements that may be achieved as, for example, to perform the LUT filling only for the needed classes. Also, we are planning to make our software architecture disconnect the automatic color calibration when time demanding tasks, as pursuing the ball, are being performed.

Acknowledgments. We would like to thank Luca Iocchi for kindly providing us the image sequence and the code for testing it with his method.

References

1. Jüngel, M.: Using Layered Color Precision for a Self-Calibrating Vision System. In: Nardi, D., Riedmiller, M., Sammut, C., Santos-Victor, J. (eds.) RoboCup 2004. LNCS (LNAI), vol. 3276, pp. 209–220. Springer, Heidelberg (2005)
2. Heinemann, P., Sehnke, F., Streichert, F., Zell, A.: Towards a Calibration-Free Robot: The ACT Algorithm for Automatic Online Color Training. In: Lakemeyer, G., Sklar, E., Sorrenti, D.G., Takahashi, T. (eds.) RoboCup 2006: Robot Soccer World Cup X. LNCS (LNAI), vol. 4434. Springer, Heidelberg (2007)
3. Anzani, F., Bosisio, D., Matteucci, M., Sorrenti, D.: On-Line Color Calibration in Non-Stationary Environments. In: Bredenfeld, A., Jacoff, A., Noda, I., Takahashi, Y. (eds.) RoboCup 2005. LNCS (LNAI), vol. 4020, pp. 396–407. Springer, Heidelberg (2006)
4. Gönner, C., Rous, M., Kraiss, K.: Real-Time Adaptive Colour Segmentation for the RoboCup Middle Size League. In: Nardi, D., Riedmiller, M., Sammut, C., Santos-Victor, J. (eds.) RoboCup 2004. LNCS (LNAI), vol. 3276, pp. 402–409. Springer, Heidelberg (2005)
5. Bruce, J., Balch, T., Veloso, M.: Fast and Inexpensive Color Image Segmentation for Interactive Robots. In: Proceedings of the 2000 IEEE/RSJ International Conference on Inteligent Robots and Systems (IROS 2000), vol. 3, pp. 116–122 (2000)
6. Sridharan, M., Stone, P.: Towards Eliminating Manual Color Calibration at RoboCup. In: Bredenfeld, A., Jacoff, A., Noda, I., Takahashi, Y. (eds.) RoboCup 2005. LNCS (LNAI), vol. 4020, pp. 673–681. Springer, Heidelberg (2006)
7. Palma, R., Guerrero, P., Ruiz del Solar, J.: Context-Dependent Color Segmentation for AIBO Robots. In: 3rd IEEE Latin American Robotics Symposium – LARS 2006 (CD Proceedings), Santiago, Chile, October 26 - 27 (2006)
8. Tieu, K., Miller, E.: Unsupervised Color Constancy. Advances in Neural Information Processing Systems (2002)
9. Press, W., Teukolsky, S., Vetterling, W., Flannery, B.: Numerical Recipes in C, 2nd edn. Cambridge University Press (1992)
10. Iocchi, L.: Robust color segmentation through adaptive color distribution transformation. In: Lakemeyer, G., Sklar, E., Sorrenti, D.G., Takahashi, T. (eds.) RoboCup 2006: Robot Soccer World Cup X. LNCS (LNAI), vol. 4434, pp. 287–295. Springer, Heidelberg (2007)

An Application of Gaussian Mixtures: Colour Segmenting for the Four Legged League Using HSI Colour Space

Naomi Henderson, Robert King, and Richard H. Middleton

School of Electrical Engineering & Computer Science
The University of Newcastle, Callaghan 2308, Australia
{naomi.henderson,robert.king,richard.middleton}@newcastle.edu.au

Abstract. In the colour coded environment of the RoboCup 4 Legged League it is crucial to extract as much colour information as possible from an image without error. To do this requires hours of manual YUV pixel mapping and testing to ensure robustness under all possible lighting conditions. The YUV colour space is a very convenient standard for transmission of video data, but for colour classification and segmentation it suffers from being non-intuitive and sensitive to changes in lighting. Alternatively, colour classification principles can be applied in an HSI colour space; one of the convenient characteristics of the HSI colour space is that the hue value, H, represents the colour wavelength information. From this concept it is easier to separate and label colour regions in an automated process as the theoretical hue and colour wavelength relationship is known. By fitting a Gaussian model using mixtures to HSI histograms we can generate boundaries of colour classes in HSI colour space.

1 Introduction

Colour classification is the first stage of image processing in many RoboCup software systems. It is also the primary, or in many cases effectively the only, sensor for the robot. For this reason the calibration of colour information requires very careful attention. Currently manual colour calibration is still the most common process used during competition for its reliability. Manual colour calibration process involves hours of mapping image pixels to a colour key defined in a look up table (LUT), a three dimensional array referenced by Y,U and V values. Automation of the colour calibration process would have significant impact on the human involvement required during RoboCup set up. Many methods have been used to automate the colour calibration process [1] [2] but often require a routine to gain the additional information needed to make a colour decision. However, by transforming the image from the YUV colour space to the HSI colour space we can use *hue* information to separate and label regions of object colours in colour space using Gaussian mixtures. This paper proposes a compromise between the common process of manual calibration and the uncertain performance

U. Visser et al. (Eds.): RoboCup 2007, LNAI 5001, pp. 254–261, 2008.
© Springer-Verlag Berlin Heidelberg 2008

of full automation by defining colour classes as a set of tunable upper and lower boundaries in the HSI colour space. The approach still allows for testing and manual tuning whilst greatly reducing calibration time and increasing reliability of the colour information.

Section 2 of this paper presents an overview of the vision system that the classification system is applied to. Section 3 presents the stages of the HSI calibration system which involves the collection of HSI histograms and the application of Gaussian mixtures to separate hue information for the generation of upper and lower boundaries in HSI colour space. Example images of the calibration system and results are outlined in Section 4 and future work involving multi dimensional Gaussian mixtures is outlined in Section 5.

2   Background

2.1   Vision Processing

Objects of the RoboCup Soccer environment are colour coded so that they can be uniquely identified by the region of colour space they occupy. In ideal situations, i.e. bright lighting and good quality camera, the YUV values for each object would remain constant and could therefore be directly mapped to a single hard colour [3]. However in the 4LL where there are hardware limitations and varied lighting of objects due to the dynamic environment, these regions of colour space tend to overlap making regions of colour space non unique and the previously defined YUV mapping becomes invalid [4]. A *soft colour classification system* was implemented as a practical solution to classifying this overlapping colour space, it classifies these regions as a 'soft' colour and delays the colour decision making process until the entire image is processed.

The input image of the robot vision system used follows the process; *colour classification referencing a YUV LUT, blob formation, soft colour blob filtering and size modification, and object recognition*[5]. This section details the colour classification system as it is an important aspect of the base vision system. The rest of the paper is focused on the off line colour calibration process involved with building the reference LUT.

2.2   Soft Colour Classification

In the 4LL we are concerned with identifying certain landmarks and objects that are uniquely colour coded. These objects and landmarks and their colour codes are; *orange* ball, *yellow* goal and beacons, sky *blue* goal and beacons, *red* robot uniforms, *navy blue* robot uniforms, *green* field, *white* robots and field lines. Additionally black occurs in images with robots, dark shadows and dark backgrounds. Certain colours are close in any colour space (YUV, HSI, RGB, etc) and hence more difficult to separate. For the purpose of the 4LL the pairs of close colours are; red and orange, orange and yellow, and dark green and dark blue. They are of concern as a shift in colour space due to a change in lighting causes an overlap of these classified regions for this reason the soft colours: *red-orange*, *yellow-orange* and *blue-green* were introduced to classify any overlap.

A number of approaches have been made to deal with the problem of overlapping colour space; from the implementation of multiple look up tables [6] to the implementation of 'maybe colours' [7]. Our method defers colour decision making when a colour value is shared between two objects. The overlapping shade is classified as the corresponding soft colour ('either' colour) until all colour information is processed. If a soft colour is suitably overlapping true colour information then the soft colour is used for additional object information (see Figure 1). The soft colour classification system has been implemented successfully for two years [5] of competition using a manual calibration system; however the classification principles can be applied to any calibration method.

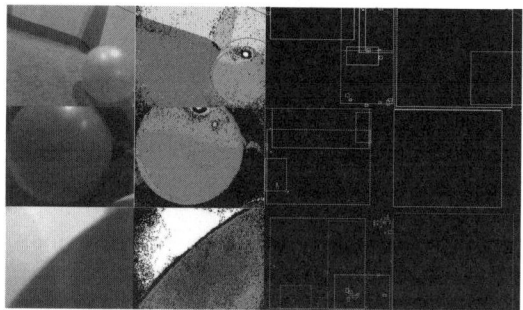

Fig. 1. Application of soft colour classification. Each row shows: Input images (taken at RoboCup 2006), colour classified image (using manually calibrated LUT), true and soft colour blobs formed and objects recognised.

2.3   YUV and HSI Colour Space

The YUV colour space is represented by; intensity (Y) and the colour information is given by chrominance values (U and V) which represent the blue-green correlation and red-green correlation. The HSI colour space is represented by; hue (H) a value that represents the predominant wavelength of colour, saturation (S) a value for the amount of colour present and intensity (I) the darkness/brightness of the colour. By transforming the YUV image to HSI we can use hue as a convenient value for separating and labeling colour keys. Additionally the HSI colour space is intuitive, this allows for three dimensional solid regions of HSI colour space to be defined as a colour class.

3   Method

The method involves a number of stages, firstly images are selected, transformed to HSI and frequency data calculated (Figure 2). The S and I data can be divided into distinct regions for colours, black and white. The H data can be and the

H frequency data is normalised for fitting Gaussian mixtures to separate colour information. Fitting of Gaussians to hue allows for upper and lower boundaries to be determined using standard deviations from the mean. Using soft colour principles the soft colour is applied to the hue values that overlap on the histogram. The upper and lower boundaries of H, S and I components are then used to fill in colour regions and convert from HSI regions to a YUV LUT to create solid classification.

3.1   HSI Histograms

We wanted a system that would be robust in the varied conditions that occur during a game; and also easily modifiable when a dramatic shift in colour space occurs (due to a change in location). Thorough manual calibration relies on image streams from hundreds to thousands of images long. From a stream of images we take a selection and convert from YUV to HSI. To reduce noise the images selected are of individual objects (colours) that fill the majority of the frame. Taking histograms of H,S and I channels (Figure 2) of test images allows for the colour space to be analysed. In the design process it was noticed that histograms of the *hue* channel showed distinct colour separation about the theoretical hue values of the object colours and a Gaussian mixture model could be applied to the frequency data. It also made apparent the problem colours and the sections of hue values that overlap.

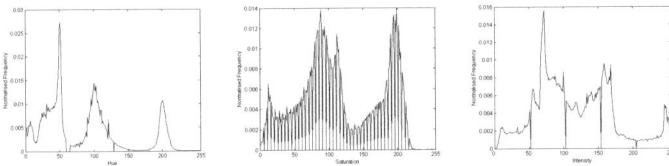

Fig. 2. Normalised histograms of hue, saturation and intensity channels from selected test images

3.2   Gaussian Mixtures

We present research into segmenting in HSI using Gaussian mixtures in one dimension. Gaussian mixtures were chosen as opposed to other mixture models (e.g. t, Gamma, generalised λ) for its ease of extension to higher dimensions. The hue frequency data was chosen as the first dimension to apply the mixtures to as it holds important colour information and requires the most attention when segmenting. Future work involving the application of multidimensional Gaussian mixtures is outlined in Section 5.

Research involving Gaussian mixtures in robot vision is extensive and has included segmenting in YUV [8] and colour learning in HSI [9]. This paper presents the application of Gaussian mixtures to ensure hue values are unique to the object colour and to detect overlapping hue values. We are trying to find unique hue values for red, orange, yellow, green and blue. Any overlapping hue values is classed as soft colour. This is different to other systems in that it is not dependent on a hard boundary, only object colour uniqueness. This ensures the integrity of the system and allows for a practical implementation.

Using the normalised hue frequency data we used an optimisation approach to fitting a Gaussian mixture. The cost function of the optimisation was the sum of the squares error between the normalised (i.e. probability density) histogram and the Gaussian mixture density. These operations were performed in Matlab using the optimisation function *fmincon*. Since the object colour codes and their theoretical hue ranges are known we can define an initial values and constraints for the Gaussian parameters including the upper and lower boundaries of the hue means. The optimisation function outputs the parameter values; mean $(\bar{h}_i)$, standard deviation (S_i) and probability (p_i) for each colour present in the histogram. Using these we calculated hue ranges, $\bar{h}_i \pm k_i S_i$, where k_i was initially interactively chosen on the basis of a histogram. Overlapping hue ranges were classified as soft colours. Figure 3 shows the optimised Gaussian mixture applied to the hue histogram data from Figure 2.

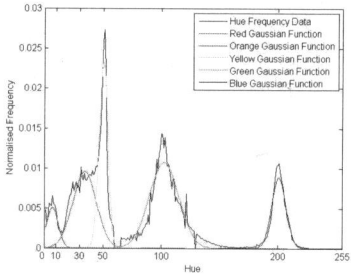

Fig. 3. Hue histogram and optimised Gaussian mixture to separate red, orange, yellow, green and blue

3.3   HSI Regions

Creating solid regions in HSI colour space can be done by defining upper and lower boundaries for H,S and I channels. The optimisation function of the Gaussian mixture outputs the mean and standard deviation for the hue values. This data is used to calculate bounday values. The number of standard deviations used for hue separation depends on the difficulty of separation of colours. For red, orange and yellow a value of one standard deviation from the mean is used to generate the upper and lower boundaries, for upper-yellow, green and blue hues are well separated standard deviations of two or three multiples are used. Saturation and intensity boundaries are interpreted from histograms as colour, black and white regions. The three dimensional region in HSI colour space can be labeled based on hue. Unique hue regions are labeled as the object colour whereas non unique hue values create the boundaries of soft colour regions. Black and white can be defined by saturation and intensity; both have low saturation but can be separated by their intensity values as they are located at either apex of the HSI cones. A larger region of intensity (brightness) values can be classified in this method due to the reliability gained from the separation of hue values. This is a step in illumination invariance [10] however a shift in colour temperature of lighting requires recalibration of hue Gaussians.

4   Results

The developed system had several notable results;

- Calibration time was greatly reduced.
- The number of required images was greatly reduced.
- The HSI regions resulted in solid classified images.
- Colours were successfully separated.
- Classification was robust and could handle change in lighting conditions.

The system has a significant reduction in calibration time from hours (approx 2 hrs) to minutes (approx 20 min). Additionally the number of images required (for this example) was reduced from 1554 random images to 10 selective images.

The effect of HSI colour region classification compared to manual classification can be seen from Figure 4. By defining solid classified regions in HSI colour has resulted in 'solid' classification of images. It can also be seen that colours were successfully separated.

This is confirmed in Table 1. The table compares performance of classification of different objects. Blob formation rate was used as a measure of successful classification, object recognition rate is a measure of success of object recognition and false positive rate is a measure of blobs incorrectly recognised as objects. Comparing the data it can be seen that the success of classification is similar between both methods of classification. Also, it can be seen that an increase in object recognition rate leads to an increase in false positive rate. This is to be expected as often a compromise must be made between object detection and false positives.

It can be seen from Figure 4 that the system results in the successful separation of red and orange hues. This is of major importance in game situations.

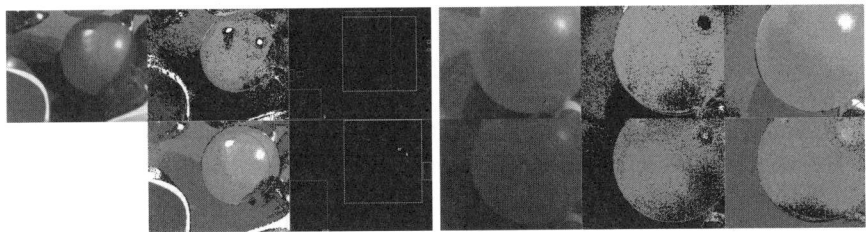

Fig. 4. (a) Separation of red and orange, manually calibrated LUT (first row) compared to HSI LUT (second row). (b) Robustness under varied lighting conditions, manually calibrated LUT (second column) compared to HSI LUT (third column).

Table 1. Blob formation rate, object recognition rate and false positive rate for objects with different classification methods

Object	Classification Method	Blob Formation Rate (%)	Object Recognition Rate(%)	False Positive Rate(%)
Ball	Manual	100	97.5	0.5
	HSI	100	98.8	1.1
Yellow Goal	Manual	99.6	97.1	0.0
	HSI	98.0	97.4	0.0
Blue Goal	Manual	100	98.0	0.0
	HSI	100	98.4	0.0

However every system has limitations, defining regions as rectangular may not be ideal and hence there is the need for extension to higher dimension Gaussian mixtures.

5 Conclusion and Future Work

Firstly this paper presented a practical colour classification method for fully separating colours, yet retaining complete colour information. Using this classification principle we have presented a solution to the hours of manual colour calibration required during setup at RoboCup events. By transforming YUV images to HSI, computing histograms of each channel and applying Gaussian mixtures to separate colour information we can define an upper and lower set of classification boundaries in HSI colour space. Then by transforming these HSI classified regions to a YUV look up table allows for an efficient calibration process that requires no additional processing to the original system. The developed solution resulted in increased reliability of object recognition due to the solid classified regions. However the reduction in misses has lead to an increase false positives. The system produced reliable results improved to that of a manually calibrated LUT whilst greatly reducing the number of required images and hence reduced image collection time, and greatly reducing the time involved with manually calibrating a LUT.

This method has proved to be successful but not optimal. Colour segmentation using Gaussian mixtures on hue alone is not sufficient in all cases. Our aim is to extend the use of Gaussian mixtures to higher dimensions. The extension of dimensions are to include saturation and intensity data and other statistically significant variables into the model to improve spatio-temporal consistency (for example edge information, IR distance sensor information, etc). To do this will require an alternate measure of goodness of fit such as maximum likelihood. Additionally we will investigate replacing the density based optimisation with either an Expectation Maximisation (EM) estimation for Gaussian mixtures or Bayesian mixture modelling.

References

1. Cameron, D., Barnes, N.: Knowledge-Based Autonomous Dynamic Colour Calibration. In: Polani, D., Browning, B., Bonarini, A., Yoshida, K. (eds.) RoboCup 2003. LNCS (LNAI), vol. 3020, pp. 226–237. Springer, Heidelberg (2004)
2. Sridharan, M., Stone, P.: Color Learning on a Mobile Robot: Towards Full Autonomy under Changing Illumination. In: The International Joint Conference on Artificial Intelligence (IJCAI 2007), Hyderabad, India (2007)
3. Bruce, J., Balch, T., Veloso, M.: Fast and inexpensive color image segmentation for interactive robots. In: Proceedings of the 2000 IEEE/RSJ International Conference on Intelligent Robots and Systems (IROS 2000), vol. 3, pp. 2061–2066 (2000)
4. Austin, D., Barnes, N.: Red is the new black - or is it? In: Proc. of the 2003 Australasian Conference on Robotics and Automation (2003)
5. Qunilan, M. J. , Henderson, N., Nicklin, S. P., Fisher, R., Chalup, S. K., Middleton, R. H., King, R.: The 2006 NUbots Team Report. University of Newcastle, EECS Technical Report (2006)
6. Quinlan, M.J., Murch, C.L., Moore, T.G., Middleton, R.H., Yung Li, L.A., King, R., Chalup, S.K.: The 2004 NUbots Team Report. University of Newcastle EECS Technical Report (2004)
7. Stanton, C., Williams, M.-A.: A Novel and Practical Approach towards Color Constancy for Mobile Robots using Overlapping Color Space Signatures. In: Bredenfeld, A., Jacoff, A., Noda, I., Takahashi, Y. (eds.) RoboCup 2005. LNCS (LNAI), vol. 4020, pp. 444–451. Springer, Heidelberg (2006)
8. Cohen, D., Hua Ooi, Y., Vernaza, P., Lee, D.D.: The University of Pennsylvania Robocup 2003 Legged Soccer Team. Technical Report University of Pennsylvania (2003)
9. van Soest, D.A., de Greef, M., Sturm, J., Visser, A.: Autonomous Color Learning in an Artificial Environment. In: Proc. 18th Dutch-Belgian Artificial Intelligence Conference, BNAIC 2006, Namen, Belgium, October 2006, pp. 299–306 (2006)
10. Sridharan, M., Stone, P.: Towards illumination invariance in the legged league. In: Nardi, D., Riedmiller, M., Sammut, C., Santos-Victor, J. (eds.) RoboCup 2004. LNCS (LNAI), vol. 3276, pp. 196–208. Springer, Berlin (2005)

Model Checking Hybrid Multiagent Systems for the RoboCup[*]

Ulrich Furbach[1], Jan Murray[1], Falk Schmidsberger[2], and Frieder Stolzenburg[2]

[1] Universität Koblenz-Landau, Artificial Intelligence Research Group, D-56070 Koblenz
{uli,murray}@uni-koblenz.de
[2] Hochschule Harz, Automation and Computer Sciences Department
D-38855 Wernigerode
{fschmidsberger,fstolzenburg}@hs-harz.de

Abstract. This paper shows how multiagent systems can be modeled by a combination of UML statecharts and hybrid automata. This allows formal system specification on different levels of abstraction on the one hand, and expressing real-time system behavior with continuous variables on the other hand. It is shown how multi-robot systems can be modeled by hybrid and hierarchical state machines and how model checking techniques for hybrid automata can be applied. An enhanced synchronization concept is introduced that allows synchronization taking time and avoids state explosion to a certain extent.

1 Multiagent Systems

Specifying behaviors for (physical) multiagent systems and multi-robot systems is a sophisticated and demanding task. Due to the high complexity of the interactions among agents and the dynamics of the environment the need for precise modeling arises. Since the behavior of agents usually can be understood as driven by external events and internal states, an obvious way of modeling multiagent systems is by state transition diagrams. Hierarchical state transition diagrams like statecharts are particularly well suited as they allow the specification of behaviors on different levels of abstraction [1].

One important aspect of physical agents and robots is that they interact with a (possibly simulated) physical environment. Such interactions typically consist of continuous actions (e.g. the movement of a robot) and perceptions like the power status of a battery. Classical state transition diagrams are not well suited for modeling this kind of interactions, as the transitions between states are discrete. However, continuous extensions to these formalisms have been proposed, e.g. hybrid automata [2].

Especially for agents employed in safety critical environments, e.g. in rescue scenarios, behavior specification has to be done very carefully in order to avoid side effects that may have unwanted or even disastrous consequences. One approach to realizing the required clarity of a specification is the use of formal design methods. Fortunately, many state transition diagram dialects like hybrid automata are equipped with a formal semantics that makes them accessible to formal validation of the modeled behavior. Thus it

---

[*] This research is supported by the grants *Fu 263/8* and *Sto 421/2* from the German research foundation *DFG* within the special priority program 1125 on *Cooperating Teams of Mobile Robots in Dynamic Environments*.

U. Visser et al. (Eds.): RoboCup 2007, LNAI 5001, pp. 262–269, 2008.

becomes possible to (semi-)automatically prove desirable features and the absence of unwanted properties in the specified behaviors, e.g. with model checking methods.

2   Hybrid Hierarchical State Machines

In this chapter, we present the combination of two concepts: hierarchical statecharts and hybrid automata. As a running example, we use a scenario from the RoboCup rescue simulation league, which is shortly described in the following subsection.

Rescue Scenario. In the RoboCup rescue simulation league [3], a large scale disaster is simulated. The simulator models part of a city after an earthquake. Buildings may be collapsed or on fire, and roads are partially or completely blocked. A team of heterogeneous agents consisting of police forces, ambulance teams, a fire brigade, and their respective headquarters is deployed. The agents have two main tasks, namely finding and rescuing buried civilians and extinguishing fires. An auxiliary task is clearing of blocked roads, such that agents can move smoothly. As their abilities enable each type of agent to solve only *one* kind of task (e.g. fire brigades cannot clear roads or rescue civilians), the need for coordination and synchronization among agents is obvious.

Consider the following simple scenario. If a fire breaks out somewhere, a fire brigade agent is ordered by its headquarters to extinguish the fire. The fire brigade moves to the fire and begins to put it out. If the agent runs out of water it has to refill its tank at a supply station and return to the fire to fulfill its task. Once the fire is extinguished, the fire brigade agent is idle again. An additional task the agent has to execute is to report any injured civilians it discovers. Part of this scenario is modeled in Fig. 1 with the help of a hierar-

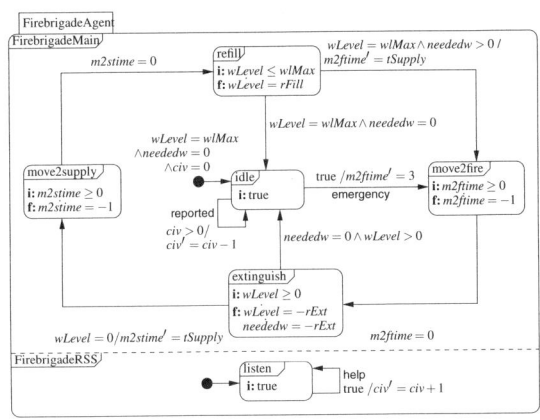

Fig. 1. A simple fire brigade agent

chical hybrid automaton [4]. In addition to the fire brigade agent the model should include a fire station, fire, and civilians as part of the environment; all this will be explained in the next section (cf. Fig. 2).

States are represented as rectangles with rounded corners and can be structured hierarchically. The specification of the fire brigade is a simple hierarchical chart, consisting of the main control structure (*FirebrigadeMain*) and a rescue sub-system (*FirebrigadeRSS*) which are supposed to run in parallel. The latter just records the detected civilians, which are not modeled in Fig. 1 (for this, see the sub-state *Civilians* in Fig. 2). *FirebrigadeMain* consists of five sub-states corresponding to movements (*move2fire*, *move2supply*), extinguishing (*extinguish*), refilling the tank (*refill*), and an idle state (*idle*). The agent can report the discovered civilians when it is in its idle state. Details from this figure will be explained in the course of this section.

Obviously, even in this simple case with few components and a deterministic environment it is difficult to see if the agent behaves correctly. Important questions like "does the fire brigade try to extinguish without water?" or "will every discovered civilian (and *only* those) be reported eventually?" depend on the interaction of all components and cannot be answered without an analysis of the whole system.

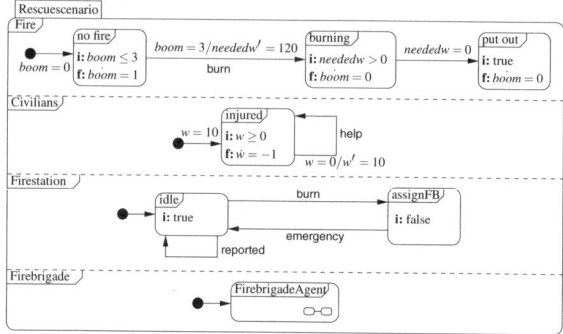

Fig. 2. A simple scenario from the RoboCup rescue simulation. The state *FirebrigadeAgent* corresponds to Fig. 1. The icon ○–○ hints at the hidden sub states.

State Hierarchies and Transitions. Statecharts are a part of UML [5,6] and a well accepted means to specify dynamic behavior of software systems. The main concept for statecharts is a state, which corresponds to an activity or behavior of a robot agent. Statecharts can be described in a rigorously formal manner [7,8], allowing flexible specification, implementation and analysis of multiagent systems [1] which is required for robot behavior engineering and modeling and simulating complex robots.

Definition 1 (basic components). *The basic components of a* state machine *are the following disjoint sets:*

S: *a finite set of states, which is partitioned into three disjoint sets:* S_{simple}, S_{comp}, *and* S_{conc} — *called simple, composite and concurrent states, containing one designated* start state $s_0 \in S_{comp} \cup S_{conc}$, *and*
X: *a finite set of (real-numbered) variables.*

In our running example, *idle*, *extinguish* or *listen* are simple states, and *FirebrigadeAgent* is a concurrent state and *FirebrigadeMain* and *FirebrigadeRSS* are composite states, called regions in this case, which are separated by a dashed line. *m2ftime* and *wLevel* are examples for real valued variables.

In statecharts, states are connected via *transitions* in $T \subseteq S \times S$, indicating that an agent in the first state will enter the second state. Transitions are drawn as arrows labeled with jump conditions over the variables in X together with actions. For example, the transition from *idle* to itself is labeled with $civ > 0/civ' = civ - 1$, with the meaning: if the value of *civ* is greater 0, the action $civ' = civ - 1$ is executed while performing the transition, i.e., the number of civilians that are found but not reported is decreased.

The label reported at the same transition is used for synchronizing the transition with another automaton working in parallel, namely the one for *Firestation* (see Fig. 2). It is only legal for the combined system if both automata take the transition labeled reported at the same time (see [2] for details). In principle, the explicit use of events and actions as in UML statecharts is not needed, as both can be expressed with the help of variables. For example the occurrence of an external event can be represented by changing the value of the corresponding variable from 0 to 1.

Since hybrid automata [2] are similar to statecharts, it makes sense to combine the advantages of both models. Statecharts have the clear advantage of allowing hierarchical specification on several levels of abstraction, while hybrid automata enable the introduction of continuous variables and flow conditions. This extension of statecharts is done by the subsequent definition. Hybrid automata are widely used for the specification of embedded systems. By reachability analyses, diagnosis tasks can be solved. We will come back to this in Sect. 4.

Definition 2 (jump conditions, flows and invariants). *In addition to the variables in X, we introduce new variables $\dot{x}$ (first derivatives during continuous change) and x' (values at the conclusion of discrete change) for each $x \in X$, calling the corresponding variable sets $\dot{X}$ and X', respectively. Then, each transition in T may be labeled by a jump condition, that is a predicate whose free variables are from $X \cup X'$, which can be split into activation condition and effect. In addition, each state $s \in S$ is labeled with a flow condition (f:), whose free variables are from $X \cup \dot{X}$, and an invariant (i:), whose free variables are from X. Flow conditions may be empty and hence omitted, if nothing changes continuously in the respective state.*

In our example we use the dotted variable $w\dot{L}evel$ to denote the change of the water level in the state *refill*. A transition from this state to the state *move2fire* is performed, if the water level reached the maximum ($wLevel = wlMax$) and water is needed ($neededw > 0$). During the transition the action $m2ftime' = tSupply$ is executed.

We will restrict our attention to linear conditions, i.e. linear equalities and inequalities among either ordinary variables in $X \cup X'$ or their first derivatives $\dot{X}$, because only then an exact reachability analysis (needed for model checking) is feasible [2]. Following the lines of [5,6], UML statecharts have a hierarchical structure which can easily be represented as a tree of states. Here, regions with cardinality greater than one must be treated as multiple composite states, which are distinguished by different indices. The behavior of the overall state machine can be described by a sequence of state tree configurations, called micro-steps. For this, the interested reader is referred to [7,9].

3   Synchronization and Cooperation

The overall performance of programmed multiagent systems heavily depends on how cooperative agents behave. Cooperation and coordination of agents can be achieved by synchronization. Hence, it is essential to implement synchronization effectively. Synchronization means that several actions must start or happen at the same time. In the rescue scenario (see Sect. 2), transition labels serve as triggers for synchronization in the formalism of hybrid automata, e.g., if an injured civilian cries for help, then the listening fire fighter hears this. However, if more complicated coordination and cooperation among agents has to be expressed, then this simple concept of synchronization may not suffice. In the following, we will therefore introduce an enhanced concept of synchronization (see [4]), which we motivate with an example from robotic soccer.

An Example of Coordination in Robotic Soccer. Since (robotic) soccer is a team sport, cooperation of agents is essential.

At best, exactly one player should go to the ball, while the others try to position themselves as good as possible. Fig. 3 shows the statechart for two players trying a coordinated behavior of going to the ball. To realize this behavior, the positions of two players (**p1, p2**), the ball (**bR**), a (stationary) opponent (**PO**) and the opponent goal **POG** are modeled. Additionally the local estimates of the ball position **b**, the player **p** and its teammate **pT** are given for each player. Finally there is the players' measurement error ME, the range DHB, within which a player has the ball, and some scaling factors $F01$–$F04$. Constant names start with capital letters, variables with lower case letters.

In this example, coordination is really important. In contrast to simple synchronization mechanisms, coordination may take some time. The time between deciding to go to the ball and actually reaching it will be almost always greater than zero. Thus, we must be able to distinguish between the allocation and the occupation of a resource (e.g. the ball) in our specification formalism. In addition, since coordination may take some time, we associate the new synchronization method with states and not with transitions. All this is comprised in the concept of *timed synchronization* introduced next.

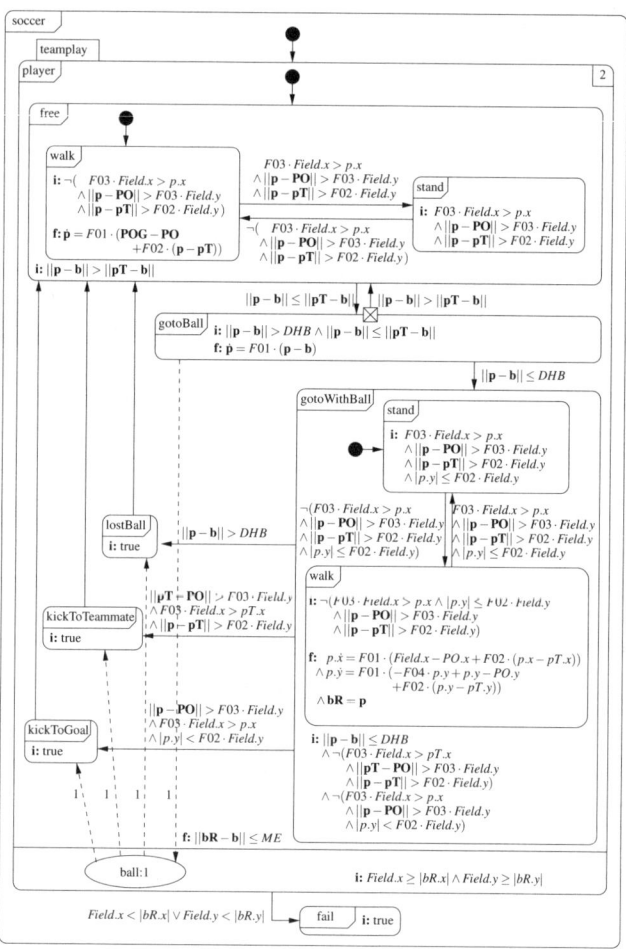

Fig. 3. Robotic soccer example

Timed Synchronization. Usually the so-called *synchrony hypothesis* is adopted for state machines, assuming that the system is infinitely faster than the environment and thus the response to an external stimulus (event) is always generated in the same step that the stimulus is introduced. However in practice, synchronization and coordination of actions cannot be done in zero time. In UML 1.5 [5], synchronization is present, but assumed to take zero time. In UML 2.0 [6] there does not seem to be a special synchronization mechanism available any longer except by join and fork transitions. Hence, it seems to be really worthwhile considering synchronization and coordination in more detail. For this, we will introduce synchronization points which are associated with states, i.e. activities that last a certain time, and not with transitions (as in UML 1.5), because the transition from one state to another takes zero time according to the synchrony hypothesis.

Definition 3 (synchronization points). *A synchronization point (represented as oval) allows the coordinated treatment of common resources. It can be identified by special synchronization variables $x \in X_{synch} \subseteq X$ with a given maximal capacity $C(x) > 0$. Each such point may be connected with several states. We distinguish two relations: $R_+ \subseteq S \times X_{synch}$ and $R_- \subseteq X_{synch} \times S$, both represented by dashed arrows in the respective direction. Further, each connection in $R_+ \cup R_-$ is annotated with a number m with $0 < m \leq C(x)$.*

As just said, according to the previous definition, synchronization is connected to states and not to transitions as in UML 1.5. In consequence, it is now possible that synchronization may take some time as desired. The process of synchronization starts when a state s connected to a synchronization variable x is entered, and it ends only after some time when s is exited. Therefore, we distinguish the allocation of (added or subtracted) resources and their (later) actual occupation.

In Fig. 3, coordination is achieved by the synchronization variable *ball* introduced in the concurrent state *teamplay*. It has capacity 1, as there is only one ball in a soccer game. The *gotoBall* state is positively connected to it, while the states *kickToGoal*, *kickToTeammate*, and *lostBall* are negatively connected to it. This means, that the ball resource is allocated during the *gotoBall* activity and deallocated after a kick. The transition marked with a crossed box indicates a failed synchronization.

4   Model Checking

As we already mentioned, hybrid automata are equipped with a formal semantics, which makes it possible to apply formal methods in order to prove certain properties of the specified systems, e.g. by model checking. However, in the context of hybrid automata the term *model checking* usually refers to *reachability* testing, i.e. the question whether some (unwanted) state is reachable from the initial configuration of the specified system. To this end, all states that can be reached by a discrete transition or evolving the continuous variables according to a flow condition are repeatedly added to the current configuration until a fixpoint R is reached. Then it can be tested, if unwanted states are reachable simply by intersecting the sets of reachable and unwanted states.

For the behavior specification shown in Figs. 1 and 2 we conducted several experiments with the standard model checkers HYTECH [10]. Both model checkers are implemented for the analysis of linear hybrid automata. They take textual representations of hybrid automata like the one in Fig. 4 as input and perform reachability tests on the state space of the resulting product automaton. This is usually done by first computing all states reachable from the initial configuration, and then checking the resulting set for the needed properties. In the remainder of this section, we present some exemplary model checking tasks for the rescue scenario.

Is it possible to extinguish the fire? When the state of the automaton modeling the fire changes from *no fire* to *burning*, the variable *neededw* stores the amount of water needed for putting out the fire (*neededw* = 120 in the beginning). When the fire is put out, i.e. *neededw* = 0, the automaton enters the state *put out*. Thus the fire can be extinguished, iff there is a reachable configuration c_{out} where fire is in the state *put out*. It is easy to see from the specification, that this is indeed the case, as *neededw* is only decreased after the initial setting, and so the transition from *burning* to *put out* is eventually forced. With the help of HYTECH's trace generation ability it is quite easy to solve the additional task of comparing different strategies, e.g. for refilling the water tanks. To this end, traces to c_{out} generated using the different strategies are compared. A shorter trace (wrt. time units) corresponds to a faster solving of the extinguishing task.

Does the agent try to extinguish with an empty water tank? The fact that the firebrigade agent tries to put out the fire without water corresponds to the simple state *extinguish* being active while *wLevel* < 0. Note that we must not test for *wLevel* ≤ 0, as the state *extinguish* is only left when the water level is zero, so including a check for equality leads to false results. Fig. 4 shows how to check this property with HYTECH. The set of reachable states is collected in the variable init_reach (l. 8), and ext_error is assigned the set of illegal states (l. 9), i.e., all states where extinguish is active and the water level is below zero. Lines 10–12 finally show the actual test. If the intersection of reachable and illegal states is not empty (l. 10), an error message is printed (l. 11).

Does the agent report all discovered civilians? This question contains two properties to be checked: (a) all discovered civilians are reported eventually, and (b) the agent does not report more civilians than he found. The discovery of a civilian is modeled by increasing the value of the variable *civ* by one. For each reported civilian one is subtracted from *civ*. From this it follows, that

```
 1 automaton Civilian
 2   synclabs: help;
 3   initially injured & w = -10;
 4   loc injured:
 5     while w<=0 wait {}
 6     when w=0 sync help do {w' = -10} goto injured;
 7 end

 8 init_reach := reach forward from init endreach;
 9 ext_error := loc[FirebrigadeMain] = extinguish & wLevel < 0;
10 if not empty(init_reach & ext_error)
11   then prints "Error: Tank empty!";
12 endif;
```

Fig. 4. HYTECH code for the civilian automaton from Fig. 2 (ll. 1–7) and analysis commands

(b) holds, iff no configuration is reachable, where *civ* < 0. To show (a), one has to ensure that from all configurations with *civ* > 0 a configuration with *civ* = 0 will be reached eventually. Testing these properties with HYTECH reveals that (b) holds in the specification, i.e. for all reachable states *civ* ≥ 0. However, the analysis also yields that (a) does not hold. As we stated earlier the fire fighter agent should report civilans when he is in the *idle* state. But as the invariant in this state (true) is never violated, the agent is not forced to take the self transition labeled reported, which corresponds to reporting a civilian. Thus, there is a legal run of the system, where no civilian is reported at all.

It should be remarked that synchronization points help us to reduce complexity. In order to see this, let us consider a composite state with cardinality *m* containing *k* (simple) states. One of them, say s, is connected to a synchronization point with capacity C. Then there are k^m different configurations, i.e. exponentially. Since at most C agents can be in s, only $\sum_{l=0}^{C} \binom{m}{l}(k-1)^{m-l}$ configurations have to be considered. This is polynomial for $k = 2$. A naïve flattening of the example in Fig. 3 e.g. leads to $8 \cdot 8 + 1 = 65$ configurations, whereas taking synchronization into account leads to only $2 \cdot 2 + 2 \cdot 2 \cdot 6 + 1 = 29$ configuration states. A translator that automatically converts hybrid hierarchical statecharts into simple flat hybrid automata is currently implemented [11].

5 Conclusions

In this paper we demonstrated the use of hybrid hierarchical state machines for the specification of multiagent systems. We presented two application scenarios from the RoboCup, one from the rescue simulation and one from robotic soccer, and we demonstrated that state-of-the-art model checkers for hybrid automata can be used for proving properties of the specified systems. We exemplified this especially with an example from the RoboCup rescue scenario. Model checking, i.e. reachability analysis helps us finding out possible paths, which could help in the pre-computation of multiagent system implementations. This point will be subject of future work as well as studies of whether the procedure scales up to more complex scenarios.

References

1. Murray, J.: Specifying agent behaviors with UML statecharts and StatEdit. In: Polani, D., Browning, B., Bonarini, A., Yoshida, K. (eds.) RoboCup 2003. LNCS (LNAI), vol. 3020, pp. 145–156. Springer, Heidelberg (2004)
2. Henzinger, T.: The theory of hybrid automata. In: Proceedings of the 11th Annual Symposium on Logic in Computer Science, pp. 278–292. IEEE Computer Society Press (1996)
3. Tadokoro, S., et al.: The RoboCup-Rescue project: A robotic approach to the disaster mitigation problem. In: Proceedings of IEEE International Conference on Robotics and Automation (ICRA 2000), pp. 4089–4104 (2000)
4. Murray, J., Stolzenburg, F.: Hybrid state machines with timed synchronization for multi-robot system specification. In: Bento, C., et al. (eds.) Proceedings of 12th Portuguese Conference on Artificial Intelligence, pp. 236–241. IEEE Inc. (2005)
5. Object Management Group, Inc.: UML Specification, Version 1.5 (March 2003)
6. Object Management Group, Inc.: UML 2.0 Superstructure Specification (October 2004)
7. Arai, T., Stolzenburg, F.: Multiagent systems specification by UML statecharts aiming at intelligent manufacturing. In: Proceedings of 1st International Joint Conference on Autonomous Agents & Multi-Agent Systems, pp. 11–18. ACM Press (2002)
8. Pnueli, A., Shalev, M.: What is in a step: On the semantics of statecharts. In: Ito, T., Meyer, A.R. (eds.) TACS 1991. LNCS, vol. 526, pp. 244–264. Springer, Heidelberg (1991)
9. Furbach, U., Murray, J., Schmidsberger, F., Stolzenburg, F.: Hybrid multiagent systems with timed synchronization – specification and model checking. In: Dastani, M., El Fallah Seghrouchni, A., Ricci, A., Winikoff, M. (eds.) ProMAS 2007. LNCS (LNAI), vol. 4908, pp. 205–220. Springer, Heidelberg (2008)
10. Henzinger, T.A., Ho, P.H., Wong-Toi, H.: HyTech: The Next Generation. In: IEEE Real-Time Systems Symposium, pp. 56–65 (1995)
11. Ruh, F.: A translator for cooperative strategies of mobile agents for four-legged robots. Master thesis, Dept. of Automation and Computer Sciences, Hochschule Harz (2007)

Physical Simulation of the Dynamical Behavior of Three-Wheeled Omni-directional Robots

Hamid Rajaie, Reinhard Lafrenz, Oliver Zweigle, Uwe-Philipp Käppeler,
Frank Schreiber, Thomas Rühr, Andreas Tamke, and Paul Levi

Institute of Parallel and Distributed Systems,
Universität Stuttgart, 70569 Stuttgart, Germany
robocup@informatik.uni-stuttgart.de
www.robocup.informatik.uni-stuttgart.de

Abstract. Hardware simulation is a very efficient way for parameter tuning. We developed a Simulink-based simulator for the navigation components of our robotic soccer team. This physical simulation has interfaces to be interconnected with the higher levels of the real control software and is therefore able to perform an overall simulation of single robots.

1 Introduction

Development of new control strategies for the driving behavior of mobile robots and extensive parameter studies on real hardware are time-consuming. To overcome this, we developed a Simulink-based simulator for the driving behavior of our soccer robots, where physical parameters of the robot and the environment can be modeled and which allows for off-line parameter studies. This physical simulation has network interfaces to be interconnected with the higher levels of the real robot control software, so that critical movements can be detected and the higher-level control can be adapted. Therefore, the system is able to perform an overall simulation of single robots.

We use Matlab and Simulink to design an integrated environment for rapid prototyping of algorithms, simulation, and modeling. Rapid prototyping is a mean for the evaluation of different control strategies, e.g., P, PID, or Fuzzy. We use the simulation environment to model the dynamic behavior of the robot and the underlying hardware components. To analyze these models, we successfully connected the physical simulator with the CoPS behavior control software, which runs on real hardware.

With the simulator, we can also analyze and visualize the different signals, e.g. motor currents and accelerations. This can be used to achieve good sets of parameters for the higher-level behavior software. For example, in robotic soccer, the driving behavior depends highly on the characteristics and conditions of the carpet. In addition, the driving behavior changes while dribbling the ball. Typical controllers like PID have several parameters, that affect the behavior. Finding a good set in the higher-order parameter space is always a critical and time-consuming procedure which can be accelerated using the simulation system.

U. Visser et al. (Eds.): RoboCup 2007, LNAI 5001, pp. 270–277, 2008.
© Springer-Verlag Berlin Heidelberg 2008

The simulator is able to interpret messages in CAN-bus format, which is used for the real hardware. In this way, the higher levels of the software don't need to be changed to switch between simulation and real hardware. The simulation is interconnected to the robot software by TCP/IP communication.

2   Three-Wheeled Omnidirectional Drive

2.1   Robot Geometry Model

Fig. 1 shows the top view of our mobile robot model and the coordinate systems we use in the model. The use of the following different coordinate systems make the analysis of the model easier.

1. The World Coordinate System (WCS), attached to the field, is an inertial Cartesian coordinate system with the unit vectors $\mathbf{i}_w, \mathbf{j}_w, \mathbf{k}$.
2. The Local Coordinate System (LCS) is also a Cartesian coordinate system, but its X axis is the heading of the robot (angle β in WCS). The unit vectors are $\mathbf{i}_l, \mathbf{j}_l, \mathbf{k}$.
3. A Local Polar Coordinate Systems or $LPCS_i, i = 1, \ldots, 3$ attached to each leg. The unit vectors are $\mathbf{e}_r^i, \mathbf{e}_\theta^i, \mathbf{k}$.

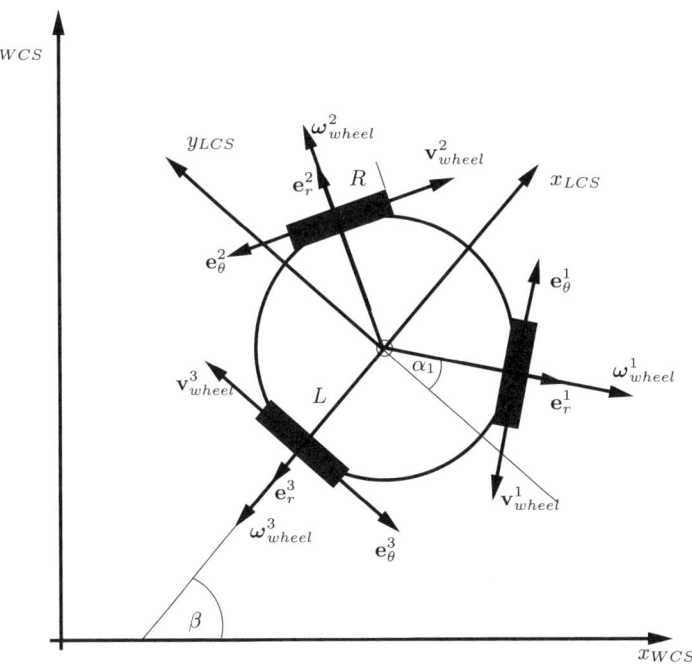

Fig. 1. Kinematics of the omnidirectional drive

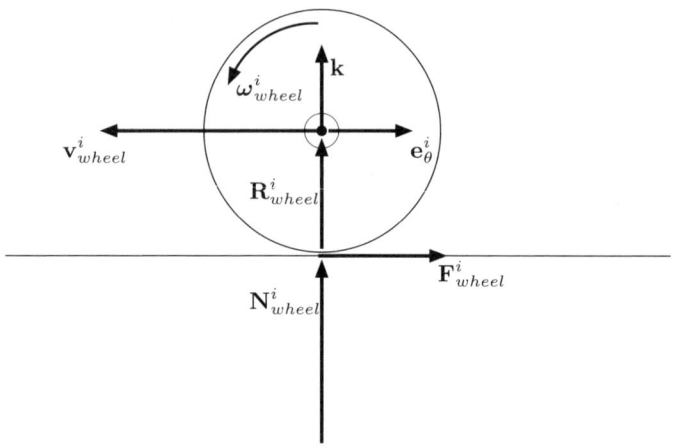

Fig. 2. Wheel model

2.2   Wheel Model

In order to access a better estimation of the dynamic behavior of the robot, we consider the effect of wheel slip. Fig. 2 shows the side view of a wheel. The view direction points to the center of the robot.

We assume that there is always a vertical force applied from the ground to each wheel, $\mathbf{N}^i_{wheel}$ as it is shown in Fig. 2. $\mathbf{N}^i_{wheel}$ is a result of the weight of the robot and its value depends on the position of the center of gravity of the robot and can be different for each wheel. We also assume that the horizontal force applied form the ground to the wheel $\mathbf{F}^i_{wheel}$ is always in the plane of the wheel, if it is not zero.

In case of a free rotation of the wheel i (no mechanical driving or breaking torque on the wheel) we can assume:

$$\mathbf{v}^i_{wheel} = \mathbf{R}^i_{wheel} \times \boldsymbol{\omega}^i_{wheel} \tag{1}$$

which leads according to Fig. 2 to

$$\mathbf{R}^i_{wheel} = R^i_{wheel}\mathbf{k}, \quad \boldsymbol{\omega}^i_{wheel} = \omega^i_{wheel}\mathbf{e}^i_r, \quad \mathbf{v}^i_{wheel} = v^i_{wheel}\mathbf{e}^i_\theta \tag{2}$$

Considering that if $\boldsymbol{\omega}^i_{wheel} \cdot \mathbf{e}^i_r$ is positive then $\mathbf{v}^i_{wheel} \cdot \mathbf{e}^i_\theta$ is negative. In this case the contact force between the wheel and the ground is zero.

$$\mathbf{F}^i_{wheel} = 0 \tag{3}$$

whereas in general

$$\mathbf{F}^i_{wheel} = F^i_{wheel}\mathbf{e}^i_\theta \tag{4}$$

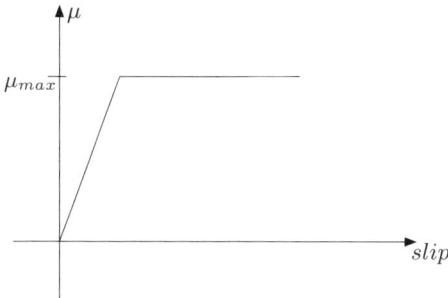

Fig. 3. Slip model

In cases where driving or breaking mechanical torques are applied on the wheel, the above equation is no longer valid, which means that slip occurs. The wheel slip velocity $\mathbf{v}_s^i$ is defined as:

$$\mathbf{v}_s^i = \mathbf{v}_{wheel}^i + \mathbf{R}_{wheel}^i \times \boldsymbol{\omega}_{wheel}^i \tag{5}$$

In order to quantify the slip, the following formula is also used as a mathematical definition.

$$\mathbf{s}^i = \frac{\mathbf{R}_{wheel}^i \times \boldsymbol{\omega}_{wheel}^i + \mathbf{v}_{wheel}^i}{\|\mathbf{v}_{wheel}^1\|} \cdot \mathbf{e}_\theta^i \tag{6}$$

In the case of a nonzero mechanical torque on the wheel axis, as a consequence of the slip, a contact force $\mathbf{F}_{wheel}^i$ between wheel and ground in the plane of wheel occurs. A simplified and useful model which relates the slip and the generated force is as follows:

$$\mathbf{F}_{wheel}^i = \mu^i \|\mathbf{N}_{wheel}^i\| \mathbf{e}_\theta^i \tag{7}$$

in which $\mathbf{F}_{wheel}^i$ is the generated force and $\mathbf{N}_{wheel}^i$ is the vertical force which is applied from the ground on the wheel i.

$$\mathbf{N}_{wheel}^i = N_{wheel}^i \mathbf{k} \tag{8}$$

In general, we assume that μ^i is a function of the slip velocity $\mathbf{v}_s^i$ according to [3]:

$$\mu^i = f(\mathbf{v}_s^i) \tag{9}$$

Using equations 7 and 9 we conclude that

$$\mathbf{F}_{wheel}^i = f(\mathbf{v}_s^i) \|\mathbf{N}_{wheel}^i\| \mathbf{e}_\theta^i \tag{10}$$

Fig. 3 shows the model we use in this paper as a relation between the slip and coefficient of friction, according to equation 9.

2.3   Kinematic and Dynamic Model

Due to the Newton's second law and according to Fig. 1 we get

$$\sum_{i=1}^{3} F^i_{wheel} \mathbf{e}^i_\theta = m_{robot} \mathbf{a}_{robot}, \qquad \sum_{i=1}^{3} r^i \mathbf{e}^i_r \times F^i_{wheel} \mathbf{e}^i_\theta = I_{robot} \boldsymbol{\alpha}_{robot} \qquad (11)$$

The above two equations are written in the Local Coordinate System LCS. Considering equation 10 we can write:

$$\sum_{i=1}^{3} f(\mathbf{v}^i_s) \|\mathbf{N}^i\| \mathbf{e}^i_\theta = m_{robot} \mathbf{a}_{robot}, \qquad \sum_{i=1}^{3} r^i \mathbf{e}^i_r \times f(\mathbf{v}^i_s) \|\mathbf{N}^i\| \mathbf{e}^i_\theta = I_{robot} \boldsymbol{\alpha}_{robot} \qquad (12)$$

2.4   DC Motor Model

The model which we used for a DC motor is the standard model with ideal motor and internal resistance. The rotation of the rotor of a DC motor generates an electromotive force U^i_{emf}. A linear algebraic equation relates the electromotive force and the rotor angular velocity of motor i:

$$U^i_{emf} = K_u \|\boldsymbol{\omega}^i_{motor}\| \qquad (13)$$

According to Ohm's law the voltage drop due to the winding's resistance is

$$U^i_{resistance} = R^i_{motor} I^i \qquad (14)$$

By applying the Kirchoff's law we get

$$U^i_{terminal} - U^i_{emf} = U^i_{resistance} \qquad (15)$$

Putting equations 13 and 14 in equation 15 we conclude that for motor i

$$U^i_{terminal} - K_u \|\boldsymbol{\omega}^i_{motor}\| = R^i_{motor} I^i \qquad (16)$$

Another very important relation in a DC motor is the linear relation between the mechanical load, T^i_{motor} and the electrical current I^i:

$$I^i = K_i T^i_{motor} \qquad (17)$$

The equations 16 and 17 are used in our model to describe mathematically the behavior of our DC motors.

2.5   Motor Controller Model

The basic task of a motor controller is to regulate the angular velocity of the motor $\omega^i(t)$ equal to a set angular velocity $\omega^i_{motor set}$. We define the velocity error as

$$error^i(t) = \omega^i_{motor set} - \omega^i_{motor}(t) \qquad (18)$$

The motor controller receives the set angular velocity from another module over the CAN bus and reads the current angular velocity from the digital encoders and calculate the error. Depending on the control strategy, in our case PID, it produces an output terminal voltage $U^i_{terminal}$ which is applied to the motor.

$$U^i_{terminal} = f^i_{PID}(error^i) \qquad (19)$$

3   Simulation Package

Based on the mathematical models introduced in part 2, we implemented the Simulink model shown in Fig. 4. It consists of the following components:

The NETWORK block provides the set motor angular velocities $\omega^i_{motor_{set}}$. As Simulink runs, the code in this block creates a listening socket and waits for a request from the CoPS control software. For each velocity message from the CoPS client, the network block updates the set motor angular velocities $\omega^i_{motor_{set}}$. The velocity messages are based on the protocol which we use in our team as the communication protocol between the higher levels of the CoPS software and the motor controller hardware.

Each POWER TRANSMISSION block models a DC motor plus the planetary gearbox as well as the tooth belt and the motor controller based on equations 17 and 16. Each block gets the corresponding set angular velocity $\omega^i_{motor_{set}}$ as input from the network block and the contact force signal $\mathbf{F}^i_{wheel}(t)$ as a feedback from the CONTACT FORCES block and outputs the wheels' angular velocities $\omega^i_{wheel}(t)$.

The SLIP block gets the $\omega^i_{wheel}(t)$ and $\mathbf{v}^i_{wheels}(t)$ as inputs and calculates $\mathbf{v}^i_s(t)$ based on Eq. 5.

The FRICTION COEFFICIENTS block calculates the $\mu^i(t)$ from the $\mathbf{v}^i_s$ based on Eq. 9.

The CONTACT FORCES block calculates $\mathbf{F}^i_{wheel}(t)$ based on Eq. 7. In this block the nonuniform weight distribution of the robot can be modeled.

The NEWTON block accepts the $\mathbf{F}^i_{wheel}(t)$ from the contact forces block as inputs and calculates the robot velocity $\mathbf{v}^i_{robot}(t)_{LCS}$ in the local coordinate system based on equations ?? and 11.

The FIELD block receives the $\mathbf{v}^i_{robot}(t)_{LCS}$ from NEWTON block and visualizes the play field and the robot on it.

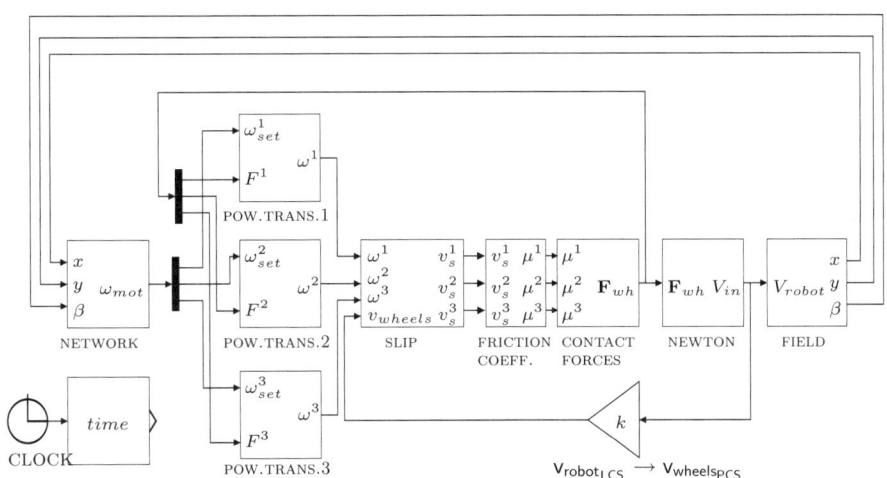

Fig. 4. Simulink model for the 3-wheel motor control

The CLOCK provides a soft real time mechanism which synchronizes the model simulation clock with the internal clock of the computer.

4   Experiments and Results

The result of the simulation for a robot with $m = 17kg, I = 0.227kgm^{-2}$ is shown in fig. 5. In the simulation we have assumed a non-uniform distribution of weight load on the wheels:

$$\|\mathbf{N}^1_{wheel}\| = 60.43N, \quad \|\mathbf{N}^2_{wheel}\| = 50.63N, \quad \|\mathbf{N}^3_{wheel}\| = 55.53N, \tag{20}$$

Also it is assumed that $\mu_{max} = 0.25$, as shown in Fig. 3. The initial velocity of the robot is zero, and the set velocity is

$$\mathbf{v}_{robot_{LCS}} = (2\,m/s,\ 0\,m/s,\ 0\,rad/s) \tag{21}$$

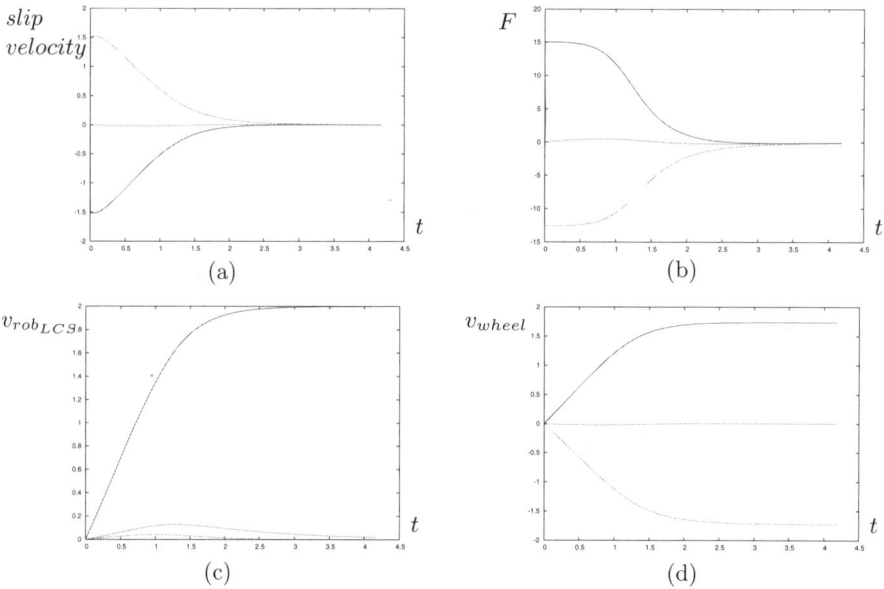

Fig. 5. Simulation result for the robot during acceleration: (a) slip velocity for the 3 wheels $[m/s]$, (b) horizontal forces for the 3 wheels $[N]$, (c) robot velocity in local coordinate system $[(m/s, m/s, rad/s)]$, (d) wheel velocities $[rad/s]$

As shown in the Fig. 5(a), at the beginning of the movement, the slip velocities of wheels 1 and 2 are high and their absolute values are not equal. The unequal slip values are a direct consequence of the non-uniform weight distribution. Therefore the traction forces on wheels 1 and 2 are not equal and therefor robot moves on a curve until the traction forces become equal. This force transition phase is shown in Fig. 5(b). As shown in Fig. 5(c),the the robot velocity reaches the set velocity $(2m/s, 0m/s, 0rad/s)$ after the transition phase.

5   Conclusion and Outlook

We presented a dynamic model for a mobile robot equipped with omnidirectional wheels, considering the wheel slip and a non-uniform distribution of weight. The network interface allows us to connect our CoPS software with the model through a the TCP/IP link. The CoPS software can communicate with the model, e.g. send *set velocity* commands and query the model for signals like velocities, positions and motor currents.

In this way, we can tune our control software for a better control over our hardware, and also implement new control strategies rapidly in our model before moving to the actual implementation in theCoPS control software.

In the near future, we want to extend the model in order to simulate not only a single robot, but a team of robots. Additionally, we want also to simulate the ball and other objects in the field, as well as components like the kicking device. Future work will also include the possibility of learning these parameters using other sensory data then odometry, e.g. the global self-localization.

References

1. Buchheim, T., Kindermann, G., Lafrenz, R., Levi, P.: A dynamic environment modelling framework for selective attention. In: Visser, et al. (eds.) Proceedings of the IJCAI 2003 Workshop on Issues in Designing Physical Agents for Dynamic Real-Time Environments (2003)
2. Craig, J.J.: Introduction to Robotics - Mechanics and Control, 3rd edn. Pearson Prentice Hall (2005)
3. Williams II, R.L., Carter, B.E., Gallina, P., Rosati, G.: Dynamic Model With Slip for Wheeled Omnidirectional Robots. IEEE Transactions on Robotics and Automation 18(3), 285–293 (2002)
4. Zweigle, O., Käppeler, U.-P., Lafrenz, R., Rajaie, H., Schreiber, F., Levi, P.: Situation recognition for reactive agent behavior. In: Artificial Intelligence and Soft Computing (ASC) (2006)
5. Zweigle, O., Lafrenz, R., Buchheim, T., Käppeler, U., Rajaie, H., Schreiber, F., Levi, P.: Cooperative Agent Behavior Based on Special Interaction Nets. In: Arai, T. (ed.) Intelligent Autonomous Systems 9 (IAS-9) (2006)

Intuitive Plan Construction and Adaptive Plan Selection

Kai Stoye and Carsten Elfers

Universität Bremen, Germany
{kstoye,celfers}@informatik.uni-bremen.de

Abstract. Typical tasks of multi agent systems are effective coordination of single agents and their cooperation. Especially in dynamic environments, like the RoboCup soccer domain, the uncertainty of an opponent's team behavior complicates coordinated team action. This paper presents a novel approach for intuitive multi agent plan construction and adaptive plan selection to attempt these tasks. We introduce a tool designed to represent plans like in tactical playbooks in human soccer which allows easy plan construction, editing and managing. Further we introduce a technique that provides adaptive plan selection in offensive situations by evaluating effectiveness of plans and their actions with statistically interpreted results to improve a team's style of play. Using experts as a concept for abstracting information about a team's interaction with another, makes fast accommodated plan selection possible. We briefly describe our software components, examine the performance of our implementation and give an example for rational plan selection in the RoboCup Small Size League.

1 Motivation and Related Work

A common hypothesis is, that the more a team's behavior is adapted to the opponent, the more effective it is. There are two major ways for adaption: One is to analyse the opponent and adapt the own team behavior to it (e.g. [2,8]). The other way, introduced in this work, is to analyse the own team effectiveness corresponding to the opponent and adapt team behavior thereby. The idea behind our approach is to enhance the B-Smart software[1], developed for the *RoboCup Small Size League*, which uses more or less only reactive behavior selection, with a deliberative component providing strategic moves and multi agent coordination. Deliberative principles become more and more one of the most challenging aspects within the focal point of artificial intelligence in RoboCup (e.g. see team descriptions to appear in [7]). This approach is restricted to offensive game situations and uses predefined stepwise action sequences, like tactical playbooks in human soccer (cf. [9]), further called plans. The approach is based on the work of Bowling et al. [3,4] who are using a play (multiagent plan)-based coordination and opponent-adaption for the CMDragons Small Size Team. However, our approach differs in the way of plan/play construction and adaptive selection. Our

---

[1] For further information see: http://www.b-smart.de/

U. Visser et al. (Eds.): RoboCup 2007, LNAI 5001, pp. 278–285, 2008.

implementation is divided into two parts. One separate component for creating and managing plans, the *Strategist*, and one for plan selection, execution and assessment, the *Coach*.

2 Plan Structure

In this section we briefly describe the structure of a plan. The highest level in our hierarchy is the *planbase*, which is a container for multiple *plans*. The planbase is used as a superior structure for classification and to provide a possibility for a (topical) subsumption of plans to improve clarity. Plans consist of *variants* and *plan steps*. A single step is defined as a tuple of *conditions* and *actions*. Conditions are restrictions which must be satisfied to enter the corresponding plan step. The variants are lists of conditions containing at least one element to describe the entry condition for a specific plan, s. Fig. 1. In order to allow different entry conditions for the same plan, different variants can be

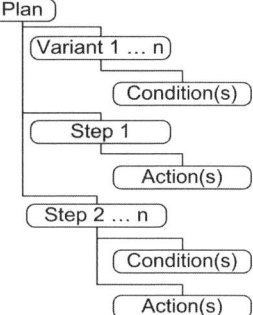

Fig. 1. Elementary plan structure

specified. Unlike all other steps, the first step only has a list of actions without any conditions, because these were already defined in the variant(s). These actions should be performed, if one of the variants matches with the actual game situation. Single conditions and actions, in contrast, must all be fulfilled. After the first plan step an arbitrary number of steps can follow. In our approach actions and conditions are predefined constructs. Conditions can be separated into different classes. *Assignment Conditions* are restrictions, which can change the world representation[2]. *Dependent Conditions* are restrictions, which are dependent on Assignment Conditions in order to determine their validity. The third class are *Independent Conditions*. In contrast to the other classes, they are independent to prior assignments and to the evaluation of other conditions. Also other constructs are needed for a complete plan step, the *actions*; they trigger agents' behaviors during plan execution. Typically a plan always ends with a goalshot, because it is the main aim in offensive situations to score for your team. Possibly it can be imagined that there also exists plans for gaining space or other intentions.

3 Plan Construction

In this section the software component for constructing plans will be briefly introduced, the *Strategist*. It is a stand-alone program designed for: Creation of plans based on the already mentioned plan elements, editing plans and managing plans and the associated planbases. One design principle for the Strategist is to

---

[2] World representation means a representation of the "world", which holds beside the world model information about internally used assignments.

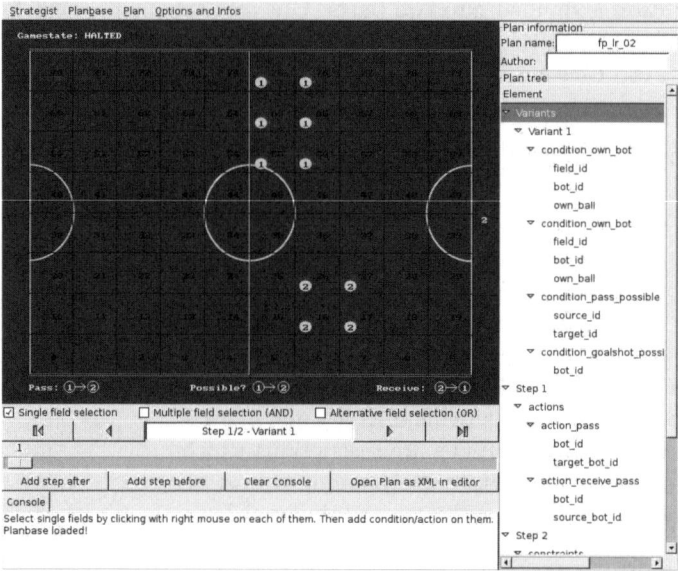

Fig. 2. B-Smart Strategist

provide an easy and intuitive useable interface for these tasks. Therefore, it uses an abstract visualization of the playing field and the plan elements, where these elements can be added or removed simply by using the mouse. This approach was inspired by, as it is common in some "real" sports, blackboard applications for explaining tactical moves. An impression of the software layout is illustrated in Fig. 2. Inspired by this, plans can be designed by human users and must not be generated automatically, like in other approaches, e.g. systems using a planner to generate possible strategies (e.g. [6]). We decided to use a qualitative world model with grid rastering. That means, the playing field is divided into equal rectangular fields, in our implementation 8 · 10 regions, where every field is numbered consecutively, so that it is explicit. This abstraction level seems to be a good trade-off between search complexity and accuracy. This concept of world representation is inspired by Schiffler et al. [11].

4   Plan Selection and Execution

4.1   Identification of Applicable Plans

An important step towards the execution of a plan is the determination of the subset of plans having a consistent variant regarding the current world model. An effective way to find this subset is to refine the search space by the position of the ball carrier. As previously described a wholly offensive plan execution has been implemented. An important property that this assumption provides, is that every variant has exactly one condition, specifying the ball carrier's position.

This feature is used for a fast refinement of the search space of plans. While loading the planbase, variants will be checked by the position of the ball carrier in the variants. This positional information is used as a key to address a list of possibly applicable plans. Doing this, the set of plans will be reduced by the plans with a ball carrier on an inconsistent position considering the current world model. In other words, we define a subset of plans, the set of possibly applicable plans with all plans containing a consistent position of the ball carrier to the current world model that can be obtained by a simple lookup.

$$Plans_{applicable} \subseteq \text{Plans}_{possiblyApplicable} \subseteq \text{Plans} \tag{1}$$

As previously described, some conditions depend on others. This makes it impossible to prove the consistency of a condition set in a commutative way. However, within one Condition Class, the consistency check is commutative. Assignment Conditions do not depend on any other conditions, Dependent Conditions only depend on Assignment Conditions and Independent Conditions do not depend on any other conditions either. To prove a set of conditions

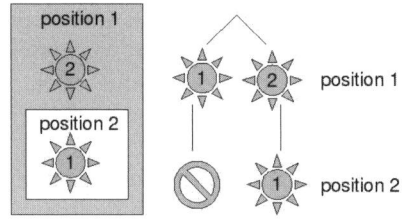

Fig. 3. Example for backtracking

of different condition classes correctly by consistency, methods to prove commutative sets with constraints (like CSPs, s. [10, ch. 5]) are not appropriate. It has to be ensured that all Assignment Conditions are proven before Dependent Conditions. To do so, all condition sets are ordered while loading the planbase by their condition class. The order of conditions allows the reduction of the branching factor for the consistency check to prove the condition set of satisfiability from $n! \cdot d^n$ to d^n (cf. [10, ch. 5]), with n conditions and d possible assignments. In order to prove a given set of conditions by consistency a modified backtracking algorithm is used to check the constraints between Dependent Conditions and Assignment Conditions. This algorithm differs from regular backtracking (cf. [10, ch. 5]) by using the given order of conditions to check consistency, assign values and solve conflicts. For every possible assignment to a condition a new node will be created and recursively checked until a consistent assignment will be found or all paths have been visited. The need for backtracking to check Assignment Conditions is exemplified in Fig. 3 with two players, player one on position 1 and 2 and player two only on position 2 on the left side of the figure. On the right side the resulting tree for the backtracking is illustrated. At first position 1 will be assigned to player one, but this assignment leads to a conflict, because no position could be assigned to player two. The conflict is solved by assigning position 1 to player two and assigning position 2 to player one.

To reduce the set of possibly applicable plans to the set of applicable plans, all variants of all possibly applicable plans are checked by the modified backtracking algorithm of consistency. Only plans with at least one consistent variant are in the set of applicable plans $Plans_{applicable}$.

4.2 Plan Assessment

Since this approach only allows one plan to be executed at a time, one plan has to be selected from the set of applicable plans. Assessing applicable plans is the first step that permits a reasonable decision. In this approach the assessment of applicable plans will be accomplished by experts. Experts are optional modules, which gather relevant information during plan execution to assess plans or aspects of plans. Most experts are able to abstract from the gathered information to evaluate similar plans. Four basic experts have been implemented, the Dribble Expert, the Goalshot Expert, the Standard Expert and the Pass Expert. Each expert evaluates one aspect of each plan[3], except the Standard Expert which evaluates the wholly plan without regarding one special aspect of the plan additionally to the individual experts. Each expert holds a table with the amount of successful executions and failures of plans or an aspect of plans.

If an expert is called to assess a given plan, it has to return a probability that contains the expert's assumption about the observed actions in the plan being successfully executed. In order to interpret a given set of samples statistically, experts need a function, which relates the given samples to a probability. The Beta distribution provides useful properties to get these probabilities to assess plans, like the mode and the probability density defined by two hyperparameters specifying the shape of the distribution (s. [10, ch. 19]). In this case the two free variables/hyperparameters are specified by the number of successes and the number of failures defining the shape of the distribution. With these two variables the distribution is fully defined and we remember that these two variables are gathered by the experts. For a low amount of samples the Beta distribution mode could result in extreme probabilities like 0 or 1. This behavior is not preferable, because it leads to excluding plans and avoids reinvestigating plans that have been failed in the beginning. Therefore we define the Beta 2-2 distribution similar to the Beta distribution by adding a fair sample, 2 successes and 2 failures. This avoids interpreting a low amount of samples inadequate funded, but still converges with a higher amount of samples to the expected value. Another feature is that the mode can be defined for unavailable a-priori evidences easily by the amount of successes $succ$ and the amount of failures $fail$ as $mode_{beta22[succ,fail]} = \frac{succ+1}{succ+fail+2}$ (cf. [1]).

In order to provide an assessment of successful executions more or less decisive than unsuccessful executions, the number of successes and failures can be factorized (cf. [3]). This allows the configuration of more optimistic or more pessimistic evaluations. The overall assessment for an expert of a given plan is defined by the product of the single assessments of each action examined by the expert, because each action is viewed as an independent event. The overall evaluation of all experts is computed in the same way, by multiplying all experts' evaluations, giving the probability about all aspects of the plan will be successful.

---

[3] E.g. the Dribble Expert evaluates dribbling actions.

4.3   Plan Selection and Execution

In order to decide which plan should be executed if more than one plan is applicable, the evaluation of each applicable plan will be normalized to 1. The normalization gives a probability distribution for the plan selection (cf. [4]), used to decide on a plan randomly, with a higher probability of choosing a high assessed plan. With a lower probability, weaker plans will be executed, which allows a reassessment of these plans. This is a helpful method to avoid an opponent predicting the plan our system is selecting and to reassess the effectiveness of our team's plans in case of the opponent team is changing their behavior or the bad luck the own team had during these plan's executions.

After selecting a plan, it will be executed. Execution means to perform the action sequences defined by the plan in the order of the steps. A transition from one step to the following is accomplished, if all conditions of the following step that are related to the ball carrier, are satisfied. If the following step contains unsatisfied conditions, but the conditions related to the ball carrier are satisfied, the plan execution will be aborted. This prevents the ball carrier from waiting for other involved players to perform their actions completely and avoids impeding the flow of the game. If a plan execution is aborted or a plan/step has been completely performed, the experts will be notified and extract relevant information. If a goal opportunity exists the plan will be stopped and the chance to score will be used. The selection, adaption and (high-level) plan execution is implemented as an optional module in the B-Smart Agent Software, called Coach. The low-level execution of plans' actions is realized by the Agent Software.

5   Preliminary Evaluation

In this section we consider performance measurements of plan selection and examine a sequence of plan executions and their resulting assessments. First we start with the performance measurements. Therefore, a series of 100 plan selections for each test with up to 100 practical plans have been performed on an AMD Athlon XP 2000+ with 1666 MHz and 1280 Mb RAM (s. Fig. 4). Each test differs in the number of loaded plans, listed in the first column. The second column, *appl.*

plans	appl. plans			required time		
	min	max	Ø	min	max	Ø
10	1	1	1	0 ms	6 ms	1.2 ms
20	1	9	3.49	0 ms	19 ms	6.26 ms
100	1	22	4.3	1 ms	22 ms	9.52 ms

Fig. 4. Performance test for plan selection

plans, lists the amount of applicable plans found by one selection request. The last column, *required time*, shows the time used for one plan selection phase. Clearly the required time depends on more factors than the amount of plans and the amount of applicable plans found, but the given criteria are of capital importance. To examine an example of a sequence of plan executions with the resulting assessments, we partly specify two plans by their executed actions in Fig. 5 and the execution sequence with the resulting plan assessments in Fig. 6

delivered by the Beta 2-2 mode. We assume that successes and failures are equally factorized by 1. Further, we assume that the three specified plans have the same variants, so they will be applicable at the same game situations.

First have a look at the a-priori assessments in column one in Fig. 6. It can be seen that plan B will be executed at 67%, this seems to be rational by the assumption that short plans have a higher success-rate than longer plans. We assume that plan B will be chosen and aborted in the first step, while performing a short pass. We notice that thereby plan A gets a higher assessment, but is assumed to be still less ef-

plan	step 1	step 2	step 3
A	*dribble*	*pass*	*goalshot*
B	*pass*	*goalshot*	

Fig. 5. Example plans

fective than plan B. This is explained by the aborted pass which exists in plan A too. Now we assume that plan A is chosen and that this plan will be executed successfully. Now we see that we cannot assume that a pass is more or less effective, because one time it was successful and the other time not. But we know that plan B failed and plan A was successful, so it is rational that our concept assumes a higher success probability for plan A than for plan B and will next choose plan A by 58%. It can be noticed, that the more promising plans become higher assessed at an accommodate speed and in a rational way. As we see, an essential feature this approach provides is to assess plans appropriately even before they have been executed the first time by using abstracted evaluations from experts, in our example plan A gets an higher assessment even before plan A have been tried the first time.

execution	plan	evl_{pass}	$evl_{dribble}$	$evl_{goalshot}$	$evl_{standard}$	evl_{all}	norm
a-priori values	A	0.5	0.5	0.5	0.5	0.06	0.33
	B	0.5	1	0.5	0.5	0.063	0.67
plan B failed	A	0.33	0.5	0.5	0.5	0.04	0.43
while passing	B	0.33	1	0.5	0.33	0.05	0.57
plan A was	A	0.5	0.67	0.67	0.67	0.15	0.58
successful	B	0.5	1	0.67	0.33	0.11	0.42

Fig. 6. Sequence of plan executions with assessments

6   Conclusions and Future Work

As shown in the evaluation, we implemented a fast and rational approach for adaptive plan selection for offensive soccer situations. Plans and actions are rated by their successes what indirectly models weakness of opponents' play and strengths of the own team abilities. This model is used to adapt the own team's play and allows to exploit the opponent's flaw. The fast adaption further allows to be prepared for unknown opponents. The intuitive plan construction software allows easy plan creation and provides useful managing features. All in all the whole system is a useful extension of the B-Smart team software and extends the current system by a deliberative concept. However, this system is not limited to the Small Size League and could be integrated in other leagues as well.

For future work more properties of the beta distribution can be used, e.g. information about the certainty of the experts' assessments could be useful for selecting plans. Another rational feature would be to evaluate positions on the field depending on the current situation, e.g. this could be done by using potential fields like Vail and Veloso [12] did, or by Voronoi diagrams used by Dylla et al. [5]. Currently, each player tries to reach the middle of the target region defined in the actions, without considering the current game situation. Potential fields or Voronoi diagrams could help to choose a proper target for these actions. We introduced the concept of experts assessing plans and their actions. This is currently done in a rudimentary way and needs to be improved. Another useful feature could be to use previously recorded assessments by the experts to support the user while creating plans in the Strategist by estimating the success-probability of the edited plans and actions.

References

1. Balakrishnan, N., Galloway, V.B., Nevzorov, V.B.: A Primer on Statistical Distributions. Wiley-IEE (2003)
2. Ball, D., Wyeth, G.: Classifying an Opponent's Behaviour in Robot Soccer. In: Proceedings of the 2003 Australasian Conference on Robotics and Automation (ACRA) (2003)
3. Bowling, M., Browning, B., Veloso, M.: Plays as effective multiagent plans enabling opponent-adaptive play selection. In: Proceedings of International Conference on Automated Planning and Scheduling (ICAPS 2004) (2004)
4. Bowling, M., Browning, B., Chang, A., Veloso, M.: Plays as Team Plans for Coordination and Adaptation. In: Polani, D., Browning, B., Bonarini, A., Yoshida, K. (eds.) RoboCup 2003. LNCS (LNAI), vol. 3020, pp. 686–693. Springer, Heidelberg (2004)
5. Dylla, F., Ferrein, A., Lakemeyer, G., Murray, J., Obst, O., Röfer, T., Stolzenburg, F., Visser, U., Wagner, T.: Towards a league-independent qualitative soccer theory for RoboCup. In: Nardi, D., Riedmiller, M., Sammut, C., Santos-Victor, J. (eds.) RoboCup 2004. LNCS (LNAI), vol. 3276, pp. 611–618. Springer, Heidelberg (2005)
6. Fraser, G., Wotawa, F.: Plan Description and Execution with Invariants. OGAI Journal 22(4), 2–7 (2003)
7. Lakemeyer, G., Sklar, E., Sorrenti, D.G., Takahashi, T.: RoboCup 2006: Robot Soccer World Cup X. LNCS (LNAI), vol. 4434. Springer, Heidelberg (2007)
8. Lattner, A.D., Miene, A., Visser, U., Herzog, O.: Sequential Pattern Mining for Situation and Behavior Prediction in Simulated Robotic Soccer. In: Bredenfeld, A., Jacoff, A., Noda, I., Takahashi, Y. (eds.) RoboCup 2005. LNCS (LNAI), vol. 4020. Springer, Heidelberg (2006)
9. Lucchesi, M.: Coaching the 3-4-1-2 and 4-2-3-1. Reedswain Publishing (2001)
10. Russell, S.J., Norvig, P.: Artificial Intelligence: A Modern Approach. Prentice Hall (2002)
11. Schiffer, S., Ferrein, A., Lakemeyer, G.: Qualitative World Models for Soccer Robots. In: Conference Paper - KI 2006 WS on Qualitative Contraint Calculi. Knowledge-based Systems Group, RWTH Aachen (2006)
12. Vail, D., Veloso, M.: Multi-Robot Dynamic Role Assignment and Coordination Through Shared Potential Fields. In: Multi-Robot System, pp. 87–98. Kluwer (2003)

Semi-autonomous Coordinated Exploration in Rescue Scenarios

S. La Cesa, A. Farinelli, L. Iocchi, D. Nardi, M. Sbarigia, and M. Zaratti

Dipartimento di Informatica e Sistemistica
Università di Roma "La Sapienza"
Via Salaria 113, 00198 Roma, Italy
last-name@dis.uniroma1.it

Abstract. In this paper we study different coordination strategies for a group of robots involved in a search and rescue task. The system integrates all the necessary components to realise the basic behaviours of robotic platforms. Coordination is based on iterative dynamic task assignment. Tasks are interesting points to reach, and the coordination algorithm finds at each time step the optimal assignment of robots to tasks. We realised both a completely autonomous exploration strategy and a strategy that involves a human operator. The human operator is able to control the robots at different levels: giving priority points for exploration to the team of robots, giving navigation goal points to team of robots, and directly tele-operating a single robot. For building a consistent global map, we implemented a centralised coordinated SLAM approach that integrates readings from all robots. The system has been tested both in the UsarSim simulation environment and on robotic platforms.

1 Introduction

The use of mobile robotic platforms for search and rescue missions is envisioned as a critical issue for society. Mobile robotic platforms can consistently help human operator in dangerous or complex tasks, providing important information for areas that cannot be directly reached.

To consistently help the human operator during rescue missions, robotic platforms should exhibit a certain level of autonomy in their behaviours. break Semi-autonomous robots can process acquired data and build a high-level representation of the surrounding environment. Moreover, robots can act in the environment (e.g. navigate) with only a limited interaction with the human operator. In this way, the human operator can easily control multiple robots providing high level commands (e.g. "explore this area", "reach this point", etc.). Moreover, in case of temporary network breakdown, the mobile bases can continue the execution of the ongoing task and return to a predefined base position.

In this paper we focus our attention on coordinated autonomous exploration by a team of mobile robots in an unstructured environment. A team of coordinated mobile robots can explore an environment faster than a single robot, moreover, using several robots the system can be robust to platform failures.

U. Visser et al. (Eds.): RoboCup 2007, LNAI 5001, pp. 286–293, 2008.
© Springer-Verlag Berlin Heidelberg 2008

Autonomous exploration has been deeply investigated in mobile robot literature [1,2]. The main problem of autonomous exploration is to choose a sequence of targets reachable by a robot, so to explore all the environment optimising an objective function. As for multi-robot exploration, the main problem is to coordinate the robot activities so to avoid conflicts in the exploration process. In particular, the goal is to spread the robots in the environment collecting all available information [3].

This paper makes two main contributions. First, we present a novel semi-autonomous coordinated exploration system for a team of mobile robots involved in a rescue scenario. The main novelty of the approach is to devise a semi-autonomous strategy that allows an operator to give high level advices to the entire team of cooperating robots. The team is responsible to coordinate and decide who should fulfil the user requirements. Second, we present an evaluation methodology for exploration strategies. Our methodology is based on an extensive evaluation of several experiments on a high fidelity simulated environment (i.e., the UsarSim environment), and then validation of the approach with real robotic platforms. We are able in this way to compare and analyse different coordination strategies for the exploration. In particular, we compared the totally autonomous coordination exploration with the semi-autonomous approach.

The idea of providing high level advices to a team of agents has been addressed in the DEFACTO system by Schurr et al. [4]. The DEFACTO system is a system built on top of the RoboCup rescue simulator used to train incident commander for intervention in large scale urban emergencies. Coordination is provided using Machinetta, a general framework for teamwork in multi-agent system [5]. With respect to the DEFACTO system our work is more focused toward specific robotic problems (e.g., cooperative SLAM, motion planning, etc.), moreover, we specifically focus on a single aspect of the coordination problem (i.e., task assignment) to have a full evaluation of this issue.

The work by Wang et al. [6] addresses the cooperation of rescue robots supervised by human operator. Also Wang et al. use the Machinetta framework for ensuring teamwork among robotic platforms. The work evaluates the cooperation between robots and human operator via a series of experiments involving different users. As before, our approach is more focused toward the specific evaluation of task assignment, and the use of real robotic platform.

2   Coordinated Exploration Strategies

The problem of coordinating the exploration of a set of robots can be conveniently formalised as a task assignment problem [3]. Task Assignment is a very well known approach to address coordination of autonomous robot activities. The classical task assignment formulation is to assign a utility value to the task to be executed. The utility function is dependent on the task and on the status of the agent that is allocated to that task. The goal of the task assignment is to maximise the sum of the utility of the allocation of agents to tasks.

We realised two coordination strategies: one is distributed and robots are not controlled by the human operator, while the second one is supervised and centralised.

2.1 Distributed Autonomous Coordination

In this strategy robots are completely autonomous.

A task assignment strategy is employed to allocate robots to different tasks. A task in our environment is a goal target to be reached by the robot Our approach extends the method proposed in [7]. In particular, with respect to [7], in our approach the number of tasks to be executed is not fixed, and there is no total priority order for the tasks.

Our approach works as follow: each robot maintains a structure containing the tasks known to all the agents. Each robot locally computes the current target points to reach, and verifies that they are not within the current tasks already known to the system. To compare the tasks a simple nearest neighbour technique is used. Each robot sends in broadcast the new tasks to all team mates, computes its utility function for all the tasks present in the system and broadcasts the function values to all other team mates. Each robot computes autonomously the best allocation of robots to targets and then execute the best task according to the chosen allocation. The best allocation is computed considering all possible assignment of robots to tasks, and choosing the one that maximises the sum of utility functions.

The algorithm used to compute the utility function is shared among all robots and is based on specific parameters that influence the task execution (e.g. distance to travel). Since all robots use the same algorithm with the same data, eventually they will all converge to the same solution, even though temporary oscillations of the algorithm might happen due to noise in the robot local estimate of the utility functions. Following the approach in [7] we use hysteresis on the allocated tasks to avoid this problem. Whenever a task is accomplished by a robot, a message is sent broadcast to all other team mates to update the task set.

2.2 Centralised Semi-autonomous Coordination

The supervised centralised coordination strategy relies on a base station common to all the robots, where all information are channeled. A human operator is in charge of supervising the robot team by monitoring the mission execution and by controlling the robots at different levels.

The central station is in charge of combining the single robot readings and provide a global comprehensible picture to the human operator. In particular, the base station combines the laser readings of different robots to build a consistent joint map. To merge the laser reading a Rao-Blackwellized particle filter method is employed. The method extends the work presented in [8], considering the joint states of all the robots as the variable to be estimated.

To ensure an effective control by the human operator a multi-robot graphic user interface has been developed within the RDK framework [9] (see figure 1). The multi-robot console allows the human operator to see the global map built by the robot team, to see interesting states of the robots, and to control them.

Robots can be controlled using one of the following operational modes: Autonomy, Navigation, Operated. When robots are in autonomy mode they will execute the exploration strategy described in [10]. Moreover, they coordinate using the algorithm described in section 2.1. When the Navigation mode is selected, the robot waits for a goal point to be provided by the operator. When a goal point is provided, the robot will autonomously navigate towards the specified point using the approach described in [11]. Finally, when the Operated mode is selected the robot will not try to act proactively, waiting for low level commands from the operator (e.g., joystick commands). Both the Navigation and the Autonomous modes accept target points by the operator. When a target point is inserted, it is sent to all the robots. Each robot will then perform the task assignment strategy specified in the previous section, treating the goal point sent by the human operator as a high priority task. In this way, the human operator does not have to decide which robot is in charge of exploring a particular area but can just signal interesting parts of the map to the robots and monitor their execution. If a robot is in the Navigation mode, when it reaches the goal point provided by the human operator, it will stop waiting for another goal point. If a robot is in Autonomy mode, after reaching a human goal-target, it will keep executing the frontier base exploration from the current position.

Notice that while the global map is estimated by a centralised process each robot maintains a local map built autonomously. Therefore if a communication breakdown interrupts the link between one robot and the central station, the robot is still able to perform its tasks reasoning on its local map. The global map is used only by the human operator to monitor the mission execution and to control the robots.

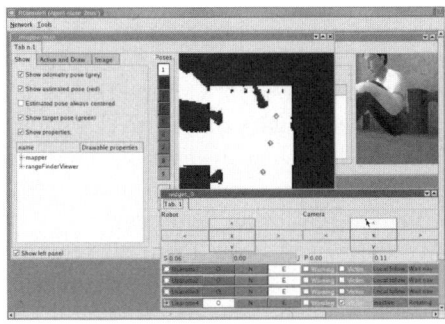 

Fig. 1. The multi robot control console   **Fig. 2.** Map used for the experiments

3   Experiments and Results

We conducted extensive experiments using the UsarSim [12] simulation environment, and then validated our approach on real robotic platforms.

3.1   Experiments in UsarSim

Our evaluation methodology is to perform experiments on a benchmark scenario and extract significant metrics for each mission, varying some interesting parameters.

To acquire quantitative data we used the UsarSim simulation environment. UsarSim is a 3D high fidelity simulation of Urban Search And Rescue (USAR) robots and environments. UsarSim is a valid tool for the study of basic robotic capabilities in 3D environment. It offers several 3D models of robotic platforms (P2DX, P2AT, Zerg, ATRVJ, etc. etc.) and sensors (Laser Range Finder, IMU, Color Camera, etc).

A sensible metric to evaluate the performance of an exploration mission is the time needed to complete the exploration.

Figure 2 reports a picture of the map used in the experiments. An important parameter to consider into the experiments is the dimension of the map.

Our reference map is composed by a central corridor (16 meters long and 3 meters wide) and a series of rooms. We divide this map into three different maps: a small size map (one third of the corridor plus the two small adjacent rooms), a medium size map (two third of the corridor and two additional rooms) and a large size map (the whole environment). Notice that, while the map used for experiments has a clear structure, the representation that robots have of the map makes no assumption of such a structure. In fact, experiments with real robots have been performed in a much less structured environment (see next section). The only assumption that must holds for our exploration method to work correctly is to have planar environment. This is due to the SLAM algorithm which is designed for such kinds of environments.

We used two wheeled robots: P2AT and P2DX, both of them equipped with a Laser Range Finder.

To have a baseline value to compare between the different coordinated strategies, we measured the time needed by a single autonomous P2AT[1] and then compute the percentage of time used by the other strategies with respect to this value.

The different strategies we compared are the following: i) Two autonomous non-coordinated platforms *AutNotCoord*; ii) Two autonomous coordinated platforms *AutCoord*; iii) Two supervised coordinated platforms *SupCoord*

Table 1 reports the obtained results. We performed several experiments for each strategy. Since each experiment involves the exploration of a consistent part of a building, the time of completion for each mission depends on several parameters. Therefore, the collected results have a consistent variance and cannot be well represented with a simple average. To take into account this issue, we report the interval between the minimum and maximum values obtained.

---

[1] The exploration time of a single autonomous P2DX is very similar to the P2AT.

Table 1. Comparison among the different coordination strategies over different environments

	Small	Medium	Large
AutNotCoord	80%	80%-85%	80%-90%
AutCoord	75%	70%-75%	65%-75%
SupCoord	65%	55%-60%	45%-55%

The results show several interesting points. First of all, the autonomous coordinated system has better performance with respect to the autonomous not coordinated one. In particular, coordination plays a crucial role for larger spaces. This is because the larger the environment the more important is to avoid that the same portion of space is covered by more than one robot. Second, the supervised coordinated strategy outperforms the autonomous coordinated strategy. This can be explained considering that the human operator has more information than the single robots. The autonomous coordinated strategy we employ, exchanges only tasks to be accomplished by the robots, trying to optimise the task allocation process. In particular, for our exploration scenario the robots exchange only the current goal points of the exploration strategy, and they do not exchange maps or map patches. When the supervised coordination strategy is used, the human operator has a more complete knowledge of the environment (given by the global merged map) and can make more informed decisions. However, notice that the supervised strategy requires a high amount of information to be transferred from the robots to the central base station. Moreover, the central base station should be always reachable through direct communication with every robots.

Since the coordination algorithms evaluated and the exploration strategy used do not make any assumption on the particular structure of the map, we believe that obtained results should be valid for other types of environments and maps. Moreover, since the computation of utility function for the Task Assignment is demanded to each single robotic agent, the coordination algorithm can be extended to take into account robots with heterogeneous sensing and mobility capabilities.

However, a deeper investigation is needed, in order to clearly understand how the obtained results relate to such situations.

3.2   Experiments with Real Robots

We validated our approach with two mobile platforms: a P2DX equipped with an Hokuyo Laser Range finder, and a P2AT equipped with a SICK Laser Range Finder. The experiments have been conducted in the arena set up in our lab.

Figure 3 shows the maps created by the robots during their mission. The environment to explore is 7×6 square meters, and the two robots completed the exploration in 10 minutes approximately. From left to right it is possible to see the initial situation, a snapshot during the exploration process and the final map. The P2DX is represented with a circle and the P2AT is represented with a square. In the maps it is possible to see the current tasks the robots are allocated to (crosses in the map). Robots performed a coordinated supervised exploration. Giving high level advices the operator was able to efficiently control the system, nicely spreading the two robots.

Fig. 3. Cooperative exploration sequence

Figure 4 shows the two maps of the single robot. These are the maps the two robots maintain locally. As it is possible to see the overlapping among the two maps is minimal, as it is desirable in a multi-robot exploration task. On the other hand, a bigger overlap between the two maps would have been beneficial for the cooperative SLAM process, and would have produced a better quality global map. In this work, we focused on minimising the exploration time rather than having a better quality map.

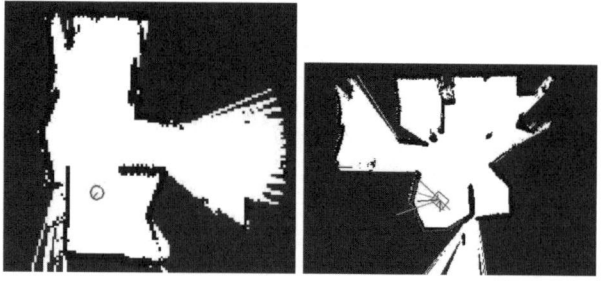

Fig. 4. Maps of each single robots: P2DX left and P2AT right

4   Conclusions and Future Work

In this work we presented an evaluation of different coordination strategies in a rescue scenario. We designed, realised and compared a totally autonomous coordination strategy and a supervised coordination strategy. Quantitative experiments have been performed using the UsarSim environment, and the system has been validated also on real robotic platforms.

An interesting future topic will be to extend our current approach to consider possible lack of communication link between robotic platforms and the human operator. This is a very important issue to consider and has a significant impact on the overall coordination strategy used in our approach.

References

1. Stachniss, C., Grisetti, G., Burgard, W.: Information gain-based exploration using rao-blackwellized particle filters. In: Proceedings of Robotics: Science and Systems (RSS), Cambridge, MA, USA (2005)
2. Yamauchi, B.: A frontier based approach for autonomous exploration. In: IEEE Int. Symp. on Computational Intelligence in Robotics and Automation (1997)
3. Zlot, R., Stenz, A., Dias, M.B., Thayer, S.: Multi robot exploration controlled by a market economy. In: Proc. of the Int. Conf. on Robotics and Automation (ICRA), pp. 3016–3023 (2002)
4. Schurr, N., Patil, P., Pighin, F., Tambe, M.: Using multiagent teams to improve the training of incident commanders. In: Proceedings of the Industry Track of the Fifth International Joint Conference on Autonomous Agents and Multiagent Systems (2006)
5. Scerri, P., Pynadath, D.V., Johnson, L.: A prototype infrastructure for distributed robot-agent-person teams. In: Proceedings of AAMAS (2003)
6. Wang, J., Lewis, M., Scerri, P.: Cooperating robots for search and rescue. In: Agent Technology for Disaster Management Workshop at AAMAS 2006 (2006)
7. Iocchi, L., Nardi, D., Piaggio, M., Sgorbissa, A.: Distributed coordination in heterogeneous multi-robot systems. Autonomous Robots 15(2), 155–168 (2003)
8. Grisetti, G., Stachniss, C., Burgard, W.: Improving grid-based SLAM with Rao-Blackwellized particle filters by adaptive proposals and selective resampling. In: Proceedings of the IEEE International Conference on Robotics & Automation (ICRA) (2005)
9. Farinelli, A., Grisetti, G., Iocchi, L.: Design and implementation of modular software for programming mobile robots. International Journal of Advanced Robotic Systems 3(1), 37–42 (2006)
10. Calisi, D., Farinelli, A., Iocchi, L., Nardi, D., Pucci, F.: Multi-objective autonomous exploration in a rescue environment. In: Proc. of IEEE Int. Workshop on Safety, Security and Rescue Robotics (SSRR), Gaithersburg, MD, USA (2006)
11. Calisi, D., Farinelli, A., Iocchi, L., Nardi, D.: Autonomous navigation and exploration in a rescue environment. In: Proc. of IEEE Int. Workshop on Safety, Security and Rescue Robotics (SSRR), Kobe, Japan, pp. 54–59 (2005) ISBN: 0-7803-8946-8
12. Carpin, S., Lewis, M., Wang, J., Balakirsky, S., Scrapper, C.: Bridging the gap between simulation and reality in urban search and rescue. In: Lakemeyer, G., Sklar, E., Sorrenti, D.G., Takahashi, T. (eds.) RoboCup 2006: Robot Soccer World Cup X. LNCS (LNAI), vol. 4434. Springer, Heidelberg (2007)

A Deeper Look at 3D Soccer Simulations

Amin Habibi Shahri, Ali Almasi Monfared, and Mohammad Elahi

Electrical & Computer Engineering Faculty
Shahid Beheshti University, Tehran, Iran
{habibiamin,almasimonfared,elahimohamad}@gmail.com

Abstract. Developing an intelligent agent requires more than an Integrated Development Environment (IDE). In multi agent environments or systems equipped with artificial intelligence it is often difficult to obtain the function or method which led to a particular behavior that is noticeable from outside. In addition to previous dilemma, the publicity that the RoboCup events get from the media provides an ideal opportunity to show the state of art of these systems during RoboCup World Cup.

This paper describes the concept and the implementation of *Team Assistant 2006* as the next generation of TA2002. The idea is to provide a tool that is able to assist developers to detect problems of their agents both in single and cooperation mode and also organizers to have better games.TA2006 won the second place in RoboCup 3D development competition 2006.

Keywords: RoboCup, Soccer Simulation, Presentation, Analyzing.

1 Introduction

Developing an intelligent agent requires more than an Integrated Development Environment (IDE). Many visualization and analysis requirements may arise at any time in the project; and developing a tool to satisfy each requirement is cumbersome and time-consuming.

In the past 10 years the simulation league was two dimensional, all players and even the ball moved on the ground. During this time numerous sophisticated tools were created for analyzing the simulated games such as Logalyzer or Team Assistant[4].

The Logalyzer provides information about detected actions like passes and several visualizations for the collected data about the game. The Team Assistant is able to display information provided via agent logfiles along with statistics about detected actions. The Team Assistant is also mentioned in the 2002 league summary[7] as the winner of the presentation tournament.

In 2003, the 3D simulation was introduced including basic tools to view and replay the simulated game. The tools used in 2D can not be used in 3D simulations because of the lack of one dimension and a different format of the logfiles.

The current monitor is capable of showing the current simulated game (at current time) and of replaying monitor logfiles. The replaying mode can be used

U. Visser et al. (Eds.): RoboCup 2007, LNAI 5001, pp. 294–301, 2008.

to watch previously simulated games again. There is also a "single step mode" which provides slow motion replay. When trying to develop a behavior for an agent or verifying behaviors acquired by machine learning methods it is hard to determine which methods or functions led to the actions observed in the game. This information is crucial when trying to debug or improve the agents in their behavior and collaboration. Especially in the case of collaboration it is tedious and time-consuming to check what each agents intention is. This is caused by the fact that every logfile has to be searched for the right record of the actual time displayed in the monitor by hand. Additionally, the agent logfiles are not numbered according to the uniform numbers of the agents which complicates finding the desired logfile.

The aim of this work as the next generation of Team Assistant[2] is to provide a rich extensible tool, providing decent log playing capabilities alongside useful agent debug and analysis information about agent's behavior and collaboration with other agents. The importance of the evaluation of agent teamwork has been addressed in many papers for 2D simulation[5][8]. It also can be used as an offline trainer to run training sessions with virtually unlimited scenario definition options.

2   Related Work

In the 2006 soccer simulation development competition just a few tools were introduced, such as "Virtual Werder Analyzer" or "UTUtd Monitor".

The Virtual Werder Analyzer offers information about ball possession, successful and unsuccessful passes and about good and bad actions[1]. In UTUtd monitor, there are some common functions with this work such as the possibility of agents to draw into the displayed scene or the displaying ability of text messages according to the current scene[6]. In 2D presentation competitions, tools like Team Assistant also provides the detection of (double-) passes, goal shots, dribbling and etc. Statistics about these detected events are shown while playing game. In Caspian monitor, there is a commentator that can comment the game according to the current detected situation. These comments were the main aspect of that work. But none of the above tools were extensible enough for new requirements.

3   Requirements

In general we can divide 3D Soccer Simulation requirements into two main parts: "Developers' requirements" and "Organizers' requirements".

3.1   Developers' Required Features

In consultation with other agent developers, providing an easy and clear way to gather information about what the agent intends and how the world looks like, according to the agent, is essentially needed. The following features are judged beneficial and essential to debug handwritten behaviors and to verify the decisions of learned behaviors.

While analyzing a special situation of the game it is obvious that forward and backward replay in different speeds is useful. With this possibility the situation can be analyzed again without starting the game from the beginning until the desired situation is reached. To gain knowledge of the agents' intentions in a situation an output of agents' logfile according to current time is needed.

In addition to the logfile displaying it is valuable to enable the agents to draw information directly into the displayed scene, like a line from the agent to the position it intends to move to. Also filtering the logfile output may be helpful to display only those information needed by the developer. This way only those information provided by the current developed behavior could be displayed. This idea can be achieved by using layered logs[3].

New camera positions, like birdview which resides directly over the agent of interest, may be beneficial. When using the single step mode of the monitor displaying the ball's and player's movements for some time forward can be an improvement, the developer could see the next movements without proceeding forward in scene display. Displaying the offside line is necessary. In some situations it could be difficult to distinguish on which side of the offside line an agent is, so marking agents that are in an offside position if the ball would be played to them may be a good feature. Detecting events and the number of their occurrence may provide essential information about what parts of the agent have to be worked on (i.e. if passes often fail there is some space for improvements). Those events are: pass success/fails, ball dribbling, lost balls, goal shot (success, out, intercepted) and kick out. By detecting event sequences more information can be extracted such as lost balls after dribbling.

3.2 Organizers' Required Features

The running of simulation league in comparison to other RoboCup fields, is so quiet and therefore it doesn't have enough visitors. So with showing a better illustration of games and final goal of RoboCup, this league could be more interesting for visitors. The publicity that the RoboCup events get from the media provides an ideal opportunity to show the state of art of these systems during RoboCup World Cup.

4 Implementation

When talking about a logplayer or a monitor this essentially means the same program in two different operation modes, logplayer means that the program replays logfiles of a previously simulated game, monitor that a game that is currently simulated is displayed.

4.1   Features

Some features of TA2006 that can be useful in developing agents are:

Forward and backward replay in different speeds even in live games is achieved by storing data into a new data structure. Playback is much faster now, it has to be slowed down artificially leading to the opportunity of different playback speeds.

The tool is also helpful for probing the agents' internals. The Layered Disclosure concept[3], first introduced in CMUnited99 simulated agents, and it proved to be very helpful in development of intelligent agents. The goal of the Layered Disclosure concept is to make various agent characteristics.

Observable, without overwhelming the human observer with data. Realizing this goal needs the agent to store a log of all relevant information from its internal state, world model, and reasoning process. Then a tool must be used to synchronize the log with a recording of the observable world and it must provide an interface to allow the developer/observer to probe a given agent's internal reasoning at any time and any level of detail. Our tool provides all the necessary functionality. The times according to the various agent outputs is also parsed at startup. When displaying, the fitting record of the agent logfile is found using the time currently displayed by the monitor. Some none-textual logging facility comes handy for the human developer/observer; the tool understands a simple notation for describing 3D or 2D geometric shapes. For example an agent may write in its log file:

```
<0><49> <Behavior> (Command) Sphere ball
( 0.023f, 0.09f, 0.111f, 0.130f, 0.5f, 0.5f, 0.5f, 0.3f );
```

And when the log-player is showing the cycle 49, the developer can select the agent, and the log- player draws a sphere with radius 0.130 at point (0.023, 0.09, 0.111). (fig. 1).

Implemented commands are: Drawing circles, spheres, cubes, lines, playing sounds, Showing a text on the screen and many more. In case of filtering the logfile output the developer just have to change the log level. New camera view modes were implemented, most of them are adjustable in a way (camera height,

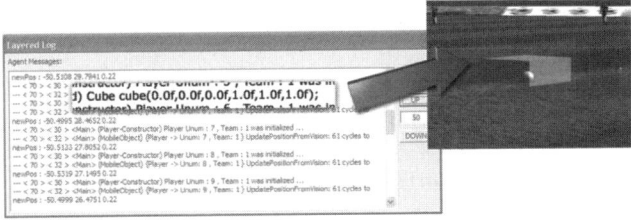

Fig. 1. Draw shapes based on output of an agent, for example agent can command the monitor to render its internal world model on the screen

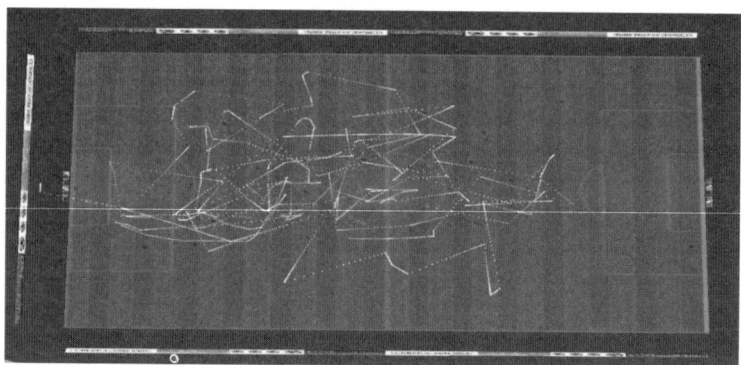

Fig. 2. Ball trajectory

Fig. 3. Robots vs. Humans, Different models can be used to represent the agents in the visualization

distance or point to look at). Some camera types like tele, top, tower views are implemented. Also cameras can reside directly over the agent of interest.

Displaying the ball's and player's movements for some time forward is solved as lines in the color of the team or white for the ball. The length of the line (how far into the future movements are shown) can be adjusted. The agents track (movements throughout the whole game) can also be displayed, this track can be colored according to the agents movement speed (fig. 2).

In case of presentation, Team Assistant 2006 provides rich game-viewing controls, plus some eye- candy features for on-site demonstration of a game, including:

- Commentator
- Automatic replays
- Customizable 3D agent model (fig. 3)

- Complete control over camera position and direction
- Several preset cameras (Tele, Top, Tower, ...)
- The ability to display different camera views side-by-side
- Visualizing Catch and Kick commands of agents

When the tool is in trainer mode, it allows defining training sessions or scenarios. A scenario definition contains a starting state and/or an end condition. Those can be specified using Angel Scripts. The trainer mode also needs a server counterpart.

4.2   Additional Features

TA2006's main power lies in its ability to be extended using *AngelScript* plug-ins. To get a glimpse of what can be done within a plug-in, it's good to mention that in current release the Commentator itself is a plug-in, all sound effects are provided by a plug-in, some training/test sessions are wrote using plug-ins, the game statistics are both calculated and rendered on screen by a plug-in. In general, plug-ins can Obtain:

- Locations of all objects
- State of the match (play mode, time,...)
- Player actions (requires new server to monitor protocol)
- Some processed values (ball and agents' speed)

And Can Perform:

- Move agents and ball
- Change play mode of the match
- Control the log player (change playback speed, jump to a specific cycle,...)
- Draw shapes in the field
- Draw markers on the field
- Write/Draw on the screen
- Control the camera
- Play audio file

4.3   Other Features Derived from the Implementation

The new data structure of the program allows the user to use single step mode, for- and backward even when watching the game "live". The server is continuing the simulation in background, even if the monitor is in single step mode. Functions that are independent from agent logfiles, such as movement display, can be shown in this mode.

5   Results

The logfile output is useful to understand the agents behavior. This program has replaced the lite monitor that comes with the simulation server. The agent may communicate its decisions to the developer. (i.e. which role it plays, which behavior of the role it has chosen and what action it selected). In occurrence of an error in the ball movement prediction the agent was changed to display its data about the ball (its position and movement vector according to the agents world model). This way the error became visible. Without displaying that data, localizing the error would have taken much longer. When creating a behavior that covers an opponent, it is mostly wanted that only one player covers an opponent at a time. To verify the opponent selection the agent can simply draw a line from himself to the opponent it intends to cover or include the chosen position into its drawings. When displaying the draw commands of all agents it becomes visible if something is wrong with the selection or the position the agent has chosen to move.

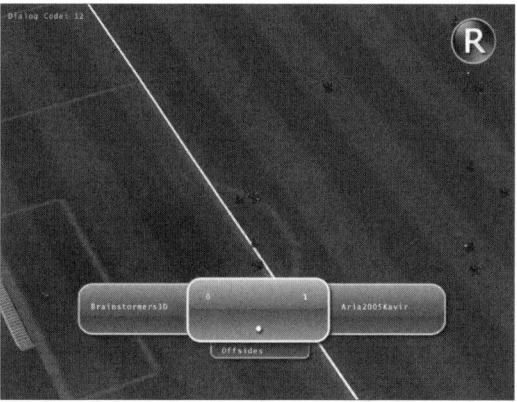

Fig. 4. Drawing offside line and automatic replay

The ability of reverse playback of the game gives the opportunity of analyzing the same scene again without watching the game from the beginning. Also slight transitions in action selection may be analyzed without much interference. This feature in addition to the display of agentlogs (in both ways, text and drawings) gives the opportunity to understand the agents decisions according to its own world model, and not only the real simulated world displayed in the monitor. The ability of delayed playback and direct analyzing of a currently simulated game provides a real speedup in development in comparison to other tools. Simulating a 3D game may take up to 10 or 20 min which is a long time the developer has to wait in order to for example analyze the previously mentioned ball approach in slow motion. With the delayed playback that situation may be analyzed while the game is still simulated in background. After the analyzing was finished the game may be watched time shifted in normal speed or the replay may be fastened to reach the most recent scene. The track display in the 3D scene along with the plotting opportunities were giving the developer essential information about the agents or ball movements throughout the game. This way the developer can find agents which are not moving much or fast, which may point to a suboptimal formation or behavior. By viewing the ball movement plot (fig. 2) of the game between Aria (left) and ZJubase (right) it is obvious that the ball was mostly located on the left side of the field and though that ZJubase dominated the game. ZJubase won this match 1:0.

6 Conclusion

The logplayer can be a useful tool when trying to resolve strange behavior of an agent or verifying the proper work of the agent. The SBCe team resolved some problems within their code using the logplayer. These resolved problems are namely the examples given in section 5. The resolving of these problems would have taken longer without the possibility of drawing or text output. Due to the parsing at the beginning the startup of the logplayer takes longer in comparison to the lite monitor/ logplayer, but the playback is much faster and the monitor does not have to be restarted to watch the game again.

Finally this tool is intended to be a general-purpose, highly customizable package. We think it has the potential to be used as the primary analyzer, visualizator, and logviewer for anyone interested in developing agents for the RoboCup 3D Soccer Simulator. TA2006 won the second place in RoboCup 3D development competition 2006.

References

1. Planthaber, S., Visser, U.: Logfile player and analyzer for RoboCup 3D simulation. In: Lakemeyer, G., Sklar, E., Sorrenti, D.G., Takahashi, T. (eds.) RoboCup 2006: Robot Soccer World Cup X. LNCS (LNAI), vol. 4434. Springer, Heidelberg (2007)
2. Nazemi, E., Kazemi, V., Habibi Shahri, A., Hosseingholizadehm, A., Nooraei B., B.: SBCE SmartSpheres 3D Development Team Description. In: Proc. of RoboCup 2006 (2006)
3. Riley, P., Stone, P., Veloso, M.: Layered Disclosure: Revealing Agents' Internals. In: Castelfranchi, C., Lespérance, Y. (eds.) ATAL 2000. LNCS (LNAI), vol. 1986. Springer, Heidelberg (2001)
4. Nazemi, E., Zareian, A., Samimi, R., Shiva, F.A.: Team assistant team description paper. In: Proc. of Robocup 2002 (2002)
5. Raines, T., Tambe, M., Marsella, S.: Automated assistants to aid humans in understanding team behaviors. In: Veloso, M.M., Pagello, E., Kitano, H. (eds.) RoboCup 1999. LNCS (LNAI), vol. 1856, pp. 85–104. Springer, Heidelberg (2000)
6. Aghaeepour, N., Bastani, M., Miri, F.D., Masoudnia, S., Radpour, S., Fotuhi, S.A.: UTUtd2006-3D Team Description Paper. In: Proc. of RoboCup 2006 (2006)
7. Obst, O.: Simulation league - league summary. In: Kaminka, G.A., Lima, P.U., Rojas, R. (eds.) RoboCup 2002. LNCS (LNAI), vol. 2752, pp. 443–452. Springer, Heidelberg (2003)
8. Takahashi, T.: Logmonitor: From player's action analysis to collaboration analysis and advice on formation. In: Veloso, M.M., Pagello, E., Kitano, H. (eds.) RoboCup 1999. LNCS (LNAI), vol. 1856, pp. 103–113. Springer, Heidelberg (2000)

Mean-Shift-Based Color Tracking in Illuminance Change

Yuji Hayashi[1] and Hironobu Fujiyoshi[1]

Dept. of Computer Science, Chubu University, Japan
yuji@vision.cs.chubu.ac.jp, hf@cs.chubu.ac.jp
http://www.vision.cs.chubu.ac.jp/

Abstract. The mean-shift algorithm is an efficient technique for tracking 2D blobs through an image. Although it is important to adapt the mean-shift kernel to handle changes in illumination for robot vision at outdoor site, there is presently no clean mechanism for doing this. This paper presents a novel approach for color tracking that is robust to illumination changes for robot vision. We use two interleaved mean-shift procedures to track the spatial location and illumination intensity of a blob in an image. We demonstrate that our method enables efficient real-time tracking of the multiple color blobs against changes in illumination, where the illuminace ranges from 58 to 1,300 lx.

1 Introduction

Tracking is a method of estimating the spatial location of a target in a camera image. It often requires real-time processing, so high-speed processing is essential. For tracking 2D blobs through an image sequence, the mean-shift algorithm is an efficient technique [1,2,3]. It seeks the nearest mode of a point sample distribution. Collins [4] proposed a method of scale change mean-shift and She [5] proposed a method of considering shape features. The mean-shift algorithm has a low calculation cost and offers high-speed execution.

Tracking is difficult when lighting changes because the RGB values from the image changes with the lighting. Thus, it is not possible to distinguish a moving object or lighting change. In addition, problems of lighting changes are usually treated as those of color transformation between different lighting conditions. Some researchers have proposed linear color transformation [6] and independent transformation [7] of each RGB component, which are derived from a physics-based color model. On the other hand, statistics-based approaches have also been proposed. Miller [8] proposed a method of non-linear color transformation using color eigenflows learned from multiple pairs of images of the same scene under different lighting conditions. It is, however, difficult for a robot vision system to get multiple reference colors in unknown lighting conditions.

This paper presents a novel approach for color tracking that is robust to lighting changes for robot vision. We use two interleaved mean-shift procedures to track the spatial location and illumination intensity of a blob in an image.

U. Visser et al. (Eds.): RoboCup 2007, LNAI 5001, pp. 302–311, 2008.

We show that our method enables real-time tracking of a color blob for varying lighting conditions.

2   Color and Illuminance

The illuminance at any surface of known color can be measured by observing the RGB values obtained by a CCD camera. Changes in the light source or meteorological effects can change the illuminance, resulting in changes in the measured RGB values. Figure 1(a) shows various color patches (Blue, Black, Green, Pink, Purple, White, Yellow) under illuminance ranging from 10 to 1400 lx. The setup used in our experiments is illustrated in Figure 1(b). The illuminance on the object's surface was obtained by an illuminance meter placed on the object, and RGB values were captured by a color CCD camera mounted at a height of 280 cm.

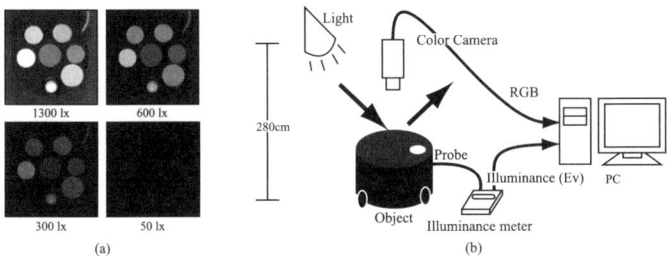

Fig. 1. Experimental setup

Using the color segmentation technique based on thresholding [9], it is difficult to distinguish color classes in RGB color space, because we cannot create a threshold criterion that specifies how the color space should be divided up into a handful of color classes. Color clustering using the HSI color system is robust to lighting changes, but it is difficult to distinguish a moving object and light change, because it does not represent illuminance on the target. To solve this problem, we augment the RGB color space to make an RGB-illuminance space, and then we use a tracking method that searches for a mode within neighboring pixels.

2.1   RGB-Illuminance Space

Our approach uses RGB-illuminance space coupled with the estimation of illuminance intensity in each frame to distinguish color classes. An example of color distributions in RB-illuminance space is shown in Figure 2. We can see that it is possible to classify color classes at each illuminance plane, as shown in Figures 2(b) and (c). However, a fixed value for thresholding does not work due to shrinking in the color space with respect to illuminance. In the RGB-illuminance space, reference-based searching such as the k-NN method for color

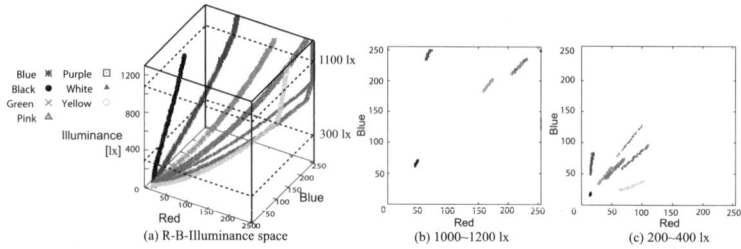

Fig. 2. RGB-illuminance space: (a) color distribution in RB-illuminance space, (b) RB value of each color class at 100 lx, and (c) RB value at 200 lx

clustering can work, but it takes a lot of time due to the number of reference patterns for each illuminance. Therefore, we use a color-illuminance model for each color class.

2.2 Color-Illuminance Model

The relationship between RGB values and illuminance is not linear. Thus, we use curve fitting on each RGB distribution over the illuminance intensity. Given the illuminance intensity Ev, we can estimate the RGB color values $\hat{I}_r, \hat{I}_g, \hat{I}_b$ using the following equations:

$$\hat{I}(Ev) = aEv^2 + bEv + c \tag{1}$$

where a, b, and c are unknowns computed by the least-squares method. Note that we assume that the object's surface has diffuse reflection.

2.3 Iris Adjustment

The RGB values is influenced by some camera parameters such as iris and white balance. To cope with special lighting situations, the iris (F-number) can be adjusted manually to let in more or less light. The F-number is given by $F = f/D$, where f is the focal length and D is the iris diameter. It affects the amount of light energy admitted to the sensor and plays a significant role in the resulting image. The relationship between intensity I and F-number is expressed by

$$I \propto \left(\frac{D}{f}\right)^2 = \left(\frac{1}{F}\right)^2. \tag{2}$$

The smaller the F-number, the more light admitted to the sensor, and hence the better the image quality achieved in low-light situations. Figure 3(a) shows RGB curves from a color-illuminance model for $F = 4$ and observed RGB values for $F = 5.6$. Using Equation (2), we can convert the RGB values observed at any F-value to the corresponding value at a desired F-value. $F = 5.6$ means $I \propto 31.36$ and $F = 4$ means $I \propto 16$, so the RGB values at 1400 lx with $F = 5.6$

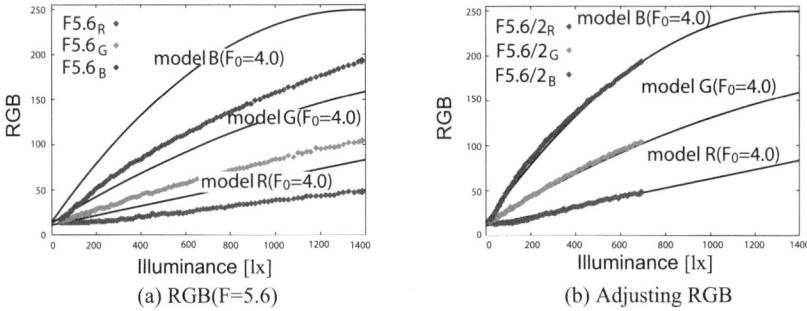

Fig. 3. Adjusting RGB color value by F-number

will be same as the RGB values at 700 lx with $F = 4$. Figure 3(b) shows an example of converted RGB values.

If we prepare a color-illuminance model for a given F-value in advance, we can estimate the color-illuminance model at the F-number corresponding to our camera's iris setting. In this paper, we assume that our light source has a constant color temperature, so we do not consider the changes in white balance of the color camera.

3   Mean-Shift Tracking through Illuminace Space

We propose a method for mean-shift-based color tracking through illuminance space, which represents the spatial location and illumination intensity of a blob in an image.

3.1   Mean-Shift in Image Space

The mean-shift algorithm is a simple nonparametric method for seeking the nearest mode of a sample distribution. It has recently been adopted as an efficient tracking technique. When the mean-shift method is used for object tracking, the gradient density is formed by weight $w(\mathbf{x})$ at each image pixel $\mathbf{x}$. The core of the mean-shift tracking algorithm is the computation of a target's motion vector from a location $\mathbf{x}$ to a new location $\mathbf{x}'$. We get the new location $\mathbf{x}' = \mathbf{x} + \Delta_{\mathbf{x}}$ from the mean-shift vector

$$\Delta_{\mathbf{x}} = \frac{\sum_{i=1}^{N} K(\mathbf{x}_i - \mathbf{x}_0, \sigma) w(\mathbf{x}_i)(\mathbf{x}_i - \mathbf{x}_0)}{\sum_{i=1}^{N} |K(\mathbf{x}_i - \mathbf{x}_0, \sigma) w(\mathbf{x}_i)|}, \tag{3}$$

where the set $\{\mathbf{x}_i\}_{i=1,...,N}$ represents the locations of pixels around the current location $\mathbf{x}$ and K is a kernel function such as the Gaussian kernel. Generally, a weight map is determined using a color-based appearance model. In [3], the weights were obtained by comparing a histogram q_u, where u is a histogram bin index, with a histogram of colors $p_u(\mathbf{x}_0)$ observed within a mean-shift window at the current location $\mathbf{x}_0$. In fact, the weight at pixel location $\mathbf{x}$ is given by

$$w(\mathbf{x}) = \sum_{u=1}^{m} \delta\,[b(\mathbf{x}) - u]\,\sqrt{\frac{q_u}{p_u(\mathbf{x}_0)}}, \tag{4}$$

where m is the total number of features, δ is the Kronecker delta function and $b(\mathbf{x})$ is feature value of the pixel at $\mathbf{x}$.

3.2   Mean-Shift in Illuminance Space for Single-Color Tracking

It is difficult to track a color blob under varying light conditions due to the limitations of color space described in Section 2. We augment the mean-shift tracker to search in illuminance space by introducing two interleaved mean-shift procedures to track the mode in image space and in illuminance space, which represent the spatial location and illumination intensity of the target blob, respectively. These two procedures are described below.

Initial input. A color-illuminance model of the target color is deformed by scaling with the current setting of the iris (F-number). The initial input is a deformed color-illuminance model of a specific color and an estimate of the blob's current illuminance intensity Ev and spatial location $\mathbf{x}_0 = (x, y)$ in the image.

Step 1: Mean-shift in image space. Given the illuminance intensity Ev in the current frame, the estimated RGB values $(\hat{I}_r, \hat{I}_g, \hat{I}_b)$ are computed using Equation (1) using the color-illuminance model for the specific color. Then, we compute a location weight map $w_{loc}(\mathbf{x})$ between the target color and the RGB values $I(\mathbf{x})$ for each pixel.

$$w_{loc}(\mathbf{x}_i) = \frac{\hat{I}_r(Ev)I_r(\mathbf{x}_i) + \hat{I}_g(Ev)I_g(\mathbf{x}_i) + \hat{I}_b(Ev)I_b(\mathbf{x}_i)}{\sqrt{(\hat{I}_r^2(Ev) + \hat{I}_g^2(Ev) + \hat{I}_b^2(Ev))(I_r^2(\mathbf{x}_i) + I_g^2(\mathbf{x}_i) + I_b^2(\mathbf{x}_i))}} \tag{5}$$

Then the spatial mean-shift vector is obtained as

$$\Delta\mathbf{x} = \frac{\sum_{i=0}^{N} K_{loc}(\mathbf{x}_i - \mathbf{x}_0, \sigma_{xy})w(\mathbf{x}_i)(\mathbf{x}_i - \mathbf{x}_0)}{\sum_{i=0}^{N} |\,K_{loc}(\mathbf{x}_i - \mathbf{x}_0, \sigma_{xy})w(\mathbf{x}_i)\,|} \tag{6}$$

where K_{loc} is a spatial kernel function given by

$$K_{loc}(\mathbf{x}, \sigma_{xy}) = \frac{1}{2\pi\sigma_{xy}^2}\exp(\frac{-(x^2 + y^2)}{2\sigma_{xy}^2}) \tag{7}$$

and the summations are performed over a local window of N pixels around the current location $\mathbf{x}$. Finally, we can get the new location $\mathbf{x}' = \mathbf{x} + \Delta\mathbf{x}$ from the mean-shift vector.

Step 2: Mean-shift in illuminance space. Our approach uses a mean-shift procedure to estimate the illuminance intensity by a local window of pixels around the new location $\mathbf{x}' = (x', y')$ obtained in step 1. First, we compute the color similarity at every illuminance $(k = 0, ..., max)$ for each pixel $\mathbf{x}$ by the following equation.

$$S(k, \mathbf{x}) = \frac{\hat{I}_r(k)I_r(\mathbf{x}) + \hat{I}_g(k)I_g(\mathbf{x}) + \hat{I}_b(k)I_b(\mathbf{x})}{\sqrt{(\hat{I}_r^2(k) + \hat{I}_g^2(k) + \hat{I}_b^2(k))(I_r^2(\mathbf{x}) + I_g^2(\mathbf{x}) + I_b^2(\mathbf{x}))}}$$

$$(k = 0, ..., max) \tag{8}$$

Then, we compute an illuminance weight map $w_{Ev}()$, which is 1D array, by the following equation:

$$w_{Ev}(k) = \sum_{i=0}^{N} K_{loc}((\mathbf{x}_i - \mathbf{x}'), \sigma_{xy})S(k, \mathbf{x}_i) \tag{9}$$

where K_{loc} is a spatial kernel function. This works as a voting mechanism from neighbor pixels using illuminance, as illustrated in Figure 4.

This mean-shift in illuminance space is performed on the 1D array of results to locate the mode. The illuminance mean-shift vector is then obtained by the equation:

$$\Delta Ev = \frac{\sum_{k=0}^{max} K_{Ev}(k - Ev)w_{Ev}(k)(k - Ev)}{\sum_{k=0}^{max} w_{Ev}(k)}, \tag{10}$$

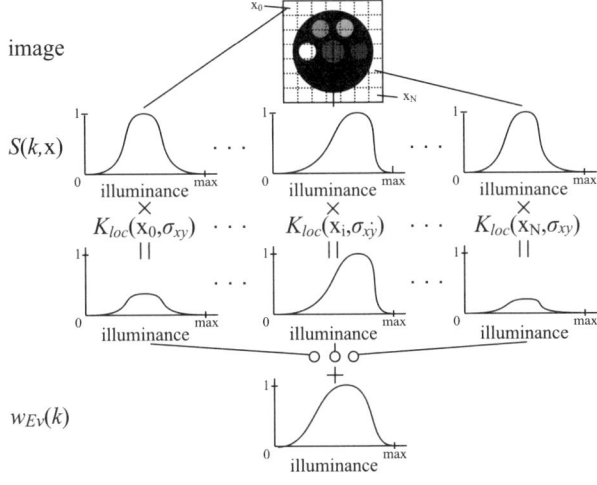

Fig. 4. Calculation of weight map for illuminance space

where Ev is the current illuminance, and K_{Ev} is a kernel function for illuminance space given by

$$K_{Ev}(k, \sigma_{Ev}) = \frac{1}{\sqrt{2\pi\sigma_{Ev}^2}} \exp\left(\frac{-k^2}{2\sigma_{Ev}^2}\right). \tag{11}$$

Finally, we can get the new illumination intensity $Ev' = Ev + \Delta Ev$ from the mean-shift vector.

Since the range of illuminance space is set as $0 < k < max$ (lx), the kernel function for illuminance space K_{Ev} limits the search to around the illuminance estimated in the last frame. However, we cannot get the illumination intensity in a rapid light change if σ_{Ev} is small and the illuminance estimation accuracy is reduced if σ_{Ev} is large. Therefore, we obtain σ_{Ev} from the maximum point k_{max} of the illuminance weight $w_{Ev}(k)$ and the difference in the front frame's illuminance Ev. It is calculated by

$$\sigma_{Ev} = (\sigma_{max} - \sigma_{min}) \times \frac{|Ev - k_{max}|}{Ev_{max} - Ev_{min}}, \tag{12}$$

where σ_{max} is the maximum value of σ_{Ev}, σ_{min} is the minimum value of σ_{Ev}, Ev_{max} is the maximum value of Ev, and Ev_{min} is the minimum value of Ev.

Step 3: Iteration. Iterate by interleaving steps 1 and 2 until both $|\Delta x| < \varepsilon_{xy}$ and $|\Delta Ev| < \varepsilon_{Ev}$.

3.3   Mean-Shift for Multiple-Color Tracking

We augment the single-color tracking method described in Section 3.2 to multiple colors. Multiple color-illuminance models and weight maps for each target color are prepared in advance. For the mean-shift in image space, we compute a spatial location weight map w_{loc}^c for each color class c by Equation (5) using each color-illuminance model. Then, the weights for spatial location are integrated into one weight by selecting the maximum value at the same pixel. The integrated weight map for spatial location w_{loc}' is obtained from each color weight map $w_{loc}^c(c = color\ variety)$ by

$$w_{loc}'(\mathbf{x}_i) = w_{loc}(\mathbf{x}_i)^{c1} \frac{w_{loc}(\mathbf{x}_i)^{c1}}{w_{loc}(\mathbf{x}_i)^{c2}}, \tag{13}$$

where $w_{loc}(\mathbf{x}_i)^{c1}$ is the 1st maximum value in multiple colors c at $\mathbf{x}_i$ and $w_{loc}(\mathbf{x}_i)^{c2}$ is 2nd one. Here, the color of $c1$ class, which has the maximum value, is stored for the next step of computing the mean-shift vector in illuminance space.

For the mean-shift in illuminance space, we compute the color similarity for each illuminance for each pixel $\mathbf{x}$ using the color-illuminance model of the target's color. Then, the weight for illuminance space (1D array) is computed according to Equation (9). This mean-shift procedure is iterated until convergence, as described in 3.2.

4   Experimental Results

The performance of the proposed method was evaluated by experiments in terms of robustness and accuracy in varying light conditions.

4.1   Experiments

A color camera was mounted at a height of 2800 [cm], as shown in Figure 1. In these experiments, the color temperature of the light source (light color) was fixed, and the white balance and iris value were not changed during the tracking task. Initial illuminance Ev and spatial location (x, y) of the colored object to be tracked were given as initial values for mean-shift tracking. To determine the accuracy of the location estimation, we compared the values estimated by the proposed method to ground truth, which was measured manually by a human. We also measured the illuminance intensity on the surface of the tracked object.

4.2   Experimental Results for Single-Color Tracking

Figure 5(a) shows tracking examples of the proposed method and the mean-shift weight map. Figure 6 shows the location errors for the proposed method and the general mean-shift method and the estimated illuminance on the object, which ranged from 50 to 1200 lx as a result of changes in the intensity of the light source. We can see that our method achieved more accurate location estimation than the general mean-shift method. Since it simultaneously computes location and illuminance, i.e., the location of the colored object while estimating the surface illuminance, our method can track the object under varying lighting conditions. When the light changes rapidly, e.g., due to flickering, our method can track a colored object by calculating σ_{Ev} of the illuminance kernel function K_{Ev} at each frame (see Figure 6(a)).

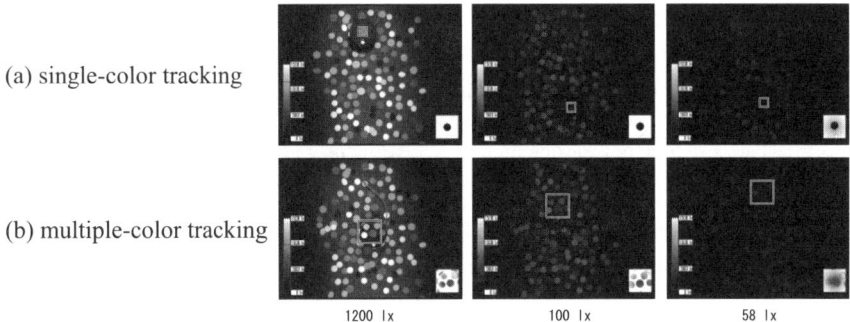

(a) single-color tracking

(b) multiple-color tracking

1200 lx   100 lx   58 lx

Fig. 5. Tracking example of the proposed method

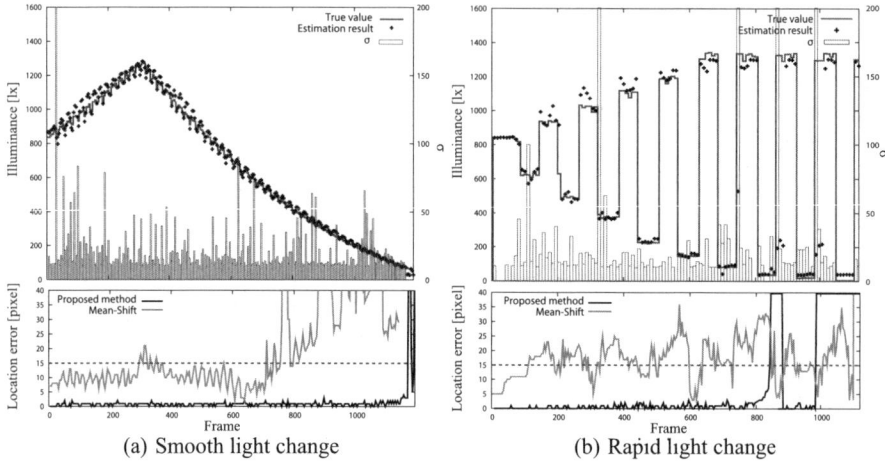

(a) Smooth light change   (b) Rapid light change

Fig. 6. Experimental results

Figure 5(b) shows examples of multiple-color tracking by our method using the integrated weight map. It is clear that our method can be easily applied to track multiple colors.

5   Conclusion

In this paper, we proposed a tracking method using two interleaved mean-shift procedures to track the mode in illuminance space, which represents the spatial location and illumination intensity of a blob in an image. We demonstrated that our method enables real-time color tracking that is robust to changes in illumination, where the illuminance ranges from 50 to 1200 lx. Since this method estimates the illuminance from the pixels of the tracked object and not by using the entire image, reliable color tracking is achieved even when the lighting changes. Color tracking when the color temperature of the light source (light color) varies is left as future work.

References

1. Fukunaga, K., Hostetler, L.: The estimation of the gradient of a density function, with applications in pattern recognition. IEEE Transactions on Information Theory 21, 32–40 (1975)
2. Comaniciu, D., Meer, P.: Mean shift analysis and applications. In: IEEE International Conference on Computer Vision, vol. 2, pp. 1197–1203 (1999)
3. Comaniciu, D., Ramesh, V., Meer, P.: Kernel-based object tracking. IEEE Transactions on Pattern Analysis and Machine Intelligence 25, 564–577 (2003)
4. Collins, R.: Mean-shift blob tracking through scale space. In: IEEE Conference on Computer Vision and Pattern Recognition, vol. 2, pp. 234–240 (2003)

5. She, K., Bebis, G., Gu, H., Miller, R.: Vehicle tracking using on-line fusion of color and shape features. In: IEEE Conference on Intelligent Transportation Systems, pp. 731–736 (2004)
6. Marimont, D.H., Wandell, B.A.: Linear models of surface and illuminant spectra. Journal of the Optical Society of America 9(11), 1905–1913 (1992)
7. Drew, M.S., Wei, J., Li, Z.-N.: Illumination-invariant image retrieval and video segmentation. Pattern Recognition 32, 1369–1388 (1999)
8. Miller, E.G., Tieu, K.: Color eigenflows: Statistical modeling of joint color changes. In: IEEE International Conference on Computer Vision, vol. 1, pp. 607–614 (2001)
9. Bruce, J., Balch, T., Veloso, M.: Fast and inexpensive color image segmentation for interactive robots. In: IEEE/RSJ International Conference on Intelligent Robots and Systems, vol. 3, pp. 2061–2066 (2000)

High Accuracy Navigation in Unknown Environment Using Adaptive Control

Fernando Ribeiro, Ivo Moutinho, Nino Pereira, Fernando Oliveira,
José Fernandes, Nuno Peixoto, and Antero Salgado

Grupo de Automação e Robótica, Departamento de Electrónica Industrial,
Universidade do Minho, Campus de Azurém, 4800-058 Guimarães, Portugal
fernando@dei.uminho.pt, ivomauro@sarobotica.pt,
ninopereira@sarobotica.pt, fremaxeei@gmail.com,
joseluisfernandes@gmail.com, nunopeixoto35242@hotmail.com,
anterosalgado@aeiou.pt

Abstract. Aiming to reduce cycle time and improving the accuracy on tracking, a modified adaptive control was developed, which adapts autonomously to changing dynamic parameters. The platform used is based on a robot with a vision based sensory system. Goal and obstacles angles are calculated relatively to robot orientation from image processing software. Autonomous robots are programmed to navigate in unknown and unstructured environments where there are multiple obstacles which can readily change their position. This approach underlies in dynamic attractor and repulsive forces. This theory uses differential equations that produce vector fields to control speed and direction of the robot. This new strategy was compared with existing PID method experimentally and it proved to be more effective in terms of behaviour and time-response. Calibration parameters used in PID control are in this case unnecessary. The experiments were carried out in robot Middle Size League football players built for RoboCup. Target pursuit, namely, ball, goal or any absolute position, was tested. Results showed high tracking accuracy and rapid response to moving targets. This dynamic control system enables a good balance between fast movements and smooth behaviour.

1 Introduction

A dynamic approach was employed to control the movement of an autonomous robot which is meant to navigate towards a moving target or goal, avoiding obstacles and collisions. Navigation direction, φ, is a behaviour variable which varies from 0° to 360° relatively to an external reference.

In Fig. 1 is represented the robots' navigation direction φ_{robot}, as well as target direction ψ_{target} and obstacle direction ψ_{obs}. The target direction is the desired value to navigation direction. The direction of the obstacle is the erroneous direction, which must be avoided by the robot [1].

The robot movement is generated by continuously calculating values for the navigation direction. The time series $\varphi(t)$ is generated by a dynamic system based on a differential equation in which the state variable is the robots' navigation direction:

U. Visser et al. (Eds.): RoboCup 2007, LNAI 5001, pp. 312–319, 2008.

$$\frac{d\varphi_{robot}}{dt} = f(\varphi_{robot}) = f_{t\arg et}(\varphi_{robot}) + f_{obs}(\varphi_{robot}) \tag{1}$$

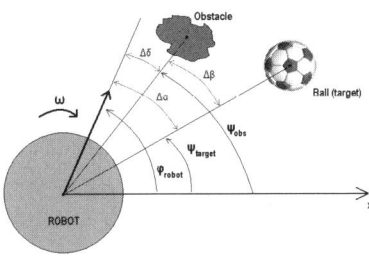

Fig. 1. Navigation direction of the robot depending on target and obstacle directions

The vector field of this dynamic system is build from a certain number of addictive forces, each one specifying a particular value (attractive or repulsive) for the navigation direction. Each of these forces is characterized by a singular direction value, ψ_{target} or ψ_{obs}, intensity of attraction or repulsion and also by the range of direction values affected by them. A force by itself establishes an attractor, state asymptotically stable, or a repulsor, state asymptotically instable in the dynamics of navigation direction.

An attractive force, f_{target}, is used to pull the system to the desired value, namely to the target direction. On the other hand, a repulsive force, f_{obs}, ensures that the system avoids moving towards the obstacle direction. Summing all these forces, results in a non-linear dynamic system [2].

Since all angles are measured related to an external reference axis, the contributions of obstacles and target to the dynamic system do not depend on the actual orientation of the robot. This navigation control system was implemented and experimental essays were carried out on an autonomous robot.

2   Navigation Direction Control

2.1   Reaching a Target

The differential equation that describes the system behaviour to pursuit a target is:

$$\frac{d\varphi_{robot}}{dt} = -k \cdot \tanh\left(\left(\varphi_{robot} - \psi_{t\arg et}\right) \cdot \gamma\right) \tag{2}$$

where:

$\varphi_{\mathbf{robot}}$ – Navigation direction of the robot, $\psi_{\mathbf{target}}$ – Target direction.
$\mathbf{k}$ – Maximum value of the attractive force, γ – Target attraction intensity

The choice of this equation is not arbitrary, because the function of hyperbolic tangent has only one zero, the attractor (fixed point), and stabilizes at maximum and minimum values for the direction variation of the robot in time.

As the robot moves, the target direction varies, pulling the attractor over the possible values for the navigation direction, as represented in Fig. 2.

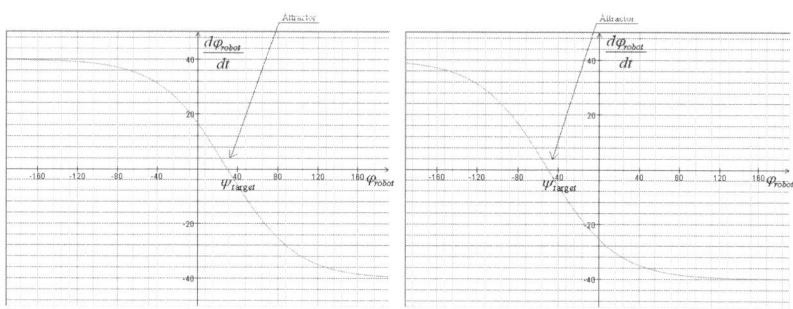

Fig. 2. Fixed point localization variation (attractor), as a function of the robot movement and target displacement

The rate of the direction variation in time corresponds to the angular velocity or robot rotational movement. The behaviour variable used for calculations was obtained from the difference between the robot direction angle and the target angle.

$$\Delta \alpha = \varphi_{robot} - \psi_{t\,\arg et} \tag{3}$$

Thus, the differential equation assumes the form:

$$\omega(t) = \frac{d\varphi_{robot}}{dt} = -\omega_{max} \cdot \tanh(\Delta\alpha \cdot \gamma) \tag{4}$$

where:

ω – Angular velocity of the robot, $\Delta\alpha$ – Target angle relative to the robot

$\omega_{max.}$ – Maximum rotation velocity, γ – Target Attraction Intensity

In this way, there isn't any shift of the attractor fixed point, since what really matters is the difference between the robot navigation direction and the target direction. This difference should always take the value zero.

No matter the navigation direction or the target direction, every time the difference assumes a non-zero value, the system acquires a positive or negative angular velocity which decreases as the difference becomes null. At that time the system stabilizes and the robot stops. This fixed point is stable and is denominated attractor since the direction variation rate is negative at that point and the system tends to the desired value. The angular velocity shown in the picture above (omega), is an entry parameter to the robot command functions that control the motors and varies from 0 to 40.

A very important parameter is γ, since it allows the specification of the intensity of the target attraction. In other words, it defines the speed variation of the angular velocity, or system time-response. A special attention must be given to the parameter γ since high values could make the system unstable.

2.2 Obstacle Avoidance

The differential equation which describes the system behaviour for obstacle avoidance is given by:

$$\frac{d\varphi_{robot}}{dt} = -\lambda \cdot \left(\varphi_{robot} - \psi_{obs}\right) \cdot e^{-\frac{\left(\varphi_{robot} - \psi_{obs}\right)^2}{2\delta^2}} \tag{5}$$

where: φ_{robot} – Robot navigation direction, ψ_{target} – Target direction
λ – Intensity of repulsive force, δ – Range of repulsion

The obstacle repulsive force magnitude depends on the distance to the robot. It can be modelled by the following equation:

$$\lambda = \beta_1 \cdot e^{-\frac{d}{\beta_2}} \tag{6}$$

where: β_1 – Maximum intensity of the repulsive force
β_2 – Decline rate of the repulsion intensity with the distance to the obstacle
d – Distance to the obstacle

The repulsion range depends on the size of the object, the bigger the obstacle the greater the range (Fig. 3).

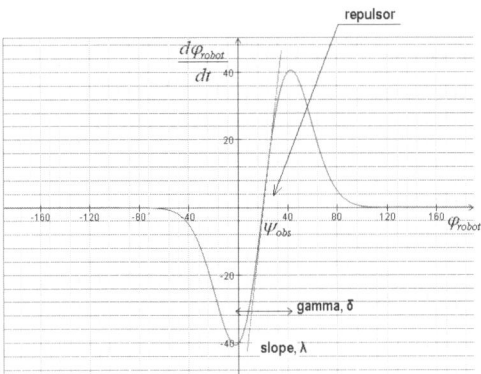

Fig. 3. Repulsive force as function of obstacle presence

Following the same technique as the target pursuing, as the robot moves also a repulsor fixed point shift occurs in the navigation direction. This brings up a repulsive force that compels the system to move away from this value due to the slope's negative value of the tangent to the curve at this point. As shown in Fig. 4, the intensity of the repulsive force varies according to the obstacle distance.

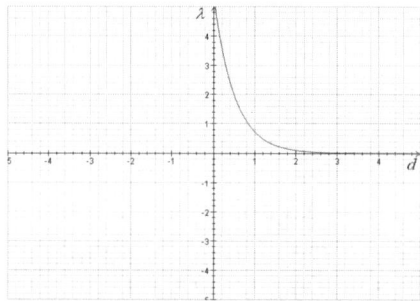

Fig. 4. Intensity of the repulsive force varying according to the obstacle distance

In this case, the behaviour variable is the angle difference between the robot direction and the target angle. The functions corresponding to obstacles will be continuously shifting along the same axis depending on their relative angles to the robot.

In the same way, the direction variation rate in time corresponds to the angular velocity or rotational movement of the robot.

The differential equation assumes the form:

$$\omega(t) = \frac{d\varphi_{robot}}{dt} = -\lambda \cdot (\Delta\alpha - \Delta\beta) \cdot e^{\frac{(\Delta\alpha - \Delta\beta)^2}{2\delta^2}} \quad (7)$$

where: ω – Robot's Angular velocity, $\Delta\alpha$ – Target angle relative to the robot
$\Delta\beta$ – Difference between obstacle and target angles

As it is possible to get an unlimited number of obstacles in an unknown environment, the global corresponding function is given by the sum of each of the individual forces, which results in the following equation:

$$\omega(t) = \frac{d\varphi_{robot}}{dt} = \sum_{i=1}^{n} -\lambda_i \cdot (\Delta\alpha - \Delta\beta_i) \cdot e^{\frac{(\Delta\alpha - \Delta\beta_i)^2}{2\delta^2}} \quad (8)$$

where: **n** – Number of obstacles

The repulsive intensity to the obstacle, i, is given by the equation:

$$\lambda_i = \beta_1 \cdot e^{-\frac{d_i}{\beta_2}} \quad (9)$$

The robot behaviour is based on the sum of all forces, including obstacles repulsive forces and the target attractive force. The resulting curve can assume many forms, as in Fig. 5. In this particular example, besides the target there are two obstacles placed in different directions. The resulting sum is a dynamic non-linear function (Fig. 6).

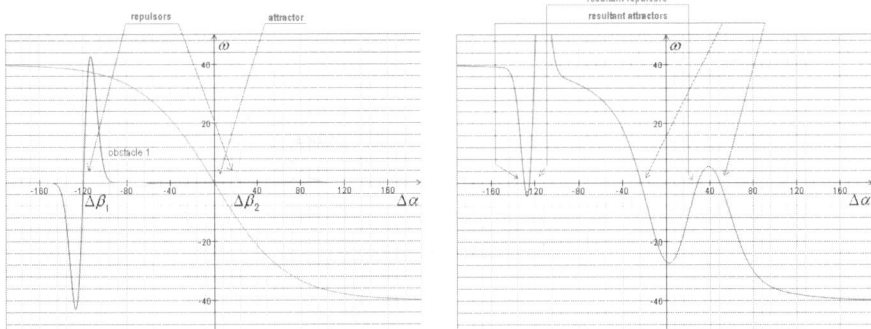

Fig. 5. Target and 2 obstacles vector fields   **Fig. 6.** Resultant vector field

Depending on the robot navigation direction, the system may adopt any of the attractor fixed points, stepping away from the repulsors. As robot moves, all vector fields vary, and attractor and repulsive points appear and disappear in a dynamic way.

This is an adaptive control to an unknown environment. The self decision is dynamic, so the behaviour is very smooth, similar to human decisions.

3 Linear Velocity Control

The robot linear velocity formula is continuously generating values so that the robot moves in a straight line to the target. This series of values, $x_{robot}(t)$, are generated by a dynamic system formulated by the following differential equation:

$$\frac{dx_{robot}}{dt} = f_{t\,arg\,et}\left(x_{robot}\right) \tag{10}$$

The linear velocity control dynamic system differential equation, is as follows:

$$v(t) = \frac{dx_{robot}}{dt} = -v_{max.} \cdot \tanh\left(\left(\left(x_{robot} - x_{t\,arg\,et}\right) - \sigma\right) \cdot \lambda\right) \tag{11}$$

where:

v –Robot linear velocity, **x_{robot}** – Robot position in a straight line to target
$v_{max.}$ – Robot max. linear velocity, **x_{target}** – Target position in a straight line to the robot
σ – Security distance to the target, **λ** – Attraction intensity to the target

Note: $\Delta d = x_{robot} - x_{t\,arg\,et}$, (distance to target)

As it can be seen in Fig. 7, when the difference between the robot and target position is equal to the safety distance the robot must stop.

The linear velocity indicated on the chart (v), is an input parameter to the function of robot motors and ranges from 0 to 100 (there is no specific unit). The distance to the target also ranges between 0 and 100.

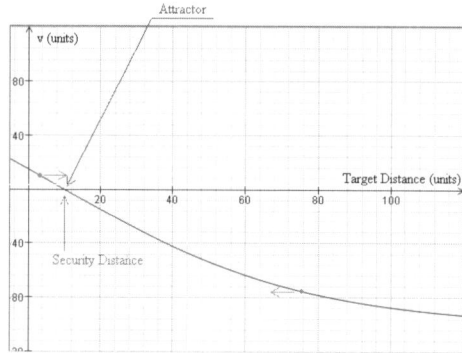

Fig. 7. Robot linear velocity graphical representation

As in the rotational movement, one important parameter is λ, because it allows specifying the target attraction intensity, also known as robot response time. But this value must not be very high otherwise the system becomes unstable.

The equation implemented on the robot and represented in figure 8 is:

$$\omega(t) = -100 \cdot \tanh\big((\Delta d - 10) \cdot 0.015\big) \tag{112}$$

4 Experimental Results

The robot's behaviour, using dynamic control, was first tested by creating a moving target (ball) in an environment without obstacles. It should be noticed that robot orientation, obstacles and target angles as well as respective distances were obtained through vision software. Fig. 8 shows results of rotational velocity variation in terms of target angle. The reference angle 180° means the robot is facing the target.

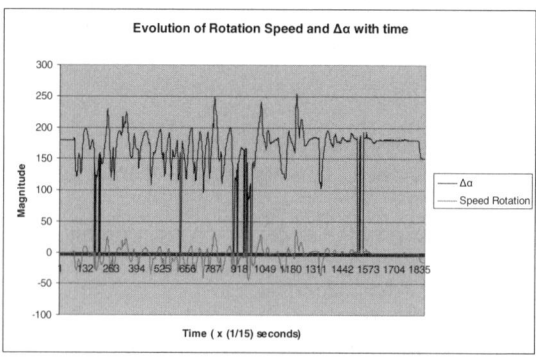

Fig. 8. Rotational velocity and corresponding angle between ball and robot

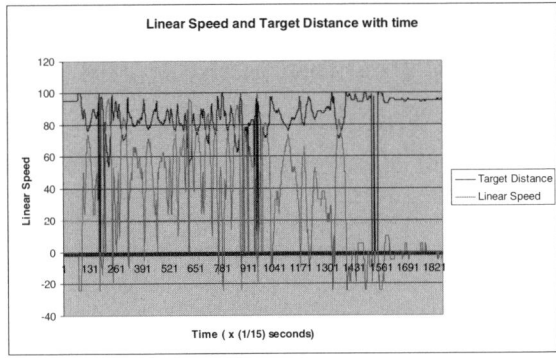

Fig. 9. Robot linear velocity versus target distance

The variation of linear velocity is represented in Fig. 9 as function of time and distance to the target. The value "95" is a reference position indicating that the robot reached the target. When the target moves away from the robot this value decreases and vice-versa. One unit of time corresponds to one frame. The processing cycle time is about 66ms.

5 Conclusions

From the experimental results shown above it can be concluded that the robot responds very fast to changes, and hence maintaining a smooth behaviour through time. This type of control doesn't need calibration parameters as opposite to PID control approach previously used [3]. Besides, the battery charge does not have any influence on the robot movement performance as happened with PID control algorithms.

This fast and stable dynamic control provides a very good precision and suits very well applications where low time-consuming algorithms are required for navigation in unknown environments.

Soon the complete control algorithm will be tested with moving obstacles included. Theoretically it is a very good adaptive control model.

The team would like to thank Prof. Estela Bicho for her support.

References

1. Bicho, E., Schöner, G.: The dynamic approach to autonomous robotics demonstrated on a low-level vehicle platform. Robotics and Autonomous Systems 21, 23–35 (1997)
2. Bicho, E., Mallet, P., Schöner, G.: Target representation on an autonomous vehicle with low level sensors. The International Journal of Robotics Research 19(5), 424–447 (2000)
3. Ribeiro, F., Moutinho, I., Silva, P., Fraga, C., Pereira, N.: Vision, Kinematics and Game strategy in Multi-Robot Systems like MSL RoboCup. In: Nardi, D., Riedmiller, M., Sammut, C., Santos-Victor, J. (eds.) RoboCup 2004. LNCS (LNAI), vol. 3276. Springer, Heidelberg (2005)

Evolutionary Design of a Fuzzy Rule Base for Solving the Goal-Shooting Problem in the RoboCup 3D Soccer Simulation League

Mohammad Jafar Abdi, Morteza Analoui, Bardia Aghabeigi, Ehsan Rafiee, and Seyyed Mohammad Saeed Tabatabaee

Intelligent Systems Lab, Computer Engineering Department, Iran University of Science and Technology, Tehran, Iran
{m.jafarabdi,m.analoui,b.aghabeigi,e.rafiee, s.tabatabaee}@gmail.com

Abstract. Most of the problems in the RoboCup soccer domain suffer from the noisy perceptions, noisy actions, and continuous state space. To cope with these problems, using Fuzzy logic can be a proper choice, due to its capabilities of inferring and approximate reasoning under uncertainty. However, designing the entire rule base of a Fuzzy rule base system (FRBS) by an expert is a boring and time consuming task and sometimes the performance of the designed Fuzzy system is far from the optimum, especially in cases that the available knowledge of the system is not enough. In this paper, a rule learning method based on the iterative rule learning (IRL) approach is proposed to generate the entire rule base of an FRBS with the help of genetic algorithms (GAs). The advantage of our proposed method compared to similar approaches in the literature is that our algorithm does not need any training set, which is difficult to collect in many cases; cases like most of the problems existing in the RoboCup soccer domain. As a test case, the goal-shooting problem in the RoboCup 3D soccer simulation league is chosen to be solved using this approach. Simulation tests reveal that with applying the rule learning method proposed in this paper on the goal-shooting problem, not only can a rule base with good performance in goal-shooting skill be obtained, but also the number of rules in the rule base can be decreased by using the general rules in constructing the rule base.

1 Introduction

Most of the problems in the RoboCup soccer domain suffer from the noisy perceptions, noisy actions, and continuous state space. To cope with these problems, the use of Fuzzy logic can be a good choice, due to its capabilities of inferring and approximate reasoning under uncertainty.

However, in the traditional design of Fuzzy systems, an expert's knowledge of the system is necessary. Due to such dependence, sometimes the performance of a Fuzzy system can be far from the optimum. To overcome these drawbacks, techniques such as neural networks (NNs), and evolutionary algorithms (EAs) are combined with Fuzzy logic to form hybrid-systems. To implement NN, lots of training data are

U. Visser et al. (Eds.): RoboCup 2007, LNAI 5001, pp. 320–328, 2008.

required which are difficult to collect in many cases. The adaptive capabilities, robust nature, and simple mechanics of EAs make them inviting tools for the development and optimization of Fuzzy rule base systems (FRBS) [1].

In this paper, we have proposed a new approach based on the iterative rule learning (IRL) method, to learn the entire Fuzzy rule base of an FRBS, which can be used for solving most of the problems in the RoboCup soccer simulation domain. In the original IRL method, a training set is needed for the algorithm. This training set should contain the correct examples, and in most cases, it should be complete enough to cover entire input space of the problem. However, obtaining such a complete training set is very difficult for some problems, because this needs supervision of an expert and this is not always possible. In our proposed approach, the entire Fuzzy rule base can be learned without needing any training set.

We chose the *goal-shooting* problem in the RoboCup 3D soccer simulation league as a test case to prove our proposed approach. We will show how the entire Fuzzy rule base for solving the *goal-shooting* problem can be generated automatically by using our proposed approach.

The remainder of this paper is organized as follows. Section 2 presents a quick overview of *Genetic Fuzzy* systems (GFS) and compares the different approaches in the literature. Section 3 describes the proposed approach for learning the Fuzzy rule base and how the Fuzzy rule base can be encoded into the chromosomes, as well as how the fitness value of chromosomes can be calculated. Section 4 introduces the *goal-shooting* problem in the RoboCup 3D soccer simulation league, and describes how the proposed algorithm is applied for this problem. Section 5 details the results of our training experiments. Conclusions and future works are summarized in Section 6.

2 Genetic Fuzzy Systems

This section describes a brief overview of *Genetic Fuzzy Systems* (GFSs) (Section 2.1). It also compares the different approaches of rule learning in the literature and explains their advantages and disadvantages (Section 2.2).

2.1 Overview of Genetic Fuzzy Systems

The general name of *Genetic Fuzzy Systems* (GFSs) refers to the systems that use the GAs for designing Fuzzy systems [2]. The most prominent types of GFSs are genetic Fuzzy rule-based systems (GFRBSs), whose genetic process learn or tune different components of a Fuzzy rule-based system (FRBS) [4]. Each FRBS usually consists of two different components.

When considering a rule-based system and focusing on learning rules, there are three main approaches that have been applied in the literature: Pittsburgh, Michigan, and iterative rule learning [4]. In Section 2.2, a comparison between these three approaches is presented.

2.2 Comparison between Rule Learning Approaches

As stated in Section 2.1, three main approaches exist in the literature for learning rules of the rule base of an FRBS. In this section a comparison between these approaches is presented.

The Michigan approach is not appropriate for our problem, since it does not consider the cooperation between rules. This problem is referred to as the cooperation vs. competition problem (CCP) [5]. If we use the Pittsburgh approach, the search space for GA becomes very large. Thus the computational time and cost is too high to choose this approach for our problem.

If we use the IRL approach, we do not have the above problems. However, in the original IRL approach, we need a training set that contains the correct examples. However, in some scenarios such as the most problems in the RoboCup soccer domain, obtaining such a training set is very difficult. Thus we changed the IRL approach, in order not to need any training set. In the next section, we will describe our proposed approach.

3   Learning a Fuzzy Rule Base

This section describes our proposed approach for learning the Fuzzy rule base of an FRBS (Section 3.1). It also describes how to encode the logic rules of a rule base (Section 3.2) and how the fitness value will be calculated for each chromosome in the population (Section 3.3). Finally the termination condition of the algorithm is explained (Section 3.4).

3.1   Overview of the Approach

Our proposal consists of a learning method based on the IRL approach in which the entire rule base is learned without needing any training set. The original IRL approach has the following steps:

1.   Use a GA to obtain a rule for the system.
2.   Incorporate the rule into the final set of rules.
3.   Penalize this rule.
4.   If the set of rules obtained is adequate to represent the examples in the training set, the system ends up returning the set of rules as the solution. Otherwise return to step 1.

All steps of the above algorithm depend on a set of input-output data pairs about the problem being solved, and even in many cases this training data set should cover the universe of discourse of all the variables in the antecedent part of the rules. However, obtaining such a complete training data set is very difficult for some problems, such as those existing in the RoboCup soccer domain. Therefore, for such problems we need a different approach that does not depend on a training set. Here we propose a new algorithm based on the IRL approach to generate the rule base of an FRBS without needing any training set. Our algorithm has the following steps:

1.   Use a GA to obtain a rule for the system.
2.   Incorporate the rule into the final set of rules.
3.   Calculate the *performance* measure and *completeness* measure for the set of rules obtained. If these two measures are acceptable, the system ends up returning the set of rules as the solution. Otherwise return to step 1.

The first two steps of the above algorithm are the same as the original IRL approach. However, the consistency criterion used in the first step of the algorithm is calculated independent of any training set. The *consistency* property of the final rule base is considered in the calculation of the fitness value for the chromosomes of the GA which were used in the first step of the algorithm. Fitness calculation is described in detail in Section 3.3. In step 3, a performance test is done to obtain the *performance* measure of the rules obtained in the final rule set so far. The *completeness* property of the rule base is checked through calculation of the *completeness* measure for the rules obtained in the final rule set. The *performance* and *completeness* measures will be described in detail in Section 3.4.

3.2 Encoding Method for Logic Rules

Suppose a Fuzzy rule base system with $X = (x_1, x_2, \ldots, x_n)$ as its inputs and $Y = (y_1, y_2, \ldots, y_m)$ as its outputs. Each input x_j ($j = 1, \ldots, n$) has $q[j]$ linguistic terms denoted as $A_j^1, A_j^2, \ldots, A_j^{q[j]}$, and each output y_j ($j = 1, \ldots, m$) has $r[j]$ linguistic terms denoted as $B_j^1, B_j^2, \ldots, B_j^{r[j]}$. V_j is defined as a vector, elements of which are the Fuzzy sets corresponding to input x_j, thus we can write: $V_j = (A_j^1, A_j^2, \ldots, A_j^{q[j]})$. Similarly U_j is defined as a vector, elements of which are the Fuzzy sets corresponding to output y_j, thus we can write: $U_j = (B_j^1, B_j^2, \ldots, B_j^{r[j]})$. A universal rule in a knowledge base has the form as

$$\textit{if } [x_{p(1)} = \bigcup_{j \in D(1)} A_{p(1)}^j] \textit{ and } [x_{p(2)} = \bigcup_{j \in D(2)} A_{p(2)}^j] \textit{ and } \ldots \textit{ and } [x_{p(s)} = \bigcup_{j \in D(s)} A_{p(s)}^j]$$

$$\textit{then } (y_1 = B_1^{k[1]}, y_2 = B_1^{k[2]}, \ldots, y_m = B_1^{k[m]}) \tag{1}$$

here, $p(\cdot)$ is an integer function mapping from $\{1, 2, \ldots, s \ (s <= n)\}$ to $\{1, 2, \ldots, n\}$ satisfying $\forall x \neq y, p(x) \neq p(y)$. $D(j)$ is an arbitrary subset of $\{1, 2, \ldots, q[p(j)]\}$ (i.e. $D(j) \subset \{1, 2, \ldots, q[p(j)]\}$), and $k[j]$ is an arbitrary number between 1 and $r[j]$. In principle, a rule base for the Fuzzy rule base can contain arbitrary rules in the form of (1).

From (1), we can see that the premise of a Fuzzy rule is characterized by sets $D(i) \subset \{1, 2, \ldots, q[p(i)]\}$. This fact suggests that a binary code be a suitable scheme for representing premises of such rules, as inclusion or exclusion of an integer in the sets $D(i)$ can be declared binary. For input variable x_j ($j = 1, \ldots, n$) with $q[j]$ linguistic terms, a segment consisting of $q[j]$ binary bits is required to encode the conditional composition for this variable. Every bit of the segment corresponds to a linguistic term with bit "1" for presence and bit "0" for absence of its Fuzzy set in forming the condition. For output variable y_j ($j = 1, \ldots, m$) with $r[j]$ linguistic terms, a segment consisting of $r[j]$ binary bits is required to encode a specific value for this variable. In contrary to the segments corresponding to the input variables, in the segments corresponding to the output variables, only one bit is allowed to be '1' and all the other bits should be '0'; because exactly one specific linguistic value should be assigned to each output variable. Therefore, each chromosome consists of L ($L = \sum_{j=1}^{n} q[j] + \sum_{j=1}^{m} r[j]$) binary genes that construct a rule.

3.3 Fitness Calculation

In the first step of our proposed algorithm, we use a GA to obtain a rule. Each GA needs a fitness function to evaluate the chromosomes. We have considered three factors in calculating the fitness value for a chromosome. The first factor is quality of the chromosome itself, i.e. the quality of the result of applying that rule in the environment. The *quality* measure is problem dependent and should be defined for each problem separately. The second factor is *generality* measure of a rule. The *generality* measure of a rule is equal to the size of the input space covered in the premise of a rule. The more general rules are used, the fewer number of rules is needed in the final rule base. The third factor is *consistency* measure of a rule with the rules inside the final rule base. The *consistency* measure of a rule is calculated as follows.

A conflict occurs in the rule base if there exist two rules which have overlapping input patterns but different linguistic consequences [3]. For example, suppose that rule R1 is defined as 'if x_1 is low and x_2 is medium then y_1 is low', and rule R2 is defined as 'if x_1 is low then y_1 is high'. In this case, there is a conflict between R1 and R2; because there is an overlapping area between the input spaces they cover, while their consequences are different. For each rule in the population, the *conflicting scale* (CS) of that rule with each rule in the final rule base is calculated separately, and summed together to construct the *conflicting amount* (CA) of that rule. The conflicting scale between two rules is equal to the cardinality of intersection set between input spaces covered in their premises. Finally, we consider a measure to evaluate the consistency of a rule with the rules in the final rule base. In the case that there are no conflicts existing, the *consistency* measure reaches its maximum value of "1". Otherwise it decreases linearly with the increment of conflicting amount, until its minimum value "0" is reached. The calculation of *consistency* measure (F_c) is given in the following formula:

$$F_c = \begin{cases} 1 - kCA & if\ kCA < 1, \\ 0 & if\ kCA \geq 1, \end{cases} \tag{2}$$

where $k \in (0,1]$ denotes the decreasing rate of F_c. The value of k should be determined in terms of particular problems to be solved.

Therefore, the final formula for calculating the fitness value of a chromosome (rule) in the population has the following form:

$$F_{total} = \alpha \cdot F_{quality} + \beta \cdot F_{generality} + \gamma \cdot F_{consistemcy} \tag{3}$$

where α, β, γ are parameters for scaling each factor, and should be determined in terms of particular problems to be solved.

3.4 Termination Condition

In the third step of our proposed algorithm, a termination condition is needed. We consider two criteria in the termination condition: The *performance* measure and the *completeness* measure. The *performance* measure is calculated through running a performance test. The performance test is a domain-dependent test, and the obtained final rule base after each iteration of the algorithm should be tested in the problem environment. The second criterion, i.e. the *completeness* measure is equal to the size of input space covered in the premises of the rules in the final rule base, divided by the size of total input space of the problem to be solved. Either of these two measures or the combination of them can be used for the termination condition of the algorithm. Also a maximum number of iterations can be considered to terminate the algorithm, if the number of iterations exceeds this value.

4 Learning the Rule Base for Goal-Shooting Problem

We chose the *goal-shooting* problem in the Robocop 3D soccer simulation league as a test case to prove our proposed approach. The *goal-shooting* is an essential skill of any successful team in the RoboCup 3D soccer simulation league. Since the goalie has the ability to catch the ball and also the height of the goal is rather short, the *goal-shooting* problem has been turned to a difficult and important problem in this league. Using our proposed approach, we will show how the entire Fuzzy rule base of the *goal-shooting* problem can be generated automatically. The input variables for this problem are as follows:

(1) *BallGoalAngle:* This variable is calculated as follows:

$$BallGoalAngle = \tan^{-1} \frac{BallY - GoalY}{BallX - GoalX}, \tag{4}$$

where *BallY* and *BallX* are the coordinates of the ball, and, *GoalY* and *GoalX* are the coordinates of the goal center.

(2) *BallGoalDistance:* This is the distance between the current position of the ball and center of the goal.

(3) *GoalieY:* This is the y-coordinate of the goalie.

(4) *GoalieX:* This is the distance between goalie and the goal line in the x-axis.

The output variables of this problem are as follows:

(1) *ShootPower:* This is the power with which a player can shoot the ball. The power is between 0 and 100.

(2) *ShootAngle:* This is the latitude angle for a shoot which is between 0 and 50.

(3) *ShootPoint:* This is the y-coordinate of the target point in the goal. This is between $-\frac{goal\_with}{2}$ and $\frac{goal\_with}{2}$, where *goal_with* is the width of the goal.

In order to apply our approach to this problem, the linguistic terms for input and output variables should be defined first. The linguistic terms and membership functions for input variables and output variables of this problem are depicted in Figure 1 and Figure 2 respectively.

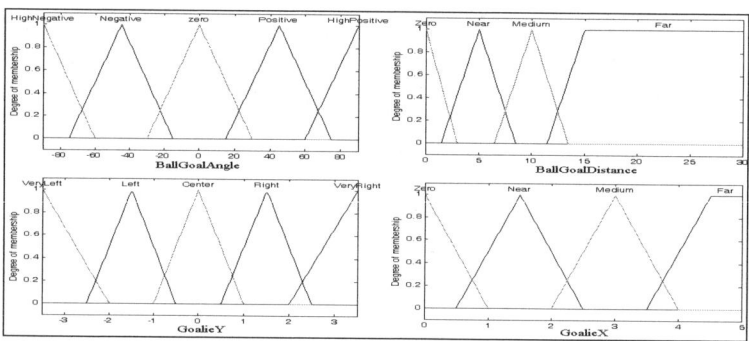

Fig. 1. Membership functions for the input variables of the goal-shooting problem

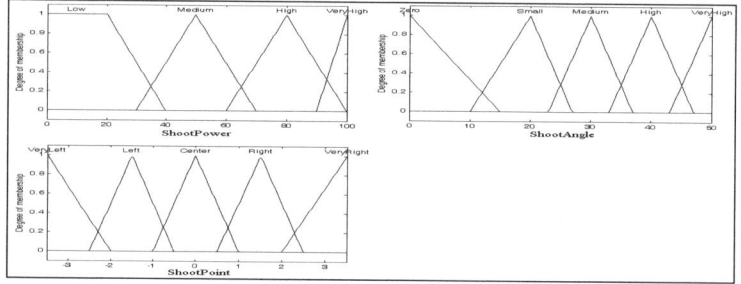

Fig. 2. Membership functions for the output variables of the goal-shooting problem

To apply our approach to the goal-shooting problem, we also need to define the fitness function for the GA used in the first step of the rule learning algorithm. As stated in Section 3.3, the fitness function consists of three parts. The only part we need to define here is the *quality* measure. The other two parts of the fitness function are general and don't depend on a specific problem, so we can use the general definition in Section 3.3 to calculate their value. To calculate the *quality* measure of a rule, we consider the set of conditions that this rule covers in its premise part. If the number of conditions this rule covers is more than *MaxEvaluation*, we pick only *MaxEvaluation* conditions randomly from them. These conditions constitute a condition set. We set the *MaxEvaluation* parameter equal to 10 in our experiments. This parameter is just for preventing the algorithm from taking very long time to run. Then for each condition in the condition set obtained in the previous step, we should create the situation described by the condition of that rule. This is accomplished by placing the ball, attacker-player, and goalie at the corresponding locations and allowing the attacker-player to shoot the ball towards the goal according to the parameters determined by the output variables of that rule. Then the *quality* measure of a rule is calculated as:

$$F_{\text{quality}} = \alpha \cdot \frac{Scores}{Total\_Shoots} + \beta \cdot \frac{Sum\,(\Delta Y)}{Total\_Shoots}, \tag{5}$$

where *Scores* is the total number of shoots that led to a score. *Total_Shoots*, is the total number of shoots towards the goal. ΔY is defined as the difference between y-coordinate of the *ShootPoint* and the y-coordinate of the goalie at the moment when the attacker-player kicks the ball. A bigger ΔY means a higher chance to score a goal. α and β are two scaling parameters that in this problem are heuristically set to 100 and 10 respectively.

5 Experiments and Results

Several tests have been carried out to show the impact of changing the parameters of the algorithm in the result. In our experiments we set the GA parameters which were used in the first step of the algorithm as described below.

The population size is fixed to 20 chromosomes. *Stochastic universal sampling* with elitism selection is adopted. One elite chromosome is reserved in each generation. The crossover operator is the uniform crossover with the probability of $P_c=0.6$. The mutation is applied to each gene of a chromosome independently with probability of $P_m=0.05$. The maximum number of generations of GA to obtain a rule is fixed to 30.

In every iteration of the algorithm, a GA runs and the best rule obtained is added to the final rule base. In step 3 of the algorithm, a performance test is carried out to evaluate the performance of the final rule base obtained so far. To carry out the performance test, 30 random situations will be created in the simulation and the attacker-player is allowed to shoot the ball towards the goal. The goalie used for our experiments is the goalie of Caspian team used in the RoboCup 2006 competitions in Bremen. The performance measure is calculated as the number of shoots that led to a score divided by the total number of shoots (i.e. 30 in our experiments). The performance measure is considered as the termination condition of the rule learning algorithm. If the performance measure of the final rule base becomes greater than a parameter named μ_p ($0 \le \mu_p \le 1$), the algorithm terminates, otherwise the algorithm goes back to step 1. Figure 3 plots the curve of performance measure of the final rule base with respect to the number of iterations required to reach a rule base with that performance measure. The curve is the average of 10 independent runs of the algorithm. Note that in Figure 3, the iteration number is equal to the number of rules in the final rule set in that iteration, because in each iteration of the algorithm, exactly one rule is added to the final rule set. Figure 3 shows that if bigger values are assigned to μ_p, more iterations will be needed to terminate the algorithm and also more rules will be obtained in the final rule set. So, the more accuracy and performance is needed for a system, the larger rule base is obtained, and also more time is needed to reach the solution.

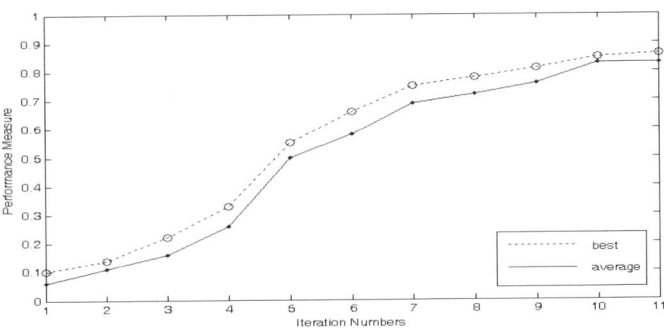

Fig. 3. Curve of performance measure of the final rule base with respect to the number of iterations

6 Conclusion and Future Work

In this paper a rule learning method was proposed to learn the entire rule base of an FRBS without needing any training set. Also a generality metric was introduced in the algorithm that can lead to a compact rule base with less number of rules. Such an FRBS can be used to solve a broad range of problems. Specially most of the problems in the RoboCup soccer simulation domain can be solved in this way. As a test case, the goal-shooting problem in the RoboCup 3D soccer simulation league was chosen and solved by applying this approach. A big advantage of this approach is using Fuzzy logic and generating a Fuzzy rule base that is expressive enough and easy to understand by a human.

In the proposed approach, membership functions should be defined by an expert. Although the most difficult part of designing an FRBS for dynamic and complicated systems is defining the rule base, the membership functions can play important roles in the performance of the system. We are planning to embed the tuning of membership functions in the algorithm, so as to an expert does not have to define them explicitly.

References

1. Vadakkepat, P., Peng, X., Kiat, Q.B., Heng, L.T.: Evolution of Fuzzy Behaviors for Multi-Robotic System. Journal of Robotics and Autonomous Systems 55, 146–161 (2007)
2. Herrera, F., Magdalena, L.: Genetic Fuzzy Systems: A Tutorial. Tatra Mountains Mathematical Publications. In: Mesiar, R., Riecan, B. (eds.) Fuzzy Structures. Current Trends. Lecture Notes of the Tutorial: Genetic Fuzzy Systms. Seventh IFSA World Congress (IFSA 1997), Prague, June 1997, vol. 13, pp. 93–121 (1997)
3. Xiong, N., Litz, L.: Reduction of Fuzzy Control Rules by Means of Premise Learning - Method and Case Study. Fuzzy Sets and Systems 132(2), 217–231 (2002)
4. Cordon, O., Gomide, F., Herrera, F., Hoffmann, F., Magdalena, L.: Ten years of genetic Fuzzy systems: current framework and new trends. Fuzzy Sets and Systems 141, 5–31 (2004)
5. Bonarini, A.: Evolutionary Learning of Fuzzy Rules: Competition and Cooperation. In: Pedrycz, W. (ed.) Fuzzy Modeling: Paradigms and Practice, pp. 265–283. Kluwer Academic Press (1996)

Behavioral Cloning for Simulator Validation

Robert G. Abbott

University of New Mexico, Albuquerque, NM, 87131, USA
rabbott@cs.unm.edu

Abstract. Behavioral cloning is an established technique for creating agent behaviors by replicating patterns of behavior observed in humans or other agents. For pragmatic reasons, behavioral cloning has usually been implemented and tested in simulation environments using a single nonexpert subject. In this paper, we capture behaviors for a team of subject matter experts engaged in real competition (a soccer tournament) rather than participating in a study. From this data set, we create software agents that clone the observed human tactics. We place the agents in a simulation to determine whether increased behavioral realism results in higher performance within the simulation and argue that the transferability of real-world tactics is an important metric for simulator validation. Other applications for validated agents include automated agent behavior, factor analysis for team performance, and evaluation of real team tactics in hypothetical scenarios such as fantasy tournaments.

1 Introduction

Accurate simulation of physical environments and human behavior is important for applications including training [9] and system design concept exploration [3]. However, specific metrics are required for otherwise vague notions of "accuracy" and "realism." For training tactics, a key aspect of accurate simulation is the transferability of tactics between real and simulated environments. If correct tactics are counter-effective in the simulation, students may learn incorrect tactics that work only in the simulator (negative training). We propose a metric for the transferability of tactics: the correlation between agent behavioral realism and agent performance in the simulation.

This approach requires creating software agents with realistic expert behaviors, a task that is often difficult. Domain experts are often unavailable for intensive consulting or are not trained to engineer automated systems. Therefore, domain-specific software is often created by researchers and engineers with only second-hand knowledge of the domain.

Behavioral cloning is one technique to address the challenge of creating agents. Instead of attempting to explain what they know, experts simply perform the task. The performance is recorded, and machine learning algorithms are used to create a model which is used to produce agent behaviors. This paper describes our implementation of behavioral cloning for soccer play and explores the correlation between human model fidelity and performance in the RoboCup

U. Visser et al. (Eds.): RoboCup 2007, LNAI 5001, pp. 329–336, 2008.
© Springer-Verlag Berlin Heidelberg 2008

soccer simulator. We find a significant correlation, proving that (1) the aspects of behavior captured by the model are significant, (2) the human team employed effective tactics, and (3) real-world tactics are effective within the simulation.

2   Related Work

The goal of software validation is to test completed software for compliance with its specifications. Gledhill and Illgen [6] present a survey of techniques for verification and validation of tactics simulators. For example, *trace validation* is a manual inspection of program state throughout the execution of a test scenario. The techniques are quite general and apply to software for almost any application. Our work focuses on a more specific criterion – tactics validation – that is necessary for tactical trainers.

Behavioral cloning [2] is an established technique for building agent behaviors. A person is observed performing a task and their actions are recorded. A computer model is created to capture patterns of behavior from these observations. The model is then used to control a software agent. Behavioral cloning has been successfully applied in simulations of tasks such as piloting an airplane and operating a crane [10]. Aler, Garcia, and Valls [1] used behavioral cloning in RoboCup to model data collected from a modified version of the simulator that allows a human to play RoboCup as an interactive computer game. However, the captured behavior was a single user manipulating a simulation through computer input devices, rather than a team of soccer players competing in a real match.

For pragmatic reasons, most behavioral cloning research has focused on non-expert subjects (often the researchers themselves) in a computer simulation. Because we are interested in validating simulators against reality, our observations must come from the real world.

3   A Model-Driven Simulated Soccer Team

In selecting an aspect of soccer play for behavior modeling, we considered two criteria: observability (which excludes mental skills such as situational awareness) and transferability from the simulation to reality (which excludes low-level skills such as trapping the ball). We decided to focus on team positioning dynamics.

For low-level skills, our team adopted the UvA Trilearn [7] software library. It is effective, well documented, and open source. We manually implemented a simple kick strategy (including passing, dribbling, and shooting) for our team. Modeling these aspects of play is left for future research.

3.1   A Data Set of Human Soccer Play

Our research needed a data set of skilled human soccer play. Creating the data set required both source material (recorded soccer play) and a tool to extract data from the source material. Televised soccer footage fails to capture team positioning dynamics because it focuses on the ball, so we recorded games at the University of New Mexico soccer pitch. A single video camera was unable to record the entire field with adequate resolution, so we used an array of four cameras along the top row of bleachers overlooking the field 1.

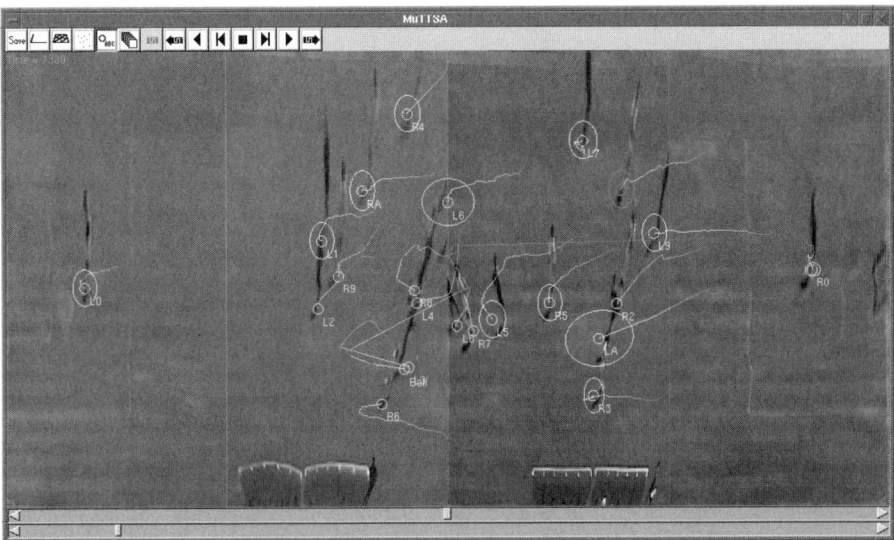

Fig. 1. A screenshot from the motion capture application. The application shows an overhead view synthesized from four cameras. Rings and trails indicate the players' current and recent positions.

Our data extraction tool is a multiple target tracking application that fuses the four viewpoints, detects players using background subtraction [5], and tracks players using Joint Probability Data Association [8] with Kalman Filtering.

The resulting dataset consists of a sequence of observations of the state of the game. Each observation consists of the positions of all 22 players (11 per team) and the ball at a sample rate of $10Hz$. Other information (e.g., velocities and distances) is derived from the sequence of positions.

3.2   Cloning Team Positioning Dynamics

We modeled team positioning dynamics using a function approximator to predict the position of each human player given the current state of the environment (e.g., the positions of players and the ball). We used nearest-neighbor matching, which is a type of instance-based learning (IBL). In the soccer domain, each instance is an observation of the state of the soccer game at a moment in time. In our implementation, an observation is an associative array of named features.

Each feature may be multidimensional. Many of the features are two dimensional because we used the 2d version of the RoboCup simulator.

3.3   Clustering and Model Context Set Size

The basic nearest-neighbor algorithm is highly susceptible to over-fitting. The model will perform poorly in the regions of nonrepresentative instances. We used clustering to prune the instance set by discarding redundant and non-representative instances; only the cluster centers are retained. We refer to retained instances as contexts.

The use of clustering adds a parameter (the number of contexts) to the model. The impact of this parameter on predictive accuracy is measured in Section 4. The number of contexts also determines the run-time memory and computation requirements of the model, which is especially important for team-oriented tasks.

3.4   Distance Metric and Feature Selection

In a complex task such as soccer, a team will never encounter precisely the same situation twice, so it is crucial to draw correct analogies between the current situation and relevant model contexts. Relevance is defined by a distance function. Our implementation calculates the weighted Euclidean distance between observation vectors. The weight vector is a parameter to the system and must be chosen to emphasize the features that most influence expert tactics. A weight of 0 causes the corresponding feature to be disregarded entirely. Feature selection algorithms [4] could be used automate the selection of weights.

Feature selection is critical for IBL, which is intolerant of irrelevant attributes. We obtained the best results for soccer field positioning with a small set of carefully chosen features as described in Section 4.

3.5   Number of Training Observations

The success of behavioral cloning for tactics modeling is indicated by the sensitivity of the model to varying amounts of training data. If varying the amount of training data has no effect, then there is no transfer of human expertise to the system, and the human model is worthless. If performance increases very slowly with increasing amounts of training data, then gathering enough data to yield a substantial performance improvement may be prohibitively labor intensive.

The feature set, number of model contexts, and amount of training data are all interdependent parameters. For example, a larger amount of training data contains more varied situations and behavior and may require more features and contexts to model effectively.

4   Experiments in Behavioral Cloning for RoboCup

This section describes two experiments. The first measures the predictive accuracy of our player-positioning model on the human data set. The second measures the outcome of RoboCup matches when the model is used to control RoboCup agents. These provide the information required to calculate the correlation between behavior modeling accuracy and agent performance in the simulation.

4.1 Predictive Accuracy of Human Player Model

The first experiment determined how prediction accuracy varies with increasing amounts of training data. We produced a data set of human soccer play by capturing the first 20 minutes of a soccer match between the University of California Irvine Anteaters and the Western Illinois University Leathernecks at the University of New Mexico Soccer Complex on September 21, 2003. This data set is available at `http://www.cs.unm.edu/~rabbott/SoccerData`

Results are reported for three feature sets: "Ball Position X" uses only the component of ball position extending from one goal to the other; "Ball Position" contains the 2-dimensional ball position; "Ball Position and Velocity" includes both the ball's position and estimated velocity. We experimented with larger feature sets and more complex features (such as the density of opponents between a player and the ball) but without significant improvement in the results.

We used 10-fold cross-validation to measure predictive accuracy. For each fold, we measured the mean prediction error (the squared distance between predicted and observed player positions in holdout data) for every combination of observation set size, context set size, and selected features.

4.2 Predictive Accuracy of Human Player Model – Results

Figure 2 shows how feature selection, observation set size, and context set size influence prediction accuracy for the human soccer data set. These results show that the x component of ball position is insufficient to accurately predict player position. In contrast, the Ball Position and Ball Position and Velocity feature sets achieve lower prediction error, with steadily improving results up to all available observations (Figure 2(a)) and the maximum tested number of contexts (Figure 2(b)). The inclusion of ball velocity yields an improvement that is very small, but consistent across varying observation and context set sizes.

4.3 RoboCup Performance

All simulations have limited fidelity. Even a perfect model of ideal human behavior on a task might not perform well in a simulation of the same task because of differences between the model and the real world. We evaluated the performance of our model-driven RoboCup team using the same parameterizations as in the previous section. The opponent team is a behavior clone of the UvA Trilearn RoboCup team [7]. Training data for this clone was generated by running several simulation matches between two instances of the UvA Trilearn.

The performance metric is the goal difference $PenaltyScore = Goals_{opponent} - Goals_{self}$ during a five minute match. As with the error metric used in the previous section, a lower penalty score indicates better performance. (The penalty

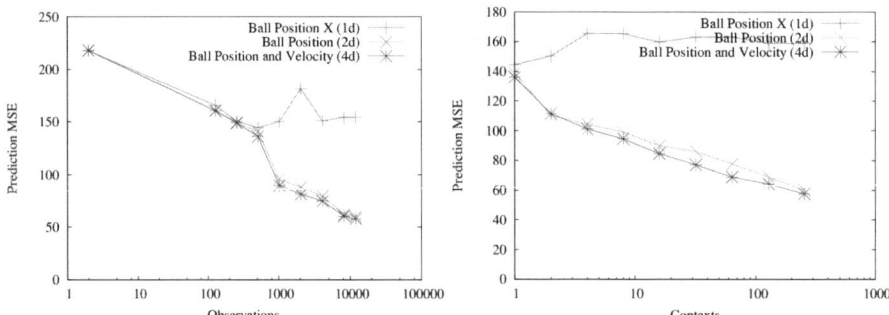

(a) Prediction error for each observation set size.

(b) Prediction error for each context set size.

Fig. 2. Summary of results from the prediction error parameter sweep

score reflects only the outcome of a match and is not related to penalties assessed by a referee for violating the rules of soccer).

As before, we tested all combinations of three feature sets and nine values each for the observation set and context set sizes. For each condition, we computed the mean of 100 trials for a total of 24,300 RoboCup matches.

4.4 RoboCup Performance – Results

The RoboCup performance for all conditions is summarized in Figure 3. The performance of Ball Position X is maximized with only 512 observations (about 50s of play) and decreases with additional observations. With this simple feature set, the model cannot adequately distinguish between different situations

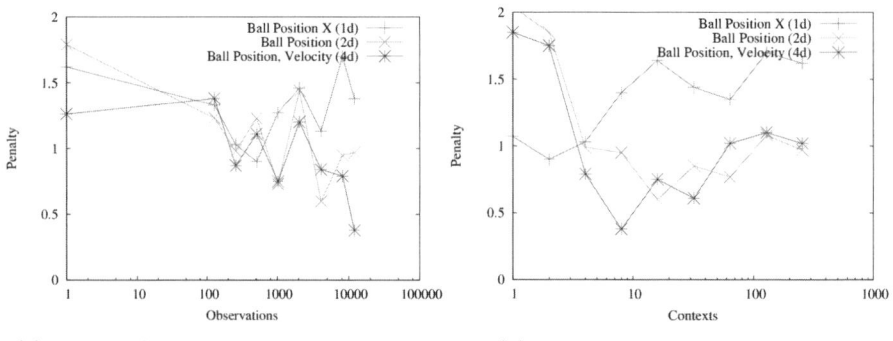

(a) Penalty for each observation set size.

(b) Penalty for each context set size.

Fig. 3. Summary of results from the RoboCup performance parameter sweep

in soccer. For the other two feature sets, performance generally improves with additional observations, and the maximum performance is achieved using both ball position and velocity with all available observations.

The result of varying the context set size is quite different; for all three feature sets, performance declines when more than 16 contexts are created (Figure 3(b)), even as the human prediction accuracy continues to increase (Figure 2(b)). This seems to indicate that clustering is an effective technique to prevent over-fitting when using IBL.

Figure 4 displays the correlation between model prediction error on the human soccer dataset and the penalty score for a RoboCup team controlled by the same model. This calculation is important because it tests the hypothesis that real soccer strategy is effective within the RoboCup simulator. No significant correlation is found for Ball Pos X ($r = 0.07$), perhaps because of the relatively small range of prediction accuracy observed for this feature. However, a significant correlation exists for Ball Position ($r = 0.41$) and Ball Position and Velocity ($r = 0.50$). When all conditions are taken together, the overall correlation is 0.43.

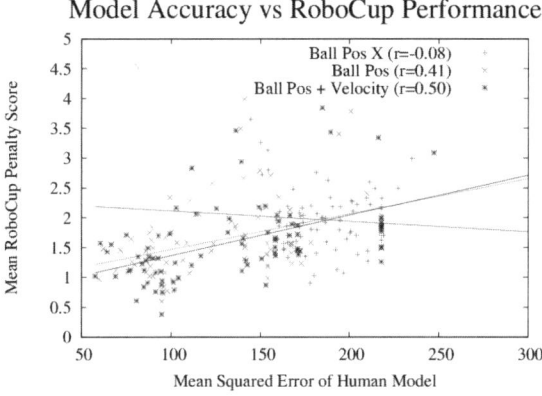

Fig. 4. Correlation between model predictive accuracy on the human soccer dataset and performance of a model-driven RoboCup team. Each data point represents a model with a unique set of parameter values. There are three point clouds, each representing a different feature set.

5   Discussion and Conclusion

The correlation between human model fidelity and performance in the simulation ($r = 0.43$) is significant, confirming the hypothesis that human soccer strategies are effective within RoboCup. What factors account for this correlation and at the same time prevent a stronger correlation? We propose three factors.

First, the correlation between predictive accuracy and performance is limited by the significance of the behaviors captured by the model. Our model captures team field positioning strategy, but does not capture other important factors such as pass selection and individual ball handling skill. In pedagogy, there is a risk of focusing on unimportant knowledge and skills simply because, for example, they are easy to explain or were important historically. The relative importance of various skills could be studied by modeling each and then measuring the impact of degrading one or more of the models.

Second, all humans (and human teams) are imperfect to varying degrees. Emulating a team of novices *should* result in worse performance than emulating a World Cup match. Thus, it may be possible to predict the outcome of a match through a contest of behavior clones.

Third, all simulations fall short of perfect realism. Ideal RoboCup tactics are distinct from ideal soccer tactics. This is a problem for training simulators because students will be discouraged from practicing proven tactics if they are ineffective in the simulator. Simulation developers should strive for a high correlation between desirable behavior and positive scenario outcomes.

Creating tests to isolate these factors is an important topic for future research. Such tests would allow instructors to re-target training, assess student performance, or focus on increasing simulator fidelity. However, any significant correlation between expert cloning accuracy and agent performance demonstrates that the modeled behavior is significant, that the example behavior is skilled, and that real-world tactics are transferable to the simulation.

References

1. Aler, R., Garcia, O., Valls, J.M.: Correcting and improving imitation models of humans for robosoccer agents. In: The 2005 IEEE Congress on Evolutionary Computation, vol. 3, pp. 2402–2409 (2005)
2. Bratko, I., Urbancic, T., Sammut, C.: Machine Learning and Data Mining: Methods and Applications. In: Behavioural Cloning of Control Skill. John Wiley & Sons Ltd. (1997)
3. Cares, J.R.: Agent modeling: the use of agent-based models in military concept development. In: WSC 2002, pp. 935–939 (2002)
4. Cost, S., Salzberg, S.: A weighted nearest neighbor algorithm for learning with symbolic features. Machine Learning 10, 57–78 (1993)
5. Fuentes, L.M., Velastin, S.A.: People tracking in surveillance applications. In: PETS 2001, Hawaii (2001)
6. Gledhill, D.W., Illgen, J.D.: 21st century verification and validation techniques for synthetic training models and simulations. In: Proc. I/ITSEC (1999)
7. Kok, J.R., Boer, R.: The incremental development of a synthetic multi-agent system: the UvA Trilearn 2001 robotic soccer simulation team. Master's thesis. University of Amsterdam, The Netherlands (2002)
8. Oh, S., Sastry, S.: A polynomial-time approximation algorithm for joint probabilistic data association. In: Proc. Am. Control Conf. (2005)
9. Stone, B.: Serious gaming. Defence Management Journal (2005)
10. Suc, D., Bratko, I.: Symbolic and qualitative reconstruction of control skill. Electronic transactions on artificial intelligence, Section B 3, 1–22 (1999)

A Model-Based Approach to Calculating and Calibrating the Odometry for Quadruped Robots*

Haitao He and Xiaoping Chen

Department of Computer Science, The University of Science and Technology of China,
HeFei, China

Abstract. This paper presents a model-based odometry calculation and calibration method (MBO) for quadruped robots. Instead of establishing the direct relation between target and actual speeds as previous methods did, MBO sets up a "parametric physical model" incorporating various properties of the robot and environment such as friction and inertia, through optimization with locomotor data. Based on this optimized model, one can compute the loci of robot legs' movement by forward kinematics and finally obtain odometric readings by analyzing the loci. Experiments on Sony AIBO ERS-7 robots demonstrate that the odometry error of MBO is generally 50% less than the existing methods. In addition, the calibration complexity is low.

1   Introduction

Odometry is very important for an autonomous robot[1,2,3,9], especially for the purpose of determining the robot's locomotory parameters (e.g., speed, position, orientation.) between fixes which are rare or cost demanding. It is relatively simple for a wheeled robot to calculate its odometry. The vehicle's offset from a known starting position can be computed with the data of encoders which monitor the wheels' revolutions and/or steering angles. Odometry calibration for wheeled robots often involves determining the values of kinematic parameters [2,9] or calibrating error model [3]. However, these methods cannot be implemented directly on legged robots, which have completely different mode of locomotion.

Two common motivations behind the current investigation into the odometry of legged robots are: (1) the odometry should be accurate and powerful enough with respect to the needs of real-world applications; (2) the odometry calibration should be as simple as possible. There is some work on meeting these two requirements jointly. Thomas Röfer [11] reports on a method for calculating odometry based on proprioception. German Team optimizes the parameters [5] of their walking engine [4] to make the actual walking speed as close as possible to that specified by the corresponding walk request. The rUNSWift team establishes the relation between raw walk command values and the corresponding actual speed values through polynomial curve fitting [6]. Lin and his colleagues propose calculating odometry for a hexapod robot from its body pose based on the kinematic configuration of its legs [7]. Stronger and Stone put forth a method for simultaneous calibration of action and sensor models autonomously [10].

---

* This work is supported by the NSFC 60275024 and the 973 program 2003CB317000.

U. Visser et al. (Eds.): RoboCup 2007, LNAI 5001, pp. 337–344, 2008.

All the previous work can be divided into two categories: one is to directly establish a mapping between the target and actual speeds – we call this "direct calibration". The other is mainly based on proprioception, we call this "indirect calibration". With the "direct" method try to establish a mapping from one space (target speeds) to another space (actual speeds), one has to calibrate lots of points of the space, and another disadvantage is the mapping lack the information of the process of state changing from one point to a another point. The "indirect" method uses the information of sensors, and generates the continuous proprioceptive state, so as to avoid the negative aspects of "direct" methods.

Based on previous work, we propose in this paper a new odometry calculation and calibration method for quadruped robots, named MBO (model-based odometry), that is the first "indirect calibration" based on a "parametric physical model" for quadruped robots. Instead of establishing the direct relation, MBO sets up a "parametric physical model" incorporating both geometrical and physical properties of the robot such as friction and inertia through optimization of a parameterized version of the "triangle model" proposed in [8] with the locomotor data of the robot. Based on this model, MBO can figure out the loci of leg movement with forward kinematics and finally obtain the odometric readings by analyzing loci. Since it is straightforward to calculate the acceleration/deceleration of the robot's movements at runtime, MBO provides a "finer-granularity" odometry for quadruped robots. In addition, the calibration complexity is rather low—only twelve walking samples are needed for optimization of the parametric physical model, although MBO runs semi autonomously at the current stage. We implemented MBO and carried out experiments on AIBO ERS-7. The results show that the accuracy of the odometry calculated by MBO is markedly higher than that by previous methods.

The remainder of this paper is organized as follows: Section 2 describes briefly the principle of MBO. Section 3 explains calibration of the parametric physical model on the AIBO platform. Section 4 discusses how to calculate odometry from loci. Section 5 presents the experimental results and the paper is concluded in Section 6.

2   The Principle

Consider a walking engine that responds to any walking request by arranging the legs' movements according to some preset gait loci. If the robot is driven by motors, the robot's architecture usually allows each joint to receive a request every t_u time and returns feedback on its corresponding angles at the same frequency. According to a given leg model, the paw positions can be calculated by forward kinematics using the joint angles fed back by joint sensors. Moreover, a calculated locus can be simply acquired by connecting the consecutive paw positions generated from the angles.

If a robot is held in the air while walking, its paw locus looks like the curve shown in Fig.1. The solid line from s to e in Fig.1 presents the trajectory of support phase of a leg. If the leg keeps in contact with the ground and has no paw slippage during the support phase, the movement of the robot is fully determined by the locus of the support phase. Therefore, the problem of odometry calculating is reduced into that of calculating the actual loci of legs.

Specifically, the displacement of one walk cycle is determined by the corresponding trajectory on the $x - y$ plane. Suppose a leg makes contact with the ground at point s and is raised in the air at point e, their projected projective points on the $x - y$ plane is marked as P_s and P_e respectively (Fig.1). The displacement S of this walk cycle can be computed by setting $S = P_s - P_e$. It follows under the assumption of no obstruction that the displacement of one walk cycle only concerns the point where the leg is placed onto the ground and the point where the leg is raised. Therefore, the error caused by discrete feedback of joint sensors is limited and can occur only within a few cycles of feedback when the leg is placed on or raised from the ground.

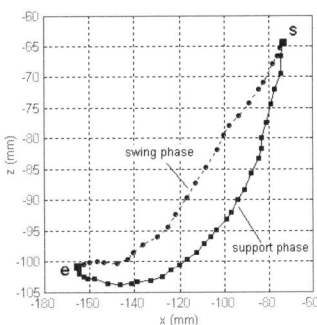

Fig. 1. One walk cycle of sampled leg locus. The sample interval t_u=8ms. The leg is in contact with the ground at point s, and raised in the air at point e.

The assumption of no slippage is problematic for real-world applications. Generally, odometry for legged robots is error-prone due to the noises and/or uncertainty inherent in the robots, their motion and even their environments. In order to simplify the modeling, one usually neglects physical properties such as friction, weight of the robot and its components, inertia, motor strengths, causing impaired performance in real-world applications [4]. As a remedy to this problem, we try to model these factors including both geometrical and physical properties as nuch as possible. We take a parameterized version of "triangle model" [8] as the starting point for our "parametric physical model" (see Fig. 2) and optimize it with the locomotor data of the robot's movements in the application environment by using a genetic algorithm. The expressive power and the adaptability of this model primarily stems from the variance of the parameters. For example, the length of some parts of the leg would be reduced by the genetic algorithm when the robot walks on a slipperier carpet. Due to this adaptability of the parametric physical model, MBO can provide more accurate odometric readings, as our experiments demonstrate.

3   Model Setup

There are two steps in MBO to set up a parametric physical model: (1) setting an temporary profile of the parameterized triangle model; (2) obtaining the final model through optimization. This section describes the detailed method on the sony AIBO robotics platform.

3.1   Parameterized Triangle Model

The AIBO robot has four legs. Each leg has three joints known as the rotator, abductor, and knee. The most widely used model is the triangle model described in [8], in which the lengths and widths of the legs are both considered. The model is simple and convenient to use. MBO uses the triangle model as the basis, but parameterizes it (Fig.2).

The following parameters describe in a related coordinate system whose origin is the position of joint rotator: 1). position correction of joint rotators six parameters.

2). length of fore and hind upper legs, L1 in Fig.2 two parameters. 3). width of fore and hind upper legs, L3 in Fig.2 two parameters. 4). length of fore and hind upper legs, L1 in Fig.2 two parameters. 5). the rotation angle correction of joint knee,θ in Fig.2. 6). the zero correction of joint rotator, abductor and knee, three parameters.

The paw position $P(x, y, z)$ can be determined using following transformations, written in matrix:

$$\begin{pmatrix} A1' \\ A2' \\ A3' \end{pmatrix} = \begin{pmatrix} A1 \\ A2 \\ A3 \end{pmatrix} + \begin{pmatrix} A1Correction \\ A2Correction \\ A3Correction \end{pmatrix} \tag{1}$$

$$\begin{pmatrix} x \\ y \\ z \\ 1 \end{pmatrix} = \begin{pmatrix} \cos(A1') & 0 & -\sin(A1') & 0 \\ 0 & 1 & 0 & 0 \\ \sin(A1') & 0 & \cos(A1') & 0 \\ 0 & 0 & 0 & 1 \end{pmatrix} * \begin{pmatrix} \cos(A2') & 0 & -\sin(A2') & 0 \\ 0 & 1 & 0 & 0 \\ \sin(A2') & 0 & \cos(A2') & 0 \\ 0 & 0 & 0 & 1 \end{pmatrix} * \begin{pmatrix} 1 & 0 & 0 & 0 \\ 0 & 1 & 0 & 0 \\ 0 & 0 & 1 & -L1 \\ 0 & 0 & 0 & 1 \end{pmatrix} *$$

$$\begin{pmatrix} 1 & 0 & 0 & 0 \\ 0 & 1 & 0 & L3 \\ 0 & 0 & 1 & 0 \\ 0 & 0 & 0 & 1 \end{pmatrix} * \begin{pmatrix} \cos(A3'+\Delta R) & 0 & -\sin(A3'+\Delta R) & 0 \\ 0 & 1 & 0 & 0 \\ \sin(A3'+\Delta R) & 0 & \cos(A3'+\Delta R) & 0 \\ 0 & 0 & 0 & 1 \end{pmatrix} * \begin{pmatrix} 1 & 0 & 0 & 0 \\ 0 & 1 & 0 & 0 \\ 0 & 0 & 1 & -L2 \\ 0 & 0 & 0 & 1 \end{pmatrix} * \begin{pmatrix} \Delta x \\ \Delta y \\ \Delta z \\ 1 \end{pmatrix} \tag{2}$$

where $(\Delta x, \Delta y, \Delta z, \Delta 1)^T$ denotes the position correction of leg joint rotator,ΔR the rotation angle correction of the joint knee.

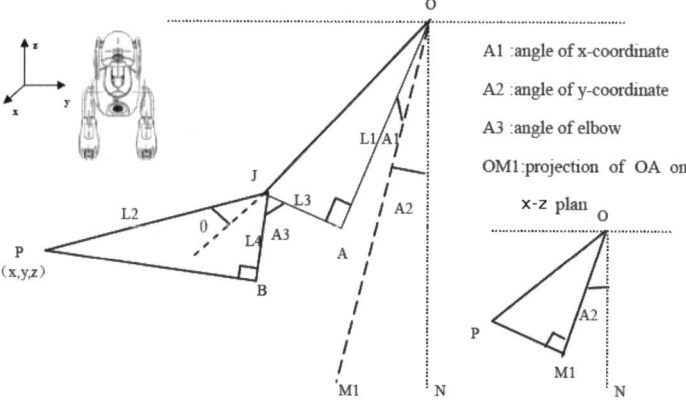

Fig. 2. The triangle leg model of AIBOs. Point O is the joint rotator position. Point P is the paw position.

3.2 Model Optimization

A position can be determined by a given sequence of all leg joints data and a given leg model. However, the calculated position is not close sufficiently to the actual position if the calculation is based on a pure leg model, because there are too many factors affecting the accuracy, (as described in Sect.2). Therefore, MBO optimizes the initial model described above with the locomotor data of the robot's movements in the application environment. For this purpose, the model is re-described by a set of parameters, denoted $M_i(p_{i1}, p_{i2}, \ldots, p_{in})$. Thus model optimization is reduced to a search of the best set of parameters in an n-dimensional space. A basic genetic algorithm is implemented here for the optimization. It first computes the displacement S_i with a given model M_i and a given sequence of consecutive joints data D according to formula (3) and determines the fitness of a model M_i according to formula (4):

$$S_i = F(M_i, D) \tag{3}$$

$$f_i = \{ \begin{pmatrix} 1 & 0 & 0 \\ 0 & 1 & 0 \\ 0 & 0 & 10 \end{pmatrix} * (S_i^o - S_i) \}^T * \{ \begin{pmatrix} 1 & 0 & 0 \\ 0 & 1 & 0 \\ 0 & 0 & 10 \end{pmatrix} * (S_i^o - S_i) \} + \omega \tag{4}$$

where ω denotes the variance between the displacements of fore legs and hind legs. The method of analyzing odometry readings from loci is used here and will be described in Sect. 4.

4   Odometry Calculation

The purpose of MBO is to obtain the odometric readings for the robot through calculating its paw loci by forward kinematics based on the optimized physical model. The robot's movement is completely determined by the movement of its support legs. From the loci of all legs, one can deduce the support legs and swing legs and calculate the odometric readings for the robot.

The detailed method of odometry calculation depends on the robot and its walking type. In this paper, we assume that the quadruped robot uses the trot gait which lifts the two diagonally opposite legs alternately [8].

Support Legs Calculation. A pair of diagonal legs are in contact with the ground as support legs and the other two legs swing in the air at any given time. Even though the robot may not have static balance and a third leg may fall on the ground, the effect of the third leg is ignored here for sake of simplicity. The two diagonal legs whose positions are lower are taken as the support ones.One plane can be determined by three arbitrarily chosen legs, the remaining leg is below or above the plane. If the remaining is under the plane, the supports legs are the diagonal legs which are composed by it and otherwise are the other pair.

Translation and Rotation Calculation. A robot walks by lifting the two pairs of diagonally opposite legs alternately. Each pair of diagonal legs swing in the air for a moment and then stay touching ground . The translation of the robot is calculated according to the movement of the legs that are touching the ground. Let R_{trans} and R_{rot} denote the translation and rotation of the robot, respectively; f_{trans} and h_{trans} denote the displacement of fore and hind support leg, respectively; $P_f(t)$ and $P_h(t)$ projective points on $x-y$ plane of the fore and hind support leg's paw position at time t.

The translation is computed by following rules:i. if support legs do not change from time $t-1$ to time t, then, $f_{trans} = P_f(t-1) - P_f(t)$, $h_{trans} = P_h(t-1) - P_h(t)$, $R_{trans} = (f_{trans} + h_{trans})/2$. ii. if support legs have changed form time $t-1$ to time t,then $R_{trans} = 0$.

The rotation is computed by following rules: i. if support legs do not change from time $t-1$ to time t, then set vector $l = P_f(t-1) - P_h(t-1)$,vector $c = P_f(t) - P_h(t)$. The rotation of robot is equal to the angle between vector l and vector c. ii. if support legs have changed form time $t-1$ to time t, then $R_{rot} = 0$.

5 Experiments

The experimental platform is the sony AIBO-ERS7.The parametric physical model is optimized using 12 samples in Table 1.

Table 1. Training samples for setting up the parametric physical model in the experiments

walk type	$C_i(x, y, \theta)$ of samples, $T_i = 5s$	number of sample
forward	$\{(200, 0, 0), (300, 0, 0)\}$	4
backward	$\{(-250, 0, 0)\}$	2
sidewalk	$\{(0, 200, 0)\}$	2
rotation	$\{(0, 0, 130), (0, 0, 180)\}$	4

These experiments are to compare the accuracy of MBO with that of German Team 2004.Both odometric readings returned by German Team 2004 and MBO are recorded and the actual displacement of the robot is measured manually. Equation (6) in Sect.3 is adopted here to evaluate the errors.We use a external camera to capture the position and orientation of the robot. The average error of position and orientation are ± 2cm and ± 5 degree.

Experiment 1. First, we test the errors caused by executing a single instruction each time, with each of these instructions sampled twice or thrice. The experiment results are shown in Table 2, which show that the odometrical readings returned by MBO are always closer to the measurements and is robust to the different actions.

Table 2. Results of experiment 1

instruction (C,T)	measured displacement	displacement of GT04	displacement of MBO	error rate of MBO	rate proportion (MBO/GT04)
$\{(300,0,0),6000\}$	(2050,90,5)	(1850,0,0)	(2046,-58,-2.7)	0.08	0.74
	(2060,200,10)	(1850,0,0)	(2040,67,0.4)	0.08	0.53
$\{(300,0,60),6000\}$	(110,80,415)	(21,0,364)	(129,72,414.2)	0.005	0.04
	(-80,100,420)	(21,0,364)	(91,92,412.6)	0.04	0.32
$\{(-250,0,0),6000\}$	(-1050,-40,0)	(-1500,0,0)	(-1198,-45,-0.7)	0.14	0.32
	(-1080,0,-10)	(-1500,0,0)	(-1180,-52,-5.9)	0.11	0.27
$\{(0,200,0),4000\}$	(0,870,10)	(0,811,0)	(-45,850,3.5)	0.09	0.70
	(-50,860,7)	(0,811,0)	(-62,833,12.5)	0.07	0.63
$\{(0,0,130),4000\}$	(-40,80,470)	(0,0,524)	(2,12,475.2)	0.02	0.18
	(10,10,475)	(0,0,524)	(-37,-1,441)	0.07	0.76
	(-10,15,445)	(0,0,511)	(-30,3,443)	0.007	0.05

Experiment 2. Let the robot start at point (0,0,0), perform one of the following sequences of instructions, and then stop: $\Psi_1 = \{\{(150,0,0),500\}, \{(200,0,0),500\}, \{(250,0,0), 500\}, \{(300,0,0),500\}\}$; $\Psi_2 = \{\{(300,0,0),1000\}, \{(250,0,0),500\}, \{(200,0,0),500\}, \{(150,0,$

0),500}}; Ψ_3={{(0,0,100),500}, {(0,0,130),500}, {(0,0,160),500}, {(0,0,190),500}}; Ψ_4={{(0,0,200),1000}, {(0,0,170),500}, {(0,0,140),500}, {(0,0,110),500}}; Ψ_5={{(200,0,0), 2000}, {(200,0,60),2000}, {(200,0,0),2000}, {(200,0,-60),2000}}.

Ψ_1 is an accelerating process in x-axis direction and Ψ_2 a decelerating process after a sudden start . Ψ_3 and Ψ_4 are similar to Ψ_1 and Ψ_2 , respectively, but in φ direction. The results (Table 3) show that MBO is more accurate than German Team 2004 except in the case Ψ_4, where both methods are equally good. It is worth noting that odometrical readings returned by MBO for rotation is much better. Ψ_5 is used to test the odometry accuracy when the robot moves along a curve and MBO is also better in this case. Moreover, the actual trajectory caused by Ψ_5 is approximated by fitting a smooth curve over several manually measured points that the robot passes, as shown in Fig. 3. The shape of the trajectory generated by MBO is much closer to the actual one than that generates by German Team 2004. To sum up, the accuracy of MBO improves by at least 50% in most cases.

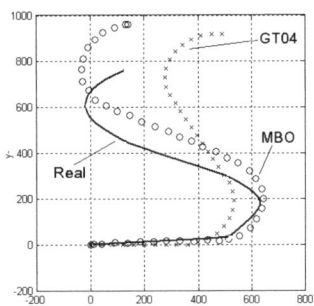

Table 3. Results of experiment 2

Order	Measured	GT04	MBO	MBO /GT04
Ψ_1	(2050,90,5)	(450,0,0)	(612,-17,-1.3)	0.43
Ψ_2	(675,30,3)	(600,0,0)	(670,47,2.2)	0.48
Ψ_3	(-10,-5,-6)	(0,0,290)	(-10,-2,-13.1)	0.02
Ψ_4	(-15,20,60)	(0,0,50)	(-12,8,50.1)	0.99
Ψ_5	(130,760,0)	(529,919,0)	(141,959,8.0)	0.46

Fig. 3. The trajectories of Ψ_5

6 Conclusions

The original notion of odometry calculation for quadruped robots basically concerns the relation between the output (the odometric readings) and the input (the actual motion data) of a movement. As far as we know, all previous methods of the odometry calculation for quadruped robots are technically based on this notion in the sense that some direct relations between the target and actual speed are established and employed to tell the odometric readings. An alternative approach is proposed in this paper. Instead of establishing the direct relation, MBO sets up a "parametric physical model" incorporating various properties of the robot and even the environment such as friction and inertia. Based on this model, MBO deduces the loci of leg movements by forward kinematics and obtains the odometric readings by analyzing loci in execution time.

We described the major steps and tested the performance of this method on the AIBO platform with a generally applicable methodology. MBO fits the quadruped robot whose swing phase occupies not greater than 50% of its whole gait trajectory. The experiments showed the calibration complexity is as low as only twelve samples, covering 4 basic motions—straight forward, straight backward, pure sideways walking, and pure rotation. The error is 50% less than existing methods and the error rate is below 8% for most motion types. MBO does not demand additional sensors, and it only employs the feedback of the leg joint sensors while returning odometric readings online. Another feature of MBO is that the instantaneous speed of the robot offered by MBO is very sensitive with steadily lower error. This would provide a new opportunity for more precise motion control of legged robots[7].

For that purpose, we need to work further on the prediction based on the odometry. In addition, achieving the fully autonomous odometry calibration is a most important future work. This implies that we need some on-line and on-board optimization methods. Another interesting problem concerns the choice of types and number of training samples for building an optimal "parametric physical model", especially when "mixed motions" (e.g., moving forward and sideways at the same time [5]) are considered more thoroughly.

Acknowledgment

We would like to thank Kai Xu, Fei Liu and Benjamin Johnston for their valuable advice and helpful discussions. We also would like to thank the members of the WrightEagle Team for their efforts in developing the software used as a basis for the work reported in this paper.

References

1. Kelly, A.: Fast and easy systematic and stochastic odometry calibration. In: International Conference on Intelligent Robots and Systems (October 2004)
2. Borenstein, J., Feng, L.: Measurement and correction of systematic odometry errors in mobile robots. IEEE Transactions on Robotics and Automation (December 1996)
3. Chong, K.S., Kleeman, L.: Accurate Odometry and Error Modelling for a Mobile Robot. In: International Conference on Robotics and Automation, April 1997, pp. 2783–2788 (1997)
4. Düffert, U., Hoffmann, J.: Reliable and Precise Gait Modeling for a Quadruped Robot. In: Bredenfeld, A., Jacoff, A., Noda, I., Takahashi, Y. (eds.) RoboCup 2005. LNCS (LNAI), vol. 4020, pp. 49–58. Springer, Heidelberg (2006)
5. Hengst, B., Ibbotson, D., Pham, S.B., Sammut, C.: Omnidirectional locomotion for quadruped robots. In: Birk, A., Coradeschi, S., Tadokoro, S. (eds.) RoboCup 2001. LNCS (LNAI), vol. 2377, pp. 368–373. Springer, Heidelberg (2002)
6. Chen, W.: Odometry Calibration and Gait Optimisation. The University of New Wales School of Computer Science and Engineering, Technical Report (2005)
7. Lin, P.-C., Komsuoğlu, H., Koditschek, D.E.: Legged Odometry from Body Pose in Hexapod Robot. Experimental Robotics IX, STAR 21, pp. 439-448 (2006)
8. Hengst, B., Ibbotson, D., Pham, S.B., Sammut, C.: The UNSW United 2000 Sony Legged Robot Software System, School of Computer Science and Engineering University of New South Wales, Technial Report (2000)
9. Antonelli, G., Chiaverini, S., Fusco, G.: An Odometry Calibration Method for Mobile Robots Based on the Least-Squares Technique. In: Proceedings of the American Control Conference (June 2003)
10. Stronger, D., Stone, P.: Simultaneous Calibration of Action and Sensor Models on a Mobile Robot. In: IEEE International Conference on Robotics and Automation (April 2005)
11. Röfer, T.: Evolutionary Gait-Optimization Using a Fitness Function Based on Proprioception. In: Nardi, D., Riedmiller, M., Sammut, C., Santos-Victor, J. (eds.) RoboCup 2004. LNCS (LNAI), vol. 3276, pp. 310–322. Springer, Heidelberg (2005)

A Framework for Learning in Humanoid Simulated Robots

Esther Luna Colombini[1], Alexandre da Silva Simões[2],
Antônio Cesar Germano Martins[2], and Jackson Paul Matsuura[1]

[1] Itandroids Research Group
Technological Institute of Aeronautics (ITA), Brazil
[2] Automation and Integrated Systems Group (GASI)
São Paulo State University (UNESP), Brazil
{esther,jackson}@ita.br,
{assimoes,amartins}@sorocaba.unesp.br

Abstract. One of the most important characteristics of intelligent activity is the ability to change behaviour according to many forms of feedback. Through learning an agent can interact with its environment to improve its performance over time. However, most of the techniques known that involves learning are time expensive, i.e., once the agent is supposed to learn over time by experimentation, the task has to be executed many times. Hence, high fidelity simulators can save a lot of time. In this context, this paper describes the framework designed to allow a team of real *RoboNova-I* humanoids robots to be simulated under *USARSim* environment. Details about the complete process of modeling and programming the robot are given, as well as the learning methodology proposed to improve robot's performance. Due to the use of a high fidelity model, the learning algorithms can be widely explored in simulation before adapted to real robots.

1 Introduction

In recent years, there has been much discussion concerning how knowledge can be acquired and used by autonomous agents. Through learning an agent can interact with an unknown environment and improve its performance over time by focusing its sensors on parts of the environment that are relevant to the task at hand.

In this scenario, the RoboCup® has created in the last years a set of realistic and simulated leagues to stimulate developments in the robotic field. One of these leagues is the *Humanoid League*, where autonomous mobile robots with a human-like appearance play soccer against each other. Humanoid League rules follow FIFA soccer laws in general lines. However, currently, some simplifications are assumed. Differently from conventional soccer, for example, each team consists of two players, in which one can be designated as a goalkeeper.

Another recently created league is the RoboCup *Rescue Simulation Virtual Robots*, in which a team of heterogeneous robots is asked to look for victims in

U. Visser et al. (Eds.): RoboCup 2007, LNAI 5001, pp. 345–352, 2008.

a urban search and rescue (USAR) task. This category also aims to fill the gap between real and simulated environments by using a high fidelity simulator, the USARSim [1], recently extended by the work of Zaratti et al. [2] to work with legged robots. It is also an important research tool for studies of learning, Human Robot Interaction (HRI) and multi-robot coordination, given the possibility of simulating commercial and self-developed robot platforms.

Recently, some simulated models of real legged robots have been proposed. The Sony Aibo and Sony QRIO [2] are some of these models. One of the main constraints for using these robots as the basis for researching learning relies in the fact that they are no longer commercially available. In fact, there is a commercial platform that has been modelled for USARSim, Robovie-M [3]. However, its model is not yet fully available.

This paper presents the details about the complete process of modeling and programming the commercially available version of *Robonova-I* humanoid robot, as well as the learning methodology proposed to improve its performance on the Humanoid league task. The rest of this paper is structured as follows. Section 2 explains the main characteristics of autonomous Learning. Section 3 presents the proposed approach for building the robot model, describing, in details, the complete high fidelity geometric model of the robot and the set of script files needed to configure this model in the USARSim RoboCup simulator. The learning framework is presented in Section 4. Finally, Section 5 summarizes with the main conclusions and presents some lines for further work.

2   Learning

Reinforcement Learning (RL) [4] is a class that lies between the extremes of supervised learning, where the policy is taught by an expert, and unsupervised learning, where there is no evaluative feedback. It is a technique that allows an agent to adapt to its environment through the development of an action policy, which determines the action that should be taken in each environmental state in order to maximize (or minimize) a function over a cumulative reinforcement. The reinforcement is a real value that defines the desirability of a state and can be expressed both in terms of rewards or punishments. In RL systems, the a priori domain knowledge incorporated by the designer is minimal and is mostly encapsulated in the reinforcement function.

Q-learning [5] is the preferred RL algorithm because it provides good experimental results in terms of learning speed and it is a model-free learning for optimal policies. It learns the values of all actions in all states, rather than only representing the policy.

3   Proposed Approach

The use of Learning, more specifically RL, is wide spread on RoboCup. In the development of the simulated robots' plan, DAMAS Rescue team used Jack

Intelligent Agent programming language [6], decision tree algorithms and reinforcement learning [7]. Moreover, F180 champions CMUDragons are known to use RL techniques. Furthermore, Soccer Simulation 3D champion FC-Portugal can be cited as another successful example [8].

However, most of the techniques known that involves RL are time expensive, i.e., it takes time to find a policy to successfully accomplish the proposed task and difficult to configure. In these cases, as the agent is supposed to learn over time by experimentation, the task has to be executed many times. Hence, high fidelity simulators can save a lot of time.

In our case, the study of real robots in a simulated environment only makes sense if the resultant study can be sent back to the real robot. For this purpose, the construction of the *Robonova-I* humanoid robot simulated model in USARSim is proposed.

4   Bulding Robonova-I Model

The construction of a robot model in the USARSim environment is a very complex process. Since documentation for this task is extremely rare, we will detail in the following sections the construction of the *Robonova-I* model. In this approach, two main steps were adopted: *i)* robot geometric model construction and *ii)* robot scripting.

4.1   Geometric Model

The construction of the geometric model of the robot makes necessary the steps mentioned next.

Creating the static meshes. In our approach a tridimentional model of the robot was made in a CAD (Computer Aided Design) environment. We adopted the *AutoCad*® *2007* software, which provides a rich set of 3D creation and management tools, necessary to reproduce the complex forms of the robot.

One important remark must be done with respect to the XYZ coordinate system. Autocad environment is well known to use the XYZ positive axis arranged according to the LHR (left hand rule). The final assembly of the robot in the Unreal engine is assumed to arrange XYZ axis according to the RHR (right hand rule). In order to convert between systems, the orientation of the X axis must be changed. This situation forces the feet of our CAD robot model to be constructed above the XY plane, with the robot front oriented to the -X axis.

Accomplishing the first step of the process, a high fidelity model of the Robonova-I robot was generated in the CAD environment using exclusively static meshes. It is also important to remark that other kinds of primitives (like surfaces or regions) are not recognized in Unreal engine. Each robot material (in this case golden metal, black plastic and servomotors plastic) was represented in a different layer. The complete robot drawing was splited, and one new drawing was created to each rigid part of the robot (without joints). The complete robot model and its parts are shown in Figure 1.

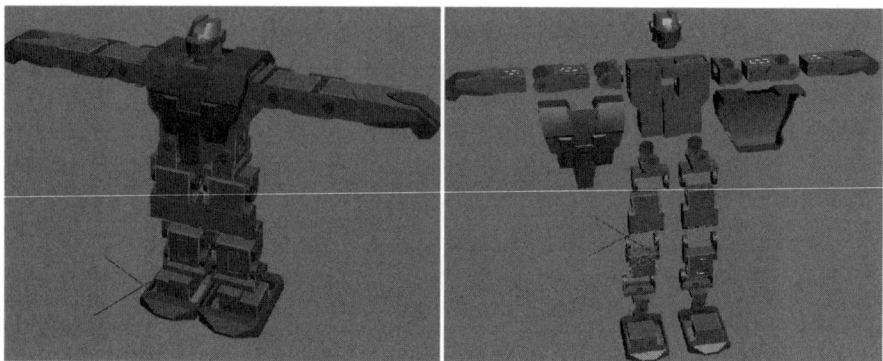

Fig. 1. Geometrical model of the Robonova-I robot. a) Assembled robot; b) Exploded main parts of the robot.

Converting the static meshes. In next step, the static meshes must be converted to its preferred file format to be imported in USARSim. The ASE (ASCII Scene Exporter) file format is the standard, since it stores identifiers for all file objects materials. In our approach, we adopted the software *3D Studio MAX®* *8* to realize this task. After a CAD file was imported, three different materials (simple color patterns) were created using the material editor tool. One material was assigned to each layer of the original CAD drawing and, so, all static meshes were attached together forming a single body composed by different materials. The body was so rendered to texture, in order to generate a texture map. The texture properly was discarded, and the file was exported to the ASE format. This process was repeated to each one of the rigid parts of the robot.

Creating textures. In order to allow the use of simple textures in the robot into the USARSim environment, in proposed approach three standard 256x256 Bitmap files with 8 colors depth were created in the *paint* software and filled with the three different colors of the textures.

Assigning textures. In the next step, the textures must be assigned to the materials specified in the ASE file. This process was realized inside of the *Unreal Editor 2004*. First, all three textures were imported and a unique UTX texture package was created. After this, each of the static meshes of the rigid parts of the robot were imported into a unique USX package. Still using the Unreal Editor, each material of each component of the meshes in the USX package was linked to one of the textures in the UTX package. The USX package now stores all information about robot geometry, except the information concerning the position to correctly assemble this parts, which will be informed in the robots configuration script.

4.2 Robot Configuration

After preparing the robot parts geometric model and textures, it is necessary to add these new models to the USARSim file structure. However, the robot

Table 1. Robonova-I parts and parameter values: weight, rotation angle, static friction rupture with robot up and down

Robot part	Quantity	Weight	Rotation	Up SFR	Down SFR
				-	-
Head	1	27g		-	-
Chest	1	337	-	-	-
Hand	2	65g	180^0	-	-
Elbow	2	65g	180^0	-	-
Shoulder	2	6g	360^0	-	-
Thigh	2	23g	90^0	-	-
Knee	2	135g	90^0	-	-
Superior ankle	2	44g	180^0	-	-
Inferior ankle 2	2	23g	180^0	-	-
Foot	2	83g	90^0	-	-
Spins	-	8g	-	-	-
Robonova-I	-	1.260g	-	260 Kgf	600Kgf

Fig. 2. Frontal view (out of scale)

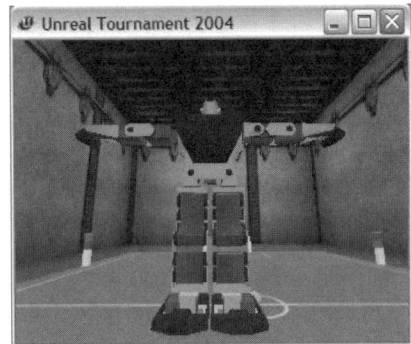

Fig. 3. Back view (out of scale)

physical parameters and dynamics still have to be configured. In this phase, scripts written in the Unreal Script language are prepared for each part of the robot, as well as for the complete robot model. For the individual parts, parameters such as torque, mass, angular velocity, friction, restitution, etc., are described. These parameters are used by the Karma engine [9] that is responsible for modeling the USARSim system dynamics. As for the complete robot model script, it contains the static meshes assembling and relative movements (i.e. axis spin) information.

To keep the fidelity of the model, some experiments were carried out with the real *Robonova-I* in order to obtain some of the Karma parameters. Based on the data of these experiments, well-known physical constants and robot geometry, one can estimate static and dynamic friction, maximum and minimum joint aperture, motor torque, etc. Some of the acquired values are shown in table 1.

Finally, scripts were compiled in order to generate the robot model into the USARSim environment. The final robot model built and imported in the virtual environment (out of scale) is presented in Figures 2 and 3.

5   *Robonova-I* Learning Framework

The design of architectures composed of very simple skills is not easy, nor is the learning of its sequence, as producing an adequate combination of these behaviours is not straight-forward. Furthermore, the controller decomposition introduces the need for determining when to trigger control, i.e. when to re-evaluate the previously selected behaviour and choose a new one.

Considering these constraints, the framework described proposes the intro-duction of learning in two levels. In the first level, the information provided by sensors (gyroscopes, camera and pressure sensors) is used to build the controllers responsible for movements. These basic controllers represent the set of sequen-tial servo commands that a robot may perform to execute a movement. We can divide the controllers in four main groups: *i)* Walking Controllers; *ii)* Pre-cise Positioning Controllers; *iii)* Special Actions Controllers; and *iv)* Goalkeeper Controllers.

In the *Walking Controllers* the three main controllers are the shift-right, the shift-left and the forward walk controllers. As the names propose, they are re-sponsible for the shift-sideways movements and for the forward walk movement of the robot. Other Walking Controllers are the backward walk, the diagonal walks (forward left, forward right, backward left and backward right), the turns (left and right) and the forward run.

In the *Precise Positioning Controllers* there are just three controllers that are smaller and more precise versions of the three main Walking Controllers. They are step-right, step-left and step-forward. The steps are small movements sideways or forward executed to allow a precise positioning of the robot.

There are four *Special Actions Controllers*, two of them responsible for the interaction with the ball, the kick right and the kick left controllers, used to kick the ball with the right leg and with the left leg respectively. The other two are the stand-up controller and the bend controller. The stand-up controller is used when the robot falls to get back to the upright condition while the bend controller is used to allow the camera to track objects near the feet of the robot.

Finally, the *Goalkeeper Controllers* are specific actions for the goalkeeper, such as: defend-right, defend-left and defend-mid. Each of them is used to defend a ball kicked to the right, left or in the direction of the robot respectively.

As for the second level, once defined the basic controllers, it is possible to apply the learning approach to automate the process of choosing a controller to execute in a specific environment situation. Figure 4a presents an overview of the proposed system architecture.

5.1   State and Action Space Modeling

For using RL algorithms, one has to guarantee that the problem can be modelled as a MDP (Markov Decision Process), i.e, the problem has to be represented as a finite set of actions and states and a discrete time model where the states should be available for measurement. However, real robot tasks have infinite state and action spaces, continuous time and due to sensorial limitation are not always

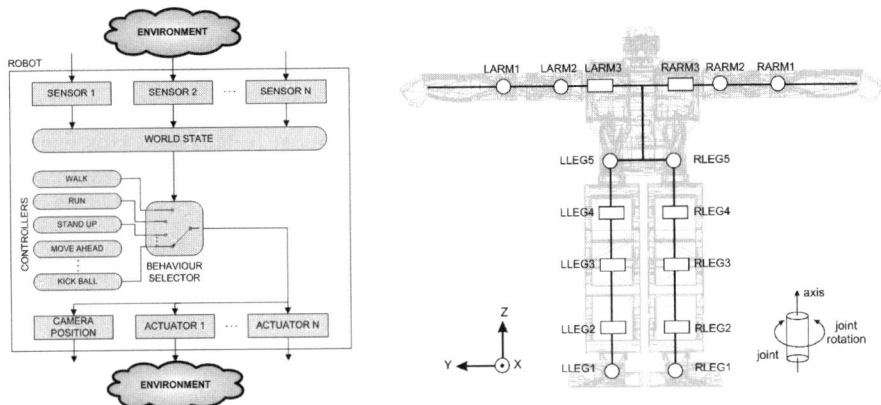

Fig. 4. a) System Architecture. b) Joints position and spin.

measurable. To deal with this problem, a discretization model for the state and action spaces is proposed.

First, consider the joints robot structure presented in Figure 4b. For each arm there are three joints with one DOF each, while for each legs these number reaches five, what give us sixteen DOFs. The state and action space discretizations are divided into two groups: 1) for low level learning or controllers learning and 2) for high level learning or switching controller determination.

Low level discretization. The low level action space is composed of sixteen elements, each representing a servo. Each action corresponds to change a servo angle by adding or subtracting 15 degrees to its actual state.

If the state vector, defined by the gyroscopes, camera and pressure sensors values indicate a falling, the robot is assigned with a null value reward and for each action performed, it receives an unitary reinforce. We work in all cases with a minimization criterium and with a step corresponding to a change in the servos configuration. In these cases, the goal is to perform the movements without much changing in the robot servos and without falling down.

High level discretization. For the high level learning, after fine tuning the individual controllers, one can apply the high level learning, using the same state vector defined for the low level phase, to decide which basic controller to execute. The reward structure considers that each action performed costs a unit to the learning agent, while accomplishing the goal gives it a null reinforcement value. This criterium can help the robot to save battery. To help configuring the learning parameters, a graphical interface was implemented.

6   Conclusion

This paper presented a complete framework for a *Robonova-I* humanoid robot, composed by: *i)* A complete high fidelity geometric model of the robot; *ii)* A set of scritp files to configure this model in the USARSim RoboCup simulator; *iii)* An architecture model to be used in robot learning, and *iv)* A graphical interface for learning parameters settings.

At the best of our known, this set of features represents the first available framework of a commercially available humanoid robot. In this way, this framework is expected to work as an important tool in robots dynamics research and also to contribute to reduce time required to test learning algorithms.

The paper also presented a detailed description of the robot modeling and configuration process for USARSim environment, filling some gaps in the related technical literature and expecting to reduce the amount of time required to create new robots and models in this environment.

As main ongoing works, there is a set of experiments to establish the confidence degree between proposed model and real Robonova-I robot with respect to dynamics, sense and acts. As future work, we point the implementation of a large number of RL algorithms in order to extend the framework capabilities.

References

1. Wang, J.: USARSim V2.0.2 Manual: A Game-based Simulation of the NIST Reference Arenas (2006)
2. Zaratti, M., Fratarcangeli, M., Iocchi, L.: A 3d simulator of multiple legged robots based on usarsim. In: Lakemeyer, G., Sklar, E., Sorrenti, D.G., Takahashi, T. (eds.) RoboCup 2006: Robot Soccer World Cup X. LNCS (LNAI), vol. 4434, pp. 13–24. Springer, Heidelberg (2006)
3. Greggio, N., Silvestri, G., Antonello, S., Menegatti, E., Pagello, E.: A 3d model of a humanoid for usarsim simulator. In: First Workshop on Humanoid Soccer Robots, Genova, pp. 17–24 (2006)
4. Sutton, R.S., Barto, A.G.: Reinforcement learning: an introduction. MIT Press, USA (1998)
5. Watkins, C.: Learning from delayed rewards. PhD thesis, King's College (1998)
6. Howden, N., Renquist, R.: Jack intelligent agents - summary of an agent infrastructure. In: 5th Int. Conf. on Autonomous Agents, Montreal, Canada (2001)
7. McCallum, A.K.: Reinforcement Learning with Selective Perception and Hidden State. PhD thesis, University of Rochester, New York (1996)
8. Reis, L.P., Lau, N.: Fc portugal team description: Robocup 2000 simulation league champion. In: Stone, P., Balch, T., Kraetzschmar, G.K. (eds.) RoboCup 2000. LNCS (LNAI), vol. 2019, pp. 29–40. Springer, Heidelberg (2001)
9. The Unreal Engine Site (2007), http://wiki.beyondunreal.com/wiki/Karma

Let Robots Play Soccer under More Natural Conditions: Experience-Based Collaborative Localization in Four-Legged League

Qining Wang, Yan Huang, Guangming Xie, and Long Wang

Intelligent Control Laboratory, College of Engineering,
Peking University, 100871, Beijing, China
qiningwang@pku.edu.cn
http://www.mech.pku.edu.cn/robot/fourleg/

Abstract. This paper presents an experience-based collaborative approach for a group of autonomous robots to localize in asymmetric, dynamic environments. To help robots play soccer under more natural conditions, we propose a Markov localization based hybrid method with integration of environment experience construction and dynamic reference object based multi-robot localization. By using this method, the robot can estimate and correct its position perception more accurately and effectively among a group of autonomous robots, taking the odometry error and other negative influence into consideration. Satisfactory results are obtained in the RoboCup Four-Legged League environment.

1 Introduction

On the move to real human soccer conditions, current localization approaches applied in RoboCup (eg. [1], [2]) seem not enough. In the human soccer, there are two aspects which may inspire the self localization of mobile robot systems. On the one hand, the features surrounding the soccer field may be exploited as the sensory information in probabilistic approaches. Inspired by the features, some systems applied image-retrieval approach in localization [5]. However, the computational cost is expensive. Besides, the requirement of building a huge database is not practical, especially in complex environments. On the other hand, collaboration among the robot team may help self localization. Previous research in localization has proven that the cooperation in self-localization among multiple robots has impressive performance in real robot systems (see [3] for overview). The limitation of such robot systems is that the robot needs to identify other one precisely. It is difficult to perform collaborative localization for robots dealing with situations where they can detect but not identify other robots.

Our work focused on applying image-retrieval approach and collaboration in self localization in RoboCup. In the following section, we describe the method for individual localization with experience. In section 3, we present how to use the sharing information to improve the Markov localization when the robot can not localize accurately by itself. In section 4, satisfactory results on localization through our approach is shown in experiments using Sony Aibo ERS-7 robots.

U. Visser et al. (Eds.): RoboCup 2007, LNAI 5001, pp. 353–360, 2008.
© Springer-Verlag Berlin Heidelberg 2008

2   Individual Localization with Experience

Based on [1]-[2], in our approach, the current position of the robot is modelled as the density of a set of particles which are seen as the prediction of the location. Initially, at time t, each location l has a belief:

$$Bel_t(l) \leftarrow P(L_t^{(0)} = l) \tag{1}$$

To update the belief of robot possible location, at first, this approach uses the new odometry reading o_t:

$$Bel_t(l) \leftarrow \int P(l|o_t, l^-)Bel_t(l^-)dl^- \tag{2}$$

Considering the mobile robot with complex motions, let the geometric center of robot body as the location vector ϕ, which contains the x/y- global coordinates of the center point. Another vector θ is defined as the heading direction. Then every particle is updated by the motion model as follows when the robot moves:

$$\phi_t = \phi_{t-1} + \Delta_t \tag{3}$$

where Δ_t represents the displacement in x/y coordinates and heading direction.

To implement image retrieval system in Markov localization, we divide the sensory update into two parts: updating position probability by landmark perception and experience matching. If the robot recognizes landmarks well enough, landmark based sensor model will update the belief of position with the new landmark reading s_t:

$$Bel_t(\phi_t) \leftarrow \beta P(s_t|\phi_t)Bel_t(\phi_t) \tag{4}$$

where β is a normalizing constant. We set $N_1(t)$ which is the amount of lasting frames of having no landmark perception from t as a condition to activate the experience system. If $N_1(t)$ is great enough, the experience based sensor model will update the probability as follows:

$$Bel_t(\phi_t) \leftarrow \gamma P(e_t|\phi_t)Bel_t(\phi_t) \tag{5}$$

where e_t is the new reading experience with γ being the normalizing constant.

2.1   Experience Construction

The feature that is exploited from images with no landmark in the view, and represents the invariant character of images obtained at positions where collisions and other negative effects more likely occur is defined as *Experience*.

In our method, we divide one image which is obtained by the robot camera into six parts. First, image features including average color value $f_{i,j}$ and color variance d_i in the divided areas are calculated by the following equations:

$$f_{i,j} = \frac{\sum\limits_{x,y} M[y][j][x]}{N_i}; \{j = 0, 2, 3; i = 1, 2, 3, 4, 5, 6\} \tag{6}$$

where $f_{i,j}$ is the average value in the color channel j of area i. $M[y][j][x]$ represents the value in the color channel j at the position (x, y) in the image. N_i is the number of the pixels in area i. Clearly, the $f_{i,j}$ is in the range from 0 to 255.

$$d_i = \frac{\sum\limits_{x,y}(|M[y][0][x] - f_{i,0}| + |M[y][1][x] - f_{i,1}| + |M[y][2][x] - f_{i,2}|)}{N_i} \tag{7}$$

where $i=1, 2, 3, 4, 5, 6$. d_i is in the range from 0 to 382.5. When the value of color variance in the certain area gets maximum, d_i is 382.5.

After calculating features in divided areas, we collect average color value F_j and color variance D in the whole image which are calculated by the following equations:

$$F_j = \frac{\sum_i f_{i,j}}{S}; \{j = 0, 1, 2\} \tag{8}$$

where F_j represents the average value in the color channel j of the whole image. S is the number of divided areas in the image.

$$D = \frac{\sum\limits_{i}(|f_{i,0} - F_0| + |f_{i,1} - F_1| + |f_{i,2} - F_2|)}{S} \tag{9}$$

where D is in the range from 0 to 382.5.

In our system, the invariant features of images include $f_{i,j}$, d_i, F_j, and D. All the features are calculated from images collected in certain places where the robot needs experience to help. We construct experience database embedded in robot's memory. This database stores the features along with the global coordinates of the position where the image is taken. All the features are calculated off-line and stored in the database as experience. When the experience module is activated, the feature of current image taken by camera is computed on-line notated as *imageFeature*. Meanwhile, the record notated as *bestRecord* whose feature is most similar to *imageFeature* is selected from the database. Fig. 1 shows the result of finding the best pose in database based on experience. The query image is on the left while its most similar image in the database is on the right. Their poses are represented by (x, y, θ). x, y are calculated in millimeter, while θ is in degree.

When the experience module is activated, difference between *imageFeature* and the feature of *bestRecord* is calculated. If the difference is small enough, the pose of *bestRecord* is transferred into *bestPose* notated as l_{best} which is in the form of world coordinates in the robot system. With such *bestPose*, probabilities of all the sample poses are updated and new pose templates which are random poses near the *bestPose* are generated to perform the resample procedure in Markov localization. It is true that the more experience in database, the more precisely the calculation is. However, building such database is expensive in time cost and even unreachable in complex environments. As a part of the sensor update module, experience can help the Markov localization converge as soon as possible, which means the robot can know own position immediately. In our approach, we only need to construct the database in those really difficult situations. This method works well in real robot applications.

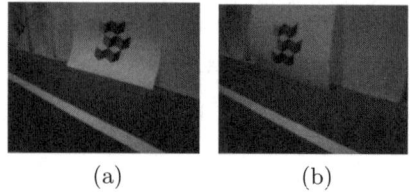

(a)   (b)

Fig. 1. Examples for finding the best pose in image database. Images in the database are collected in the areas of the field where the robot can not see any landmark every $100mm$ in x, $100mm$ in y and $45°$ in θ. (a) is the current image taken by robot's camera when its real position is $(-1660, 1520, 135°)$. (b) is the most similar picture to image (a) in the experience database which the corresponding position of the robot is $(-1600, 1500, 135°)$. The location error is $60mm$ in x, $20mm$ in y, and $0°$ in θ.

2.2   Self Learning in Experience Collection

One of the difficulties in applying image-retrieval system into real robot localization is how to collect the experience efficiently and correctly. We create a self learning method for experience collection. The robot can collect images along with corresponding positions autonomously. When construct the experience database, we use the black-white stripes to adjust robot body which is similar to the one used in gait optimization mentioned in [4]. In the self learning procedure, at first, the robot adjusts its own body to the initial position which is preset by our control system. By using the stripes, the robot walks to the next position and stops to capture images in left and right view respectively as shown in Fig. 2. The black-white stripes help robot go to the preset position precisely.

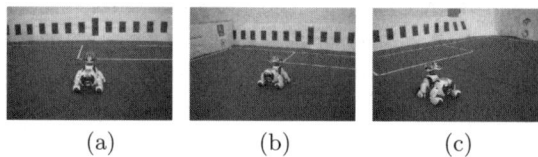

(a)   (b)   (c)

Fig. 2. Self learning procedure in experience collection. (a) shows the Black-white stripes for body adjusting. The robot captures image in the left view and right view as shown in (b) and (c) respectively.

3   Collaborative Localization

In RoboCup, static reference objects like beacon, and goal can be used to help localize in complex environments. However, global coordinates of such objects need to be known beforehand. Those static reference objects are not applicable in an unknown environment. To solve this problem, we propose the concept of *Dynamic Reference Object*. The object that can be detected by more than one robots among the team will be the candidate dynamic reference object. If the frequency of clearly recognizing the object is high enough, it may be set as the

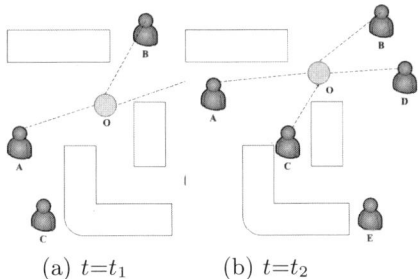

(a) $t=t_1$ (b) $t=t_2$

Fig. 3. A simple system with five mobile robots and a dynamic reference object: (a) At time t_1, robot A, B and E can see the dynamic reference object O. If at this time robot A, for example, needs the reference object to help, A will use the calculated position of the object from B or E. Querying the most possible position in team message shown in Table 1, A will take the calculated result by B as the reference. (b) At time t_2, C and D have not detected any landmark or experience for a period. Thus their answers to the object position is relatively unreliable. Position possibilities of them are shown to be low in Table 1. The reference object position will be set as B percepts.

dynamic reference object. There is no need to know the object's position as a precondition. If a robot can localize itself accurately, the position of the dynamic reference object calculated by this robot is reliable. Meanwhile, another robot that has seen the reference object can use this calculated position of the object to measure own location. This information is useful for decreasing the time cost of Markov localization convergence and improve the result of position estimate especially for multiple robots collaboration.

With the assumption that robots can communicate with each other, our approach integrates *Reference Object Position Possibility* in the team message which will be broadcasted to every robot. The item which is relevant to the object position in team message includes calculated position, robot ID, time, and position possibility. This position possibility is due to the accuracy of the robot self localization. In our system, the object position possibility is notated as P_r is measured by the following equation:

$$P_r = P_l e^{-\mu^2} + P_e e^{-\omega^2} \tag{10}$$

where P_l and P_e are certain probabilities for landmark and experience update respectively. μ is the sum of lasting frames after detecting the latest landmark, while ω is the sum of lasting frames after exploiting good experience. In real robot application, P_r will be normalized less than 1. If P_r is high enough, the calculated result by this robot will be the most reliable one among different robots perception. A robot that needs help always uses the most possible position of the reference object at the same time when it detects the object by itself. To illustrate the method, a common robot system is shown in Fig. 3 with five mobile robots. Object O is supposed to be the dynamic reference object. Table 1 is the real-time information in team message of the system in Fig. 3.

Table 1. Team message relevant to dynamic reference object

Calculated Position	Robot ID	Time	Position Possibility
(2388, 700)	A	t_1	0.71
(2264, 658)	B	t_1	0.92
(2530, 710)	E	t_1	0.86
(2368, 803)	A	t_2	0.81
(2401, 801)	B	t_2	0.91
(2103, 743)	C	t_2	0.32
(2215, 725)	D	t_2	0.43

In our approach, collaboration is a part of probability update modules in Markov localization. There is a problem that robots should known when to activate the collaboration module using the dynamic object as a reference. To improve Markov localization using our collaborative approach, the collaboration module will be activated in two situations. We set $N_2(t)$ by using as the sum of lasting frames of having no landmark perception or experience as a condition to activate the collaboration system. If $N_2(t)$ is great enough and the robot has detected the dynamic reference object, the collaboration module will update the probability of every poses. In addition, if the robot has a perception of the object which has a relatively high position possibility, the robot will use this reference to improve the Markov localization in a collaborative way.

4   Experimental Results

The experience-based collaborative approach presented above has been implemented on the Sony Aibo ERS7 legged robot in RoboCup environment. Fig. 4(a) shows the environment in 2007. In our localization experiment field, we use the field similar to the standard field in four-legged soccer field 2007. However, we remove the beacons. As shown in Fig. 4(b), our field is surrounded by colorful advertisement which simulates the real human soccer environment.

(a)   (b)   (c)

Fig. 4. Experimental field. (a) is the soccer field with two colorful beacons in 2007. (b) shows field with no beacon which is used to test our localization approach. (c) is the colorful advertisement placed around our test field.

4.1   Individual Robot Localization

We randomly select 8 points to test the self localization results. The robot is expected to go to the preset positions through localization. When it stops, we

Table 2. Results of self localization in randomly walking

Point Number	Expected Position (x, y, θ)	Real Position (x, y, θ)	Error (x, y, θ)
1	$(-1290, -440, 15)$	$(-1496, -713, 147)$	$(206, 273, 132)$
2	$(-1450, -300, 0)$	$(-1410, -150, 0)$	$(40, 150, 0)$
3	$(-180, -670, 45)$	$(-230, -610, 9)$	$(50, 60, 36)$
4	$(1430, -250, 55)$	$(-1909, -1162, 132)$	$(461, 912, 76)$
5	$(-650, 170, 0)$	$(-404, -427, 5)$	$(246, 597, 5)$
6	$(270, -480, -90)$	$(102, -402, -48)$	$(168, 78, 42)$
7	$(-1440, -340, 10)$	$(-1322, -332, 5)$	$(78, 8, 5)$
8	$(-2160, -390, 0)$	$(1979, -454, 8)$	$(181, 64, 8)$

calculate the real positions on the ground. Table 2 shows the results in detail. x, y are calculated in millimeter, while θ is in degree.

4.2 Collaborative Localization

In this experiment, the orange ball used in the four-legged league is considered as the *dynamic reference object*. We use three robots to perform multi-robot localization. Every robot uses the hybrid system tested in the individual experiment mentioned above. We set one of the three robots as a sample to estimate our collaborative approach. The other two robots move randomly to catch the ball and broadcast the ball position with position possibilities mentioned in section 3. We receive the calculated result from the sample robot. Only experience and collaboration can help the robot localize. The localization result of the sample robot which has used the collaborative approach is shown in Fig. 5. The probability distribution can converges quickly after 3-9 seconds when the dynamic reference object is taken into account.

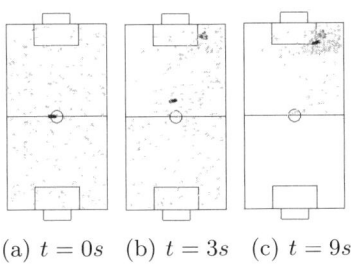

(a) $t = 0s$ (b) $t = 3s$ (c) $t = 9s$

Fig. 5. The localization result of applying collaborative approach with dynamic reference object. Solid arrows indicate MCL particles(100). The calculated robot position is indicated by the solid symbol. (a) is the initial uniform distribution. (b) is the calculated result after 3 seconds. (c) is the well localization result after 9 seconds.

5   Conclusion

In this paper, we have demonstrated an experience-based collaborative approach that combines image database for experience without landmarks and real-time sensor data for vision-based mobile robots to estimate their positions under more natural conditions towards real human soccer environment. On the one hand, our approach presented a fast and feasible system for vision-based mobile robots to localize in the dynamic environment even if there is no artificial landmark to help. On the other hand, we showed the collaborative method with introduction of Dynamic Reference Object to improve the accuracy and robustness of self localization, even in the circumstance that the robot can not localize individually or has no idea of who is nearby.

Acknowledgments

This work was supported by the 863 Program of China (No. 2006AA04Z258) and 985 Project of Peking University.

References

1. Fox, D., Burgard, W., Dellaert, F., Thrun, S.: Monte Carlo localization: Efficient position estimation for mobile robots. In: Proc. of the National Conf. on Artificial Intelligence, pp. 343–349 (1999)
2. Röfer, T., Jüngel, M.: Vision-Based Fast and Reactive Monte-Carlo Localization. In: Proc. of the IEEE ICRA, pp. 856–861 (2003)
3. Arkin, R.C., Balch, T.: Cooperative multiagent robotic systems. In: Kortenkamp, D., Bonasso, R.P., Murphy, R. (eds.) Artificial Intelligence and Mobile Robots. MIT/AAAI Press, Cambridge (1998)
4. Röfer, T., et al.: GermanTeam RoboCup 2004, tech. rep. (2004), http://www.germanteam.org/GT2004.pdf
5. Wolf, J., Burgard, W., Burkhardt, H.: Robust vision-based localization by combining an image-retrieval system with Monte Carlo localization. IEEE Trans. on Robotics 21(2), 208–216 (2005)
6. Fox, D., Burgard, W., Kruppa, H., Thrun, S.: A probabilistic approach to collaborative multi-robot localization. In: Autonomous Robots, vol. 8(3), pp. 325–344 (2000)

Strategic Layout of Multi-cameras Based on a Minimum Risk Criterion

Ryota Narita, Kazuhito Murakami, and Tadashi Naruse

Graduate School of Information Science and Technology, Aichi Prefectural University,
Nagakute-cho, Aichi 480-1198 Japan
im071018@cis.aichi-pu.ac.jp, {murakami,naruse}@ist.aichi-pu.ac.jp

Abstract. This paper proposes a method to allocate multiple cameras to a better or the best positions. In RoboCup Small Size League(SSL), two or more cameras are used, and we have to decide the layout of them at the venue. This paper gives a criterion which minimizes the risk, for example, the occlusion of a ball by robots, and solves it by using Fletcher-Reeves conjugate gradient algorithm. Experimental result shows the effectiveness of the proposed method.

1 Introduction

Figure 1 shows an example of multi-camera's layout in RoboCup SSL, and in this case, the cameras are placed to cover whole of the game field. These kinds of cameras are set at the suitable places based on the human experiences where the cameras could catch an object around the center of the image.

Kono et al. have reported an assist system which utilizes several cameras attached on the human's body[1]. Although they allocate multiple cameras based on a subjective criterion, there is no objective criterion. When we use multi-camera system, we are requested to decide the number and the places of the cameras. It is necessary to solve the most suitable number of cameras. How to solve the minimum number of security cameras is expressed in Art Gallery Problem[2]. This research provides only the cost minimum criterion. On the other hand, Kato et al. have treated the data traffic in the network and showed how to solve the maximum number of cameras[3].

As shown in Fig. 2, according to the number of the cameras increases, the computing cost or the traffic on the network increases, on the other hand, the

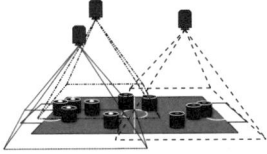

Fig. 1. An example of multi-camera's layout

U. Visser et al. (Eds.): RoboCup 2007, LNAI 5001, pp. 361–368, 2008.

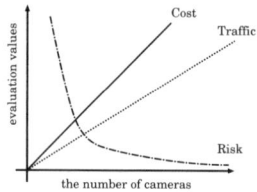

Fig. 2. The relation between the number of cameras and some evaluation values

Fig. 3. Cameras attached on the truss(interruption to other team's view is not allowed)

risk such as missing a ball decreases. An answer for the former is obtained by solving the trade-off problem between the computing cost and the risk, for example. This answer gives us one of the most suitable number of cameras. For the latter problem, a kind of criterion such as minimum cost, minimum data transfer traffic in the network, minimum risk, and so on, is required to decide a better or the best positions of them.

As for the places of multi-cameras, some criterion have been provided. There is a system which measures the shape of insects for the electronic museum. The paper resulted that it is better to place many cameras near the thready places than to place them at regular intervals around the measured insects[4].

Furthermore, there sometimes exists spatial restriction as shown in Fig. 3. It is not allowed to interrupt other camera's view. We have to decide a better or the best positions of cameras under these conditions or the restrictions at the venue.

This paper proposes a method to decide a better layout of multiple cameras under the condition that the risk is minimum. In the following, sections 2. and 3. describe how to model and calculate the risk and how to decide the best positions of cameras, respectively. Section 4. shows the effectiveness of the proposed method based on the experimental simulation results.

2 Risk Model in RoboCup Competition

Figure 4 shows a robot in RoboCup SSL. It is limited less than $18cm$ diameter and less than $15cm$ height. In RoboCup SSL, the robots game on a field of size $5.5m \times 4.0m$ including $0.3m$ width technical area. In the competition, each team uses 5 robots and an orange color golf ball on the field.

Fig. 4. Robot's overview

(a) A robot is close to the ball.   (b) In the image through a camera,
the ball can't be observed.

Fig. 5. Occlusion by a robot

There are two methods to take the coordinates of robots and a ball. Global vision overlooks robots and ball from the ceiling and local vision detects circumjacent robots and ball from a robot. Multiple cameras in the global vision system are placed on the truss built on $4m$ high. Many teams use multi-camera global vision system in RoboCup SSL because it is difficult to overwatch the whole field with one camera.

One of the risk in RoboCup SSL is the occlusion by the robots. The global vision loses a ball if the robot moves closer to or many robots close up the ball. Figure 5 is a typical case of the occlusion in RoboCup SSL. Figure 5 (a) shows that a robot is near the ball. Figure 5 (b) is an occluded image through a camera. There appears only robots in the image and the ball is not observed in it. If the occlusion occurs like this, it affects on the strategic planning of robots.

The area S of occlusion caused by a robot is defined as a function of the coordinates of camera $C(x, y, z)$, robot $R(x, y, z)$ and ball $B(x, y, z)$. Let the probability distribution of a robot be $P_{(R)}(x, \ y)$ in the field F. Then, $P_{(R)}(x, \ y)$ satisfies

$$\iint_F P_{(R)}(x, \ y)dxdy = 1. \tag{1}$$

So, the evaluation value E of occlusion is defined and expressed as

$$E = \iint_F S \ P_{(R)}(x, \ y) \ dxdy. \tag{2}$$

3   How to Decide Multi-cameras Layout

This paper calculates camera's coordinates $C_{(i)}(x, \ y, \ z) \ (1 \leq i \leq N)$ that minimize E. E is minimum means that the risk is minimum, so, this solution is one of the optimal positions of cameras.

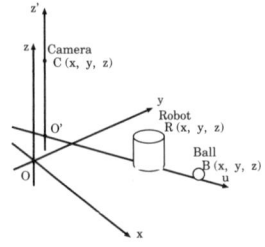

Fig. 6. Definition of u-z' plane

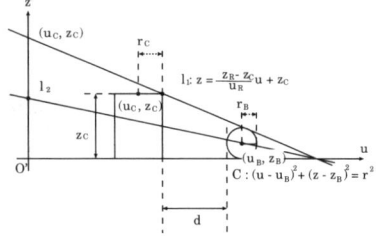

Fig. 7. Modeling to calculate the distance between a robot and a ball

3.1   Simulation of Occlusion with a Cameras

When camera, robot, ball are in the same plane, a ball is frequently occluded by a robot. In order to discuss simply, let a line connected with the centers of them on the x-y plane be u-axis, and its vertical line which passes C be z'-axis, respectively as shown in Fig. 6, then the following equations are devised.

$$u = \sqrt{(x - x_C)^2 + (y - y_C)^2} \qquad (3)$$

$$x = u \times \cos\left\{ \tan^{-1}\left(\frac{y_B - y_C}{x_B - x_C} \right) \right\} + x_C \qquad (4)$$

$$y = u \times \sin\left\{ \tan^{-1}\left(\frac{y_B - y_C}{x_B - x_C} \right) \right\} + y_C \qquad (5)$$

Figure 7 demonstrates how to calculate the distance d between the robot and the ball. d is calculated on u-z' plane using the relation between the circle and the tangential line in Fig. 7.

A tangent line is drawn from the camera to the robot (displayed as a rectangle in Fig. 7). The equation of this line(L_1) is obtained as:

$$z = \frac{z_R - z_C}{u_R + r_R} u + z_C. \qquad (6)$$

The coordinates $U(u_0, 0)$ is obtained as the point at the intersection of this line and u-axis. L_2 is a line passing through (u_B, z_B) and $(u_0, 0)$. The following equation

$$\frac{z_B}{u_0 - u_B} \simeq \frac{1}{2}\frac{z_C}{u_0} \qquad (7)$$

satisfies the gradient of L_1 and that of L_2. In consideration of robot and ball radius, d is calculated as

$$d = |u_B - u_R| - r_R - r_B. \qquad (8)$$

If d is less than 0, it means that a ball is not occluded by a robot. So, in this case, let it be $d = 0$. By using this value d instead of S, a new evaluation value E', replacing to Eq.(2), is obtained as:

$$E' = \iint_F d \times P_{(R)}(x,\ y)\ dxdy. \qquad (9)$$

3.2   Simulation of Occlusion with Multiple Cameras

When multiple cameras are used, it is necessary to consider the overlapped area. $d_{(i)}$ denotes d of the i-th cameras($1 \leq i \leq N$). Considering that at least one camera could catch the ball, Eq. (9) is rewritten as:

$$E'' = \iint_F min_i\{d_{(i)}\} P_{(R)}(x, \, y) \, dxdy \qquad (10)$$

due to evaluate multiple cameras. This paper solves all the positions $C_{(i)}(x, \, y, \, z)$ $(1 \leq i \leq N)$ of all cameras which minimize E''.

4   Experiment

4.1   Simulation Environment

Simulation experiment was done with parameters using RoboCup SSL. Based on the official regulation, Laws of the F180 League 2006, the field size is 5500mm × 4000mm including 4900mm × 3400mm court and 300mm width technical area, the maximum size of the robot height is 150mm and the radius is 90mm, respectively. The radius of a ball is 21.5mm, the heights of the cameras is 4000mm because the height of truss shown in Fig. 3 is $4000mm$. So the coordinates of the cameras are $C_{(i)}(x, \, y, \, 4000)$ $(1 \leq i \leq N)$. Based on Eq.'s (3), (8) and (10), the occlusion probability is obtained.

We have reported the result of multi-camera's layout which minimizes the occlusion for SSL. This was calculated under the condition that the robots existed equally in the field[5]. Figure 8 shows the probability distribution of our robots. It was given by analyzing the five logs of the past RoboCup SSL competitions. The logs recorded the coordinates and the velocity information of robots and ball, referee signal, time and so on. The coordinates information of our team's robots is used to make this distribution. According to the Laws of the F180 League 2006, the teams change their attack side in each half. And E'' is regarded as an approximate solution as

$$E''' = \sum_F min_i\{d_{(i)}\} P_{(R)}(x, \, y). \qquad (11)$$

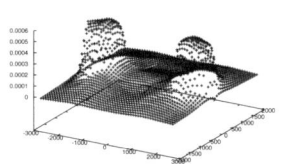

Fig. 8. Probability distribution of our robot

Table 1. Strategic camera coordinates (N = 9)

Camera Number	$C^*_{(i)}(x,y)$		$C_{(i)}(x,y)$		Camera Number	$C^*_{(i)}(x,y)$		$C_{(i)}(x,y)$	
1	(-1955,	-948)	(-1877,	27)	6	(1851,	1969)	(1393,	1215)
2	(310,	700)	(2044,	15)	7	(-411,	-1058)	(-421,	-833)
3	(-2148,	693)	(-1784,	1258)	8	(-1679,	-1634)	(-1734,	-1196)
4	(679,	-1879)	(1496,	-1216)	9	(-1618,	521)	(-471,	782)
5	(-243,	-862)	(707,	-32)					

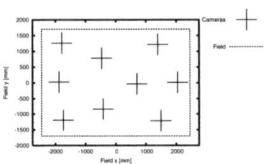

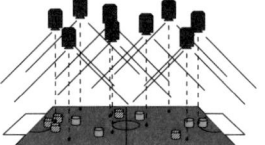

Fig. 9. Strategic multi-cameras layout (N = 9)

Fig. 10. Layout example (N = 9)

It takes about a day to calculate the minimal solution with full search in the range of $-2750 \leq x \leq 2750$, $-2000 \leq y \leq 2000$. So, we utilized Fletcher-Reeves conjugate gradient algorithm to obtain an optimized solution.

4.2 Algorithm

The algorithm used in this experiment is shown below.

Step 1. Set the initial position for (x, y) with pseudo random number in the range $-2750 \leq x \leq 2750$ and $-2000 \leq y \leq 2000$.

Step 2. Move a robot in the raster scan procedure on the field F. $\Delta x = \Delta y = 50$mm for the reduction of computing time.

Step 3. Calculate the distance $d_{(i)}(1 \leq i \leq N)$ for all i. Search $\min_i\{d_{(i)}\}$ and calculate E''' by Eq.(11).

Step 4. By using Fletcher-Reeves conjugate gradient algorithm, update camera coordinates $C_{(i)}(x, y)$ for all i to the direction that E''' decreases.

Step 5. Repeat from Step 2 to Step 4 while E''' decreases. If this is not satisfied, terminate this algorithm.

4.3 Experimental Result

Table 1 shows the result of strategic multi-cameras layout that nine cameras $(N = 9)$ are used. Figure 9 demonstrates the result $C_{(i)}(x, y)$ plotted on the 2-dimensional field of 4900mm × 3400mm. Figure 10 is an example of multi-cameras layout on the 3-dimensional space. Figure 11 shows the simulation results for other numbers.

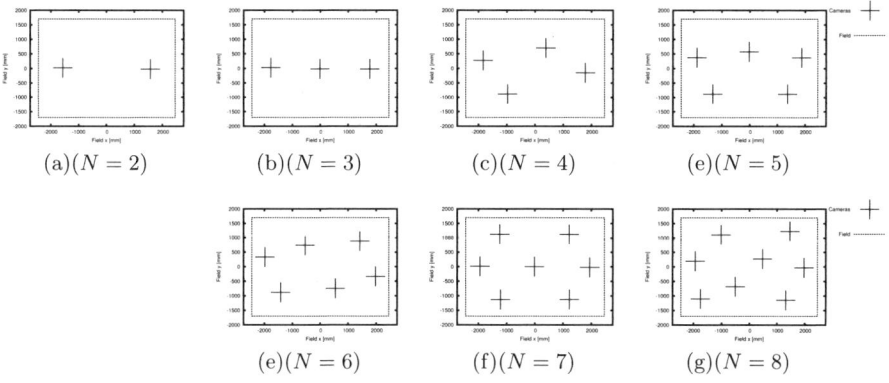

(a)$(N = 2)$ (b)$(N = 3)$ (c)$(N = 4)$ (e)$(N = 5)$

(e)$(N = 6)$ (f)$(N = 7)$ (g)$(N = 8)$

Fig. 11. Strategic multi-cameras layout $(N = 2, 3, \cdots 8)$

Table 2. Results of E''' $(N = 2, 3, \cdots 9)$

N	E''' [mm]	N	E'''
2	7.809	6	1.813
3	5.558	7	1.263
4	4.026	8	0.951
5	2.668	9	0.636

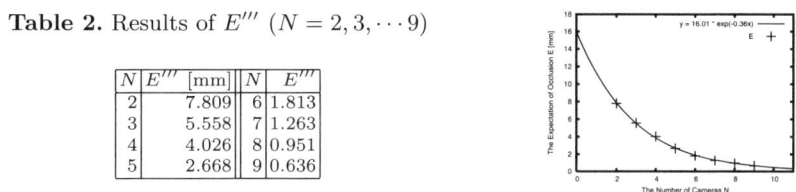

Fig. 12. Relation between N and E'''

5   Discussion

In order to obtain the best number of cameras, we experimented by changing the number of cameras N from 2 to 9 and calculated each E'''. Table 2 and Fig. 12 show the relation between N and E'''. Here, the values(N, E''') in Table 2 are plotted in it, and the curves are solved approximately as $y = a * exp(-bx)$ with least square method. In this experiment, the data are fitted as $y = 16.01 * exp(-0.38x)$. From this result, it is considered that the risk (occlusion) obeys exponentially.

6   Conclusion

This paper proposed optimal cameras' layout for global vision in RoboCup SSL. By giving a criterion that minimizes the risk of occlusion by the robots, a method to decide the camera's coordinates is realized. The occlusion was modeled from the positional relations of cameras, robots, a ball and probability distribution of the robots. This method was applied to the concrete parameters used in RoboCup SSL and solved the optimal camera's positions.

Though the simulation was done on the condition that cameras looked down directly below, it is concretely difficult to satisfy this condition. To improve simulation parameters like direction and angle of view, and to realize high speed simulation are coming subjects.

References

1. Kono, Y., Miyake, Y., Saiwaki, N., Kawamura, T., Kidode, M.: Evaluation of Waist-Mounted Camera for Object-Finding in Everyday Life. In: IPSJ-SIGHI-118, pp. 31–38 (2006) (in Japanese)
2. Mikuri, H., Mukai, N., Watanabe, T.: Monitoring Arrangement by Using Inclusive Relation Based on Visibility. The Database Society of Japan (DBSJ) Letters 5(2), 57–60 (2006) (in Japanese)
3. Katoh, K., Hibino, S., Murakami, K., Naruse, T.: On the Layout of Multi-Cameras for RoboCup Small-Size League. In: 10th SSII, pp. 445–460 (2004) (in Japanese)
4. Chen, S., Iiyama, M., Kakusho, K., Minoh, M.: Camera Arrangement Search for Satisfying Required Resolution in Shape Reconstruction. In: Proceedings of the 50th Annual Conference of the Institute of Systems, Control and Information Engineers(ISCIE), pp. 603–604 (2006) (in Japanese)
5. Narita, R., Murakami, K., Naruse, T.: A Better Solution of Multi-camera's Layout for RoboCup Small Size League. In: Proceeding of 13th Japan-Korea Joint Workshop on Frontiers of Computer Vision(FCV 2007), pp. 351–356 (2007)

Region-Based Segmentation with Ambiguous Color Classes and 2-D Motion Compensation*

Thomas Röfer

Deutsches Forschungsinstitut für Künstliche Intelligenz GmbH,
Sichere Kognitive Systeme, Enrique-Schmidt-Str. 5, 28359 Bremen, Germany
Thomas.Roefer@dfki.de

Abstract. This paper presents a new approach for color segmentation, in which colors are not only mapped to unambiguous but also to ambiguous color classes. The ambiguous color classes are resolved based on their unambiguous neighbors in the image. In contrast to other approaches, the neighborhood is determined on the level of regions, not on the level of pixels. Thereby, large regions with ambiguous color classes can be resolved. The method is fast enough to run on a Sony AIBO in real time (30 Hz), leaving enough resources for the other tasks that have to be performed to play soccer. In addition, the paper discusses the problem of motion compensation, i. e. reversing the effects of a rolling shutter on the images taken by a moving camera.

1 Introduction

In most RoboCup leagues, cameras are the central sensor of the robots. As the environment is color-coded, image segmentation is one of the most important topics in soccer robot's image-processing systems. Especially in leagues with limited on-board computing power, such as the Four-Legged League and the Humanoid League, extremely fast and robust image segmentation algorithms are a necessity. A new approach to this problem is presented in Section 2. Also typical in these leagues is the use of inexpensive CMOS cameras, e. g., the Sony AIBO, the standard platform used in the Four-Legged League, is equipped with such a sensor. Section 3 deals with the central problem of this kind of sensor, the so-called rolling shutter.

2 Image Segmentation

2.1 Current Methods

There are two general image-processing approaches in RoboCup. The first one is the *blob-based* approach (e. g. CMVision [1]), the second one is the *grid-based*

---

* This work has been funded by the Deutsche Forschungsgemeinschaft in the context of the Schwerpunktprogramm 1125 (*Kooperierende Teams mobiler Roboter in dynamischen Umgebungen*).

U. Visser et al. (Eds.): RoboCup 2007, LNAI 5001, pp. 369–376, 2008.

approach (e. g. the vision system of the GermanTeam [2]). While blob-based approaches such as CMVision entirely rely on color segmentation, the vision system of the GermanTeam still does it for the most part. The general problem is that a color classification is hard to find in which, e. g., the ball is completely orange, but orange is detected nowhere else on the field (neither in yellow goals, nor in red uniforms). Hence, teams such as the GermanTeam or the NUbots [3] have started to use additional color classes such as *yellow-orange* or *red-orange* that delay the decision about which color the pixels actually have. Quinlan *et al.* [3] call these color classes *soft colors*. They resolve these soft colors after the blob formation based on relations between the bounding boxes of the blobs. The relations are object-specific. Palma-Amestoy *et al.* [4] present a more general approach for defining and resolving soft colors. From example images that are labeled manually with the unambiguous color classes that the objects in the images should have, a color table with soft colors is automatically generated. Images are segmented using the soft colors and afterwards a mode filter is applied that assigns the color class to each pixel that is the most frequent in its 3×3 neighborhood.

2.2   Ambiguous Color Tables

A color table is a mapping from image colors to color classes. Typically 256^3 possible colors exist in an image (one byte for each color channel; Y, Cr, and Cb in case of the AIBO), and they have to be reduced to only a few color classes. In case of the Four-Legged-League in 2007, these are the seven classes orange, yellow, sky-blue, red, blue, green, and white. An *ambiguous color class* is a color class that contains more than one of these base classes, e. g. yellow-orange or red-orange. Ambiguous color classes are represented as a bit-set, i. e. each bit of a byte stands for one of the color classes. A color table can be implemented as a simple lookup table, i. e. the color values function as indices into a 3-D array of color classes. Since a color table of 16 MB would be quite big, the table takes only the six highest bits of each color channel into account. Thus the size of the color table is $64^3 = 262144$ bytes. In contrast to the work described in [4], the color tables are created manually, using a color table editor.

2.3   Image Segmentation with Ambiguous Color Classes

In general, image segmentation with ambiguous color classes uses the same base algorithms as segmentation with unambiguous color classes [1], but it applies them several times with slight differences to resolve the ambiguities. Since the whole image is processed, no color correction is performed to eliminate the bluish corners of the camera images of the AIBO. Instead it is assumed that this problem is handled by the ambiguous color classification.

In a first step, all pixels are segmented determining their (ambiguous) color class from the color table. Successive pixels with the same color class are combined to *runs*. Each run contains its color class, its *y*-coordinate, and its first and last *x*-coordinate. For the image size of 208×160 pixels used in the AIBO, all

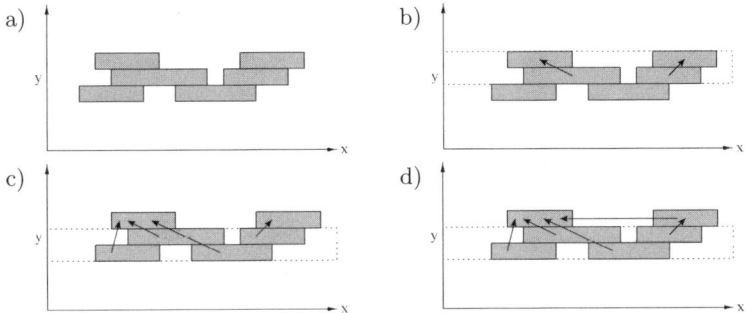

Fig. 1. Grouping runs to regions (taken from [1]). a) Runs start as a fully disjoint forest. b) Scanning adjacent lines, neighbors of the same color class are merged. c) New parent assignments are to the furthest parent. d) The first run in a region is always the parent of all others.

these values can be represented by bytes. Since the camera images of the AIBO are rather noisy when taken with the fastest shutter setting, the run length encoding typically reduces the image size only by factor 10, i. e. around 3000 runs will be generated. Please note that an ambiguous segmentation creates more runs than an unambiguous one, because there are simply more color classes.

The collection of runs is traversed five times. In the first phase, only runs of the same ambiguous color class are grouped together, because these regions are required for the next two phases. Runs with unambiguous color classes are simply skipped. Grouping runs to regions is based on the work of Bruce et al. [1], who describe it as a *union-find* problem that can be solved highly efficiently (cf. Fig. 1). The collection of runs is processed from top to bottom and from left to right. There are always two current runs in two adjacent rows. Whenever they overlap and have the same color class, they are grouped together. Groups are defined as a tree structure with each run having a pointer to its parent. During merging, path compression is applied, and it is ensured that the first run in a group is always the parent of all others.

Regions of ambiguous color classes inherit the unambiguous color class of which they are surrounded most. This extends the mode filter approach described in [4] to the level of regions. Therefore, a histogram has to be calculated for each region with an ambiguous color class on which unambiguous color class is neighbored how often, i. e. how many pixels on the region's perimeter are neighbored to which color class. Therefore, a second pass over the runs is performed. This time it is focused on the neighborhood between runs of unambiguous and ambiguous color classes, and for each such pair, the histogram of the ambiguous region involved is updated. In this pass, the neighborhood between runs in the same row has also to be considered.

After all the histograms have been calculated, the ambiguous color classes can be resolved. This is done in a third pass. Each ambiguous color is replaced by the unambiguous class that reached the highest score in the histogram of the region,

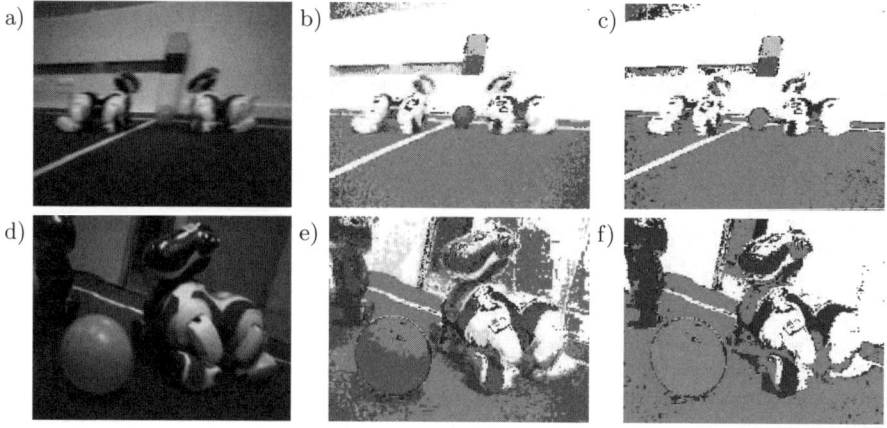

Fig. 2. Image segmentation with ambiguous color classes. a, d) Original camera images. b, e) Segmentation with ambiguous color classes. c, f) Resolution of ambiguous color classes.

but only if that score is above a threshold (e. g. more than two pixels). Regions that do not satisfy this criterion are deleted. Thus, ambiguous regions that are only surrounded by other ambiguous regions are removed. It would be possible to delay the resolution of the color class of such a region until its neighbors have unambiguous color classes, but this would take considerably more computing time, and therefore it was not implemented.

After the ambiguous color classes are resolved, a final merging pass over the runs is performed to group all regions. Again, some special treatment is required, because now there are successive runs within the same row that have the same color class. They are grouped together to single runs on the fly while merging the regions. In the final phase, the blobs are collected.

2.4   Results

Figure 2 shows some examples of segmented images. The first column shows two images as they are taken by the camera of the AIBO with the fastest shutter setting. The camera was turning quickly while the image shown in Figure 2a was taken, resulting in rather blurry colors. The column in the middle shows the result of the segmentation using an ambiguous color table. Please note that large parts of the ball and the uniform of the red robot are segmented as the ambiguous color class red-orange. Since the blue uniform is very dark, it was decided to use the same color class for black and blue. Hence, not every blue region on the field is an indication for a robot of the blue team, but this problem exists as long as the Four-Legged League, and cannot be solved on the level of color segmentation. However, as can be seen in the right column, other ambiguities can be resolved very well. The red uniforms are entirely red and the ball is always completely orange. The robots and the walls are white and the field is green, even in the

Module	Min	Max	Avg	Freq
AmbiguousRLEColorClassImage	1.0	9.0	2.7	30.0
RLEColorClassImage	2.0	15.0	6.6	30.0
groupAmbiguousRegions	0.0	7.0	0.9	30.0
calcPerimeterStats	0.0	6.0	2.6	30.0
resolveAmbiguousColors	0.0	4.0	1.3	30.0
groupRegions	0.0	4.0	1.1	30.0
Blobs	0.0	3.0	0.5	30.0

Fig. 3. Runtime measurements on an AIBO in ms

corners of the image. Figure 3 shows the runtime of the individual parts of the system.

3   Motion Compensation

The AIBO is equipped with a simple CMOS camera. Such cameras can also be found in PDAs such as the Siemens Pocket Loox 720 that is used by several teams in the Humanoid League, e.g. by the BredoBrothers [5]. Such cameras have a central weakness, the so-called rolling shutter. Instead of taking images at a certain point in time, a rolling shutter takes an image pixel by pixel, row by row. Thus the last pixel of an image is taken significantly later than the first one. The general problem is depicted in Figure 4. The AIBO is equipped with a camera in its head that takes 30 images per second. By moving its head, the AIBO can point the camera in different directions. Since an image is not taken all at once, the camera may point to a different direction when the first pixel is recorded than when the last pixel is taken. In Figure 4a the AIBO turns its head from right to left. Thus the head is pointing further right when the upper image part is taken and further left when the lower part is taken. This results in a distorted image as depicted in Figure 4b. In fact, the effect is not only present during panning the camera, but also when it is tilted or even rolled.

3.1   State of the Art

The problem of the rolling shutter was first mentioned by Nistico and Röfer in [6], and later analyzed in more detail by Nicklin *et al.* [7]. The approach of compensating the effect of the rolling shutter described by Nistico and Röfer is the most general one. However, it requires calculating the positions of all percepts twice, and in merging the pairs of locations or bearings a rather heuristic approach is followed. Nicklin *et al.* concentrate on the compensation of horizontal distortions resulting from panning the camera [7]. They determined that the camera of the AIBO actually takes the full 33.3 ms to take a picture, i.e. there is basically no delay between recording the last pixel of one image and the first pixel of the next one. The camera images and the joint angles on which the calculation of the camera position is based do not arrive at the same time, but they are time-stamped. As is indicated in the team report of the Microsoft Hellhounds [8], it seems that the timestamp of the joint angles matches the time when 3/4 of the image is recorded.

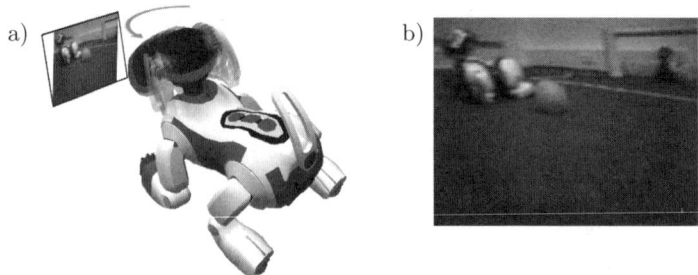

Fig. 4. Image distortion by a rolling shutter. a) Robot is turning its head while taking an image. b) Resulting camera image.

3.2   2-D Motion Compensation

Nicklin *et al.* [7] only compensate for horizontal distortions of the image. However, as can be seen in Figure 5b/e and c/f the rolling shutter shrinks the image if the head is turning downwards and stretches it when turning upwards. Because of the gravity, the head can be tilted faster downwards than upwards, resulting in the stronger effect shown in Figure 5b/e. The distortion in vertical direction impedes the calculation of distances to objects if the calculation is based on their position in the image, e. g. the distance to field lines can be determined this way. The distortion can add systematic errors to these calculations. For instance if the head is always turning upwards when looking to the right and turning downwards when looking to the left, as it is often done when scanning for the ball, distances to field lines are overestimated in one direction and underestimated in the other, resulting in a bad self-localization.

The motion of the camera between the previous image and the current image can be determined from the rotation matrices R_t and $R_{t-\Delta t}$ that describe the orientations of the camera as $\Delta R = R_t^{-1} R_{t-\Delta t}$. From ΔR the relative pan angle α and relative tilt angle β can be calculated and used when applying the equation from [7] to the 2-D case:

$$\begin{pmatrix} x' \\ y' \end{pmatrix} = \begin{pmatrix} cx - f\tan\left(\arctan\frac{cx-x}{f} - d\alpha\right) \\ cy + f\tan\left(\arctan\frac{y-cy}{f} - d\beta\right) \end{pmatrix}$$

$$\text{where } d = \frac{\frac{y}{imageHeight} - 0.75}{\Delta t 30\text{Hz}}$$

(1)

(x', y') is the corrected position of the pixel (x, y). (cx, cy) is the image center and f the focal length of the camera. The decimal number 0.75 is used because the camera position was measured when 3/4 of the image were taken (cf. previous section). The factor d is only dependent on the y-coordinate, because the effect of the x-coordinate on the distortion of the image is far below a single pixel. The equation also works if Δt is more than the delay between two successive images, e. g., if an image was skipped.

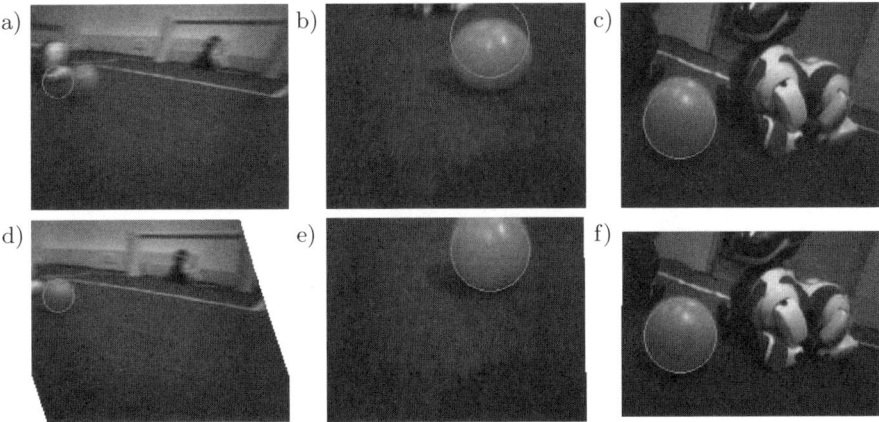

Fig. 5. Images taken will the camera was moving a) right, b) down, c) up. d–f) Corrected images. The orange circles represent the ball positions, also projected back to the original images.

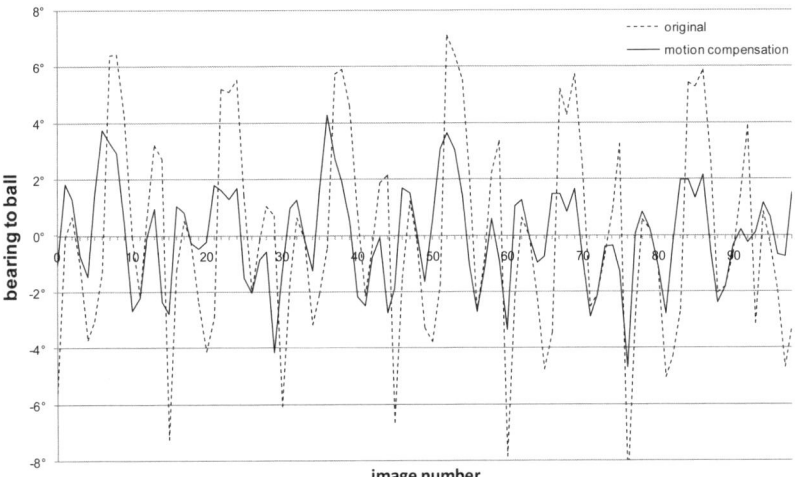

Fig. 6. Errors in horizontal bearings to the ball when turning the head with $300°/s$

3.3 Results

The simplest way to determine the improvements achieved by motion compensation is to let the AIBO turn its head quickly while the image-processing system determines the bearing to a static object in the environment. Here, as well as in [7] and [8], the static object is a ball. Figure 6 shows the results of the experiment using a ball detector based on the approach presented in Section 2, and the current head motion of the GermanTeam, which has a slight stop in the middle. The standard deviation of the bearings without motion compensation is 3.59° (without the stop it is 5.54°). When the ball position is determined from the distorted image using a Levenberg-Marquardt least-square fitting [9] and the position of its center is corrected afterwards, the standard deviation of the bearings is 1.81°. If the edge points of the ball are corrected before they are used in the fitting algorithm, the standard deviation is reduced to 1.76°. The influence of stopping or not stopping the head on these results was negligible.

4 Conclusions

This paper describes two methods to improve image-processing in RoboCup. The image segmentation using ambiguous color classes is a general solution to blob-based image-processing. It is fast and simplifies post-processing significantly, because many of the problems that result from noisy camera images are already dealt with during image segmentation. The section on motion compensation brings together all the findings of different teams on this topic and adds a compensation for up and down rotations of the camera. Thus the precision of bearings to objects could significantly be improved.

References

1. Bruce, J., Balch, T., Veloso, M.: Fast and inexpensive color image segmentation for interactive robots. In: Proceedings of the IEEE/RSJ International Conference on Intelligent Robots and Systems (IROS 2000), vol. 3 (2000)
2. Röfer, T., Jüngel, M.: Vision-based fast and reactive Monte-Carlo localization. In: IEEE International Conference on Robotics and Automation, Taipei, Taiwan, pp. 856–861. IEEE (2003)
3. Quinlan, M.J., et al.: The 2006 NUbots Team Report (2007), http://robots.newcastle.edu.au/publications/NUbotFinalReport2006.pdf
4. Palma-Amestoy, R.A., Guerrero, P.A., Vallejos, P.A., del Solar, J.R.: Context – dependent color segmentation for Aibo robots. In: Proc. of the 3rd Latin American Robotics Symposium. IEEE, Robotics and Automation Society (2006)
5. Laue, T., Röfer, T.: Getting upright: Migrating concepts and software from four-legged to humanoid soccer robots. In: Pagello, E., Zhou, C., Menegatti, E. (eds.) Proceedings of the Workshop on Humanoid Soccer Robots in conjunction with the 2006 IEEE International Conference on Humanoid Robots (2006)
6. Nistico, W., Röfer, T.: Improving percept reliability in the Sony Four-Legged League. In: Bredenfeld, A., Jacoff, A., Noda, I., Takahashi, Y. (eds.) RoboCup 2005. LNCS (LNAI), vol. 4020, pp. 545–552. Springer, Heidelberg (2006)
7. Nicklin, S.P., Fisher, R.D., Middleton, R.H.: Rolling shutter image compensation. In: Lakemeyer, G., Sklar, E., Sorrenti, D.G., Takahashi, T. (eds.) RoboCup 2006: Robot Soccer World Cup X. LNCS (LNAI), vol. 4434. Springer, Heidelberg (2007)
8. Hebbel, M., et al.: Microsoft Hellhounds Team Report 2006 (2007), http://www.microsoft-hellhounds.de/fileadmin/pub/MSH06TeamReport.pdf
9. Seysener, C.J., Murch, C.L., Middleton, R.H.: Extensions to object recognition in the four-legged league. In: Nardi, D., Riedmiller, M., Sammut, C., Santos-Victor, J. (eds.) RoboCup 2004. LNCS (LNAI), vol. 3276, pp. 274–285. Springer, Heidelberg (2005)

Multi-agent Positioning Mechanism in the Dynamic Environment

Hidehisa Akiyama and Itsuki Noda

Information Technology Research Institute,
National Institute of Advanced Industrial Science and Technology
Ibaraki, Japan
{hidehisa.akiyama,I.Noda}@aist.go.jp

Abstract. In this paper, we propose a novel agent positioning mechanism for the dynamic environments. In many problems of the real-world multi-agent/robot domain, a position of each agent is an important factor to affect agents' performance. Because the real-world problem is generally dynamic, a suitable positions for each agent should be determined according to the current status of the environment. We formalize this issue as a map from a focal point like a ball position in a soccer field to a desirable positioning of each player agent, and propose a method to approximate the map using Delaunay Triangulation. This method is simple, fast and accurate, so that it can be implemented for real-time and scalable problems like RoboCup Soccer. The performance of the method is evaluated in RoboCup Soccer Simulation environment compared with other function approximation method like Normalized Gaussian Network. The result of the evaluation tells us that the proposal method is robust to uneven sample distribution so that we can easily to maintain the mapping.

1 Introduction

In many problems of the multi-agent/robot domain in real-world environment, the position of each agent is a significant factor to affect the agent's performance. For example, in a multi-robot transportation system, a cordinatin of positionings of robots are decided carefully by given tasks and status inputs in order to realize efficient convey. Similarly, in the games which teams play in the dynamic environment like a soccer, each agent should make a decision about its positioning continuously according to the current game status in order to fulfill its role or duty. Otherwise, the performance of the team will be decreased significantly.

In order to realize effective positionings or formations of agents, most of RoboCup teams generally use rule-based and/or numerical-function-based position definition systems. However, this approach meets a difficulty when we like to tune the positionings for a certain situations like case when a ball is in a penalty area. Because penalty areas are very important region in soccer, small miss of the position may cause critical situation. Therefore, rules and functions to determine the position in such cases becomes so complex that it becomes difficult to maintain.

U. Visser et al. (Eds.): RoboCup 2007, LNAI 5001, pp. 377–384, 2008.
© Springer-Verlag Berlin Heidelberg 2008

Another approach for the positioning is to utilize machine learning methods. In this approach, we just need to give examples of desireble positioning or training schema for agents to learn suitable positioning rules/functions. For example in the penalty-area cases above, we just give many examples for penalty area to tune sensitive positioning of the dangerous region. However, this learning approach has a general problem how to give enough example or training for agent. Especially in a complex domain like RoboCup, it is difficult to give enough number of examples for learning by hand. It is also difficult to provide reasonable evaluation method for trainings to provide suitable index of the learning.

In this paper, we propose a novel positioning mechanism which uses Delaunay Triangulation and the linear interpolation method. In this method, we formalize learning problem of the positioning as a function approximation. The mechanism used in the model is effective to acquire the interpolation function, which a focal point in the environment is used as input and agents' move target position is output. In the experiment, we used the RoboCup Soccer Simulator[6,1] as an experiment environment. We compared our method with other function approximation method, Normalized Gaussian Network.

2   Positioning Problem in the RoboCup Soccer Simulation

In the RoboCup Soccer Simulation domain, Situation Based Strategic Position (SBSP)[8] is well known as the agents' positioning mechanism. SBSP uses a ball position as a focal point and does not consider other agents. But, if we assume that all agents always pay attention to the ball, the cooperative behavior can be done indirectly. Because it is easy to implement SBSP, almost all teams in the RoboCup Soccer Simulation League use SBSP or the similar model.

SBSP uses a simple function that uses the the ball coordinate value as an input value and outputs the player agent's basic move position. And the attraction parameters and the movable region are defined for each agent and are used to calculate the final output value. Fig.1 shows a basic SBSP algorithm.

BasePos means the agent's position when the ball is located at the center of field. AttractionToBall means the rate that agent follows the ball move.

```
GetSBSPPosition( Num, BallMove )
  Num : agent number
  BallMove: ball move vector from the center of field
1. BasePos := BasePosition( Num );
2. AttractionToBall := BallAttract( Num );
3. SBSPPos := BasePos + BallMove * AttractionToBall
4. Region := MovableRegion( Num );
5. if Region does not contain SBSPPos
6.    adjust SBSPPos into Region
7. return SBSPPos
```

Fig. 1. The basic algorithm of SBSP

SBSPPos means the agent's move position. If SBSPPos is not contained by Region, the SBSPPos is adjusted into Region.

The problem of SBSP is that the output value depends on the used function. In the basic SBSP algorithm, the characteristic of agent's move also becomes simple because the simple linear function is used. If we want the more complicated agent's move, we need to prepare the several parameter set for each situation. But, it is difficult for us to manage such many parameters and the relation of the situations correctly.

In our previous research[2], we used a non-linear function instead of the SBSP algorithm. To acquire the non-linear function, we used a traditional three layer perceptron as a function approximator and the training data set is created by human supervisor using a GUI tool. As a result, we could acquire the good approximation function that can be used in the real game. However, it is difficult to acquire the completely desired result because the overfitting is a critical problem. So, we concluded that we need a locally adjustable function approximation model.

3   Positioning Mechanism Using Delaunay Triangulation

We propose a novel positioning mechanism that uses Delaunay Triangulation and the linear interpolation algorithm. In this model, the region is divided into several triangles based on the given training data, and each training data affects only the divided region where it belongs to. So, the proposal model is locally adjustable.

3.1   Delaunay Triangulation

Delaunay Triangulation is one of the method to triangulate tha plane region based on the given point set. Delaunay Triangulation for a set P of points in the plane is a triangulation $DT(P)$ such that no point in P is inside the circumcircle of any triangle in $DT(P)$. If the number of given points are more than 3, we can get a unique Delaunay Triangulation. Fig. 2(a) shows the example of Delaunay Triangulation. In this figure, Voronoi Diagram is also shown. There is a duality between Voronoi Diagram and Delaunay Triangulation. In our programs, we implemented the incremental algorithm[4] that is one of the fastest algorithm to calculate Delaunay Triangulation and the time complexity is $O(n \log n)$.

In our proposal method, the ball positions in training data are used as the vertices of triangles and each vertex means the given training data. Each vertex has the output value as an agent's move position for that vertex(=ball) position. When the ball is contained by one triangle, agent's move position is calculated by interpolation algorithm described in next section.

3.2   Linear Interpolation Algorithm

We use the simple linear interpolation algorithm to calculate the agent's move position. This algorithm is same as Gouraud shading algorithm[3]. Gouraud

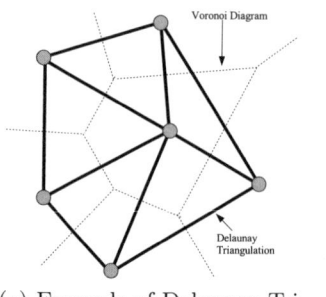
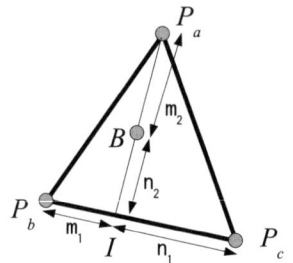

(a) Example of Delaunay Tri-   (b) Liner   interpolation   by
angulation   Gouraud shading algorithm

Fig. 2. Delaunay Triangulation and Linear Interpolation

shading algorithm is a method used in the computer graphics domain to simulate
the differing effects of light and color across the surface of an object.

Fig. 2(b) shows the process of Gouraud shading algorithm. The output values
from vertices P_a, P_b and P_c are $O(P_a)$, $O(P_b)$ and $O(P_c)$ respectively. Now,
we want to calculate $O(B)$, the output value of the point B contained by the
triangle $P_a P_b P_c$. The algorithm is as follows:

1. Calculates I, the intersection point of the segment $P_b P_c$ and the line $P_a B$.
2. The output value at I, $O(I)$, is calculated as:

$$O(I) = O(P_b) + (O(P_c) - O(P_b))\frac{m_1}{m_1 + n_1}$$

where $|\overrightarrow{P_b I}| = m_1$ and $|\overrightarrow{P_c I}| = n_1$.
3. $O(B)$ is calculated as:

$$O(B) = O(P_a) + (O(I) - O(P_a))\frac{m_2}{m_2 + n_2}$$

where $|\overrightarrow{P_a B}| = m_2$ and $|\overrightarrow{B I}| = n_2$.

3.3   Advantages

The proposal positioning mechanism has following advantages:

- High approximation accuracy. Our method realizes higher accuracy than
  other function approximator.
- Locally adjustable. Even if new training data is added or existing data is
  modified, the triangle region where that data is not contained is never affected.
- Simple and fast running. It is possible to use it in real time.
- High scalability. Even if the considered region is extended or shrinked, it is
  easy to correspond to the new region without any changes.

– Complete reproducibility. If the training data is same, we can acquire the completely same result.

Especially, the complete reproducibility is an important advantage. Because there is no limitation of the input order of the training data, any training data can be added at any time. This means that human can intervene at any time.

4   Experiment

In order to evaluate our method, we compare our method with other function approximator. We developed the simulated soccer team which can use our method and Normalized Gaussian Netrowk(NGnet)[5].

4.1   NGnet

This section describes the brief definition of NGnet. NGnet is one of the extended methods of Radial Basis Function(RBF) Network[7]. The network structure of NGnet and RBFNetwork is almost same as the three layer perceptron. But, the Gaussian basis function is used as an activation function instead of the sigmoid function. These methods are known as a locally adjustable function approximator.

The difference between NGnet and RBFNetwork is whether the output of each unit is normalized by the sum of units' output or not. The normalization guarantees the activation of at least one unit for any input value. On the other hand, in the RBFNetwork, if the distance between units is big, output values become almost 0. In the agent positioning problem, it is not preferable that the output values become 0. Because NGnet can solve this, we adopt NGnet as the compared method in this paper.

The output of the unit i in output layer is

$$f_i(x, w) = \frac{\sum_{j=1}^{N} w_{ij} \phi_j(x)}{\sum_{j=1}^{N} \phi_j(x)} \tag{1}$$

where x is an input vector, w_{ij} is the connection weight for unit i from hidden layer to output layer. $\phi_j(x)$ is the output of basis unit j in hidden layer:

$$\phi_j(x) = \exp(-\frac{||x - c_j||^2}{2\sigma_j^2}) \tag{2}$$

where c_j is the center position of basis unit j and σ_j is the variance parameter of basis unit j. In NGnet, we need to acquire three parameters, w_i, c_j and σ_j.

The connection weight w_i is learned by the following gradient descent method:

$$w(t + 1) = w(t) - \eta \frac{\partial \epsilon}{\partial w} + \alpha(w(t) - w(t - 1)) \tag{3}$$

The error value of agent's position at each training data point is used as ϵ . In this experiment, the learning rate η is set to 0.1 and the rate of moment method

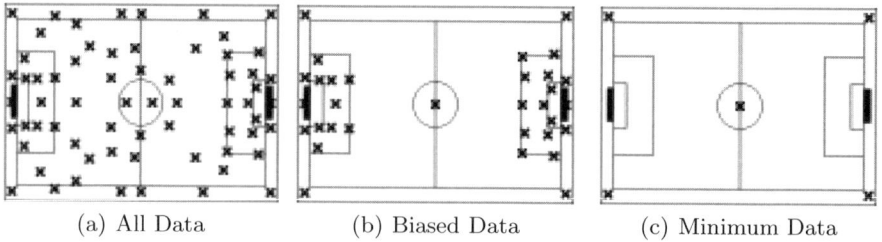

| (a) All Data | (b) Biased Data | (c) Minimum Data |

Fig. 3. Training data set used in the experiment

α is set to 0.5. The ball positions of training data are just used as the center of basis unit c_j. So, the number of training data is same as the number of basis unit. The following heuristic value is used as the variance parameter σ:

$$\sigma = \frac{1}{N} \sum_{i=1}^{N} ||c_i - c_j|| \tag{4}$$

where c_j is the nearest basis unit to the unit c_i. All basis units use a same variance parameter.

4.2 Experiment Settings

We prepare three training data set which are shown in Fig.3. In each sub figure, 'X' marked points mean the given training data. We used them as the training data for NGnet and the proposal method.

Fig.3(a) shows the normal training data set. In this data set, the training data are given for the whole region. This data set is used by our team TokyoTechSFC in the RoboCup2006 Soccer Simulation League and our team won the 4th place. Fig.3(b) shows the biased training data set. In this biased training data set, several training data are removed from Fig.3(a) and we set the big difference of the density between the field middle area and the penalty area. Fig.3(b) shows the minimum training data set. Almost all training data are removed from Fig.3(b). This is the minimum training data set for the simulated soccer domain.

We developed a simulated soccer team that uses both methods and played games against the fixed team. As an opponent team, we use UvA Trilearn 2004[1] which participated RoboCup2004 and was ranked 7th. For each training data set, one half match[2] is played 50 times respectively.

4.3 Results

Table 1 and 2 show the statistics data extracted from log files. Table 1 shows big difference in the average conceded goal between NGnet and the proposal

---

[1] http://www.science.uva.nl/~jellekok/robocup/
[2] One half is about 3000 cycles.

Table 1. This table shows the average scored goal and the average conceded goal. The values in the parenthesis mean the standard deviation for each average value.

	Goal Scored		Goal Conceded	
	NGnet	Proposal method	NGnet	Proposal method
All	0.2(0.4)	0.08(0.27)	0.98(0.89)	0.76(0.9)
Biased	0.24(0.52)	0.06(0.22)	1.76(1.24)	0.6(0.74)
Minimum	0.02(0.14)	0.02(0.14)	3.7(1.97)	3.16(1.67)

Table 2. This table shows the average number of successful passes, and the average number of successful intercepts. The values in the parenthesis mean the standard deviation for each average value.

	Successful Pass		Successful Intercept	
	NGnet	Proposal method	NGnet	Proposal method
All	99.06(19.08)	110.6(18.79)	20.2(5.06)	17.18(6.06)
Biased	67.68(15.07)	91.44(21.35)	18.8(5.52)	19.32(5.29)
Minimum	83.86(19.1)	84.3(18.48)	12.06(4.17)	14.5(5.21)

method with the biased data set. And Table 2 shows that the average number of successful pass becomes minimum in the case of NGnet with the biased data set.

We guess this is because the positions acquired by NGnet with the biased data set has the unique characteristic. When agents uses NGnet with the biased data set, they do not move so much in the field middle area. Even if a teammate agent has the ball, other agents do not move according to the ball move. So, the ball owner agents might not be able to find tha pass courses. On the other hand, in the case of NGnet with all data and minimum data set or the proposal method with any data set, the acquired positions show the smooth move in the whole field area. These characteristic are caused by the variance parameter of NGnet.

These results shows that the density of the training data significantly affects the positioning characteristic of NGnet. As a result, the performance of the team is also easily affected by that characteristic. This means that it is difficult to tune the variance parameter of each basis function unit because the positioning characteristic of NGnet depends on the variance parameter. Therefore, we can say that it is necessary to prepare the uniformly distributed basis function units for NGnet in order to get the stable team performanc, and the proposal method is robust to uneven sample distribution so that we can easily to maintain the mapping.

5   Conclusion and Future Directions

Although the statistics data from the experiment may not show the meaningful quantitative result, some criteria show the important characteristic of each method.

We can say that our proposal method has many advantages obviously. However, the proposal method has the following disadvantages:

- The proposal method needs to store all training data. So, it requires many memories.
- The proposal method requires the high cost to keep the consistency of the training data. If one training data is modified, the near training data may be also required to modify.

We plan to develop the method to adjust the training data automatically in order to reduce the management cost of the training data set. At least, we need the method to help us to find the inconsistent training data. And also, we should consider about the multiple dimensional input and output. Now, our proposal method can handle only two dimensional input and output. If we can handle other agents' positions as an input and the other decision making parameters as an output, they will be very useful. We have to consider about the method to compress the dimension or to overlap the information.

References

1. The RoboCup Soccer Simulator, `http://sserver.sourceforge.net/`
2. Akiyama, H., Katagami, D., Nitta, K.: Team formation construction using a gui tool in the robocup soccer simulation. In: Proceedings of SCIS & ISIS 2006 (2006)
3. Gouraud, H.: Continuous shading of curved surfaces. In: Wolfe, R. (ed.) Seminal Graphics: Pioneering efforts that shaped the field. ACM Press (1998)
4. van Kreveld, M., de Berg, M., Schwarzkopf, O., Overmars, M.: Computational Geometry: Algorithms and Applications, 2nd edn. Springer (2000)
5. Moody, J., Darken, C.: Fast learning in networks of locally-tuned processing units, vol. 1, pp. 289–303 (1989)
6. Noda, I., Matsubara, H.: Soccer server and researches on multi-agent systems. In: Kitano, H. (ed.) Proceedings of IROS 1996 Workshop on RoboCup, pp. 1–7 (November 1996)
7. Poggio, T., Girosi, F.: Networks for approximation and learning, vol. 78, pp. 1481–1497 (1990)
8. Reis, L.P., Lau, N., Olivéira, E.: Situation Based Strategic Positioning for Coordinating a Simulated RoboSoccer Team. Balancing Reactivity and Social Deliberation in MAS, 175–197 (2001)

Layered Learning for a Soccer Legged Robot Helped with a 3D Simulator

A. Cherubini, F. Giannone, and L. Iocchi

Dipartimento di Informatica e Sistemistica
Università di Roma "La Sapienza"
Via Ariosto 25, 00185 Roma, Italy
{cherubini,iocchi}@dis.uniroma1.it

Abstract. Mobile robots can benefit from machine learning approaches for improving their behaviors in performing complex activities. In recent years, these techniques have been used to find optimal parameter sets for many behaviors. In particular, layered learning has been proposed to improve learning rate in robot learning tasks. In this paper, we consider a layered learning approach for learning optimal parameters of basic control routines, behaviours and strategy selection. We compare three different methods in the different layers: genetic algorithm, Nelder-Mead, and policy gradient. Moreover, we study how to use a 3D simulator for speeding up robot learning. The results of our experimental work on AIBO robots are useful not only to state differences and similarities between different robot learning approaches used within the layered learning framework, but also to evaluate a more effective learning methodology that makes use of a simulator.

1 Introduction

In order for robots to be useful for many real-world applications, they must be able to effectively perform complex tasks. For this purpose, a popular research activity is the annual RoboCup soccer competition[1]. In this domain, the robot should be able to respond to changes in its surroundings by adapting both its low-level skills (e.g., the walking style) and the higher-level skills (e.g., the behaviour) which depend on them. Firstly, creating effective motion for walking and kicking the ball is a challenging task, since there are many parameters to be set, and since successful motions strongly depend on many factors, including: playing surface, robot hardware, and game situation. In recent years, machine learning techniques have been used to find optimal parameter sets. Secondly, in Robocup matches, the correct choice of the best behaviours required to accomplish a certain task (e.g., to score a goal) is fundamental for success. Machine learning techniques have been used in this field as well, in order to adapt the behaviours to the given game situation.

---

[1] Here, we focus on the RoboCup Four-Legged league (see www.tzi.de/4legged/).

U. Visser et al. (Eds.): RoboCup 2007, LNAI 5001, pp. 385–392, 2008.
© Springer-Verlag Berlin Heidelberg 2008

Indeed, robot learning has been used both for fine-tuning of the parameters used by the low level algorithms – *parameter learning* – and for finding the optimal composition of simple behaviors for accomplishing a certain task – *behavior learning*. In the first class, an important application area is robot motion control, such as gait optimization for legged robots. For example, a Genetic Algorithm has been used for optimization of the vector of quadruped walk parameters in [3]. Kohl and Stone [6], empirically compared four machine learning algorithms for quadruped walk optimization. Parameter learning has proved very effective for improving other motion control tasks, such as grasping. This task is achieved in [5] by applying the *layered learning* paradigm [9]: grasping parameters rely on previously learned walk parameters.

On the other hand, researchers proved the utility of behavior learning for improving high level tasks, e.g. navigation [4], path-planning, and multi-robot cooperation [8]. To our knowledge, parameter and behaviour learning have rarely been joined in a single framework. In [1], this has been done by gradually developing cognitive capabilities, starting from abilities for detecting and recognizing objects, up to task execution.

Besides, experimental comparison of different learning methodologies has been rarely addressed. Here, we focus on learning a task for a AIBO soccer robot. Behavior learning and parameter learning are integrated, and we present an experimental evaluation of a layered learning approach [9] using three different learning techniques: genetic algorithm (GA), Nelder-Mead (NM), and policy gradient (PG). Finally, we study how to use a 3D simulator for speeding up robot learning. The results of our experimental work can be summarized as follows: 1) layered learning is very effective in the complex scenario considered in this paper; 2) using a 3D simulator can speed up the learning process on real robots; 3) the learned low-level parameters are strongly related to the desired behavior. We have successfully experimented the presented learning methodology in preparation of RoboCup 2006, showing a notable improvement of performance of the robots in our team.

2   Problem Definition

In this paper, we consider learning a complex task as a composition of different behaviors. More specifically, we consider situations in which a task $\mathcal{T}$ can be accomplished by applying different strategies, where each strategy is a composition of different behaviors. Each behavior B is characterized by a set of parameters $\Theta_B = \{\theta_1, \ldots, \theta_{k_B}\}$. Notice that a behavior can be present in different strategies and possibly requires the definition of different parameters depending on the situation in which it is used. The learning problem is thus twofold: on one hand, behavior learning is needed to select the best strategy, i.e., the best composition of basic behaviors; on the other hand, each behavior needs fine tuning of each parameter. In the next section we present a learning methodology that integrates behavior and parameter learning for a complex robot task.

To make this problem more clear we will present the application example in which we have tested our method. Consider a robot playing soccer within the RoboCup Four-Legged league competitions. One of the main tasks to be accomplished is to approach the ball and kick it to the opponent goal. This is a complex task that requires the integration of different behaviors in different ways (i.e., strategies). Besides, each behavior has several parameters to be tuned: walking gait parameters, kick parameters, etc. Learning such a complex task requires to define a strategy (as a combination of behaviors) and tune the parameters of the behaviors involved in such a strategy.

3   System Description

In this section, we describe learning the *attacking* task for a soccer robot in the Four-Legged League.

3.1   Behaviors and Parameters

For learning the *attacking* task, a set of six behaviors $\mathcal{B}_P = \{B_1, \ldots, B_6\}$ has been considered. These behaviors (see Figure 1) can be classified in three subsets:

- *ball approaching* behaviors: B_1: *fast ball approach* (the path length is minimized by an omnidirectional walk), B_2: *aligning ball approach* (while the robot approaches the ball, its heading is oriented towards the goal);
- *ball carrying* behaviors (aimed at grasping the ball with the robot chin, and orienting the robot heading towards the opponent goal): B_3: *rotational ball carrying* (alignment is achieved by pure rotation), B_4: *rototranslational ball carrying* (alignment is achieved by a rototranslational movement);
- *ball kicking* behaviors (realized by direct kinematics control of the 15 joint positions): B_5: *head straight kick* (the robot 'dives' on the ball and hits it with the head), B_6: *head spanning kick* (the robot 'dives' on the ball and hits it with varying head pan).

A strategy is a combination of behaviors. Here, we consider only a simple form of behavior composition: chaining. Thus, we consider *strategies* as sequences of behaviors. For example, $\{B_1; B_4; B_5\}$ is a possible strategy for the attacking task. Each behavior B is characterized by a set of parameters Θ_B. Speed and

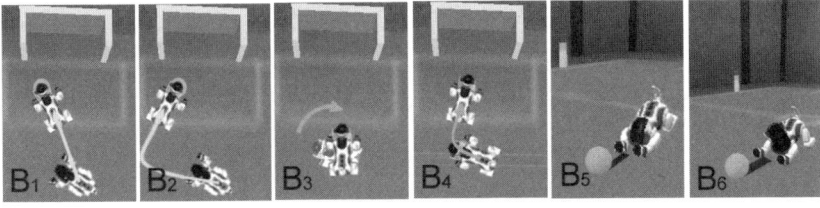

Fig. 1. The six behaviours that can be used in the attacking task

stability of the ball approach are mainly characterized by the **eleven** *walking gait parameters* Θ_{WG}, which define the kinematic characteristics of the walk. The **four** *ball carrying parameters* Θ_{BC} characterize the way the robot slows down and eventually stops near the ball. These parameters also influence the quality of ball control in attacking strategies with no ball carrying (e.g., strategy $\{B_1; B_5\}$). Finally,the robot kicks are generated by a sequence of fixed joint positions, characterized by **nine** *ball kicking parameters* Θ_{BK}.

3.2 Learning to Attack

Here, we present the optimization problem that must be solved in order to improve the attacker performance, based on the above system characteristics. As aforementioned, we present a learning approach that allows for the concurrent search of optimal behaviors and optimal parameters. The first issue deserves some clarification. We represent the attacker strategy P_i with a string of three integers, instead of symbolic expressions. We use the coding function:

$$c : P_i \rightarrow \{\phi_1 \ \phi_2 \ \phi_3\} \in \mathbf{Z}^3 \qquad i = 1, \ldots, 12$$

such that: ϕ_1 indicates the *ball approaching* behaviour used in strategy P_i (1 for B_1, or 2 for B_2), ϕ_2 the *ball carrying* behaviour (1 for B_3, 2 for B_4, 0 for no ball carrying), and ϕ_3 the *ball kicking* behaviour (1 for B_5, 2 for B_6). For example, the strategy $\{B_1; B_4; B_5\}$ is coded with the string $\{122\}$.

  With this approach, the search for the optimal composition of simple behaviors amounts to a parameter optimization problem in the discrete set $S_\phi = \{1, 2\} \times \{0, 1, 2\} \times \{1, 2\} \subset \mathbf{Z}^3$. In practice, since we consider parameter and behavior learning together, a candidate solution of the optimization problem is:

$$\underline{X} = [\theta_{WG,1} \ \ldots \ \theta_{WG,11} \ \theta_{BC,1} \ \ldots \ \theta_{BC,4} \ \theta_{BK,1} \ \ldots \ \theta_{BK,9} \ \phi_1 \ \phi_2 \ \phi_3]^T \in \mathbf{R}^{24} \times S_\phi$$

The very large dimensions of the search space, and the system characteristics, suggest the use of the *layered learning* paradigm [9] for optimizing this problem. In fact, given a hierarchical task decomposition, layered learning allows for learning at each level of the hierarchy, with learning at each level directly affecting learning at the next higher level. The *incremental learning* approach that we use is inspired by the layered learning paradigm; however, in contrast with classic layered learning, we utilize the same learning method for each layer of the hierarchy. Specifically, we can decompose optimization of the attacker task in the following four optimization subtasks (layers):

- L_1: find $\underline{X}_{1,opt} = [\theta_{WG,1} \ \ldots \ \theta_{WG,11}]_{opt}^T \in \mathbf{R}^{11}$ for 'best' walk;
- L_2: find $\underline{X}_{2,opt} = [\theta_{BC,1} \ \ldots \ \theta_{BC,4}]_{opt}^T \in \mathbf{R}^4$ for 'best' ball carrying, given $\underline{X}_{1,opt}$ found by L_1;
- L_3: find $\underline{X}_{3,opt} = [\theta_{BK,1} \ \ldots \ \theta_{BK,9}]_{opt}^T \in \mathbf{R}^9$ for 'best' ball kicking, given $\underline{X}_{1,opt}$ and $\underline{X}_{2,opt}$ found by L_1 and by L_2;
- L_4: find $\underline{X}_{4,opt} = [\phi_1 \ \ldots \ \phi_3]_{opt}^T \in \mathbf{Z}^3$ for 'best' attacking strategy, given $\underline{X}_{1,opt}$, $\underline{X}_{2,opt}$, and $\underline{X}_{3,opt}$, found by the three previous layers.

We note k_i the dimension of $\underline{X}_i$ at each level L_i ($k_1 = 11$, $k_2 = 4$, $k_3 = 9$, $k_4 = 3$). The solution can be obtained as:

$$\underline{X}_{opt} = [\underline{X}_{1,opt}\ \underline{X}_{2,opt}\ \underline{X}_{3,opt}\ \underline{X}_{4,opt}]^T$$

The appropriate choice of the objective function for optimization is fundamental. The function for evaluating the quality of an *attacking performance*, must take into account: the quality (speed and precision) of the robot motion for approaching the ball, the quality of ball carrying, and the quality (power and precision) of the kick. Hence, we adopt the following objective function:

$$F(\underline{X}) = k_{WG} f_{WG}(\underline{X}) + k_{BC} f_{BC}(\underline{X}) + k_{BK} f_{BK}(\underline{X})$$

where k_{WG}, k_{BC}, k_{BK} are positive weights indicating the significance desired for each of the three aspects in the learning process, and f_{WG}, f_{BC}, f_{BK} indicate respectively the quality of the walking gait, of ball carrying, and of ball kicking. These functions are derived with heuristics on the robot and ball positions at various stages of the attacking task.

3.3   Learning Techniques

Our objective is to maximize $F(\underline{X})$ in a space of dimension k. This is not trivial, since $F(\underline{X})$ is 'black box', analytical computation of its derivatives is impossible, and all the parameters are box-constrained due to the physical characteristics of the system (we note : Δ_j the range size for each θ_j). Hence, conventional derivative-based optimization methods cannot be utilized, and convergence analysis of a method is impossible. The selected approach must handle non differentiable search space, have high convergence rate, and be resistent to noise in $F(\underline{X})$. Many algorithms possess these characteristics. In particular, we will focus on three different machine learning algorithms: Genetic Algorithms, Nelder-Mead Simplex Method, and Policy Gradient.

In *Genetic algorithms* [2], a *population* of q parameter sets (*individuals*) is used to find a solution for the optimization problem. To evolve a new population from the tested one, the q_e best individuals are preserved (*elitism*) and the remaining individuals are generated by applying *crossover* (q_c individuals) and *mutation* (q_m individuals) operators.

The *Nelder-Mead simplex algorithm* [7] explores a search space of dimension k by moving a *simplex* of $v = k + 1$ vertices via four possible geometrical transformations: reflection, expansion, contraction, and shrinkage. From an initial parameter set $^0\underline{X}$, the other $^l\underline{X}$ vertices ($l = 1, \ldots, k$) of the first simplex are generated as: $^l\underline{X} =^0 \underline{X} + [0, \ldots, \pm\zeta_l, \ldots, 0]^T$.

The *Policy Gradient algorithm* has been used for robot learning in [6]. From an initial parameter set $^0\underline{X}$, p randomly generated *policies* $^m\underline{X}$ ($m = 1, \ldots, p$), are evaluated, such that: $^m\underline{X} =^0 \underline{X} + [\rho_1, \ldots, \rho_k]^T$, and each ρ_j is chosen randomly in the set $\{+\epsilon_j, 0, -\epsilon_j\}$. Each $^m\underline{X}$ is grouped into one of three sets for each j: $S_{+\epsilon,j}$, $S_{0,j}$ or $S_{-\epsilon,j}$ depending on the variation applied to its jth parameter. $-\epsilon_j$. The average objective functions $\bar{F}_{+\epsilon,j}$, $\bar{F}_{0,j}$, and $\bar{F}_{-\epsilon,j}$ are computed for $S_{+\epsilon,j}$,

$S_{0,j}$ and $S_{-\epsilon,j}$. These are used to estimate the *gradient*, which is then scaled by a *step-size* η, and added to $^0\underline{X}$, to begin the next iteration.

4   Experimental Results

Here, we report the experimental results obtained by applying the proposed learning methodology in the described robot soccer scenario. More specifically, we comment on three results: 1) the effectiveness of the incremental approach; 2) the comparison among the three learning methods 3) the effectiveness of using a 3D simulator for speeding up the learning process. We have executed the incremental learning approach presented in the previous section, but without considering the fourth layer L_4, since this would require additional input to the system. In fact, selecting the best strategy depends on other variables, such as the position of the robot and of the ball with respect to the target goal, the position of other robots in the field, etc. For layer L_4 it is necessary to learn a function that maps the current situation with a suitable strategy. This aspect needs to be further investigated and it is beyond the scope of this paper. In practice, we applied incremental learning, focused on layers L_1 to L_3, for optimizing the strategy $P_1 = \{B_1; B_5\}$, i.e., the robot approaches the ball as fast as possible, and kicks it forward, without grasping it, with fixed head pan. Learning this task has been initially developed and configured within the 3D AIBO simulator [10] embedded in USARSim[2], before experimentation on the real robot.

Let us briefly outline the configurations of the three learning algorithms. For the genetic algorithm, we use: $q = 10$, $q_e = 1$, $q_c = 6$, $q_m = 3$. *Selection* of individuals from the original population is based on the popular roulette wheel scheme, and mutation is obtained by altering the jth parameter with an offset chosen randomly in the set $[-0.2\Delta_j, +0.2\Delta_j]$. For the Nelder-Mead algorithm, at each layer L_i: $v = k_i + 1$, and $\zeta_l = \pm 0.2 \, \Delta_j$, with the sign of ζ_l chosen randomly. For the policy gradient, we use for the three layers: $p_1 = 8$, $p_2 = 4$, $p_3 = 6$, $\epsilon_j = 0.1 \, \Delta_j$, $\eta = 3$. The same initial parameter set $^0\underline{X}$ is used for starting each learning technique. Since there is significant noise in each experiment, each set of parameters is evaluated three times, and the resulting fitness *evaluated* for that set, is computed by averaging over the three experiments. For each layer, and each learning technique, we terminate learning after $10 \, k_i$ *evaluations* (e.g. for L_1, 110 evaluations, i.e., 330 experiments). We choose to use the same amount of learning time, since this is usually a given specification in learning problems.

The results of incremental learning are shown in Fig. 2(a): Nelder-Mead slightly outperforms GA and PG. A similar plot is shown in Fig. 2(b), where we ran the same number of evaluations by learning all 24 parameters at the same time. Comparison between the two plots confirms the quality of the incremental approach: for instance, for the GA, the final fitness obtained by the incremental approach is 34% higher than that obtained by learning all parameters together.

Other experiments were carried out to show how the optimal low-level parameters are strongly related to the desired behavior. To emphasize this aspect, we

---

[2] `usarsim.sourceforge.net`

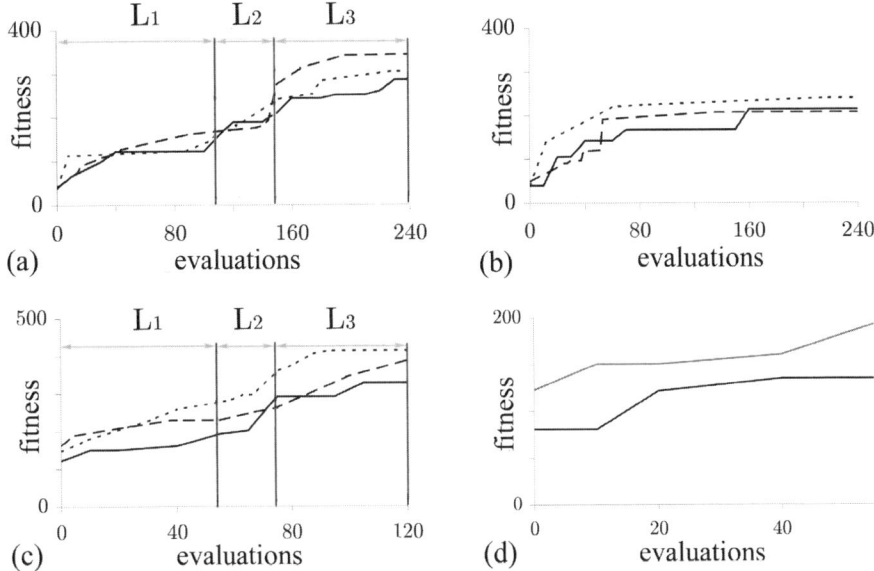

Fig. 2. Fitness values at each evaluation: (a) incremental learning with USARSim, (b) learning all parameters with USARSim, (c) incremental learning with AIBO (GA solid, NM dashed, PG dotted) (d) comparing GA for L_1: with initial population derived from USARSim (grey), and with random initial population (black)

used the same learning configuration used for optimizing strategy P_1, to optimize strategy $P_2 = \{B_1; B_3; B_5\}$ (the robot approaches the ball as fast as possible, rotates while grasping it, and kicks it with fixed head pan). This experiment was carried out with the genetic algorithm, in USARSim: starting from the optimal walking gait parameters $\underline{X}_{1,opt}$ learned for the previous strategy, we proceeded with the other two levels to derive $\underline{X}_{2,opt}$ and $\underline{X}_{3,opt}$. Comparison between the optima learned for P_1 and P_2 showed major differences. Specifically, for P_2, the robot must slow down farther away from the ball (parameters $\theta_{BC,1}$ and $\theta_{BC,2}$ are different in the two cases) and the head movement for kicking the ball (which in P_2 has been grasped) is different and depends on parameters $\theta_{BK,7}$ and $\theta_{BK,8}$. These results outline the dependency of the behavior parameter sets on the chosen strategy.

After having configured the algorithm for learning strategy P_1 in USARSim, we ported it on the real robot. The initial parameter set for each technique is the set derived at the end of simulator optimization. This time, for each layer, and each learning technique, we terminate learning after $5\,k_i$ evaluations. The results of the incremental learning algorithm are shown in Fig. 2(c). On AIBO, policy gradient slightly outperforms the two other approaches.

To emphasize how the use of a 3D simulator speeds up the learning process on real robots, we ran another experiment where the GA was used to learn L_1 for the same strategy P_1, starting from a random population, different from the one derived at the end of USARSim optimization. The results of this experiment are shown in Fig. 2(d), where the fitness of GA walking gait learning starting from different populations are plotted: the figure clearly shows the advantage of the simulator for deriving the GA initial population.

5   Conclusions

In this paper we presented a layered-like approach for learning optimal parameters, and strategy selection. We compared three different methods in the different layers: genetic algorithm, Nelder-Mead, and policy gradient. Moreover, we showed how the use of a 3D simulator speeds up robot learning. The proposed learning methodology has been applied to the soccer attacking task, on both simulated and real robots, and it has shown a notable improvement in the performance of the robot basic behaviours. The main results of the experiments are: the utility of the layered approach for this complex scenario, and the effectiveness of the 3D simulator for configuring the learning algorithms before porting on the robot. Incremental learning has been executed without considering the strategy selection layer. In fact, this depends on the 'game situation' (e.g., positions of robot and ball with respect to the goal). Learning a function that maps the game situation with a suitable strategy will be the object of further work.

References

1. Arsenio, A.M.: Developmental learning on a humanoid robot. In: IEEE International Joint Conference on Neural Networks, Budapest (2004)
2. Back, T.: Optimization by means of genetic algorithms. In: 36th International Scientific Colloquium, pp. 163–169 (1991)
3. Chalup, S.K., Murch, C.L., Quinlan, M.J.: Machine learning with AIBO robots in the Four-Legged League of Robocup. IEEE Trans. on Systems, Man and Cybernetics, Part C (2006)
4. Millan, J.d.R.: Rapid, safe, and incremental learning of navigation strategies. IEEE Trans. on Systems, Man and Cybernetics, Part B 26(3), 408–420 (1996)
5. Fidelman, P., Stone, P.: The chin pinch: A case study in skill learning on a legged robot. In: Proc. of 10th International Robocup Symposium (2006)
6. Kohl, N., Stone, P.: Machine learning for fast quadrupedal locomotion. In: The 19th National Conference on Artificial Intelligence (2004)
7. Lagarias, J.C., Reeds, J.A., Wright, M.H., Wright, P.E.: Convergence properties of the Nelder Mead simplex algorithm in low dimensions. SIAM Journal on Optimization 9, 112–147 (1998)
8. Martinson, E., Arkin, R.C.: Learning to role-switch in multi-robot systems. In: Proceedings of International Conference on Robotics and Automation (2003)
9. Stone, P., Veloso, M.: Layered learning. In: Lopez de Mantaras, R., Plaza, E. (eds.) ECML 2000. LNCS (LNAI), vol. 1810, pp. 369–381. Springer, Heidelberg (2000)
10. Zaratti, M., Fratarcangeli, M., Iocchi, L.: A 3D simulator of multiple legged robots based on USARSim. In: Proc. of 10th International Robocup Symposium (2006)

Self-localization Using Odometry and Horizontal Bearings to Landmarks

Matthias Jüngel and Max Risler

[1] Humboldt-Universität zu Berlin, Künstliche Intelligenz
Unter den Linden 6, 10099 Berlin, Germany
`matthias@matthias-juengel.de`
[2] Technische Universität Darmstadt, Simulation and Systems Optimization Group
Hochschulstr. 10, 64289 Darmstadt, Germany
`risler@sim.informatik.tu-darmstadt.de`

Abstract. On the way to the big goal - the game against the human world champion on a real soccer field - the configuration of the soccer fields in RoboCup has changed during the last years. There are two main modification trends: The fields get larger and the number of artificial landmarks around the fields decreases. The result is that a lot of the methods for self-localization developed during the last years do not work in the new scenarios without modifications. This holds especially for robots with a limited range of view as the probability for a robot to detect a landmark inside its viewing angle is significantly lower than on the old fields. On the other hand the robots have more space to play and do not collide as often as on the small fields. Thus the robots have a better idea of the courses they cover (odometry has higher reliability). This paper shows a method for self-localization that is based on bearings to horizontal landmarks and the knowledge about the robots movement between the observation of the features.

Keywords: Self-Localization, Constraints, Aibo, Bearing-Only.

1 Introduction

Localization is one of the most important challenges for a mobile robot. There are a lot of researchers developing new methods each year. In the last years the Monte-Carlo Localization has been the standard approach to the localization problem. A lot of improvements have been suggested to overcome limitations in the processing power and to address the limited angle of view of robot that are not equipped with omni-vision.

There are a lot of suggested improvements to the sensor model. Sensor-resetting reseeds new position templates obtained from observations [1] and there are improvements that build short-time history of observations to create more accurate position templates [2]. Other approaches try to incorporate negative information [3]. A lot of improvements has also been suggested for the motion model for example using the detection of collisions.

U. Visser et al. (Eds.): RoboCup 2007, LNAI 5001, pp. 393–400, 2008.
© Springer-Verlag Berlin Heidelberg 2008

This work was motivated by the experiences we collected with the localization method we use in RoboCup for our Aibo robots. We use a standard Monte-Carlo localization as described in [4,5,6]. As with the latest rule changes in the RoboCup Sony Four-Legged League besides the goals there are only two artificial landmarks on the field, the distance-based sensor resetting method does not give the desired results any longer. Size-based distance measurements give a too large error when the objects are too far away.

In this paper we provide a bearing-only method for localization that incorporates odometry and can be used as a template generator for MonteCarlo-Localization. This paper shows an approach to bearing-only self-localization that incorporates odometry in a new way. Section 2 describes the method in detail. Section 3 describes the experiments we performed with our Aibo robots.

2   Bearing-Only Localization Using Odometry

In this section we show a method that allows a robot to localize based on two inputs. The first input are observations. The vector $\boldsymbol{\alpha} = (\alpha_{l_1}, \alpha_{l_2}, ..., \alpha_{l_n})$ contains the measured bearings to the landmarks $l_1, l_2, ..., l_n$. These angles were measures at different times $t_1, t_2, ..., t_n$. The second input is the knowledge about the motion of the robot. The vector $\boldsymbol{u} = (u_1, u_2, ..., u_n)$ contains the robot's odometry at times $t_1, t_2, ..., t_n$.

A robot can obtain these vectors $\boldsymbol{\alpha}$ and $\boldsymbol{u}$ by storing its observations and the according odometry in a buffer. Figure 1 shows a visualization of such a buffer.

In this section we define a function $F(x, y, \boldsymbol{\alpha}, \boldsymbol{u})$ which describes the likelihood for the robot of being at position (x,y) on the field. This function can be used

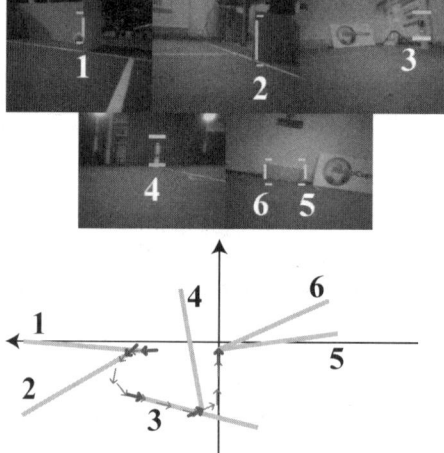

Fig. 1. Odometry and horizontal bearings. Top: Five images with six horizontal bearings (1: right goal post, 2: left goal post, 3 and 4: center landmarks, 5 and 6 goal posts) Bottom: Gray arrows show the robots odometry at different times, bold arrows show the odometry associated with the horizontal bearings.

to calculate a robot position (the maximum of the function) or to generate templates for Monte-Carlo localization.

2.1 Localization with Three Simultaneously Seen Horizontal Bearings

In this subsection we show two methods to determine the position of the robot when the robot is not moving. The first one uses well-known simple geometry, the second one is a constraint-based approach.

Using simple geometry. When a robot perceives three landmarks without moving between the observations, the calculation of the position is straightforward. With the known position of the landmarks a circle can be constructed for each pair of bearings. The radius of the circle is determined by the difference of the angles and the distance between the landmarks. The intersection point of the circles is the only possible position for the robot.

Pose estimation using angular constraints. When the position is determined by intersecting circles, there is nothing known about the influence of errors in the measurement of the bearings. This influence can be determined using a constraint-based approach. A single observation of a landmark l at a certain relative angle constrains the angle ϑ_l the robot can have at a certain position (x, y) on the field. This angle is given by

$$\vartheta_l(x, y, \alpha_l, x_l, y_l) = \arctan\left(\frac{y_l - y}{x_l - x}\right) - \alpha_l$$

where (x_l, y_l) is the position of the landmark on the field and α_l is the relative angle to the landmark. When two bearings to two landmarks are given, the function

$$D_{l_1, l_2}(x, y) = (\vartheta_{l_1}(x, y) - \vartheta_{l_2}(x, y))^2$$

describes the likelihood for being at position (x, y). The shape of the function represents how good a certain pair of landmarks is suited to constrain the position on the field. For example a plateau in this function means that a small error in an observation leads to a large error in the resulting position.

The function $D_{l_1, l_2}(x, y)$ introduced above describes for each position (x, y) how good the angles ϑ_{l_1} and ϑ_{l_2} obtained from two different horizontal bearings match. To use more than two observations $\alpha_{l_1}, \alpha_{l_2}, ..., \alpha_{l_n}$, we can calculate the average angle of all resulting $\vartheta_{l_1}, \vartheta_{l_2}, ..., \vartheta_{l_n}$ for each position (x, y) using this formula

$$\vartheta_{average}(x, y) = \arctan\left(\frac{\sum_{i=1}^{n} \sin(\vartheta_{l_i}(x, y))}{\sum_{i=1}^{n} \cos(\vartheta_{l_i}(x, y))}\right)$$

a)   b)

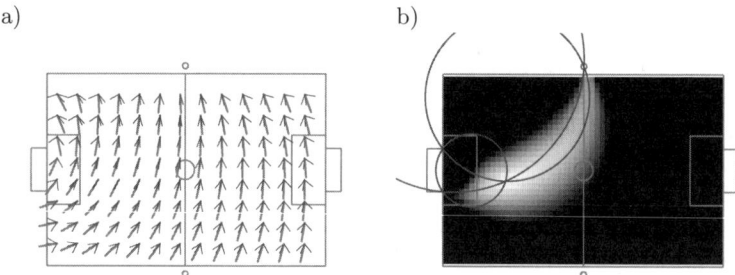

Fig. 2. Similarity of angles. a) The thin lines show the angles $\vartheta_l(x, y)$ for three different observations. The bold line shows the average angle. The robot's position is constrained to the positions where the angles are similar. b) Function $G(x, y)$ displayed as height map. White: small difference between the angles, black: large difference between the angles. The red circles are obtained from the method using simple geometry described above.

Figure 2 a) shows function $\vartheta_l(x, y)$ for three different landmarks and the resulting average angle. Using $\vartheta_{average}(x, y)$ we can define the function

$$G(x, y) = \sum_{i=1}^{n} (\vartheta_{average}(x, y) - \vartheta_{l_i}(x, y))^2$$

which describes how similar the angles ϑ_l are. This function has its maximum at the position (x,y) that best fits with all observations $\alpha_{l_1}, \alpha_{l_2}, ..., \alpha_{l_n}$. Furthermore the function provides an estimation of the position error for known errors in the observation. Figure 2 b) shows this function for three observations.

2.2   Incorporating Odometry

To incorporate odometry we define a function $v_l(x, y, \alpha_l, \Delta_{odometry_l}, x_l, y_l)$ which determines the angle of the robot at position (x, y) when the landmark l was seen at angle α_l and the robot moved $\Delta_{odometry}(\Delta_x, \Delta_y, \Delta_\phi)$ since the observation. Figure 3a) illustrates these parameters and the resulting angle v_l. To determine v_l we define a triangle with its corners at the position (x_l, y_l) of the landmark l (angle β), at the position (x, y) (angle γ) and at the position (x_0, y_0) where the observation was taken (angle δ). Figure 3b) shows this triangle. Note that in this triangle (x_l, y_l) and (x, y) are fixed. The position of (x_0, y_0) can be calculated using the angle ω from (x, y) to (x_l, y_l) and the distance Δ_d the robot walked:

$$x_0 = x + \cos(\omega + \gamma) \cdot \Delta_d; y_0 = y + \sin(\omega + \gamma) \cdot \Delta_d$$

where γ follows using sine rule:

$$\gamma = \pi - \delta - \beta$$
$$= \pi - \alpha_l - \arctan\left(\frac{\Delta_y}{\Delta_x}\right) - \arcsin\left(\frac{\Delta_d \cdot \sin(\delta)}{d_l}\right).$$

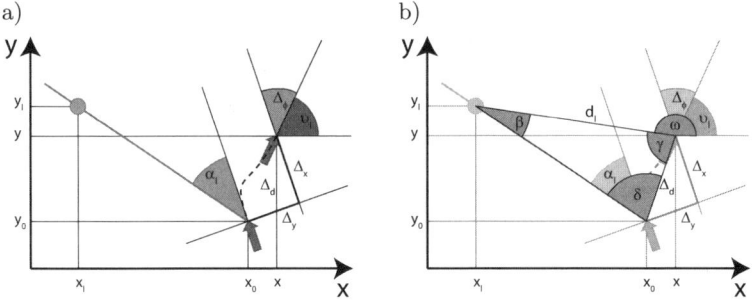

Fig. 3. Bearing + odometry define the robots angle for a given position (x_0, y_0)

With the known position (x_0, y_0) follows

$$v_l(x, y, \alpha_l, \Delta_{odometry_l}, x_l, y_l) = \vartheta(x_0, y_0) + \Delta_\phi.$$

When the robot is at position (x, y), has seen the landmark l at angle α_l some time ago, and has moved by $\Delta_{odometry_l}$ since that observation, the function v_l gives the angle the robot must have. Similar to the function G from section 2.1 we define a function

$$F(x, y, \boldsymbol{\alpha}, \boldsymbol{u}) = \sum_{i=1}^{n} \left(v_{average}(x, y) - v_{l_i}(x, y)\right)^2$$

which describes the likelihood of the robot for being at position (x, y). This function can incorporate an arbitrary number of observations from the past and does not need any internal representation of the position that is updated by alternating sensor and motion updates. The selected sensor information $\boldsymbol{\alpha}$ and the according motion information $\boldsymbol{u}$ are processed at once.

2.3   Calculating the Robot Pose

The maximum of function F given in the last section is the position of the robot. The rotation of the robot can immediately be calculated using $v_{average}$ or the angle v_{l_0} that is defined by the last observation. When a fast and rough estimation of the robot pose is wanted, the maximum can be determined by an iteration through the domain of the function. When a more accurate estimation is wanted it can be obtained by means of standard methods as Gradient Descent with only a few iterations. Note that such methods usually find only local maxima of the function.

2.4   Generating Templates for Monte-Carlo Localization

Often there is more information than the horizontal bearings to unique landmarks to determine the pose of the robot. Especially when there is ambiguous

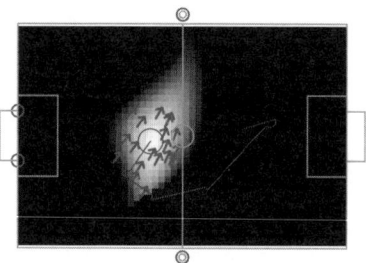

Fig. 4. The Function $F(x, y)$: white - high likelihood, black - low likelihood, Arrows: position templates that can be used for sensor resetting in Monte-Carlo localization - note that usually only a small number of these templates will be used. Small circles: the landmarks that were used for position calculation. Large circle: the robot pose (known from the simulation). Path: the way the robot walked.

information like distances to walls or field lines a localization method that is able to track multiple hypotheses might be preferred. In such a case the function F described in section 2.2 can be used to create template poses for sensor resetting. Which is in particular useful when only a small number of particles can be used due to computational limitations. To obtain a fixed number of samples you can normalize F such that all values are between 0 and 1 using function $F' := 1/(1 + F^2)$ and create a template pose at each position (x, y) with $random() < F'(x, y)^n$. Where n is a parameter to adjust how much the sample poses can deviate from the maximum. Figure 4 shows templates obtained from function F.

3   Experimental Results

We developed the bearing-only localization approach as a replacement of the distance based sample template generation that we use for our Monte-Carlo self localization [7,8,9]. The old method was not usable any longer as with the 2007 rule change in the Sony Four Legged league two more beacons were removed and thus there are less beacons and the beacons have a higher average distance to the robots.

Thus we added the method described in section 2.2 as a sample template generator in a way described in section 2.4. The particle filter uses 200 particles.

To measure the quality of our improvements we steered a real robot via remote control over the soccer field in our lab performing an s-like shape on the field. The head of the robot performed the typical Aibo scan motion which looks around searching for the ball and the landmarks. During this process log data was recorded containing camera images, head joint values, odometry data, and ground truth robot positions obtained by a ceiling mounted camera. Such log-files can be played back off-line to feed our algorithms with data. The angles to the landmarks needed for our location approach were extracted from images and

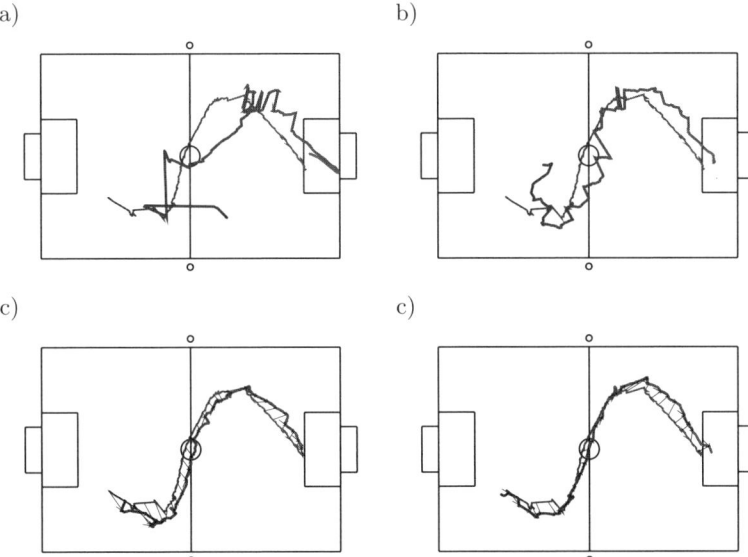

Fig. 5. blue line: ground truth robot position, red line: result of self localization a,b) no sample templates used. c,d) sample templates used.

joint sensor data. We used the recorded log data to compare different parameterizations of the approach.

Figure 5 shows a visualization of the path the robot walked and the pathes obtained by our method. We also tested the influence the number of samples used for reseeding has. Table 1 gives the results.

The result of the experiments is that without template generation there were random jumps and a large deviation from the ground truth robot pose. With sample template generation (using one sample per frame) the resulting trajectories were smoother and closer to the ground truth.

Table 1. Results of Localization tests. In our experiment the position obtained by the approach introduced in this paper was compared with the one obtained by the ceiling camera. The table shows the average distance between the two positions for the whole run repeated six times. To show how reseeding influences the localization quality we conducted the experiment with different re-sampling rates (top row). The table shows that even a single reseeded particle in each frame improves self-localization drastically. Adding more samples has almost no effect.

num. of reseeded samples	0	1	2	5	10
position error in cm	54.8±21.6	21.7±13.1	19.1±13.0	19.5±12.6	22.4±17.0
position error percentage	9,13%	3,63%	3,18%	3,25%	3,74%

4   Conclusion

In this paper we presented an approach for bearing-only self-localization incorporating odometry. The method does not need an internal representation of the position estimate which is updated by alternating sensor and motion updates. The history of observation and motion information (stored in a small buffer) is processed directly. A big advantage is that no wrong model from the past can disturb the current pose estimation.

However, we showed that our method also provides good positions for sensor resetting in the well known Monte-Carlo localization. Tests in simulation and on a real Aibo robot with ground truth by a ceiling camera showed the robustness of our approach. Further experiments have to show whether the localization method can cope with larger errors in odometry caused by strong influence of opponents in RoboCup games.

References

1. Lenser, S., Veloso, M.M.: Sensor resetting localization for poorly modelled mobile robots. In: Proceedings of the 2000 IEEE International Conference on Robotics and Automation (ICRA 2000), pp. 1225–1232. IEEE (2000)
2. Sridharan, M., Kuhlmann, G., Stone, P.: Practical vision-based monte carlo localization on a legged robot. In: IEEE International Conference on Robotics and Automation (April 2005)
3. Hoffmann, J., Spranger, M., Goehring, D., Juengel, M.: Making use of what you don't see: Negative information in markov localization. In: Proceedings of the 2005 IEEE/RSJ International Conference of Intelligent Robots and Systems (IROS), IEEE (2006)
4. Fox, D., Burgard, W., Dellaert, F., Thrun, S.: Monte Carlo localization: Efficient position estimation for mobile robots. In: Proc. of the National Conference on Artificial Intelligence (1999)
5. Dellaert, F., Burgard, W., Fox, D., Thrun, S.: Using the condensation algorithm for robust, vision-based mobile robot localization. In: Proc. of the IEEE Computer Society Conference on Computer Vision and Pattern Recognition (CVPR) (1999)
6. Wolf, J., Burgard, W., Burkhardt, H.: Robust vision-based localization for mobile robots using an image retrieval system based on invariant features. In: Proc. of the IEEE International Conference on Robotics and Automation (ICRA) (2002)
7. Roefer, T., Brunn, R., Dahm, I., Hebbel, M., Hoffmann, J., Juengel, M., Laue, T., Loetzsch, M., Nistico, W., Spranger, M.: GermanTeam 2004: The German national RoboCup team. In: Nardi, D., Riedmiller, M., Sammut, C., Santos-Victor, J. (eds.) RoboCup 2004. LNCS (LNAI), vol. 3276. Springer, Heidelberg (2005)
8. Roefer, T., Juengel, M.: Fast and robust edge-based localization in the sony four-legged robot league. In: Polani, D., Browning, B., Bonarini, A., Yoshida, K. (eds.) RoboCup 2003. LNCS (LNAI), vol. 3020, pp. 262–273. Springer, Heidelberg (2004)
9. Roefer, T., Juengel, M.: Vision-based fast and reactive monte-carlo localization. In: Polani, D., Bonarini, A., Browning, B., Yoshida, K. (eds.) Proceedings of the 2003 IEEE International Conference on Robotics and Automation (ICRA), pp. 856–861. IEEE (2003)

Incremental Generation of Abductive Explanations for Tactical Behavior

Thomas Wagner, Tjorben Bogon, and Carsten Elfers

Center for Computing Technologies (TZI), Universität Bremen, D-28359 Bremen
{twagner,tbogon,celfers}@tzi.de

Abstract. According to the expert literature on (human) soccer, e.g., the tactical behavior of a soccer team should differ significantly with respect to the tactics and strategy of the opponent team. In the offensive phase the attacking team is usually able to *actively select* an appropriate tactic with limited regard to the opponent strategy. In contrast, in the defensive phase the more passive *recognition* of tactical patterns of the behavior of the opponent team is crucial for success. In this paper we present a qualitative, formal, abductive approach, based on a uniform representation of soccer tactics that allows to recognize/explain the tactical and strategical behavior of opponent teams based on past (usually incomplete) observations.

1 Introduction

The quality and success in human soccer is determined to a large extent by the individual abilities (high-level as well as low-level) of the acting agents and by the strategical and tactical abilities on the individual- and team level [9]. Tactics and strategies can be used in at least two ways. In the offensive phase a team/coach can actively choose an appropriate tactic while in the defensive phase a team has to recognize the opponents tactical behavior in order to adopt defensive play. A first attempt to recognize opponent behavior has been made in [4] which allowed us to detect online rule-based pattern of behavior. In order to achieve the demanding goal to defeat the human world champion in 2050 we claim that the recognition of (simple) rules will not be sufficient. Instead the recognition of more complex pattern in terms of strategies and tactics is required.

According to the expert literature on (human) soccer the tactical behavior of a soccer team should differ significantly with respect to the tactics and strategy of the opponent team [9]. In the offensive phase the attacking team is usually able to *actively select* an appropriate tactic with limited regard to the opponent strategy. In contrast, in the defensive phase the more passive *recognition* of tactical patterns of behavior of the opponent team is crucial for success. E.g., depending on wether one team is playing a *4-4-2-* and the other team is playing a *3-4-1-2*-strategy the defensive behavior should differ significantly with respect to the opponents strategy. In the case that the attacking opponent is playing a *3-4-1-2*-strategy while the defending team relies on a *4-4-2*-strategy the defending team has to face e.g., *overlapping movements*, which are typical

U. Visser et al. (Eds.): RoboCup 2007, LNAI 5001, pp. 401–408, 2008.

for the offensive phase of the *3-4-1-2*-strategy. Without adopting the *4-4-2*-strategy the attacking team has a very good chance e.g., to reach numeric superiority on the wing sides [9].

In this paper we present an abductive approach based on the basic algorithms of [5]. Although the original algorithm has been proved to be efficient [5] in contrast to other abductive approaches, efficiency is still a problem for many practical real world problems. Although efficiency is essential plan-recognition method should furthermore support configurable, context-dependent (e.g. role dependency) and incremental recognition.

2   Motivation

In order to apply strategy and tactics from the theory of soccer [9] successfully in the RoboCup-domain we do not only have to consider the offensive play. Strategy and tactic is also highly relevant in the defensive phase. The role of strategy changes significantly depends on whether a team is in the offensive or in the defensive phase. In the offensive phase a team can actively *choose* a tactic to a large extent independently of the opponents strategy[1], in contrast in the defense phase a team has to *recognize* both the more general strategy as well as the concrete tactic in order to adopt the overall defensive team behavior (e.g., avoiding a superior number of opponent players on wing (i.e., in *4-4-2*)).

Depending on the specific context we are interested in different types of explanations for observed behavior. E.g., the *coach* is interested in recognizing more general pattern of behavior in order to adopt the strategy and choose more appropriate tactics. In contrast, *players* have a significantly stronger interest in more precise explanation which allow them (at least) to some extent to predict the opponent behavior. Both types of observations differ not only with respect to the level of granularity. While the coach usually relies on a large set of observations on a larger time scale, a player is usually restricted to a limited set of observations (which belong to the current opponent attack) and is required to recognize the opponent tactics fast. As a consequence not only a single inference is needed but instead a framework of inferences which allow for context sensitive explanations.

An additional important aspect of plan recognition is the underlying role-dependency. Depending on the specific role (i.e., *left side defender*, *central attacker*, ...) and the context (e.g., *defensive phase*, *final phase of the game*, ...) the focus and the level of granularity of the plan recognition process can differ significantly. E.g., usually a *left side defender* is not interested in an exakt prediction (which is usually impossible due to the non-deterministic nature of the game) but instead in the set of possible future opponent behaviors at varying levels of granularity, depending his specific capabilities (e.g., whether it is reasonable to play offside).

---

[1] Of course, the choice of a specific tactic should be done with respect to the specific strength and weaknesses of the opponent teams.

3   Related Work

The generation of explanations of observed events/behavior has gained much interest within the research community and resulted in various applications like diagnosis [11], natural language understanding [6] and plan recognition [2]. The methods under consideration vary from probabilistic methods like bayes networks e.g., [1], classification-based approaches e.g., [3] to the already mentioned logic-based abductive methods e.g., [2]. One aspect most of these approaches have in common are the constraints of their application: most approaches focus on static scenarios with an precise model of behavior (e.g., *closed world assumption*). A popular exception is the work of *Albrecht et al.* [1] which applys probabilistic reasoning to an online-dungeon game but also grounded on a complete (predefined and also limited) number of actions and valid combinations. An additional approach that overcomes these limitations at least to some extent is the work of Intille and Bobick [7] who apply their classification-based approach to the multi-agent scenario of *American Football*-domain. Although the *American Football*-domain appears to be highly related to the *RoboCup*-scenario the differences (within the Intille-approach) are significant. The *American football*-domain provides a complete, predefined taxonomic playbook which specifies all possible (allowed) patterns of behavior and therefore allows for classification-based approaches. Additionally, *Bobic et al.* are able to use manually generated data without noise. In contrast, in the soccer domain tactics and strategies describe behavior on much higher level that allows a wide range of variations that cannot be specified in any detail. Instead, the given observations have to be assumed to be uncomplete due to sensor limitation.

Abduction has been introduced by C.S. Pierce [12] as a third kind of logic inference next to induction and deduction and has gained much interest in the late 80'th and early 90'th. Abduction does not rely on complete observations but instead supports to infer missing knowledge i.e., premises. The abductive inference process can generally be decomposed in two steps:

1. generation of all abductive explanation
2. selection of the most appropriate explanation

Several proposals have been made for the time consuming generation process depending on the underlying representation[2]. A serious problem for the use of abduction especially in time critical applications like RoboCup is that the generation of abductive explanations has been proofed to be NP-hard (in the general case) [13]. Nevertheless, more recently *Eiter et al.* [5] developed an efficient algorithm that allows to generate all explanations of positive queries based on a logic horn-clause representation.

4   Abduction-Based Generation of Explanations of Tactical Behavior

Observations are essential elements for plan recognition. In this approach the necessary information are recognized actions from a team of soccer players. We use an egocentric, online version of the approach presented by [10].

---

[2] Abduction does not necessarily have to rely on an logic representation.

4.1   Generation of Tactical Knowledgebase

Our knowledge base is based on the book of soccer tactics from Lucchesi [9]. The key assumption which is essential for the use of abductive reasoning is that the (logical) implication can be interpreted not only as an inference from *cause to effect*, but also as an inference from *effect to cause*[3]. Following this pattern each single action within a complex tactical pattern can be interpreted as the cause for a possible sequence of successional actions and may itself be the effect of a previous action and therefore sequences

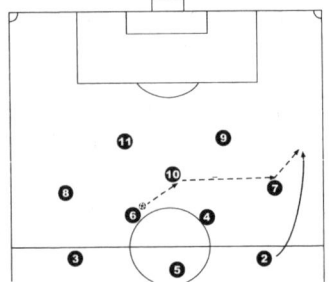

Fig. 1. A left-right sidechange of a human soccer tactic

of actions are modeled strictly as sequences of implications. The situation depicted in figure 1 is in the first step described as a sequence of implications (see figure 2) which is in the second step transfered into a *horn* representation. Two limitations had to be considered (1) no cyclic horn theories and (2) no expressions like $x \to 0$ or $1 \to x$ are allowed[4]. The resulting horn-clause knowledge base is described in figure 2.

(1) PassPlayer6ToPlayer10 $\land$ MovePlayer2ToRightOppMid $\to$ RecievePassPlayer10
(2) RecievePassPlayer10 $\to$ PassPlayer10ToPlayer7
(3) PassPlayer10ToPlayer7 $\to$ RecievePassPlayer7
(4) RecievePassPlayer7 $\to$ PassPlayer7ToPlayer2
(5) PassPlayer7ToPlayer2 $\to$ RecievePassPlayer2
(6) HaveBallPlayer6 $\land$ IsFreePlayer10 $\to$ PassPlayer6ToPlayer10
(7) IsFreeRightOppMid $\to$ MovePlayer2ToRightOppMid
(8) HaveBallPlayer10 $\land$ IsFreePlayer7 $\to$ PassPlayer10ToPlayer7
(9) HaveBallPlayer7 $\land$ IsFreePlayer2 $\to$ PassPlayer7ToPlayer2

Fig. 2. Extracted Implication from the Tactic

4.2   The Basic Algorithm of Eiter and Makino

The basic algorithm can be decomposed into two main steps: (1) calculation of *prime implicants* and (2) calculation of abductive explanations[5] The first step results in a knowledge base of *prime implicants*.

In our example all clauses are also *prime implicants*. The following generations of abductive explanations will be done on this representation. The general idea is quite simple: in the first step the positive request clause σ is used to look up in the *consequences* of the set of all *prime implicants* (in the following *pi*). If σ is found in some

---

[3] It should be mentioned that reasoning from *effect* to *cause* is *non-monotonic*.
[4] The first condition is a prerequisite of the algorithm of [5], the latter helps to ensures a consistent knowledge base.
[5] For more details please refer to [5].

prime clause ρ the corresponding *antecedents* of ρ as annotated as the first solution. Based on the first (simple) solution the algorithm tries systematically to find a more fundamental explanation by trying to find a new resolvent between ρ and some different clause in the set of *pi*. Given a new resolvent ρ is found *pi* is expanded by ρ and the same procedure is applied to *pi'* until no changes occur and therefore all solutions have been calculated.

4.3   Optimizing the Algorithm of Eiter and Makino

Although the described algorithm has not only been proved to be complete and correct it also is the only abductive algorithm known to be efficient (non NP-hard). In contrast to various approaches to the generation of explanations the abductive algorithm is very robust with respect to redundant actions which is a serious problem in soccer in general[6]. Furthermore, an abductive algorithm can easily be applied to different knowledge bases at different levels of granularity. Nevertheless, efficiency is still a very serious problem for the application in a *RoboCup*-domain[7].

In order to improve efficiency various optimizations have been applied. It should be noted that the following optimizations may not be reasonable in all domains. I.e., although the optimized algorithm is (of course) domain-independent the optimizations can only be applied with certain restrictions. Therefore they are optional with respect to different domains[8].

The complexity of the algorithm is given by $O(e * m * n * \|\varphi\|)$, whereas e denotes the number of solutions, m the number of clauses in φ and n the number of literals. The first significant improvement can be achieved by separating the complete knowledge base into different separat knowledge bases. As a matter of consequence, we get a special knowledge base for *counter attack on the right wing*, *counter attack on in the center*, ... This modularisation has an interesting advantage: as assumed in the motivation (see section2) a player is usually only interested in explanations that leads to an improved/adopted behavior: E.g., a *right wing defender* is specially interested in *counter attack on the right wing* and not on the left one. The modularisation allows him to focus on context sensitive, role specific tactic explanations.

Additionally, four different optimizations have been applied:

1. Pre-calculation of prime implicants: Independently of the specific request the basic algorithm calculates all prime implicants. This process is done before runtime in our realization and will be uploaded together with the knowledge-base. The level of improvement is strongly dependent on the complexity of the causal relations and will lead at least in complex models to significant improvements.

2. Use of additional observations: In most of the cases a player has made different observations that account to a specific tactic. These observations can be used to improve the abductive reasoning process by skipping proofs. Since a player has already observed

---

[6] Even if a team is strictly using declarative tactical patterns of behavior, redundant actions result due to necessary adaptations as a result of unexpected opponent behavior.

[7] On a more complex knowledge base the basic algorithm took 21 sec.! As we will see in section 5, with all optimization and with some restrictions the performance can be improved to 22ms.

[8] And they are also optional in our implementation.

an action it is not necessary to find out under which conditions the observation is/was true.

3. Skipping satisfiability-test: The described handling of observations has an additional interesting side effect: In the classic abductive approach new observations are expanded in the knowledge base. In the case of false observations this *may* lead to inconsistency. In order to avoid to an inconsistent knowledge base a satisfiability-test would be needed. But since observation are never expanded in the knowledge base the satisfiability-test can be skipped in our case.

4. Pre-calculation of possible solutions: A static knowledge base offers additional advantages: it allows the pre-calculation of all possible solutions! A possible disadvantage can be the increase in memory usage which is minimal in this domain due to the modularisation of the knowledge base.

In addition to the improvements in efficiency the basic algorithm had to be adopted in order to increase robustness. Although the tactical patterns described in Lucchesi [9] are strictly associated with specific tactical roles these assumption does rarely hold in the *RoboCup*-domain. Therefore we provided the abductive algorithm with a flexible role association method. The use of this method allows to detect tactical behavior independently from specific player numbers or roles and increases the robustness significantly. The obvious drawback is a decrease in efficiency since all player-number configurations have to be considered in the role assignment. The results of our extentions are described in the following section.

5   Experiments and Preliminary Results

Before we tried to integrate the modified algorithm in our 3D-team we evaluated the efficiency in different scenarios. In the first test scenario described in table 1 we wanted to evaluate the effect of pre-calculated solutions under (1) the varying condition of request complexity: simple vs. complex and (2) under varying goal: whether the agent wants to know if a plan is possible at all or whether he wants to get a list of all possible (opponent-) plans. The request complexity is simply changed by the action selection. In the case we observe an action that appends very late in a possible plan, the algorithm will find significant more solutions than in the inverse case. In the following tests we used 26 different plans in each case.

The table 1 presents some interesting results. First, the calculation whether a single plan is possible is highly efficient and can be done in 4,7 ms to 10,5 ms with respect to

Table 1. Finds all solution and search all plans with a possible solutions

Test 1	with Pre-Calculation		without Pre-Calculation	
	simp. Query	komp. Query	simpl. Query	komp. Query
all Time	62,8 ms	55,7 ms	69,3 ms	74,2 ms
⌀ Time for one Plan	4,7 ms	4,5 ms	6,3 ms	10,5 ms
processed Plans	4 P.	3 P.	4 P.	3 P.
⌀ compute Possible Plans	44,2 ms	42 ms	43,8 ms	42,2 ms

Table 2. Used cycles in 2D-league, watch a game with a *trainer-agent* in 3d-league

| | iterativ Test | | | | | | | | | |
| | all Queries | | | | | Queries with Solutions | | | | |
	min	max	∅	∅ over	Tests	min	max	∅	∅ over	Tests
P28	2	7	4,66	0,6	22,6	2	7	4,66	0,6	22,6
P53	0	4	1,06	6	182,2	1	4	2,62	6	55,8
P99	0	7	2,24	1,2	49,2	1	7	3,96	1	24,8
P130	0	6	2,6	0,6	53,4	2	6	4,6	0,2	16,6

the specific conditions. Interestingly, the pre-calculation of results is significantly more efficient but interestingly the difference is surprisingly small. Two main reasons can be found: (1) the complexity of our tactical model is quite low (in terms the capability of the algorithm). The efficiency decreases in the case of no pre-calculations only for complex requests. (2) The modularisation of the knowledge base appears to be highly efficient. In the case that all possible plans should be calculated (which represents the case of non-modularisation) the run-time requirements are significant higher, but is still efficient enough to be used e.g., in the *3D-simulation league*, as it can be seen in table 2.

In table 2 we used a 3D-trainer agent who has been restricted to use at maximum 80ms in order to simulate cycles. In the test the coach was required to detect whether a single plan is possible and to find all possible solutions. These conditions have been tested on four different varying plans (P28, P53, P99, P130). The test mainly showed two important results: (1) The modularisation of the knowledge base provides a basis for incremental abductive reasoning. The algorithm in our implementation can interrupt the calculation process at (relative) fixed time steps in order to allow other tasks within a single cycle (instead using complete cycles). (2) Depending on the specific condition the algorithm requires at most between 4 to 7 cycles for all solutions. These results can also be approved under different conditions e.g., used by a 3D- or a 2D-player.

6   Summary and Discussion

The role of strategy and tactic is becoming more and more important especially in the simulation- and the small-size league. The use of more complex tactics of an offensive team will require that defensive teams are at least to some extent able to detect the set of possible opponent tactical patterns in order to coordinate defensive behavior. In this paper we presented an approach to symbolic plan recognition (more precisely genera-tion of explanation for opponent behavior) at all relevant stages: from the generation of qualitative action- and world descriptions based on [8] to the generation of abductive explanations of these observed behavior. The algorithm of Eiter and Makino [5] has been adopted to the specific requirements of the *RoboCup*-domain. We showed that the adopted algorithm can efficiently be used for explanation generation. Furthermore, the algorithm can be used in an incremental fashion (which is especially useful for the sim-ulation leagues). An additional positive characteristic is the robustness with respect to redundant/false observations/actions. The modularisation of the knowledge base allows for role- and context sensitive requests but does not prohibit the generation of complete solutions, i.e., without respect to role and context e.g., for the trainer.

Besides the application of the modified algorithm in different domains the hypotheses generation is still an open task. Although we claim that it will be sufficient in many situations to identity possible tactical behavior (with respect to role and context) it is clear that there also exits situations where we would like to find a single prediction, i.e., the selection of a single hypothesis out of the set of possible explanations. Various solutions may be considered, varying from probabilistic to symbolic approaches proposed in the abduction community [11].

References

1. Albrecht, D.W., Zukerman, I., Nicholson, A.E.: Bayesian models for keyhole plan recognition in an adventure game. User Modeling and User-Adapted Interaction 8(1-2), 5–47 (1998)
2. Appelt, D.E., Pollack, M.E.: Weighted abduction for plan ascription. User Modeling and User-Adapted Interaction 2(1-2), 1–25 (1991)
3. Avrahami-Zilberbrand, D., Kaminka, G.A.: Fast and complete symbolic plan recognition. In: IJCAI, pp. 653–658 (2005)
4. Drücker, C., Hübner, S., Visser, U., Weland, H.-G.: As time goes by - using time series based decision tree induction to analyze the behaviour of opponent players. In: Birk, A., Coradeschi, S., Tadokoro, S. (eds.) RoboCup 2001. LNCS (LNAI), vol. 2377, pp. 325–330. Springer, Heidelberg (2002)
5. Eiter, T., Makino, K.: On computing all abductive explanations. In: Eighteenth national conference on Artificial intelligence, Menlo Park, CA, USA, pp. 62–67. American Association for Artificial Intelligence (2002)
6. Hobbs, J.R., Stickel, M., Martin, P., Edwards, D.D.: Interpretation as abduction. In: 26th Annual Meeting of the Association for Computational Linguistics: Proceedings of the Conference, Buffalo, New York, pp. 95–103 (1988)
7. Intille, S., Bobick, A.: Recognizing planned, multi-person action. In: Computer Vision and Image Understanding, vol. 81, pp. 414–445 (2001)
8. Lattner, A.D., Miene, A., Visser, U., Herzog, O.: Sequential pattern mining for situation and behavior prediction in simulated robotic soccer. In: Bredenfeld, A., Jacoff, A., Noda, I., Takahashi, Y. (eds.) RoboCup 2005. LNCS (LNAI), vol. 4020, pp. 118–129. Springer, Heidelberg (2006)
9. Lucchesi, M.: Coaching the 3-4-1-2 and 4-2-3-1. Reedswain Publishing (2001)
10. Miene, A., Visser, U.: Interpretation of spatio-temporal relations in real-time and dynamic environments. In: Birk, A., Coradeschi, S., Tadokoro, S. (eds.) RoboCup 2001. LNCS (LNAI), vol. 2377, pp. 441–447. Springer, Heidelberg (2002)
11. Ng, H.T., Mooney, R.J.: On the role of coherence in abductive explanation. In: National Conference on Artificial Intelligence, pp. 337–342 (1990)
12. Peirce, C.S.: Collected Papers of Charles Sanders Peirce. Harvard University Press (1931)
13. Selman, B., Levesque, H.J.: Support set selection for abductive and default reasoning. Artif. Intell. 82(1-2), 259–272 (1996)

Implementing Parametric Reinforcement Learning in Robocup Rescue Simulation

Omid Aghazadeh[1], Maziar Ahmad Sharbafi[1,2], and Abolfazl Toroghi Haghighat[1]

[1] Mechatrronic Research Lab, Azad University of Qazvin, Qazvin, Iran
[2] Electrical and Computer engineering Department, University of Tehran,Tehran, Iran
aghazadeh@mrl.ir, m.sharbafi@ece.ut.ac.ir

Abstract. Decision making in complex, multi agent and dynamic environments such as Rescue Simulation is a challenging problem in Artificial Intelligence. Uncertainty, noisy input data and stochastic behavior which is a common difficulty of real time environment makes decision making more complicated in such environments. Our approach to solve the bottleneck of dynamicity and variety of conditions in such situations is reinforcement learning. Classic reinforcement learning methods usually work with state and action value functions and temporal difference updates. Using function approximation is an alternative method to hold state and action value functions directly. Many Reinforcement learning methods in continuous action and state spaces implement function approximation and TD updates such as TD, LSTD, iLSTD, etc. A new approach to online reinforcement learning in continuous action or state spaces is presented in this paper which doesn't work with TD updates. We have named it Parametric Reinforcement Learning. This method is utilized in Robocup Rescue Simulation / Police Force agent's decision making process and the perfect results of this utilization have been shown in this paper. Our simulation results show that this method increases the speed of learning and simplicity of use. It has also very low memory usage and very low costing computation time.

Keywords: Reinforcement Learning, Multi Agent Coordination, Decision Making.

1 Introduction

Rescue simulation environment as a disaster space and a branch of RoboCup competitions, models a city after an earthquake occurrence. Its main purpose is to provide emergency decisions supported by integration of disaster information, prediction, planning, and human interface. In such a multi agent system, the coordination between heterogeneous agents is the main problem.

Reinforcement Learning (RL) is one of the most powerful strategies in dynamic and time variant environments. Adaptation with changes according to the results of actions is the basic property of RL which is needed in these situations. RL-based techniques with an adaptive behavior use interactions with the system to optimize the policy used to generate the decisions.

U. Visser et al. (Eds.): RoboCup 2007, LNAI 5001, pp. 409–416, 2008.

RL has many outstanding characteristics in using a feedback to improve the policy in discrete spaces. However, in continuous spaces, RL still has some drawbacks in adapting to huge state space. The most famous RL methods like Q-learning and Sarsa are defined for discrete spaces and their traditional methods are not practical in continuous ones [8-10].

The best solution in such a complex system is to use learning methods which would solve the curse of dimensionality as the main challenge in continuous and large discrete spaces. In this paper we represent a simple approach which is very low costing in computation time and memory needs.

In this method, like other on policy RL methods which use function approximation, there is a Function Approximator (FA) which presents the Q-values and works as the behavior generation policy. After taking an action, using the observed reward, the policy (FA) is updated using an innovative update process.

Our test bed for evaluating the ability of this method was Rescue Simulation. We evaluated our learning method by comparing it with our earlier algorithm. The latter was our team (MRL) algorithm which we used it in RoboCup2006. The better operation of the new system compared with the MRL algorithm -which was the first team in RoboCup Bremen 2006- shows the performance of the new RL algorithm. We have also won the championship of Robocup 2007 US's agent competitions using this method in Police Force agent's decision making process. We have named this method Parametric Reinforcement Learning (PRL) which is useful in continuous spaces and discrete with huge action/state spaces.

We arranged this paper as follows: In section 2, a summarized review of RL is presented in continuous spaces. Section 3 explains about RoboCup and Rescue Simulation as the test bed for PRL. The results of this implementation are shown in section 4. Finally, section 5 concludes this paper.

2 Reinforcement Learning in Continuous Spaces

2.1 Continuous Reinforcement Learning

In systems having continuous state and action spaces, the value function must operate with real-valued variables representing states and actions, which means that it should be able to represent the value for infinite states and action pairs. Choosing the value function's structure is a real challenge. RL methods should use memory resources efficiently, support learning without too much computational burden and generalize the immediate outcome of specific state-action combinations to other regions of the state and action spaces [7].

Function Approximation provides the estimations of the expected returns of every state-action pair for an agent. FAs are useful because they can generalize the expected return of state-action pairs that the agent actually experiences to other regions of the state-action space. In this way, the agent can estimate the expected return of state-action pairs that it has never experienced before.

However, note that a FA may not be able to accurately represent the Q-function for the entire state and action space due to its finite resources [6].

2.2 Our Approach: Parametric Reinforcement Learning

2.2.1 Action Evaluation Function (AEF)

Action Evaluation Function is the implementation of a linear FA in PRL. This is a linear combination of some parameters which affect the importance of an action. In fact, the aim of the learning is discovering the importance of each parameter comparing with other ones. In other words finding a sub-optimal AEF is the objective of our learning process. AEF is denoted by $V(s_i, a_i)$. The importance of parameter $P_k(s_i)$ is determined by its coefficient α_k shown in (1). In this formula n is the number of parameters in AEF which are chosen by the designer of an implementation.

$$V(s_i, a_i) = \sum_{k=1}^{n} \alpha_k P_k(s_i) \tag{1}$$

2.2.2 Update Process

After the reward is computed by the "Rewarder", parameters are changed in order to gain better rewards in the similar states i.e. moving AEF toward an optimal policy. The update process is described in (2):

$$\alpha_k = \alpha_k + \gamma \frac{\alpha_k P_k(s_i)}{\sum_{m=1}^{n} |\alpha_m P_m(s_i)|} sign(V(s_i, a_i)) \cdot R \tag{2}$$

In this formula γ is the learning factor, and the sign function prevents diverging coefficients when the value becomes negative and R is the observed reward. This definition for updating is inspired from Takagi-Sugeno coefficients in fuzzy Q-learning algorithm. To make this issue more obvious, assume that the value is positive, and then every reward should be divided linearly according to the effect of the parameter and its related coefficient in the decision i.e. $\alpha_k P_k(s_i)$. The bigger is the $\alpha_k P_k(s_i)$ the more portion of the reward it takes.

2.2.3 Descriptions of PRL

PRL is used in actor-critic configuration as. It is an on-policy Reinforcement Learning for continuous action or state spaces and discrete problems with huge action or state spaces. On-policy characteristic of PRL leads to a fast learning method. Actions can be generated using greedy or ε-soft methods with AEF.

To initialize the AEF some notifications are required:

1. Each P_k must be normalized to have similar initial effects on the value function. Otherwise, from the beginning bigger parameters will effect the decisions more and this will make the learning process instable.
2. Considering an initial knowledge (policy) which is close to the optimal policy, will speed up the learning process. The nearer the initial policy to the optimal policy, the faster the learning process converges to the desired area.

Since we do not know about the optimal policy in different problems, it is reasonable to initialize AEF with a policy which is not far from any other possible policy. In such a policy, importance of the parameters should be equal and this can be represented with each α_k set to 1.

3 RoboCup and Rescue Simulation

Our main reasons for choosing RoboCup Rescue Simulation as a test bed are:

1. The ability of evaluating system in its perfect manner is with participating in RoboCup competitions. When the algorithm is better than 19 other teams' who were working on AI, there is no doubt about its performance.
2. RoboCup Rescue is used throughout the international research community as a platform for testing aspects of integrated information fusion and agent systems.
3. The RoboCup Rescue scenario is based on real-world scenarios, with detailed simulators modeling different parts of the system.
4. RoboCup Rescue is particularly pertinent to exploring coordination at different levels of granularity, and coordination processes which interact with each other. It models different scenarios which are well suited to a combination of local and global coordination.

In the rescue simulation environment, a map is simulated 72 hours after an earthquake occurrence in 300 time steps. In this simulation the buildings collapse which causes some ignitions, obstruct the roads and injure the people. There are three groups of rescuers. The fire brigades try to put out the fire, ambulances can rescue injured people from damaged buildings and Police Force (PF) agents should clear the blocked roads and make them passable for others.

The quality of these agents operations are evaluated by the score which its formula is computed by (3).

$$Score = \sqrt{\frac{UBA}{TBA}}\left(NAH + \frac{LHP}{THP}\right) \tag{3}$$

Where UBA is the total unburned area, TBA is the total buildings area, NAH is the number of alive humanoids, LHP is the summation of living humanoid's HPs (showing the health of people with an integer number) and THP is the total HP of all humanoids at the beginning of the simulation.

4 Simulation Results

4.1 Definition of the Problem

As mentioned in chapter 3, Police Force (PF) agent's goal is to open blockades for other agents. At the beginning of the simulation, if fire sites are not reachable for the fire brigades because of some blocked roads, the fire will spread out quickly and will not be controllable which would lead to worse results. If blocked roads which are blocking buried civilian buildings are not opened quickly enough, they will die and the score will decrease. Hence, in maps with heavy blockades the role of the PF agent, especially their earliest actions and adapting their strategies to the map condition would become too important to the final result of the simulation. Because the agents do not know anything about the condition of map at the beginning steps, pre-designed decision makers can not solve the problem effectively, so learning during operation can help the agents to promote their action selectors.

The descriptions and assumptions of the PRL method, which we have implemented, are presented below:

Action space is split to areas that we call paths. Paths are made of one or more roads having no junctions. In VC map we have about 400 paths in each cycle to choose from.

In this problem, AEF is a function of many parameters. These parameters are from three different groups. The first group is distance containing distances of the path to the police force, the nearest burning building and the nearest refuge. The second group is the buried humans including a number of the civilians, the fire brigades, the police forces and the ambulance team agents buried in the buildings connected to the path. The third and forth sets are reported blockades and majority factor of the path. Reported blockade parameter of a path increases when an agent requests PF agents to open it.

Rewards are only positive and are computed according to the number of agents encountered the target, the number of agents that have requested opening of the target and some other factors.

All methods implement the same agents as MRL agents except the PF agents. The only difference between α_1 and PRL methods is the target selection policy of PF, and all other factors are exactly the same.

Greedy behavior generation policy is used.

4.2 Results

We tested three different methods in two different maps with heavy blockades and many fire sites. Both maps are based on "Virtual City" map with different distribution of civilians, agents and ignition points. The first method is the implementation of police force agent of the RoboCup Rescue 2006 Bremen's champion: MRL. The second method α_1 implements no learning with AEF's α_i s set to 1 which is the initializing policy of AEF in learning methods. The third method is PRL in which γ shows the learning speed constant.

Table 1 shows that police force agents in PRL $\gamma=0.05$, on average will lead to better results than MRL and α_1 police forces. Results show that the big learning rates are more risky, they could lead to best results but they also might lead to a poor result which again shows the trade-off between the accuracy and the speed of learning. Police forces in PRL $\gamma=0.05$ have the most stable behavior of all (smallest variance) and PRL $\gamma=0.2$ has the best average score of all. Table 2 shows the same facts in a different condition.

Table 1. Results for implementing 3 methods in VC map Runs

MRL	α_1	$\gamma = 0.2$	$\gamma = 0.05$	Method
90.48	93.95	100.76	93.13	
95.78	90.68	89.87	98.20	
95.47	84.69	93.14	97.28	
95.05	96.08	91.26	96.17	
91.20	83.58	98.01	93.86	
93.96	85.12	100.61	98.87	
91.89	96.22	98.41	94.20	
91.71	95.28	99.60	96.10	Scores
90.39	95.71	99.65	93.74	
82.65	83.55	98.83	92.03	
93.07	95.71	92.74	96.80	
93.63	90.46	99.78	94.95	
93.44	90.65	90.71	96.86	
90.52	97.07	91.83	96.32	
92.23	99.05	102.70	93.03	
92.09	91.85	96.52	95.43	Mean
10.03	28.61	19.09	4.16	σ^2 (Var)

The mean values for Table 2 are 92.1, 90.7 and 94.8 for MRL, α_1 and $\gamma=0.05$ respectively. The variance values for the mentioned methods are 4.9, 31.5 and 3.1 and these values show the same characteristics that the results from Table 1 show i.e. PRL is the most robust and satisfactory method and α_1 is the worst.

Police Forces work in different areas of the map. Each area has different conditions. As it is expectable, learning from different conditions will lead to different learned policies. For example, an area might have fire in it and other ones might have not, an area might have lots of buried civilians while another area may have several agents locked in blockades. This issue is depicted in Fig. 1. We have empirically learned that cooperation of agents having their own achieved knowledge will solve the problem more effectively.

Since in these simulations the learning speed is of a major concern, we should use a fast and accurate learning method, which should work properly in such complex and time-critical situations. The presented algorithm has these properties.

Table 2. Results for implementing 3 methods in VC3 map

Method	Scores									
MRL	93.1	91.8	91.6	89.5	92.6	91.6	91.4	95.2	88.7	95.9
α_1	95.7	87.0	86.2	96.9	96.1	93.2	86.7	95.4	80.1	90.0
$\gamma=0.05$	96.1	96.8	95.1	96.1	96.2	92.7	93.5	91.3	95.6	95.0

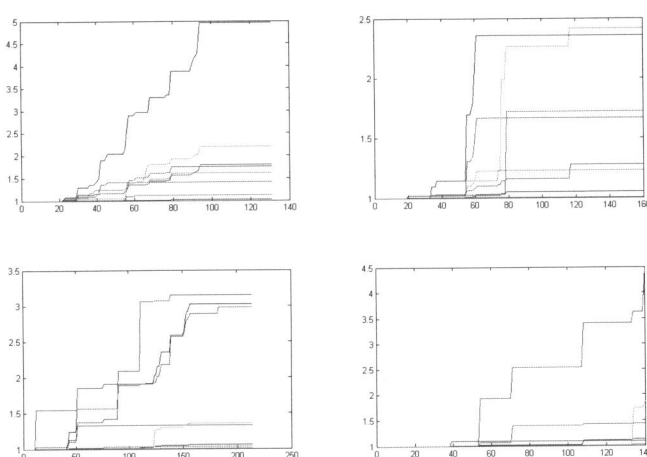

Fig. 1. Variations of learned policies for 4 PF agents acting in 4 different partitions

5   Conclusion

Reinforcement Learning as a powerful solution to the unknown dynamic multi agent systems [13-15], is the focus of this paper. Although RL has many advantages which make it a favorite method in a complex environment, it has some drawbacks especially in continuous spaces or discrete ones with a large action or state space. In this paper, we presented a useful method which is inspired from a linear Function Approximation and an innovative updating technique. We named this method Parametric Reinforcement Learning (PRL). Our method was evaluated in a very challenging problem (Robocup Rescue Simulation). Excellent results of utilizing PRL in these cases show its capabilities in intricate situations. Low usage of memory and simplicity of implementation are two additional advantages of this technique. Its only drawback is that it is suboptimal which is inevitable in environments with curse of dimensionality. Applying PRL in many other complicated continuous space problems can solve their difficulties too. Perhaps extending it to the nonlinear function approximation can reduce its distance to the optimal solution. It is obvious that this will take more time for the agents to learn and it is not suitable for time-critical situations like rescue simulation.

References

1. Ahmad Sharbafi, M., Lucas, C., AmirGhiasvand, O., Aghazadeh, O., Toroghi Haghighat, A.: Using Emotional Learning in Rescue Simulation Environment, Transactions on Engineering, Computing and Technology. 13, 333–337 (2006)
2. Allen-Williams, M.: Coordination in multi-agent systems, PhD thesis, University of Southampton (2006)
3. Dorais, G., Bonasso, R., Kortenkamp, D., Pell, P., Schreckenghost., D.: Adjustable autonomy for human-centered autonomous systems on Mars. In: Mars Society Conference (1998)
4. Schurr, N., Marecki, J., Lewis, J.P., Tambe, M., Scerri, P.: The defacto system: Coordinating human-agent teams for the future. In: Multi-Agent Programming, pp. 197–215. Springer, New York (2005)
5. Scerri, P., Sycara, K., Tambe, M.: Adjustable Autonomy in the Context of Coordination. In: AIAA 1st Intelligent Systems Technical Conference, Chicago, Illinois (2004)
6. Santamaria, J.C., Sutton, R.S., Ram, A.: Experiments with reinforcement learning in problems with continuous state and action spaces. Adaptive Behavior 6(2), 163–218 (1998)
7. Sutton, R., Barto, A.: Reinforcement Learning: An Introduction. MIT Press (1998)
8. Baird, L.: Reinforcement learning in continuous time: Advantage updating. In Neural Networks. IEEE World Congress on Computational Intelligence 4, 2448–2453 (1994)
9. Doya, K.: Temporal difference learning in continuous time and space. In: Advances in Neural Information Processing Systems, pp. 1073–1079. The MIT Press (1996)
10. van Kampen, E.-J.: Continuous Adaptive Critic Flight Control using Approximated Plant Dynamics, Master of Science Thesis Faculty of Aerospace Engineering, Delft University of Technology (2006)
11. Martin Appl.: Model-Based Reinforcement Learning in Continuous Environments, PhD thesis Technical University of Munich (2000)
12. Precup, D., Sutton, R., Dasgupta, S.: Off-Policy Temporal-Difference Learning with Function Approximation. In: ICML 2001, pp. 417–424 (2001)
13. Sutton, R.: Open Theoretical Questions in Reinforcement Learning. In: Fischer, P., Simon, H.U. (eds.) EuroCOLT 1999. LNCS (LNAI), vol. 1572, pp. 11–17. Springer, Heidelberg (1999)
14. Habibi, J., Ahmadi, M., Nouri, A., Sayyadian, M., Nevisi, M.: Utilizing Different Multiagent Methods in Robocup Rescue Simulation. In: Polani, D., Browning, B., Bonarini, A., Yoshida, K. (eds.) RoboCup 2003. LNCS (LNAI), vol. 3020. Springer, Heidelberg (2004)
15. Kitano, H., Tadokoro, S., Noda, I., Matsubara, H., Takahashi, T., Shinjou, A., Shimada, S.: RoboCup-Rescue: Search and Rescue in Large Scale Disasters as a Domain for Autonomous Agents Research. In: IEEE Conference on Man, Systems, and Cybernetics (1999)

Obtaining the Inverse Distance Map from a Non-SVP Hyperbolic Catadioptric Robotic Vision System

Bernardo Cunha, José Azevedo, Nuno Lau, and Luis Almeida

LSE-IEETA/DETI, Universidade de Aveiro, Portugal
{mbc,jla,lau,lda}@det.ua.pt

Abstract. The use of single viewpoint catadioptric vision systems is a common approach in mobile robotics, despite the constraints imposed by those systems. A general solution to calculate the robot centered distances map on non-SVP catadioptric setups, exploring a back-propagation ray-tracing approach and the mathematical properties of the mirror surface is discussed in this paper. Results from this technique applied in the robots of the CAMBADA team (Cooperative Autonomous Mobile Robots with Advanced Distributed Architecture) are presented, showing the effectiveness of the solution.

Keywords: Omnidirectional vision, robot vision, visualization.

1 Introduction and Related Work

The use of a catadioptric omni-directional vision system based on a regular video camera pointed at a hyperbolic mirror is a common solution for the main sensorial element found in a significant number of autonomous mobile robot applications. This is the case of the Middle Size Robocup Competition, where most of the teams adopt this approach for their robots vision sub-system [1-5]. This ensures an integrated perception of all major target objects in the robots surrounding area, allowing a higher degree of maneuverability at the cost of higher resolution degradation with growing distances away from the robot [6] when compared to non-holonomic setups. For most practical applications, this setup requires the translation of the planar field of view, at the camera sensor plane, into real world coordinates at the ground plane, using the robot as the center of this system. To simplify this non-linear transformation, most practical approaches choose to create a mechanical geometric setup that ensures a symmetrical solution by means of single viewpoint (SVP) approach [1][2][5]. This calls for a precise alignment of the four major points comprising the vision setup: the mirror focus, the mirror apex, the lens focus and the center of the image sensor. It also demands the sensor plane to be both parallel to the ground field and normal to the mirror axis of revolution, and the mirror foci to be coincident with the effective viewpoint and the camera pinhole respectively [7]. This approach generally precludes the use of low cost video cameras, due to the commonly found problem of translational and angular misalignment between the CCD sensor and the lens plane and focus. In this paper we describe a general solution to calculate the robot centered distances map on non-SVP catadioptric setups, exploring a back-propagation ray-tracing approach, also known as *"bird's eye view"*, and the mathematical properties of

U. Visser et al. (Eds.): RoboCup 2007, LNAI 5001, pp. 417–424, 2008.

the mirror surface [8][9]. This solution effectively compensates for the misalignments that may result either from a simple mechanical setup or from the use of low cost video cameras. Results from this technique, applied to the robots of the CAMBADA team (Cooperative Autonomous Mobile Robots with Advanced Distributed Architecture), are presented.

2 The Framework

In the following discussion we will assume a specific setup comprising a catadioptric vision module mounted on top of a mechanical structure (figure 1a)). It includes a low cost Fire-I BCL 1.2 Unibrain camera with a 3.6mm focal distance inexpensive lens. The main characteristics of this sensor can be depicted in figure 1b).

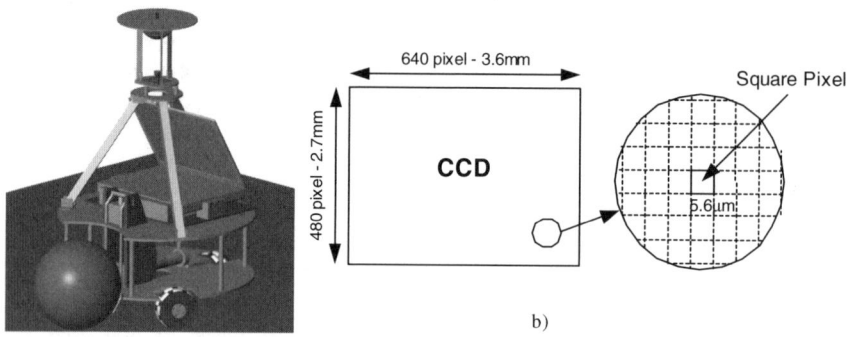

Fig. 1. a) The robot setup with the top catadioptric vision system. b) The Unibrain camera CCD main characteristics.

The used mirror has a hyperbolic surface, described by the following equation:

$$\frac{y^2}{1000} - \frac{(x^2 + z^2)}{1000} = 1 \ \ (\text{mm}) . \tag{1}$$

where y is the mirror axis of revolution and z is the axis parallel to a line that connects the robot center to its front. Height from the mirror apex to the ground plane is roughly 650mm. Some simplifications will also be used in regard with the diffraction part of the setup. The lens has a narrow field of view and must be able to be focused at a short distance. This, together with the depth of the mirror, implies a reduced depth of field and therefore an associated defocus blur problem [6]. Fortunately, since spatial resolution of the acquired mirror image is significantly reduced with distance, this problem has a low impact in the solution when compared with the low-resolution problem itself. A narrow field of view, on the other hand, also reduces achromaticity aberration and radial distortion introduced by the lens. Camera/lenses calibration procedures are a well-known problem and are widely described in the literature [10][11] – e.g Zhang's method. We will also assume that the pinhole model can provide an accurate enough approach for our practical setup, therefore disregarding any radial distortion of the lens.

3 Discussion

3.1 Initial Approach

Lets assume a restricted setup as in fig. 2. Assumptions of this setup are as follows:

- The origin of the coordinate system is coincident with the camera pinhole through which all light rays will pass;
- **i**, **j** and **k** are unit vectors along axis *X*, *Y* and *Z*, respectively;
- The *Y* axis is parallel to the mirror axis of revolution;
- CCD major axis is parallel to the *X* system axis;
- CCD plane is parallel to the **XZ** plane;
- Mirror foci do not necessarily lie on the *Y* system axis;
- The vector that connects the robot center to its front is parallel and have the same direction as the positive system *Z* axis;
- Distances from the lens focus to the CCD plane and from the mirror apex to the **XZ** plane are *htf* and *mtf* respectively and can be readily available from the setup and from manufacturer data.
- Point *Pm*(m_{cx}, *0*, m_{cz}) is the intersection point of the mirror axis of revolution with the **XZ** plane;
- Distance unit used throughout this discussion will be the millimeter.

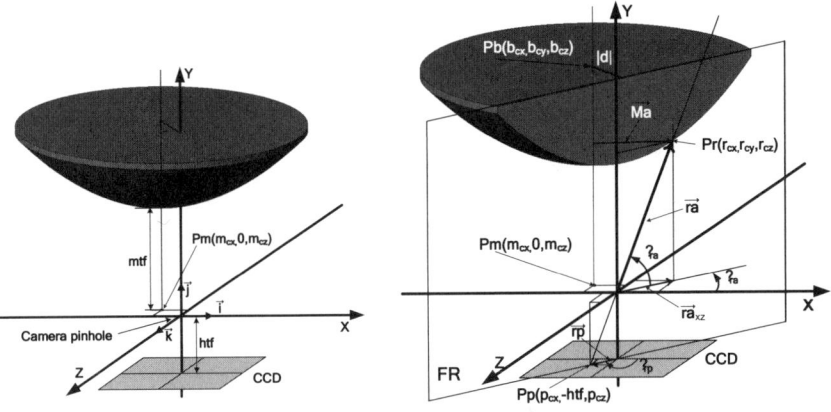

Fig. 2. left) The restricted setup with its coordinate system axis (*X, Y, Z*), (*mirror*) and (*CCD*). The axis origin is coincident with the camera pinhole (figure objects are not drawn to scale). right) A random pixel in the CCD sensor plane is the start point for the back propagation ray.

Mapping equation (1) it into the defined coordinate system, we get

$$y = \sqrt{1000 + (x - m_{cx})^2 + (z - m_{cz})^2} + K_{off} \text{ where } k_{off} = mtf - \sqrt{1000} . \quad (2,3)$$

Assuming a randomly selected CCD pixel (X_x, X_z), at point *Pp*(p_{cx}, *-htf*, p_{cz}) (fig. 2 b)), the back propagation ray that crosses the origin, may or may not intersect the mirror surface. This can be evaluated from the ray vector equation, solving for *y=mtf+md*, where *md* is the mirror depth, and obtaining the distance module from the mirror center. If this module is greater than the mirror maximum radius then the ray will not intersect the mirror and the selected pixel will not contribute to the distance map.

Assuming now that this particular ray will intersect the mirror surface, we can then define a plane **FR**, normal to **XZ** and containing this line, equated by

$$z = x \tan(\alpha_{ra}).$$
(4)

The line containing position vector **ra,** can then be expressed as a function of X as

$$y = x \tan(\beta_{ra})/\cos(\alpha_{ra}).$$
(5)

Substituting (4) and (5) into (2) we get the equation of the line of intersection between the mirror surface and plane **FR**. **Pr**, can then be determined from the equality

$$\frac{x \tan(\beta_{ra})}{\cos(\alpha_{ra})} = \sqrt{1000 + (x - M_{cx})^2 + (x \tan(\alpha_{ra}) - M_{cz})^2} + K_{off}.$$
(6)

which can be transformed into a quadratic equation of the form

$$ax^2 + bx + c = 0.$$
(7)

where

$$a = \left(1 + k_{in}^2 - k_{tc}^2\right). \quad b = 2\left(k_{tc} k_{off} - k_{in} M_{cz} - M_{cx}\right).$$
(8,9)

$$c = 1000 + M_{cz}^2 + M_{cx}^2 - K_{off}^2.$$
(10)

and

$$k_{tc} = \frac{\tan(\beta_{ra})}{\cos(\alpha_{ra})} \quad k_{in} = \tan(\alpha_{ra}).$$
(11)

Having found **Pr**, we can now consider the plane **FN** (fig. 3 a)) defined by **Pr** and by the mirror axis of revolution. The angle of the normal to the mirror surface at point **Pr** can be equated from the derivative of the hyperbolic function at that point, as a function of |**Ma**|,

$$\frac{\partial h}{d|M_a|} = \frac{|M_a|}{\sqrt{1000 + |M_a|^2}} \quad \beta_{tm} = \tan^{-1}\left(\frac{|M_a|}{\sqrt{1000 + |M_a|^2}}\right) - \frac{\pi}{2}.$$
(12)

This normal line intercepts the **XZ** plane at point **Pn**. The angle between the incident ray and the normal at the incidence point can be obtained from the dot product between the two vectors, -**ra** and **rn**. Solving for ϕ_{rm}:

$$\phi_{rm} = \cos^{-1}\left(\frac{r_{cx}(r_{cx} - n_{cx}) + r_{cy}(r_{cy} - n_{cy}) + r_{cz}(r_{cz} - n_{cz})}{|ra||rn|}\right).$$
(13)

The reflection ray vector, **rt**, starts at point **Pr** and lies on a line going through point **Pt**. Its line equation will therefore be

$$P = (r_{cx}i + r_{cy}j + r_{cz}k) + u((t_{cx} - r_{cx})i + (t_{cy} - r_{cy})j + (t_{cz} - r_{cz})k).$$
(14)

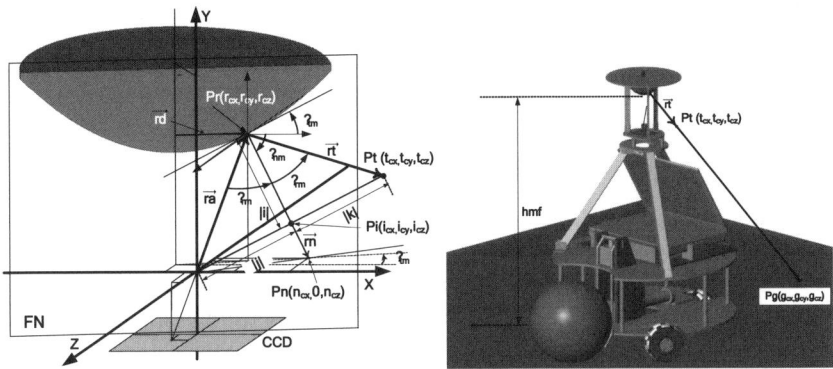

Fig. 3. left) Determining the normal to the mirror surface at point *Pr* and the equation for the reflected ray. right) (*Pg*) will be the point on the ground plane for the back-propagation ray.

The point *Pg* can then be obtained from the mirror to ground height *hmf*, and from the ground plane and **rt** line equations (fig. 3 b)), which, evaluating for *u*, gives

$$u = \frac{(mtf - hmf) - r_{cy}}{t_{cy} - r_{cy}}.$$ (15)

3.2 Generalization

To generalize this approach we must now consider the following misalignment factors: 1) The CCD plane may not be parallel to the **XZ** plane; 2) The CCD minor axis may not be correctly aligned with the vector that connects the robot center to its front; 3) The mirror axis of rotation may not be normal to the ground plane.

The first of these factors may result from two different sources: the CCD plane not being parallel to the lens plane; and the mirror axis of rotation being not normal to the CCD plane. Since both effects result in geometrical transformations of the setup, we will integrate these two contributions in the CCD plane, therefore providing a simpler solution. The second of the misalignment factors, on the other hand, can also be integrated as a rotation angle around the *Y* axis. To generalize the solution for these two correction factors, we will assume a CCD center point translation offset given by (*-dx, 0, -dy*), and three rotation angles applied to the sensor: γ, ρ and θ, around the *Y'*, *X'* and *Z'* axis respectively. These four geometrical transformations upon the original *Pp* pixel point can be obtained from the composition of the four homogeneous transformation matrices,

$$R_x(\rho) \bullet R_y(\gamma) \bullet R_z(\theta) \bullet T = \begin{bmatrix} t1_{\rho\gamma\theta} & t2_{\rho\gamma\theta} & t3_{\rho\gamma} & d_x \\ t1_{\rho\theta} & t2_{\rho\theta} & t3_{\rho} & 0 \\ t1_{\rho\gamma\theta} & t2_{\rho\gamma\theta} & t3_{\rho\gamma} & d_z \\ 0 & 0 & 0 & 1 \end{bmatrix}$$ (16)

The new start point $Pp'(p'_{cx}, p'_{cy}, p'_{cz})$, already translated to the original coordinate system, can therefore be obtained from the following three equations:

$$p'_{cx} = p_{cx}(\cos(\gamma)\cos(\theta) + \sin(\rho)\sin(\gamma)\sin(\theta)) + p_{cz}(\sin(\gamma)\cos(\rho)) + d_x$$
$$p'_{cy} = p_{cx}(\cos(\rho)\sin(\theta)) + p_{cz}\sin(\rho) - htf \qquad (17)$$
$$p'_{cz} = p_{cx}(-\sin(\gamma)\cos(\theta) + \sin(\rho)\cos(\gamma)\sin(\theta)) + p_{cz}(\cos(\gamma)\cos(\rho)) + d_z$$

Analysis of the remaining problem can now follow from (5) substituting Pp' for Pp.

Finally we can also deal with the third misalignment pretty much in the same way. We just have to temporary shift the coordinate system origin, assume the original floor plane equation defined by its normal vector $\mathbf{j}$, and perform a similar geometrical transformation to this vector. This time, however, only rotation angles ρ and θ need to be applied. The new unit vector $\mathbf{g}$, will result as

$$g_{cx} = -\sin(\theta)$$
$$g_{cy} = \cos(\rho)\cos(\theta) - mtf + hmf \qquad (18)$$
$$g_{cz} = \sin(\rho)\cos(\theta)$$

The rotated ground plane can therefore be expressed in Cartesian form as

$$g_{cx}X + g_{cy}Y + g_{cz}Z = g_{cy}(mtf - hmf) \qquad (19)$$

Replacing the **rt** line equation (14) for the X, Y and Z variables into (19), the intersection point can be found as a function of u. Note that we still have to check if **rt** is parallel to the ground plane – which can be done by means of the **rt** and $\mathbf{g}$ dot product. This cartesian product can also be used to check if the angle between **rt** and $\mathbf{g}$ is obtuse, in which case the reflected ray will be above the horizon line.

3.3 Obtaining the Model Parameters

Some of the parameters needed to obtain the distance map can be measured from the setup itself, e.g., the ground plane rotation relative to the mirror base. A half degree and 0.5mm precision has been proven enough for practical results. Other parameters can be extracted from algorithmic analysis of the image or from a mixed approach. Consider, for instance, thin lens law

$$f = \frac{g}{1 + G/B} . \qquad (20)$$

G/B is readily available from the diameter of the mirror outer rim in the sensor image; g can be easily obtained from the practical setup while f and the actual pixel size are defined by the sensor and lens manufacturers. Since G/B is also the ratio of distances between the lens focus and both the focus plane and the sensor plane, the g value can also be easily obtained. The main image features used in this automatic extraction are the mirror outer rim diameter and eccentricity, the center of the mirror image, the center of the robot image, and both the radius, distance and eccentricity of the game field lines – mainly the mid-field circle, lateral and area lines.

4 Support Visual Tools and Results

Although misalignment parameters can actually be obtained from a set of features in the acquired image, the resulting map can still present minor distortions. This is due to the fact that spatial resolution on the mirror image greatly degrades with distance. Since parameter extraction depends on feature recognition on the image, degradation of resolution actually places a bound on feature extraction fidelity. To allow further trimming of these parameters, two simple image feedback tools have been developed.

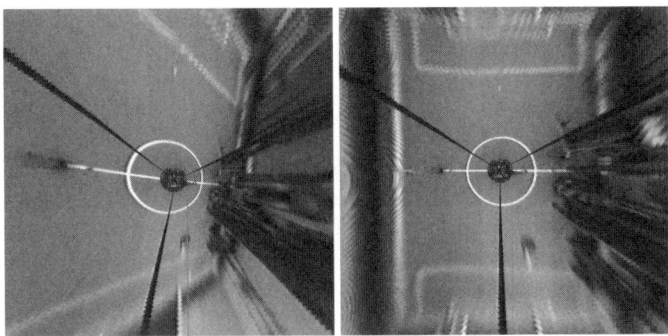

Fig. 4. Acquired image after reverse-mapping into the distance map. On the left, the map was obtained with all misalignment parameters set to zero. On the right, after automatic correction.

The first one creates a reverse mapping of the acquired image into the real world distance map. A fill-in algorithm is used to integrate image data in areas outside pixel mapping on the ground plane. (fig. 4).

The second generates a visual grid with 0.5m distances between both lines and columns, which is superimposed on the original image. This provides an immediate visual clue for the need of possible further distance correction (fig. 5). Since the mid-field circle used in this setup has exactly an outer diameter of 1m, incorrect distance map generation will be emphasized by grid and circle misalignment.

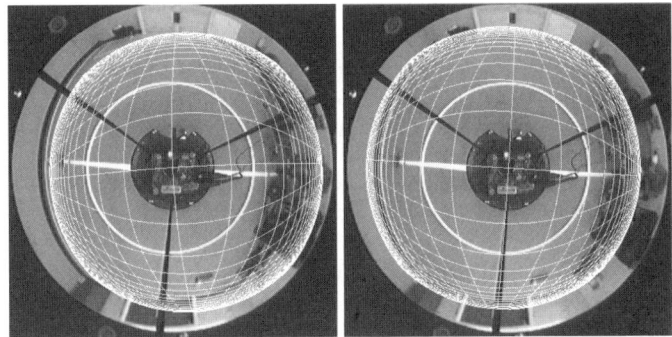

Fig. 5. A 0.5m grid, superimposed on the original image. On the left, with all correction parameters set to zero. On the right, the same grid after geometrical parameter extraction.

Comparison between real distance values measured at more than 20 different field locations and the values taken from the generated map, have shown errors always bellow twice the image spatial resolution. These results are perfectly within the required bounds for the robot major tasks, namely object localization and self-localization on the field.

5 Conclusions

Use of low cost cameras in a general-purpose omni-directional catadioptric vision system, without the aid of any precision adjustment mechanism, will normally preclude the use of a SVP approach. To overcome this limitation, this article explores a *"birds eye view"* algorithm to obtain the ground plane distance map in the CAMBADA football robotic team. Taking into account the intrinsic combined spatial resolution of mirror and image sensor, the method provides viable and useful results that can actually be used in practical robotic applications. This method is supported by a set of image analysis algorithms that can effectively extract the parameters needed to obtain a distance map with an error within the resolution bounds. Further trimming of these parameters can be manually and interactively performed, in case of need, with the support of a set of visual feedback tools that provide the user with an intuitive solution for analysis of the obtained results.

References

1. Zivkovic, Z., Booij, O.: How did we built our hyperbolic mirror omni-directional camera - practical issues and basic geometry. Intelligent Systems Laboratory Amsterdam, University of Amsterdam, IAS technical report IAS-UVA-05-04 (2006)
2. Wolf, J.: Omnidirectional vision system for mobile robot localization in the Robocup environment. Master's thesis, Graz, University of Technology (2003)
3. Menegatti, E., Nori, F., Pagello, E., Pellizzari, C., Spagnoli, D.: Designing an omnidirectional vision system for a goalkeeper robot. In: Birk, A., Coradeschi, S., Tadokoro, S. (eds.) RoboCup 2001. LNCS (LNAI), vol. 2377, pp. 78–87. Springer, Heidelberg (2002)
4. Menegatti, E., Pretto, A., Pagello, E.: Testing omnidirectional vision-based Monte Carlo localization under occlusion. In: Intelligent Robots and Systems (IROS 2004). IEEE/RSJ, vol. 3, pp. 2487–2493 (2004)
5. Lima, P., Bonarini, A., Machado, C., Marchese, F., Marques, C., Ribeiro, F., Sorrenti, D.: Omni-directional catadioptric vision for soccer robots. Robotics and Autonomous Systems 36(2-3), 87–102 (2001)
6. Baker, S., Nayar, S.K.: A theory of single-viewpoint catadioptric image formation. International Journal of Computer Vision 35(2), 175–196 (1999)
7. Benosman, R., Kang, S.B.: Panoramic Vision. Springer, Heidelberg (2001)
8. Blinn, J.F.: A Homogeneous Formulation for Lines in 3D Space. In: SIGGRAPH 1977, pp. 237–241 (1977)
9. Foley, J.D., van Dam, A., Feiner, S.K., Hughes, J.F.: Computer Graphics: Principles and Practice in C, 2nd edn. Addison-Wesley Professional (1995)
10. Zhang, Z.: A flexible new technique for camera calibration. IEEE Transactions on Pattern Analysis and Machine Intelligence 22(11), 1330–1334 (2000)
11. Hartley, R., Zisserman, A.R.: Multiple View Geometry in Computer Vision, Cambridge University Press (2003)

Tailored Real-Time Simulation for Teams of Humanoid Robots

Martin Friedmann, Karen Petersen, and Oskar von Stryk

Simulation and Systems Optimization Group
Technische Universität Darmstadt
Hochschulstr. 10, D-64289 Darmstadt, Germany
{friedmann,petersen,stryk}@sim.tu-darmstadt.de
http://www.sim.tu-darmstadt.de

Abstract. Developing and testing the key modules of autonomous humanoid robots (e.g., for vision, localization, and behavior control) in software-in-the-loop (SIL) experiments, requires real-time simulation of the main motion and sensing properties. These include humanoid robot kinematics and dynamics, the interaction with the environment, and sensor simulation. To deal with an increasing number of robots per team the simulation algorithms must be very efficient. In this paper, the simulator framework `MuRoSimF` (Multi-Robot-Simulation-Framework) is presented which allows the flexible and transparent integration of different simulation algorithms with the same robot model. These include several algorithms for simulation of humanoid robot motion kinematics and dynamics (with $O(n)$ runtime complexity), collision handling, and camera simulation including lens distortion. A simulator for teams of humanoid robots based on `MuRoSimF` is presented. A unique feature of this simulator is the scalability of the level of detail and complexity which can be chosen individually for each simulated robot and tailored to the requirements of a specific SIL test. Performance measurements are given for real-time simulation on a moderate laptop computer of up to six humanoid robots with 21 degrees of freedom, each equipped with an articulated camera.

1 Introduction

Besides suitable hardware the performance of an autonomous humanoid robot is mainly determined by the software modules applied for cognition, behavior and motion control. Efficient performance measuring and debugging of these modules on the real robot is very difficult in general, because physical robots are expensive and only limitedly available and may suffer from many experiments. Additionally an observed, undesired robot performance can be caused by any of the used software modules, which in physical experiments are difficult to test in isolation. Replacing the robot by a real-time simulation of its main physical characteristics including the kinematics and dynamics of humanoid motion with many actuated joints enables SIL-testing, monitoring and debugging of software modules under repeatable and controllable conditions.

U. Visser et al. (Eds.): RoboCup 2007, LNAI 5001, pp. 425–432, 2008.

In this paper a real-time simulator for teams of humanoid robots is presented which is closely integrated with the robot control framework `RoboFrame` [1]. The simulation can transparently be interfaced with all software modules (see Fig. 1). Many robot designs include controller hardware for motion generation, e.g., [2]. The firmware of the controller can also be integrated into the simulation.

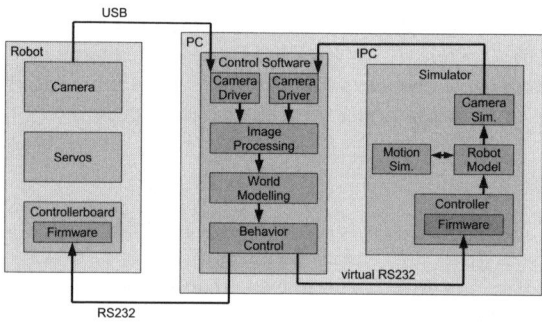

Fig. 1. The modules of the robot control software may be connected transparently either to the real robot or a real-time simulation

While SIL-tests of motion generation or image processing require physical accuracy, the needs for testing the behavior control are different: For investigating how the behavior control is affected by exact or inexact localization, computer vision with ideal or blurred camera images or walking without or with possible slipping of the feet the simulator must support different levels of simulation detail. Thus, different approaches for behavior-based humanoid robot control can be tested by controlling the representations between stimulus and action.

To meet these requirements the simulator must support a flexible exchange and combination of different simulation algorithms for the same purpose, e.g., robot dynamics, on different levels of detail. When exchanging algorithms there should be no need to change the models of the robots or the environment. The framework presented in this paper fulfills these requirements. Several algorithms for real-time motion-, contact- and sensor-simulation are considered. Results for real-time simulation of soccer playing humanoid robots are presented.

2   Overview

2.1   Modeling and Simulation of Robot Motion

Humanoid and four-legged robots are modeled as tree-structured kinematic chains of (usually rigid) links and (usually rotational) joints. The forward kinematics model describes the 3D position and orientation of the robot's bodies depending on the current joint angles. To build the kinematical model the geometrical data of the robot (link lengths, type and position of joints etc.) is needed.

A physics-based modeling of legged locomotion describes the nonlinear relationships between the forces and moments acting on each joint and the feet etc. and the position, velocity and acceleration of each joint. The high dimensional nonlinear multibody system dynamics (MBS) results in second order differential equations which can be formulated in various ways differing in terms of efficiency, modularity and flexibility [3,4]. In addition to the geometrical data a dynamics model requires kinetical data as mass, center of mass and inertia matrix for each link and joint, max/min motor torques and joint velocities. To simulate interaction with the environment detection and handling of collisions as well as suitable models of foot-ground contacts are required.

In the context of simulation of autonomous robots and RoboCup most often the open source Open Dynamics Engine (ODE) [5] is applied. ODE provides collision detection for several geometric primitives and a simulation of MBS dynamics. Only a one-step integrator with constant time step length and without integration error monitoring is available.

2.2 Simulation of Sensors

For closed loop simulation artificial readings of the robot's sensors are necessary. While joint position sensors can be simulated using a robot kinematics model, simulation of inertial or contact-force sensors requires a dynamics model. Cameras can be simulated using real-time rendering based on OpenGL or Direct3D. Only few simulators (cf. Sect. 2.3, e.g., [6,7]) enable the simulation of projections beyond the standard pinhole model.

2.3 Overview of Existing Robot Simulators

UCHILSIM [8] is a simulator for the RoboCup Four-legged League limited to the AIBO robots. Motion simulation is based on ODE. OpenGL is used for visualization and simulation of camera images.

SimRobot [6] is a general simulator for different mobile robots. The current version uses ODE for motion simulation. Simulation of four legged and wheeled robots have been demonstrated in [6]. Simulated sensors include tactile sensors, distance sensors and cameras, the later using accelerated hardware rendering.

USARSIM [9] is a simulator for different types of vehicles, including legged robots [10]. It provides a variety of sensors with the simulation of noise. The simulation is based on the Unreal Engine, a game engine providing a physics simulation and realistic visualization through high accuracy rendering.

Gazebo [11] is a general 3D multi-robot simulator with graphical interface and dynamics simulation. It is part of the Player/Stage project [12]. Motion simulation is based on ODE. Several sensor systems can be simulated including distance sensors and cameras. Additional sensors can be added.

Webots [13] is a commercially available general robot simulation package providing a wide variety of legged and wheeled robots. The motion simulation is based on ODE and the visualization uses OpenGL. Webots provides several types of sensors including cameras, distance and contact sensors.

2.4   Discussion

All simulators mentioned above use either ODE or a proprietary engine as motion simulator which all derive from physics engines in games. In games, however, only a physically plausible appearance based on simplified robot dynamics is required and not an accurate simulation of moments and forces acting in each of the robot's joints. The latter would require a full robot MBS dynamics model as well as error monitoring integration methods. Furthermore, an adaption of the level of detail in robot motion simulation or a flexible exchange or combination of different motion simulators for robots in the same environment is not possible.

MuRoSimF overcomes this problem by introducing a flexible interface allowing the combination of different simulation algorithms. By using the concept presented in this paper, simulation algorithms can be recombined easily for different simulators while maintaining efficient exchange of data.

3   Simulator Framework: Concept and Implementation

3.1   Modeling and Integration of Algorithms

The simulated robots as well as the environment are described as collections of objects. Each thereof is a collection of constant properties (like mass, size, shape, etc.) and variable properties (like position, velocity, etc.) which are assigned at runtime. During creation of robots and environment only constant properties are assigned. The objects are linked to the algorithms used for simulation next. In this phase, variable properties are assigned to the objects if they are needed by an algorithm. When linking algorithms to an object, it is checked if the object provides the necessary constant properties thus avoiding unnecessary calculation.

Robots are modeled as kinematic trees, consisting of one base, forks, static and variable translations, and static and variable rotations. Each of these elements is a specialized object having the necessary properties like length for a translation or angle and axis for a fixed rotation. During creation of the kinematic tree, additional constant properties may be added to the object, depending on the desired level of detail. All objects belonging to one robot are stored in one container, providing access to several subsets of the objects as the *joints* (denoting all variable translations and rotations) or the *bodies* (denoting all objects which may interact physically with the environment).

3.2   Simulation of Humanoid Robot Motion

Two alternative algorithms for humanoid robot motion simulation are currently utilized. Both have $O(n)$ computational complexity, with n the number of joints of the humanoid robot and can be individually selected for different humanoid robots in a multi-robot simulation.

Kinematic walking is a computationally cheap algorithm based on the assumption that during walking always (at least) one foot of the robot is in contact with the ground. During each time-step of the numerical integration, the direct kinematics of the humanoid robot is computed with the constraint, that the stance foot is not tilted or lifted off the ground unless the swinging foot touches or penetrates the ground plane. Then, the roles of the feet are exchanged. Besides its low computational complexity this algorithm has the benefit that the simulated robot can not fall over. If highly accurate simulation of the humanoid robot motion is not crucial, this algorithm can be used successfully, e.g., for testing behavior-based control or self localization modules. The algorithm is restricted to purely humanoid walking applications.

A *simplified robot dynamics algorithm* has been developed to overcome the drawbacks of the purely kinematic simulation. During each time-step, all external forces and resulting torques are summed up at the total center of mass of the robot. Only for the center of mass dynamics calculations are performed. Relative motions of all other elements of the robot are calculated using direct kinematics. The basic version of the algorithm uses a center of mass and an inertia matrix of the robot which are in a constant position relative to the robot's base. For more realistic motions, the center of mass and the inertia tensor can be calculated for each time step based on masses assigned to the robot's bodies. Both versions allow for a rich variety of motions. The second version is especially useful for humanoid robot motions during which the robot's overall mass distribution significantly changes as for falling down, standing up or balancing. As these algorithms do not rely on too specific assumptions on the robot's kinematical structure, they are not limited to humanoid robots.

A special strength of MuRoSimF is, that *algorithms for full MBS forward dynamics* like composite rigid body or articulated body algorithms [3,4] can be incorporated as well in case a higher level of detail in physical motion simulation. Also for numerical integration not only the common Eulers method but also higher order methods with variable step size may be employed.

3.3 Collision Detection and Handling

To simulate the interactions of a robot with its environment the collisions between the robot's bodies and the environment must be detected and handled. Collision detection and handling are treated as separate modules for which different algorithms of different complexity or level of detail can be applied easily.

Collision detection is based on an object's shape, position and orientation. Currently the primitive objects box, ellipsoid, cylinder and plane are supported. To avoid the $O(n^2)$ complexity of checking each object for collision with any other object, collision detection can be activated during setup of the simulation for each pair of objects individually. This process can be automated by defining sets of objects which do not need to be intersected with each other. For specialized simulations it is also possible, to use only selected bodies of a robot for collision detection (e.g., only the feet, if no other application than humanoid walking is considered). A collision is described by the penetration depth c_{depth} of the two bodies, the position $\mathbf{c}_{pos} \in \mathbb{R}^3$ of the contact point and the direction $\mathbf{c}_{normal} \in \mathbb{R}^3$ of the normal force pushing the bodies apart.

Collision handling is based on a *soft* collision model, allowing bodies to penetrate. Collisions are handled by calculating the resulting normal and frictional

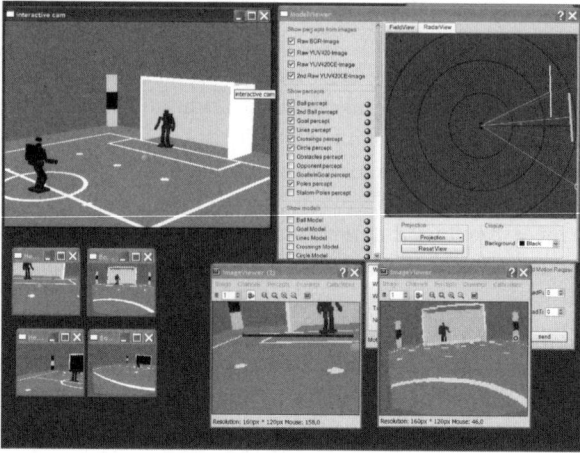

Fig. 2. Left side: Simulation of motion and sensing of two humanoid robots, each equipped with an articulated camera in the head and a wide angle camera in the chest whose images are displayed below. Right: GUI of control application displaying the striker robot's percepts in robot centric coordinates and within the camera images.

forces. To allow for different kinds of surfaces, additional parameters describing the contact situation are used. These parameters depend on the two colliding surfaces and are stored for each pair of surface types once. The normal force

$$\mathbf{f}_{normal} = \begin{cases} 1 \cdot s_c \cdot c_{depth} \cdot \mathbf{c}_{normal} & \text{if objects are getting closer} \\ s_b \cdot s_c \cdot c_{depth} \cdot \mathbf{c}_{normal} & \text{if objects are separating} \end{cases} \quad (1)$$

is calculated using a spring model with the spring constant s_c. The scaling factor $0 \leq s_b \leq 1$ is used to model different forces depending on the objects' relative velocity, allowing for different kinds of impact. Frictional forces $\mathbf{f}_{fric}$ are calculated using a viscous friction model depending on the relative velocity of the colliding bodies and the constant μ. As only one contact point is calculated for each pair of bodies, a pseudo-friction $\mathbf{n}_{fric}$ depending on the relative angular velocity is calculated which is used to stabilize standing bodies

$$\mathbf{f}_{fric} = \mathbf{v_{rel}} \cdot s_\mu \cdot f_{normal}, \quad \mathbf{n}_{fric} = \omega_{\mathbf{rel}} \cdot s_\nu \cdot f_{normal}. \quad (2)$$

3.4 Visualization and Camera Simulation

Real time rendering for visualization of the simulated scene is based on OpenGL. It is possible to display any property of any simulated object: all objects which need to be displayed are registered at the rendering module which keeps a set of renderers for the properties of interest and matches any object having a specific property with the respective renderer. Complexity and realism of the rendering process can be adjusted easily by exchanging the respective renderers. Therefore, scene rendering can be adjusted to different needs and levels of detail.

Camera simulation is based on the visualization module. After rendering the scene from the camera's point of view, the created image can be post-processed. Distortion caused by a camera lens (cf. Fig. 2) is simulated by moving the pixels according to the camera model from the well known camera calibration toolbox for Matlab [14]. This approach enables easy modeling of real cameras by calibrating them with the freely available toolbox. By changing the level of detail (and therefore realism) of the rendering algorithms used, the camera simulation can be adjusted to different cases and purposes.

4   Results

Using `MuRoSimF` various simulators for specific applications, e.g., for teams of humanoid robots or mixed teams of humanoid and wheeled robots can be realized. For each robot the simulation algorithms for motion, sensing, collision detection and handling can be chosen individually. Thus a high scalability of the level of detail can be achieved for tailoring the real-time simulation accuracy and computational complexity to the current needs.

The simulator has been used successfully for testing the software modules for vision, behavior control and self localization of robots in different scenarios from the RoboCup Humanoid League (Fig. 3). Further images and videos can be obtained from **www.dribblers.de/murosimf**. The performance has been measured using a standard laptop computer (IntelCentrinoDuo CPU (1.66 GHz), 1GB of RAM, Intel 945GM graphics chip set) with the simulation running single-threaded (see table 1). Each robot model consists of 21 joints [2].

Fig. 3. Left: Penalty kick. Middle: 3 versus 3 soccer game. Right: Slalom challenge.

Table 1. Simulation performance

Real-time simulation of	*frame rate*	*robots*
Kinematic motion	100	10
Simplified dynamics motion	1000	8
Kinematic motion and one camera	100 resp. 20	6
Dynamic motion and one camera	1000 resp. 20	5
Kinematic motion and one camera with distortion	100 resp. 20	3
Dynamic motion and one camera with distortion	1000 resp. 20	3

5   Conclusions

The new simulator framework `MuRoSimF` enables real-time simulation of motion, sensing and interaction with the environment for SIL-testing of onboard control software modules for teams of humanoid robots. It supports a scalable level of detail and a flexible exchange of algorithms for different simulation subtasks for different robots in the same multi-robot simulation. Results have been presented for scenarios of the RoboCup humanoid league. The source code of the simulator will be made available for other researcher upon request.

Acknowledgement. Parts of this research have been supported by the German Research Foundation (DFG) within the Research Training Group 1362 "Cooperative, adaptive and responsive monitoring in mixed mode environments".

References

1. Friedmann, M., Kiener, J., Petters, S., Thomas, D., von Stryk, O.: Modular software architecture for teams of cooperating, heterogeneous robots. In: Proc. IEEE Intl. Conf. on Robotics & Biomimetics, December 17-20, 2006, pp. 613–618 (2006)
2. Friedmann, M., Kiener, J., Petters, S., Sakamoto, H., Thomas, D., von Stryk, O.: Versatile, high-quality motions and behavior control of humanoid soccer robots. In: Proc. Workshop on Humanoid Soccer Robots of the 2006 IEEE-RAS Int. Conf. on Humanoid Robots, Genoa, Italy, December 4-6, 2006, pp. 9–16 (2006)
3. Featherstone, R., Orin, D.: Robot dynamics: Equations and algorithms. In: Proc. IEEE Intl. Conf. on Robotics and Automation (ICRA), April 2000, pp. 826–834 (2000)
4. Hardt, M., von Stryk, O.: The role of motion dynamics in the design, control and stability of bipedal and quadrupedal robots. In: Kaminka, G.A., Lima, P.U., Rojas, R. (eds.) RoboCup 2002. LNCS (LNAI), vol. 2752, pp. 206–223. Springer, Heidelberg (2003)
5. Smith, R.: ODE - Open Dynamics Engine (2007), http://www.ode.org
6. Laue, T., Spiess, K., Röfer, T.: SimRobot - a general physical robot simulator and its application in RoboCup. In: Bredenfeld, A., Jacoff, A., Noda, I., Takahashi, Y. (eds.) RoboCup 2005. LNCS (LNAI), vol. 4020, pp. 173–183. Springer, Heidelberg (2006)
7. Otsuka, F., Fujii, H., Yoshida, K.: Development of 3D dynamics simulator with omnidirectional vision model. In: Lakemeyer, G., Sklar, E., Sorrenti, D.G., Takahashi, T. (eds.) RoboCup 2006: Robot Soccer World Cup X. LNCS (LNAI), vol. 4434. Springer, Heidelberg (2007)
8. Zagal, J.C., Ruiz-del-Solar, J.: UCHILSIM: A dynamically and visually realistic simulator for the robocup four legged league. In: Nardi, D., Riedmiller, M., Sammut, C., Santos-Victor, J. (eds.) RoboCup 2004. LNCS (LNAI), vol. 3276, pp. 34–45. Springer, Heidelberg (2005)
9. Carpin, S., Lewis, M., Wang, J., Balakirsky, S., Scrapper, C.: USARSim: a robot simulator for research and education. In: Proc. of the 2007 IEEE Intl. Conf. on Robotics and Automation (ICRA) (2007)
10. Zaratti, M., Fratarcangeli, M., Iocchi, L.: A 3D simulator of multiple legged robots based on USARSim. In: Lakemeyer, G., Sklar, E., Sorrenti, D.G., Takahashi, T. (eds.) RoboCup 2006: Robot Soccer World Cup X. LNCS (LNAI), vol. 4434. Springer, Heidelberg (2007)
11. Koenig, N., Howard, A.: Gazebo - 3D multiple robot simulator with dynamics (2003), http://playerstage.sourceforge.net/gazebo/gazebo.html
12. Website: Player/stage project (2006), http://playerstage.sourceforge.net/
13. Website: Cyberbotics Webots, http://www.cyberbotics.com/products/webots/
14. Website: Camera Calibration Toolbox, http://www.vision.caltech.edu/bouguetj/calib_doc/

Evolution of Biped Walking Using Neural Oscillators and Physical Simulation*

Daniel Hein, Manfred Hild, and Ralf Berger

Humboldt University Berlin, Department of Computer Science
{dhein,hild,berger}@informatik.hu-berlin.de

Abstract. Controlling a biped robot with a high degree of freedom to achieve stable movement patterns is still an open and complex problem, in particular within the RoboCup community. Thus, the development of control mechanisms for biped locomotion have become an important field of research. In this paper we introduce a model-free approach of biped motion generation, which specifies target angles for all driven joints and is based on a neural oscillator. It is potentially capable to control any servo motor driven biped robot, in particular those with a high degree of freedom, and requires only the identification of the robot's physical constants in order to provide an adequate simulation. The approach was implemented and successfully tested within a physical simulation of our target system - the 19-DoF *Bioloid* robot. The crucial task of identifying and optimizing appropriate parameter sets for this method was tackled using evolutionary algorithms. We could show, that the presented approach is applicable in generating walking patterns for the simulated biped robot. The work demonstrates, how the important parameters may be identified and optimized when applying evolutionary algorithms. Several so evolved controllers were capable of generating a robust biped walking behavior with relatively high walking speeds, even without using sensory information. In addition we present first results of laboratory experiments, where some of the evolved motions were tried to transfer to real hardware.

Keywords: Biped Walking, Humanoid Robot Simulation, Evolutionary Algorithms, Walking Controllers, Neural Oscillators.

1 Introduction

Making a biped robot walk is a complex task. Describing and calculating joint trajectories is a common way to control servo motor driven humanoid robots. In the majority of the cases, the trajectory describing coefficients are calculated based on a model of the robot and a stability criteria. As an example, Takanishi's research group in Waseda University presented the humanoid robot WABIAN, where the trajectories of the arms, legs and ZMP were described by Fourier series [1]. The coefficients were determined in simulation in a way to ensure the Zero Moment Point (ZMP,[2]) conditions. As a drawback of this approach, a detailed and valid model of the target system has to be identified, and changes in the target system require a redesign of this model.

---

* Our work is granted by the German Research Foundation (DFG) in the main research program 1125 "Cooperating teams of mobile robots in dynamic environments".

U. Visser et al. (Eds.): RoboCup 2007, LNAI 5001, pp. 433–440, 2008.

Another well-established approach of gaining the reference trajectories, which is emerged from studying vertebrate animals are the Central Pattern Generators (CPG, [3,4,5]). Central pattern generators are circuits which are able to produce periodic signals in a self-contained way, i.e. without having any rhythmic input into themselves. In order to build structures with similar properties to the neural oscillators found in animals, several mathematical models have been proposed (e.g. [6,7,8]). Matsuoka proposed a mathematical model of CPGs and demonstrated that the combination of simple neural models can generate the neural activities for biped locomotion [9]. This model has been applied across several biped simulations (e.g. [10]), as well as used for real robots (e.g. [11]). One of the difficulties in the application of the CPG model to real robots is to determine the weights of neural connections. This is the main reason why genetic algorithms have often been used to solve this problem [12,13].

Within this paper we present a model-free approach of biped motion generation, based on a neural oscillator. The neural architecture has a biological analogy which is particularly interesting from a cognitive point of view. Furthermore it provides a very easy and natural way to incorporate arbitrary sensory input. We demonstrate the use of physical simulation and evolutionary algorithms to identify appropriate parameter sets of the presented motion generation model. This methodology is independent of a certain robot instance and does not require the detailed physical analysis of the target system. The application of simulation and artificial evolution permits an easy adaption of the motion generation to any modifications in the target system itself or in the requirements of the motion.

2   Simulation Environment

The target system of our study is a 19-DoF *Bioloid* robot with a shoulder height of 34*cm* and a weight of approx. 2.2*kg*. Due to the natural limits of real hardware experiments a physical simulation of this robot was developed. The simulation is based on the Open Dynamics Engine library (ODE, [14]) and simulates a simplified model of the real robot, consisting of 59 body parts and 19 servo motor joints. The time-integrated simulation is processed with a resolution of 100 simulation steps per second. Several isolated motor characteristic experiments were accomplished, in order to adequately simulate the servo motors torque and friction (see Fig. 1). Finally, as a weak validation of the simulation behavior, several real robot motions were transferred to the simulated one and could reproduce almost identical behavior. As an example, the handcrafted stand-up motion of the real robot is simulated accordingly (see Fig. 2).

The modular structured simulation environment was designed for exploring appropriate non-model based control structures which are potentially able to generate *robust* biped motions of our target system. Within this paper, a robust motion denotes a motion that is capable to compensate small environmental disturbances (e.g. small obstacles, impacts, rough floors, etc.). Regarding the simulation, the simulated robot had to pass at least 120*s* without falling or visible tumbling, while facing the ODE's simulated environmental noise.

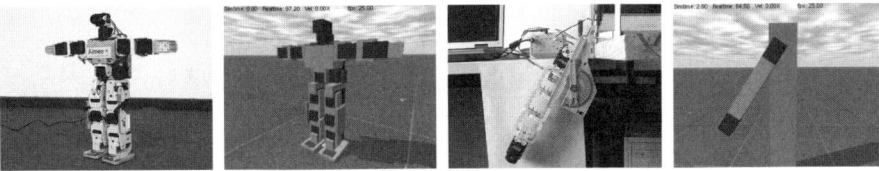

Fig. 1. Real and simulated world (left to right): Real Bioloid, Simulated Bioloid, Real servo motor torque and friction experiment setup, Simulated servo motor torque and friction experiment setup

Fig. 2. A first weak validation of the simulation: The stand-up motion is based upon interpolated keyframes and was developed on the real Bioloid. The (raw) transfer of the identical keyframe structure to simulation shows almost identical behavior.

3   Motion Generation – Neural Oscillator Approach

The neural oscillator approach generates a core oscillation with the use of the discrete-time dynamics of a two neuron network. Aspects of discrete-time dynamics with recurrent connectivity have been studied extensively, e.g. in [15,16]. The basic idea behind this approach is formulated by Pasemann, Hild and Zahedi in [17], which is also a good address for its mathematical background. The network update formula is as follows:

$$a(t + 1) := \tanh\left(\Omega\ a(t)\right), \quad \Omega = \begin{pmatrix} \omega_{11} & \omega_{12} \\ \omega_{21} & \omega_{22} \end{pmatrix} \tag{1}$$

It is demonstrated, that certain configurations of the weight matrix Ω cause periodic or quasi-periodic attractors in the phase space of the network [17,18]. These types of networks are able to generate different types of oscillations which in turn can be used for generating reference trajectories. An example of such a quasi-periodic orbit is displayed in Fig. 4.

The oscillations of the presented two neuron networks are now used for generating the joint's reference trajectories. The reference trajectory of a single joint is represented by the output of a dedicated (standard additive) neuron. The neuron derives its activation by two synapses coming from the two neurons oscillator and a bias term which

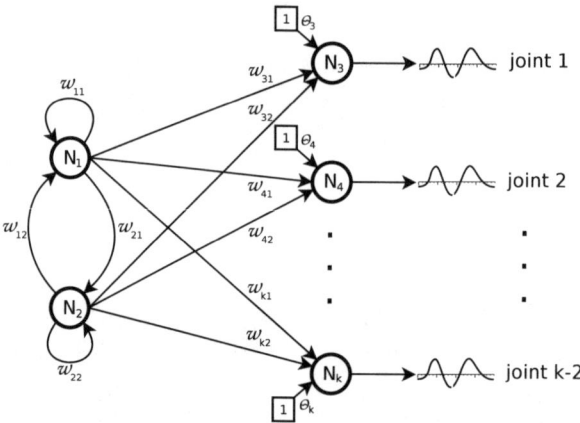

Fig. 3. Topology of the neural net controller. Each joint's reference trajectory is given by a dedicated neuron, which derives its activation by the two oscillating neurons N_1, N_2 and a bias term θ_j.

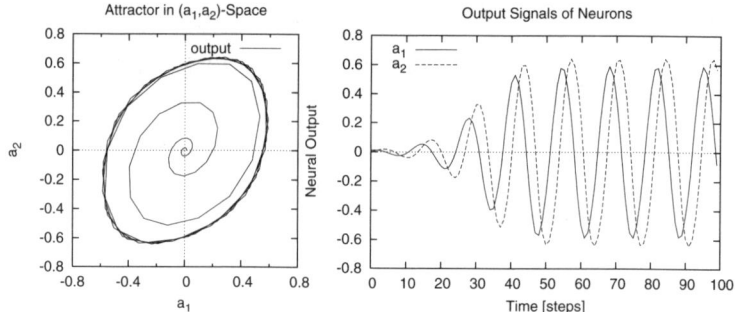

Fig. 4. Example dynamics of a two neuron network output: Phase trajectory in (a_1, a_2)-space (left), and output signals of neuron 1 and 2 (right) for $\omega_{11} = 1.17$, $\omega_{12} = 0.61$, $\omega_{21} = -0.47$, $\omega_{22} = 0.83$. Graphs show the initial phase up to reaching the quasi-periodic attractor within the first 100 time steps. The initial activation was set to $a_1 = 0.01$, $a_2 = 0.0$.

represents the offset of the trajectory's amplitude. In this way, the reference trajectory of a single joint is described by three parameters, ω_{j1}, ω_{j2} and θ_j, where ω_{ji} denotes the synaptic weight coming from neuron $i = 1, 2$ and θ_j the bias of joint j. Figure 3 illustrates the neural topology of the controller's network.

In order to reduce this parameter space, we further made use of a sagittal symmetry assumption, which states same movements between corresponding left and right sided joints with a half-period phase shift. In doing so, all trajectories are described by 10 output neurons, and the parameter space has a dimension of 34 synaptic weights.

4   Evolution of Walking Motions

Within this simulation environment, artificial evolutions were processed for identifying applicable parameters sets of the neural net controller. The primary object was to

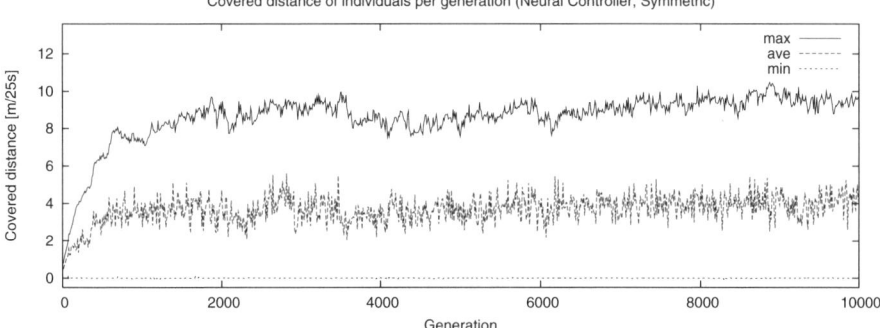

Fig. 5. Fitness development of an exemplary evolution experiment using the neural oscillator approach

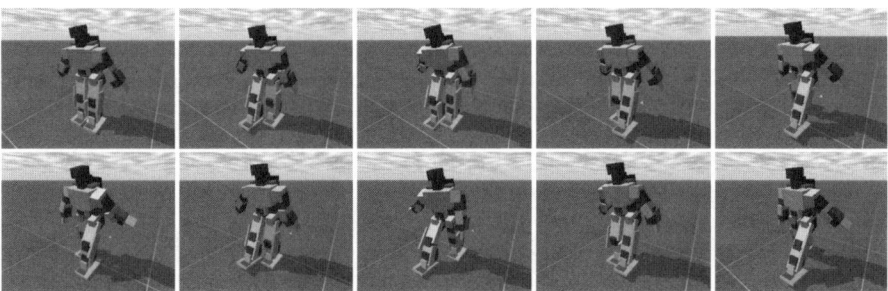

Fig. 6. Evolution of Walking Pattern: Example of an evolved walking pattern applying the neural oscillator approach. Pictures illustrate the start of walking and first steps. The displayed motion reaches a walking speed of about $0.45m/s$, which corresponds to a human walking speed of approx. $7km/h$.

identify motion patterns, that could pilot the robot a maximum possible distance within a certain time. Each individual has to pass an episode, in which the corresponding distance is measured. An episode starts with the relocation of the robot to its initial position. Subsequently the robot is given time to adopt its starting pose, in order to pass the episode run. The episode run is aborted if either the maximum episode duration is exceeded, if the robot falls or if it loses the desired path. The fitness value of an individual was set to its covered distance in stated walking direction. The actual 'position' of the robot was defined as the center of both feet. In doing so, the fitness is defined as follows:

$$fitness = min\left(\Delta y_{rfoot}, \Delta y_{lfoot}\right) \quad (2)$$

$$\Delta y_{rfoot} = y_{rfoot_{end}} - y_{rfoot_{start}} \quad (3)$$

$$\Delta y_{lfoot} = y_{lfoot_{end}} - y_{lfoot_{start}} \quad (4)$$

where $y_{xfoot_{start}}$ denotes the y-coordinate of the right/left foot at the beginning of the episode, and $y_{xfoot_{end}}$ the y-coordinate of the right/left foot at the end of the episode.

We already processed several hundreds of evolutions experiments, and the present results are the outcome of about 210,000 (simulated) hours.

Figure 5 shows the fitness development of such an evolution experiment.

The genotypes of the first generations were initialized with a (weak) Gaussian distribution ($\sigma = 0.01$) around $m = 0.0$. Only the synaptic weights of the two neuron network were chosen in a way, that the two neurons had already oscillating dynamics, which could significantly speed up the evolution progress. The chosen parameters were: $\omega_{11} = 1.1$, $\omega_{12} = 0.7$, $\omega_{21} = -0.7$ and $\omega_{22} = 1.1$, which corresponds to a oscillating frequency of approx. 8 periods per 100 net-update steps. The net-update frequency was set to $10Hz$ (10 updates per simulated second), hence the initial overall step frequency was $0.8Hz$.

5   Motion Transfer to Real Robot

Subsequently to the simulated evolutions, we transfered and tested several of the evolved motions patterns on the real robot. In general, the real robot was capable to reproduce all motions with a similar visual motion phenotype - as long as the robot acts free and does not touch the floor. Actually, none of the transfered motions could reproduce a robust walking motion. All walking motions need manual stabilization to avoid a fall down of the robot (see Fig. 7).

Fig. 7. Transfer of motion pattern to hardware: The 'grounded' real robot shows similar behavior compared to its simulated counterpart, but still needs manual support for walking

We identified two major issues that raise serious gaps between simulation and real world behavior. One refers to the considerable gears tolerances. Due to these (currently not simulated) tolerances, the actual trajectory of a joint crucially diverges from the controlled reference trajectory. As a result, whole-body motions are not reproduced with the required accuracy. To exemplify the problem: The present bodywork of the *Bioloid* robot does not even allow for standing on one foot due to the joint tolerances.

The other issue refers to the complicated motion characteristics of the servo motors. The simplified motion model of a servo motor does not sufficiently match the real servo motor behavior. This again results in significant differences between the actual whole-body movement and the desired one.

6   Conclusion and Outlook

Physical simulation is an effective and practical method, to study and explore motion generation techniques of complex biped robots. We presented a neural net controller, that could generate several robust biped walking motions for the simulated robot. The parameters of the neural net structure were identified by processing artificial evolutions within the simulation environment. Finally the simulated robot could walk with relatively high walking speeds of up to $0.51m/s$, which corresponds to a human walking speed of about $8km/h$. Interestingly while identifying the motor coupling weights, the evolution slightly modified the frequency of the neural oscillator from initially $0.8Hz$ to $0.75Hz$.

In laboratory experiments, several evolved motions were then transfered to the real robot. However, due to discrepancies between simulated and real world behavior, none of these transfered motions could actually generate a robust biped walking pattern on the real robot. Nevertheless, this paper outlines how simulation may enhance real robot motions. Generally, the presented approach may be applied to any biped robot with trajectory driven joints. In particular it can be applied to the new simulator of the 3D-Soccer-Simulation-League, that employs a physical model of the *Fujistu HOAP-2* humanoid robot.

The presented work comprises of just the first step, involved in using simulation to explore and optimize different controller models of biped robots. Several points could further expand on the completed work: For the first instance, we are currently engaged in enhancing the simulation in order to reduce the gap between real and simulated behavior. Primarily, this includes developing an enhanced servo motor joint model which describes all relevant characteristics of the applied AX-12 servos.

For the second instance we are studying the use of sensor feedback. At present, the implemented walking controller does not incorporate any sensory information. In generating a robust biped motion, the system has to be sensitive to external environmental influences, such as obstacles or various impacts, and must be able to react appropriately. This issue includes the exploration of the appropriate sensors (e.g. touch or acceleration sensors) as well as how sensor information is incorporated into the generation of motion. The synaptic architecture of the presented controller allows for several sensor coupling techniques. In conjunction with evolutionary algorithms the physical simulation enables exploring appropriate coupling structures as well as alternative neural net architectures. Regarding this point, first successful sensor coupling experiments were accomplished which we will present in a forthcoming paper.

References

1. Hashimoto, S., Narita, S., Kasahara, H., et al.: Humanoid Robots in Waseda University–Hadaly-2 and WABIAN. Auton. Robots 12(1), 25–38 (2002)
2. Vukobratović, M., Borovac, B.: Zero-Moment Point – Thirty Five Years of its Life. International Journal of Humanoid Robots 1(1), 157–173 (2004)
3. Grillner, S.: Neurobiological Bases of Rhythmic Motor Acts in Vertebrates. Science 228(4696), 143–149 (1985)
4. Grillner, S.: Neural Networks for Vertebrate Locomotion. j-SCI-AMER 274(1) (January 1996)
5. Ijspeert, A.J.: The Handbook of Brain Theory and Neural Networks. In: Arbib, M. (ed.) Locomotion, Vertebrate, 2nd edn., pp. 649–654. MIT Press (2002)

6. Ijspeert, A.J., Cabelguen, J.-M.: Gait Transition from Swimming to Walking: Investigation of Salamander Locomotion Control Using Nonlinear Oscillators. Technical report, Swiss Federal Institute of Technology (2002)
7. Chiel, H.J., Beer, R.D., Gallagher, J.C.: Evolution and Analysis of Model CPGs for Walking: I. Dynamical Modules. Journal of Computational Neuroscience 7(2) (1999)
8. Beer, R.D., Chiel, H.J., Gallagher, J.C.: Evolution and Analysis of Model CPGs for Walking: II. General Principles and Individual Variability. Journal of Computational Neuroscience 7(2), 119–147 (1999)
9. Matsuoka, K.: Mechanisms of Frequency and Pattern Control in the Neural Rhythm Generators. Biological Cybernetics 56(5–6), 345–353 (1987)
10. Miyakoshi, S., Taga, G., Kuniyoshi, Y., Nagakubo, A.: Three Dimensional Bipedal Stepping Motion Using Neural Oscillators — Towards Humanoid Motion in the Real World. In: Proceedings of IEEE/RSJ International Conference on Intelligent Robots and Systems (1998)
11. Endo, G., Nakanishi, J., Morimoto, J., Cheng, G.: Experimental Studies of a Neural Oscillator for Biped Locomotion with QRIO. In: IEEE 2005: International Conference on Robotics & Automation (2005)
12. Fujii, A., Ishiguro, A., Eggenberger, P.: Evolving a CPG Controller for a Biped Robot with Neuromodulation. In: Proceedings of the 5th International Conference on Climbing and Walking Robots, Paris, France, pp. 17–24 (2002)
13. Endo, K., Maeno, T., Kitano, H.: Co-Evolution of Morphology and Walking Pattern of Biped Humanoid Robot Using Evolutionary Computation - Consideration of Characteristic of the Servomotors. In: IEEE/RSJ International Conference on Intelligent Robots and Systems (IROS 2002), pp. 787–792 (2002)
14. Smith, R.: Homepage of Open Dynamics Engine project, http://www.ode.org
15. Chapeau-Blondeau, F., Chauvet, G.A.: Stable, Oscillatory, and Chaotic Regimes in the Dynamics of Small Neural Networks with Delay. Neural Networks 5(5) (1992)
16. Haschke, R., Steil, J.J., Ritter, H.: Controlling Oscillatory Behavior of a Two Neuron Recurrent Neural Network Using Inputs. In: Proc. of the Int. Conf. on Artificial Neural Networks (ICANN), Wien, Austria (2001)
17. Pasemann, F., Hild, M., Zahedi, K.: SO(2)-Networks as Neural Oscillators. In: IWANN (1), pp. 144–151 (2003)
18. Thompson, J.M.T., Stewart, H.B.: Nonlinear Dynamics and Chaos, 2nd edn. John Wiley & Sons, Chichester (2002)

Robust Object Recognition Using Wide Baseline Matching for RoboCup Applications*

Patricio Loncomilla[1,2] and Javier Ruiz-del-Solar[1,2]

[1] Department of Electrical Engineering, Universidad de Chile
[2] Center for Web Research, Department of Computer Science, Universidad de Chile
{jruizd,ploncomi}@ing.uchile.cl

Abstract. As the RoboCup leagues evolve, higher requirements (e.g. object recognition skills) are imposed over the robot vision systems, which cannot be fulfilled using simple mechanisms as pure color segmentation or visual sonar. In this context the main objective of this article is to propose a robust object recognition system, based on the wide-baseline matching between a reference image (object model) and a test image where the object is searched. The wide baseline matching is implemented using local interest points and invariant descriptors. The proposed object recognition system is validated in two real-world tasks, recognition of objects in the RoboCup @Home league, and detection of robots in the humanoid league.

1 Introduction

In the RoboCup soccer leagues robot vision systems are mostly based on basic color segmentation algorithms, and in some cases on the use of visual sonar (analysis of scan lines) for detecting lines. The main advantage of these vision mechanisms is their high processing speed. However, as the soccer leagues evolve, higher requirements are imposed over the vision systems, which cannot be fulfilled using those simple vision mechanisms. For instance, nowadays some teams are looking for advanced features such as: use of natural landmarks without geometrical and color restrictions, pose independent detection and recognition of teammates and opponents, detection of the teammates and opponents pose, automated refereeing tools, etc. Neither of those features can be achieved by pure color segmentation and/or using a visual sonar. Moreover, some non-soccer leagues (e.g. @Home) require robust, fast, easy trainable and general-purpose object recognition methodologies for recognizing complex objects like newspapers, bottles and soda cans (see @Home 2007 rules definition in [18]). In some tests, the object detector must be trained in runtime using only a few images as it cannot be trained before the test starts (i.e. the "lost & found" @Home test).

In this context, the main objective of this article is to propose a robust and versatile object recognition system, based on the wide-baseline matching between a reference image (object model) and a test image where the object is searched. Under this

---

* This research was funded by Millennium Nucleus Center for Web Research, Grant P04-067-F, Chile.

U. Visser et al. (Eds.): RoboCup 2007, LNAI 5001, pp. 441–448, 2008.

paradigm, local interest points (local maxima/minima in a filtered image set) are extracted independently from both the test and the reference image, then characterized using invariant descriptors (each one describes the gradient distribution in a region around an interest point), and finally several matches between similar descriptors from both images are used to get an affine transformation between the two images. Several verification stages are introduced to test the correctness of the transformation. If the object model has a known pose in the reference image, the obtained transformation allows determining the object's pose in the test image.

Object recognition based on wide-baseline matching has the following desired features: (i) no training requirements: only one image for each relevant view of the object is required; (ii) general purpose: any given object can be recognized, given that an example image of that object is available; and (iii) near real-time operation: depending on the exact characteristics of the implemented system and in the number of object classes, a processing speed of up to 3-9 frames per second can be achieved.

In the paper we describe the implemented object recognition system (section 2), and we show its use for recognizing objects in the RoboCup @Home league (section 3), and for detecting robots in the humanoid league (section 4). Finally, some conclusions of this work are given in section 5.

2   Object Recognition Based on Wide Baseline Matching

In the wide baseline matching problem formulation, the images to be compared are allowed to be taken from widely separated viewpoints, so that a point in one image may have moved anywhere in the other image, generating a hard matching problem.

Wide baseline matching approaches have become increasingly popular, experiencing an impressive development in the last years [1][4][9][12][16]. Local interest points are extracted independently from both a test and a reference image, characterized using invariant descriptors, and finally the descriptors are matched. By processing the matches, a transformation between the images is obtained.

The most employed interest point detectors are the single-scale Harris detector [2] and the multi-scale Lowe's sDoG+Hessian detector [4]. The best performing interest point detectors are the Harris-Affine and the Hessian-Affine [11], but they are slow for runtime applications. In the other hand, the most popular and best performing descriptor [10] is the SIFT (Scale Invariant Feature Transform) [4].

Lowe's system [3][4] uses the SDoG+Hessian detector, SIFT descriptors, a Hough transform to accumulate evidence from the matches for the possible similarity transformations, and a probability test to discard Hough transform bins which have few votes (then they could be generated only by random matches). This system has great recognition capabilities and near real-time operation. However, Lowe's system main drawback is the use of just a simple voting-based probabilistic hypothesis rejection stage, which cannot successful reduce the number of false positives when the true positive detection rate is prioritized. This is a serious problem in real world applications as, for example, robot self-localization [14], robot head pose detection [5] or image alignment for motion detection in video [15]. In [6][7] we proposed a system that reduces largely the number of false positives by using several hypothesis-based rejection stages. In this work, we extend this system by including the following new features: a fast probabilistic hypothesis rejection stage, a new linear correlation verification stage, a better organization of the hypothesis rejection tests into several stages, and the use of the RANSAC algorithm and a semi-local constraints test. Although RANSAC and the semi-local constraints tests have being use by many authors, Lowe's system does not use them. The proposed system is described in the following subsections.

2.1 Generation of the Matches between SIFT Descriptors for Each Image Pair

Local descriptors (SIFT descriptors) are extracted from both images, and matches between pairs of these descriptors belonging to different images are generated. This process is described in detail in [5][4].

2.2 Transformation Computation and Hypothesis Rejection Tests

This computation method (L&R – Loncomilla & Ruiz-del-Solar) considers several stages that are described in the next paragraphs.

1. Similarity transformations are determined using the Hough transform (see description in [3]). After the Hough transform is computed, a set of bins, each one corresponding to a similarity transformation, is determined. Then:
   a. Invalid bins (those that have less than 4 votes) are eliminated.
   b. Q is defined as the set of all valid candidate bins, the ones not eliminated in 1.a.
   c. R is defined as the set of all accepted bins. This set is initialized as a void set.
2. For each bin B in Q the following tests are applied (the procedure is optimized for obtaining high processing speed by applying less time consuming tests first):
   a. If the bin B has a direct neighbor in the Hough space with more votes, then delete bin B from Q and go to 2.
   b. Calculate r_{REF} and r_{TEST}, which are the linear correlation coefficients of the interest points corresponding to the matches in B that belong to the reference and test image. If the absolute value of any of these two coefficients is high, delete bin B from Q and go to 2. This numerical-robustness verification stage is explained in detail in the appendix.
   c. Calculate the fast probability P_{FAST} to B. If P_{FAST} is lower than a threshold P_{TH1}, delete bin B from Q and go to 2. This probability test is described in [7].
   d. Calculate an initial affine transformation T_B using the matches in B.
   e. Compute the affine distortion degree of T_B using a geometrical distortion verification test (described in [5]). If T_B has a strong affine distortion, delete bin B from Q and go to 2.
   f. Top down matching: Matches from all the bins in Q who are compatible with the affine transformation T_B are cloned and added to bin B. Duplication of matches inside B is avoided.
   g. Calculate Lowe's probability of bin B (see description in [3]). If this probability is lower than a threshold P_{TH2}, delete bin B from Q and go to 2.
   h. Apply RANSAC for finding a more precise transformation. In case that RANSAC success, a new transformation T_B is calculated.
   i. Accept the candidates B and T_B, what means delete B from Q and include it in R (the T_B transformation is accepted).
3. For all pairs (B_i, B_j) in R, check it they may be fused into a new bin B_k. If the bins may be fused and one of them is RANSAC-approved, do not fuse them and delete

the other in order to preserve accuracy. If the two bins are RANSAC-approved, delete the least probable. Repeat this until all possible pairs (including the new created bins) have been checked. This fusion procedure is described in [5].

4. For any bin B in R, apply semi-local constraints procedure to all the matches in B. The matches from B who are incompatible with the constraints are deleted. If some matches are deleted from B, T_B is recalculated. This procedure is described in [13].

5. For any bin B in R, calculate the pixel correlation r_{pixel} using T_B. If r_{pixel} is below a given threshold t_{corr}, delete B from R. This correlation test is described in [6].

6. Assign a priority to all the bins (transformations) in R. A more probable bin (in the Lowe's probability sense) has better priority than a less probable one, but any RANSAC-approved bin has better priority than any non RANSAC-approved one.

3 Solving RoboCup @Home Tests

The RoboCup @Home league defines seven tests to be solved in the 2007 competitions [18]. In three of them, complex and versatile visual object recognition abilities are required:

- In the "Lost & Found" test, an object is shown just one time to the robot, then the object is hidden somewhere in the environment and the robot should be able to find it within a limited amount of time [18].
- In the "Manipulate" test the robot must manipulate some specified objects (open a door, a refrigerator, get a soda can, grab a newspaper, etc) [18].
- In the "Navigate" test the robot has to safely navigate toward some specified objects in a living room environment [18].

These three tests put the following requirements to the object recognition system:

- General purpose. The objects to be recognized are of different types and in general complex: a TV, a door handle, a newspaper, a soda can, a bottle. Therefore a general-purpose object recognition system is required.
- No/Less training. In at least one of the test ("lost & found"), the objects to be recognized are not known by the robot before the test starts, while in the other two cases, the objects are not known by the participants before the RoboCup competitions start. Therefore, just one or two images of each object should be enough for a fast training and a robust characterization of the objects.
- Near real-time processing. The tests need to be solved in a short time, and for solving them the object recognition system need to be applied several times (e.g. hundreds of frames before finding an object in an arbitrary position in a complex environment).Then, the images need to be analyzed in near real-time.

These three requirements can be fulfilled using an object recognition system based on wide baseline matching, as the one described in the former section. As mentioned, this object recognition system outperforms similar ones terms of recognition rate, number of false positives and speediness, as it is shown in [7]. Therefore it will be used for implementing object recognition in the RoboCup @Home tests.

We implemented the described object recognition system in our RoboCup @Home robot [19]. We have carried out several experiments for solving the "Manipulate",

Fig. 1. Examples of object recognition results when the robot looks for different objects in different frames. In each case is shown the pair reference (left) - test (right) image.

"Navigate", and "Lost & Found" tasks, concentrating ourselves in solving the corresponding object recognition subtask. Some examples of object recognition, when the robot looks for different objects in different frames, are shown in figure 1. As in can be observed, our object recognition systems can successfully recognize in cluttered backgrounds a wide variety of objects which can appear in the *lost & found*, the *manipulate* and the *navigate* @Home tests.

4   Robot Detection in the RoboCup Humanoid League

In the RoboCup soccer competitions, the detection of teammates and opponent robots present in the scene is a key skill for good playing (e.g. passing, robot avoidance, goal kicking). Most existing vision systems, which use colors and depend on the illumination conditions, are not robust enough for solving this task. We aim at reverting this situation by using the L&R system in the detection of soccer robots, specifically humanoid robots.

We carried out several tests using our humanoid Hajime HR18 robot [20], and real video sequences processed in a notebook. The results are summarized in table 1. As it can be observed acceptable detection rates are obtained, however the processing speed should be increased, because in the humanoid league most of the robots are equipped with low-speed Pocket PCs as main processors (not notebooks). One possibility for achieving this reduction is applying this detector not in each frame, or using features that can be evaluated in less time (e.g. SURF [16]).

For exemplifying the detection of humanoid robots, in figure 2 we show some video frames where the robot is successfully matched against the reference image.

Table 1. Detection of a humanoid Hajime HR18 robot, 221 frames. Results were obtained with the system running in a notebook core-duo @ 1.66 GHz, 1GB RAM, running Windows XP.

Flavor	DR (%)	Number of False Positives	Processing Speed (fps)
Original image size: 320x240,	80.1%	14	4.4
Sub-sampled image: 240x170	75.1%	7	4.7
Sub-sampled image: 160x120	64.3%	3	11.5

Frame 24   Frame 36

Frame 49   Frame 54

Fig. 2. Some examples of humanoid robot detection in a video sequence. In each frame is shown the pair reference image (left) - test image (right).

5 Conclusions

In this article we have described a robust object recognition system, based on the wide-baseline matching between a reference image (object model) and a test image where the object is searched. The wide baseline matching is implemented using local interest points (sDoG+Hessian detector) and invariant descriptors (SIFTs). The main novelty of the described system is the inclusion of several hypothesis rejection tests that reduces largely the number of false positives, allowing the use of the system in real-world applications.

The proposed object recognition system is validated in two real-world tasks, recognition of objects in the RoboCup @Home league, and detection of robots in the humanoid league. The obtained results are satisfactory in terms of detection rate and number of false positives, although for an application in the humanoid league, where most teams employ Pocket PCs as main processors, the processing speed of the system should be increased. We are working in this direction using some novel features that can be evaluated in less time, as for example SURF features [16].

References

1. Ferrari, V., Tuytelaars, T., Van Gool., L.: Simultaneous Object.Recognition and Segmentation by Image Exploration. In: Pajdla, T., Matas, J(G.) (eds.) ECCV 2004. LNCS, vol. 3021, pp. 40–54. Springer, Heidelberg (2004)
2. Harris, C., Stephens, M.: A combined corner and edge detector. In: Proc. 4th Alvey Vision Conf., Manchester, pp. 147–151 (1998)
3. Lowe, D.: Local feature view clustering for 3D object recognition. In: IEEE Conference on Computer Vision and Pattern Recognition, Hawaii, pp. 682–688 (2001)
4. Lowe, D.: Distinctive Image Features from Scale-Invariant Keypoints. Int. Journal of Computer Vision 60(2), 91–110 (2004)
5. Loncomilla, P., Ruiz-del-Solar, J.: Gaze Direction Determination of Opponents and Teammates. In: Bredenfeld, A., Jacoff, A., Noda, I., Takahashi, Y. (eds.) RoboCup 2005. LNCS (LNAI), vol. 4020. Springer, Heidelberg (2006)
6. Loncomilla, P., Ruiz-del-Solar, J.: Improving SIFT-based Object Recognition for Robot Applications. In: Roli, F., Vitulano, S. (eds.) ICIAP 2005. LNCS, vol. 3617, pp. 1084–1092. Springer, Heidelberg (2005)
7. Loncomilla, P., Ruiz-del-Solar, J.: A Fast Probabilistic Model for Hypothesis Rejection in SIFT-Based Object Recognition. In: Martínez-Trinidad, J.F., Carrasco Ochoa, J.A., Kittler, J. (eds.) CIARP 2006. LNCS, vol. 4225, pp. 696–705. Springer, Heidelberg (2006)
8. Ruiz-del-Solar, J., Loncomilla, P., Vallejos, P.: An Automated Refereeing and Analysis Tool for the Four-Legged League. LNCS. Springer (2006)
9. Mikolajczyk, K., Schmid, C.: Scale & Affine Invariant Interest Point Detectors. Int. Journal of Computer Vision 60(1), 63–96 (2004)
10. Mikolajczyk, K., Schmid, C.: A performance evaluation of local descriptors. IEEE Trans. Pattern Anal. Machine Intell. 27(10), 1615–1630 (2005)
11. Mikolajczyk, K., Tuytelaars, T., Schmid, C., Zisserman, A., Matas, J., Schaffalitzky, F., Kadir, T., Van Gool, L.: A Comparison of Affine Region Detectors. Int. Journal of Computer Vision 65(1-2), 43–72 (2005)
12. Schaffalitzky, F., Zisserman, A.: Automated location matching in movies. Computer Vision and Image Understanding 92(2-3), 236–264 (2003)
13. Schmid, C., Mohr, R.: Local grayvalue invariants for image retrieval. IEEE Trans. Pattern Anal. Machine Intell. 19(5), 530–534 (1997)
14. Se, S., Lowe, D., Little, J.: Mobile robot localization and mapping with uncertainty using scale-invariant visual landmarks. Int. Journal of Robotics Research 21(8), 735–758 (2002)
15. Vallejos, Left blank for blindness purposes (2007)

16. Bay, H., Tuytelaars, T., Gool, L.V.: SURF: Speeded Up Robust Features. In: Leonardis, A., Bischof, H., Pinz, A. (eds.) ECCV 2006. LNCS, vol. 3951, pp. 404–417. Springer, Heidelberg (2006)
17. UCH100 database. Electronically, http://vision.die.uchile.cl/
18. RoboCup @Home 2007 competition rule book. Electronically, http://www.ai.rug.nl/robocupathome/documents/rulebook.pdf
19. Ruiz-del-Solar, J., Norambuena, S., Bernuy, F., Cubillos, S., Mascaró, M., Olavaria, I., Solís, C., Toro, C., Vargas, J.: UChile HomeBreakers, Team Description Paper (2007), http://www.robocup.cl
20. Ruiz-del-Solar, J., Vallejos, P., Parra, I., Testart, J., Ravest, P., Briones, R., Avilés, M.I.: UChile RoadRunners, Team Description Paper (2007), http://www.robocup.cl

Appendix: Linear Correlation Test

An affine transformation can be calculated from a set of matches between points (x, y) in the reference image and points (u, v) in the test image. The affine transformation can be represented in the following two ways:

$$\binom{u}{v} = \begin{pmatrix} m_{11} & m_{12} \\ m_{21} & m_{22} \end{pmatrix}\binom{x}{y} + \binom{t_X}{t_Y} \Rightarrow \binom{u}{v} = \begin{pmatrix} x & y & 1 & 0 & 0 & 0 \\ 0 & 0 & 0 & x & y & 1 \end{pmatrix}\begin{pmatrix} m_{11} & m_{12} & t_X & m_{21} & m_{22} & t_Y \end{pmatrix}^T \quad (1)$$

From the last expression, and using least squares, the parameters of the transformation can be calculated from matches between points (x_i, y_i) and (u_i, v_i):

$$\begin{pmatrix} m_{11} \\ m_{12} \\ t_X \\ m_{21} \\ m_{22} \\ t_Y \end{pmatrix} = \left(X^T X\right)^{-1} X^T \begin{pmatrix} u_1 \\ v_1 \\ u_2 \\ v_2 \\ \cdots \end{pmatrix} \quad ; \quad X = \begin{pmatrix} x_1 & y_1 & 1 & 0 & 0 & 0 \\ 0 & 0 & 0 & x_1 & y_1 & 0 \\ x_2 & y_2 & 1 & 0 & 0 & 0 \\ 0 & 0 & 0 & x_2 & y_2 & 1 \\ \cdots & \cdots & \cdots & \cdots & \cdots & \cdots \end{pmatrix} \quad (2)$$

The parameters are calculable only if the 6-by-6 $X^T X$ matrix is invertible, and this is possible only if X has rank 6. If the points in the reference image lay on a straight line, the relations $y_K = a\, x_K + b$ holds, then the second and fifth columns in X become linearly dependent, and the matrix X gets at most rank 4. Then, if the points in the reference image lay on a straight line, the parameters of a transformation from the reference to the test image cannot be successfully calculated. In the symmetric case, if the points in the test image lay on a straight line, a transformation from the test to the reference image cannot be calculated. Then, to get a numerically-stable and invertible transformation, the points in the reference and the test image cannot lie on a straight line, i.e., the correlation coefficients of the points in both images must be low. Then the following test can be done to reject numerically unstable transforms:

1. Calculate r_{REF}, the linear correlation of the interest points in the reference image
2. If r_{REF} > threshold, reject the transformation
3. Calculate r_{TEST}, the linear correlation of the interest points in the test image
4. If r_{TEST} > threshold, reject the transformation

Detection of AIBO and Humanoid Robots Using Cascades of Boosted Classifiers*

Matías Arenas, Javier Ruiz-del-Solar, and Rodrigo Verschae

Department of Electrical Engineering, Universidad de Chile
{marenas,jruizd,rverscha}@ing.uchile.cl

Abstract. In the present article a framework for the robust detection of mobile robots using nested cascades of boosted classifiers is proposed. The boosted classifiers are trained using Adaboost and domain-partitioning weak hypothesis. The most interesting aspect of this framework is its capability of building robot detection systems with high accuracy in dynamical environments (RoboCup scenario), which achieve, at the same time, high processing and training speed. Using the proposed framework we have built robust AIBO and humanoid robot detectors, which are analyzed and evaluated using real-world video sequences.

1 Introduction

In robot soccer scenarios, the detection of teammates and opponent robots is a key skill for good playing (e.g. passing, robot avoidance, goal kicking). However, most existing systems are not robust enough in the detection of other players, mainly because they are based on pure color analysis, which is very dependent on the illumination. To revert this, we have adapted our previously developed framework for face analysis system [8] to the task of building fast robot detector systems. This framework uses nested cascades of classifiers [10], the Adaboost boosting algorithm [6], and domain-partitioning based classifiers [6]. To our knowledge these statistical learning techniques have not been used before in robot detection applications.

Using the proposed framework we have built three AIBO robot detectors (ERS7 model), each one tuned for a different pose (frontal, profile and back), and also a humanoid robot detector. The main strengths of the developed robot detection systems are: the ability of working at multiple scales, being illumination invariant to a larger degree (they work in grey scale images and no preprocessing is needed for photometric normalization), and being near real-time.

The article is structured as follows. In section 2 some related work is outlined. In section 3 the robot detection framework is described. The training procedures for building AIBO and Humanoid robot detectors are described in section 4. In section 5 an evaluation of the developed robot detectors is presented. Finally, in section 6, some conclusions of this work are given.

---

* This research was partially supported by FONDECYT (Chile) under Project Number 1061158.

U. Visser et al. (Eds.): RoboCup 2007, LNAI 5001, pp. 449–456, 2008.

2   Related Work

Several approaches have been proposed to tackle the object detection problem. In the case of the RoboCup competition, most approaches for detecting robots are based on pure color segmentation and on the detection of contrast changes using scan lines (see for example [3][4]). These simple approaches are not robust enough; they are highly dependent on the illumination and background. In [2] is proposed a detection system for AIBO robots based on the use of local image descriptors and SIFT features, but its main limitations are its low processing speed and its reduced performance when highlights are present in the image, which are common in AIBO robots. However, if we consider other object detection problems, there are many robust approaches that are based on statistical classifiers [1], including systems based on neural networks, PCA projections, decision trees, SVM classifiers, and cascades of boosted classifiers.

Generally, one of the main drawbacks of detection systems based on statistical classifiers is that they are not real-time. The systems based on cascades of boosted classifiers, however, are an exception; they are very fast and accurate at the same time. The Viola&Jones classifier [9] use a cascade of filters for a fast classification, where each filter is trained using Adaboost, and the integral image for fast computation of the features, which are based on simple, rectangular features (a kind of Haar wavelets). This kind of classifier allows obtaining fast processing speed and high detection rates. These ideas are further improved in [10], where *nested* cascades are introduced. Nested cascades reuse the confidence output of a given layer, in the next layer of the cascade, which allows obtaining more compact (faster) cascades and more accurate classifications. It also uses domain-partitioning weak classifiers [6], which, compared to [9], achieves an improvement in the representation power of the weak classifiers and reduces the processing and training time. In [8] a procedure to train nested cascades of boosted classifiers that allows to considerably reduce the training time (from months in [9] to a few days) is proposed. A second improvement proposed in [8] is the use of both internal and external bootstrap for the training of the cascade. A third improvement corresponds to a criterion to automatically select the number of weak classifiers in each layer of the cascades, which aims to minimize the processing time and at the same time assures a high detection rate and a very low false positive rate. This learning framework [8] has been extended in this work to the task of robot detection.

3   Robot Detection Framework

We briefly describe the developed multiscale robot detection framework (see block diagram in figure 1). First, to detect the robots at different scales, a multiresolution analysis of the images is performed, by downscaling the input image by a fixed scaling factor --e.g. 1.2-- (*Multiresolution Analysis* module). This scaling is performed until images of about 24x24 pixels are obtained. Afterwards, windows of 24x24 pixels are extracted in the *Window Extraction* module for each of the scaled versions of the input image. The extracted windows could then be pre-processed to obtain invariance against changing illumination, but thanks to the used of illumination invariant features we do not perform any kind of preprocessing. Afterwards, the windows are analyzed by a nested cascade classifier (*Cascade Classification Module*). Finally, in the *Overlapping Detection Processing* module, the windows classified as positive

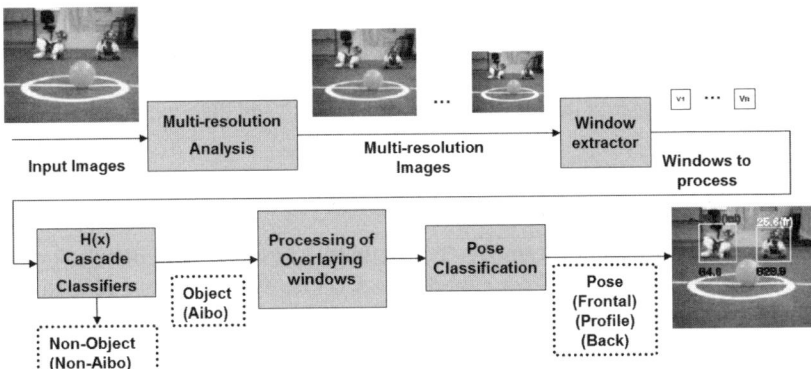

Fig. 1. Block diagram of the detection system

(they contain a robot) are fused (normally a robot will be detected at different scales and positions) to obtain the size and position of the final detections.

Using the described framework it is also possible to detect the robots pose. To achieve this, detectors tuned to different robot poses/views (e.g. frontal, profile and back) should be trained and applied. In general terms there are two possible forms of applying the detectors. The first one consists in applying the detectors in parallel. Then, the robot pose will be given by the detector having the largest confidence value. The second form consists in applying first a generic detector (not tuned to any pose) and then, in the *pose classification module,* verifying the detection, and also obtaining the pose of the robot applying the pose-specific detectors in parallel.

3.1   Learning Using Nested Cascades of Classifiers

A nested cascade of boosted classifiers is composed by integrated layers, each one containing a boosted classifier. The cascade works as a single classifier that integrates the classifiers of every layer H_C^k , defined as:

$$H_C^k(x) = H_C^{k-1}(x) + \sum_{t=1}^{T_k} h_t^k(x) - b_k \tag{1}$$

with $H_C^0(x) = 0$, h_t^k the weak classifiers, T_k the number of weak classifiers in layer k, and b_k a threshold (bias) value that defines the operation point of the strong classifier. The class assigned to the output corresponds to the sign of $H(x)$. The output of H_C^k is a real value that corresponds to the confidence of the classifier, and its computation makes use of the already evaluated confidence value of the previous layers. For details on the handling of the tradeoff between the speed and the accuracy of the cascade classifier see [8].

Domain-partitioning weak hypotheses make their predictions based on a partitioning of the domain X into disjoint blocks $X_1,...,X_n$, which cover all X, and for which $h(x)=h(x')$ for all $x, x' \in X_j$. Thus, the weak classifiers prediction depends only on which block X_j a given sample instance falls into. Herein the weak classifiers are applied over features, with each feature domain F being partitioned into disjoint blocks $F_1,...,F_n$, and a weak classifier h having an output for each partition block of its associated feature f: $h(f(x)) = c_j \ni f(x) \in F_j$.

For each classifier, the value associated to each partition block (c_j) is set to minimize a loss function on the margin [6]. This value depends on the number of times that the corresponding feature, computed on the training samples (x_i), fall into this partition block (histograms), and on the class of these samples (y_i) and their weight $D(i)$:

$$c_j = \frac{1}{2}\ln\left(\frac{W_{+1}^j + \varepsilon}{W_{-1}^j + \varepsilon}\right), \quad W_l^j = \sum_{i:f(x_i)\in F_j \wedge y_i = l} D(i) = \Pr\left[f(x_i)\in F_j \wedge y_i = l\right], \text{ where } l = \pm 1 \tag{2}$$

where ε is a regularization parameter. The outputs, c_j, from each of the weak classifiers, obtained during training, are stored in a LUT to speed up its evaluation. The Adaboost learning algorithm is employed to select the features and the weak classifiers $h_t^k(x)$. We use simple, rectangular features (a kind of Haar wavelets) [9].

3.2 Selection of the Training Examples

Every window of any size in any image that does not contain an object (e.g. an AIBO robot) is a valid non-object training example. Obviously, to include all possible non-object patterns in the training database is not an alternative, therefore non-object patterns that look similar to the object are selected using the bootstrap procedure [7]. This procedure corresponds to iteratively train the classifier, each time adding to the negative training set, negative examples that were incorrectly classified. According to our experience, it is important to use bootstrap in both situations: before starting the training of a new layer and for re-training a layer that was just trained. The *external* bootstrap is applied just one time for each layer, before starting its training, while the *internal* bootstrap can be applied several times during the training of the layer. The bootstrap procedure in both cases is the same with only one difference, before starting an external bootstrap all negative samples collected for the training of the previous layer are discarded (see [8] for details).

4   Training of the AIBO and Humanoid Robot Detectors

During the training of the cascades, validation and training sets are used. The procedure to obtain both sets is analogous, so only the training dataset is explained. To obtain the training set used at each layer of the cascade classifier, two types of databases are needed: one of cropped windows of positive examples (e.g frontal AIBOs) and one of images not containing the object to be detected. The second type of database is used during the bootstrap procedure to obtain the negative examples. The training dataset is used to train the weak classifiers, and the validation database is used to decide when to stop the training of a layer and to select the bias values of the layer. To obtain positive examples (cropped windows) a rectangle bounding the robot was annotated and a square of size equal to the largest size of the rectangle was cropped and downscaled to 24x24 pixels. In the case of the humanoid robots, two windows were cropped from each robot used during training, one corresponding to the upper half of the robot (torso and head) and the other to the lower part (mostly legs). This was made to allow the detection of either the upper or the lower part of the robot independently (using only one detector). This information should be sufficient for a successful detection under partial occlusions.

In the case of the databases used to train the AIBO detectors, the positive examples were obtained from videos captured using the AIBOs cameras and using external cameras. The videos were acquired under real-world playing conditions (variable illumination, occlusions, etc.). The sources used to build the humanoids training and validation sets were videos obtained using the same camera employed in our humanoid robots (Philips ToUCam III - SPC900NC), and videos from other humanoids obtained from the the RoboCup Humanoid league website (Hajime, Artisti, BreDo Brothers, DarmstadtDribbler and ToinPhoenix). The number of images used in each database is shown in Table 1.

Table 1. Summary of the databases used for training

Class	# Positive examples		# Negative images	
	(Training)	(Validation)	(Training)	(Validation)
Frontal AIBOs	3115	3115	5946	2550
Left AIBOs	4263	3624	5946	2550
Back AIBOs	1528	1528	5958	2562
Humanoids	3506	3500	5958	2562

5 Evaluation of the Detectors

The detection results are presented in terms of Detection Rate (DR) versus Number of False Positives (FP) in the form of ROC curves (Receiver Operation Characteristic curves) and tables, while the pose estimation results are presented using the confusion matrix. An analysis of the processing speed of the system is also presented. To evaluate the proposed system, two databases were used: one for the AIBOs (called AIBODetUChileEval) and one for the Humanoids (called HDetUChileEval). These databases were made available in http://vision.die.uchile.cl for future comparisons. No image of the training or the validation set are part of these databases. The AIBODetUChileEval database contains AIBOs in three poses (frontal, profile, back), while the HDetUChileEval database consists of images containing humanoids (from videos dribblers2006communication and dribblers2006Kicktrick). These images are from real world scenarios; containing changes in illumination, contrast, and background (see Table 2 for datils).

The performance of the proposed robot detection systems are presented in terms of DR versus FP (se Table 3 and Figure 2), and percentage of correct pose classification (Table 4). In figure 3 selected images with detection results are shown. In the AIBOs database, the first test consisted in evaluating each detector independently on the specific class it was trained to detect (e.g. "Frontal" detecting "Frontal" AIBOs). In this evaluation, AIBOs appearing under poses different to the ones being detected were not counted as false positives or correct detections. The best performing detector was the profile detector with a 90.7% DR and 70 FP (from all 724 images). The second test consisted in evaluating the performance of a particular detector when detecting all poses, including the ones they were not trained to detect. In this case the detectors were able to find AIBOs in all poses, showing a reasonably good detection rate; e.g. the Frontal detector obtained a 90% DR of AIBOs under all poses with 392 FP. The third test (*Multiple detectors in all AIBOs*) consisted in running all AIBOs detectors (Frontal, Profile and Back) in parallel. Given that in some cases the three detectors detected the same AIBOs, the final detections were obtained by selecting all non-overlaying detections, and merging overlaying detections by choosing the one with highest confidence. It is important to notice that in this case the number of false positives slightly increased, e.g. a DR of 94.8% was obtained with 392 FP. In other words, it is possible to arbitrate among the output of the detectors without increasing considerably the number of FP, although it is about 3 times slower than the individual detectors. The humanoid detector also shows high detection rates. A 92.2% detection rate was obtained with 123 false positive in a total of 244 images. This is quite high considering that the system was training using examples corresponding to different humanoid robot models than the ones used in the evaluation.

The last test made was a pose classification of the AIBOs. For this, the frontal detector was used as a generic detector (using the same parameters that obtained a 90% DR 392 FP), followed by a verification of the detections using the specific detectors. Afterwards, the pose was estimated by taking the output of the specific detector that gave the largest confidence value. Out of the 912 detected AIBOs, 657 were "pose estimated", from which 519 were correctly estimated (79% correct classification rate). Table 4 shows the confusion matrix of the pose estimation for these AIBOs. The "Frontal" and "Profile" classifiers show the best results, classifying correctly 90% and 80% of the "Frontal" and "Profile" AIBOs, respectively.

Table 2. Summary of the evaluation databases

Test database	#Images	#Frontal AIBOs	#Profile AIBOs	#Back AIBOs	#Humanoids	Image size
AIBODetUChileEval	724	344	489	180	-	208x160
HDetUChileEval	244	-	-	-	493	640x480

Table 3. Selected operation points (Detection Rate versus Number of False Positives) of the evaluated AIBO and Humanoids detectors

Detector / Target	DR %	FP	DR %	FP	DR %	FP	DR %	FP	DR %	FP
Frontal /Frontal AIBOs			89.4	254	84.4	57			74.5	18
Profile / Profile AIBOs	94.7	98	90.4	70			81.3	42		
Back / Back AIBOs			89.9	166	85.6	76	79.8	27		
Frontal / All AIBOs			90.0	392			82.9	183	73.4	95
Multiple / All AIBOs	94.8	392	88.6	183	84.3	114	80.1	52		
Humanoids	94.8	590	92.2	123					75.9	3

The processing time of the proposed detectors in the AIBO ERS7 robots was evaluated. ERS7 robots have a 64bit RISC Processor (MIPS R7000) from 576 MHz, 64MB RAM, and a color-camera of 208x160 pixels that delivers 30fps. Table 5 shows the average frame rate delivered by the "Frontal" AIBO detector in an ERS7 robot running the full four-legged Uchile1 control library [5], and in a 1.73 GHz Intel Core Duo laptop with 1GB of RAM, running Windows XP. The frame rate depends mainly on the scaling factor, and the number scales skipped by the detection system. The detector still works fine with a scaling factor of 1.2 and skipping 1 or 2 of the first scales, which allows obtaining 6.3 fps in the AIBOs. This allows using the detector in our four-legged team, considering that it is not necessary to detect the robots in each frame, but every 3-7 for frames (every 90-210 milliseconds).

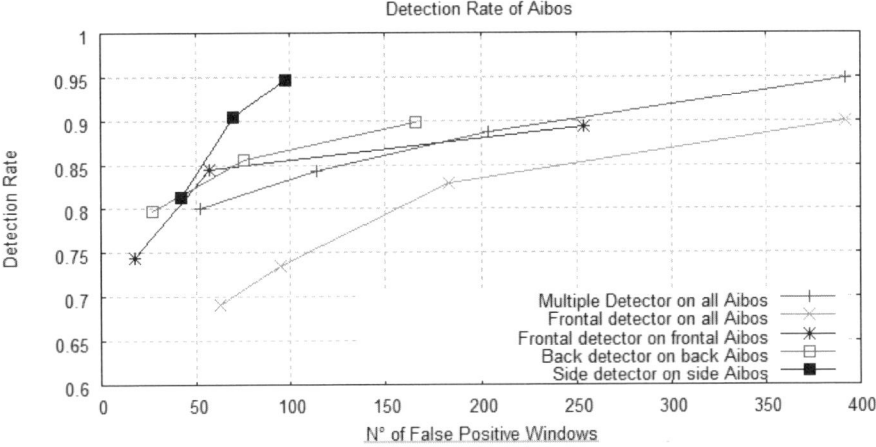

Fig. 2. ROC curves (Detection Rate versus Number of False Positives) on the AIBODetU-ChileEval database for the Frontal, Back and Profile detectors. See text for details

Table 4. Confusion Matrix: AIBO pose estimation using the detection system

True Class / Predicted Class	Frontal AIBOs	Profile AIBOs	Back AIBOs
Frontal AIBOs	91.63 %	11.64 %	33.87 %
Profile AIBOs	3.72 %	81.45 %	15.32 %
Back AIBOs	4.65 %	6.92 %	50.81 %

(a)

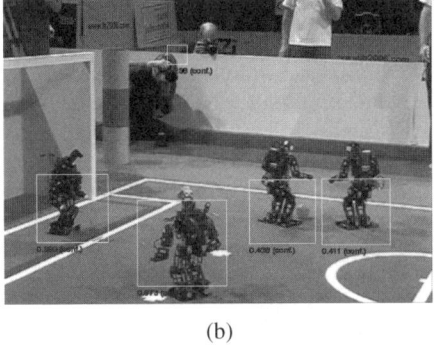
(b)

Fig. 3. Detection results of both detectors on the HDetUChileEval database are shown

Table 5. Processing time of the frontal AIBO detector

Configuration	Frame Rate (in fps) in Laptop PC		Frame Rate (in fps) in AIBO CPU	
	scaling 1.15	scaling 1.2	Scaling 1.15	scaling 1.2
no scale skipped	3.4	4.8	1.7	2.1
skip 1st scale	6.7	9.1	3.5	4.9
skip 1st,2nd scale	9.1	12.5	4.9	6.3
skip 1st,2nd,3rd scale	11.1	16.7	6.1	7.8

6 Conclusions

A framework for the robust detection of mobile robots using nested cascades of boosted classifiers was proposed. This framework was used to build robot detectors (Humanoids, and Frontal-, Profile- and Back-AIBOs). The main module of the system corresponds to a nested cascade of boosted classifiers, which is designed to perform fast detections with high DR and very low FPR. Using this cascade classifier, an exhaustive multiscale search is performed to be able to detect the robots appearing at different scales and positions. The detection rate of the obtained systems is quite high; for example a 90% DR with an average of 0.1 false positives per frame (208x160 pixels) is obtained for the "profile" AIBO detector, and a 92.2% DR with 123 false positives in 244 images (640x480 pixels) is obtained for the Humanoid detector. This shows that the detectors are working with high performance in difficult environment, and still maintain good results. Even thought the detection system was not designed to estimate the pose of the AIBO robots, it was possible to estimate it with a good accuracy in the case of the AIBOs. For example, the system correctly estimated the pose in 79% percent of the detected and verified AIBOs.

The main disadvantage of the detectors is that they achieve relatively low frame rates (e.g. 6.3 fps running in the AIBO robots). Nevertheless they can be improved in several ways. First, it is not necessary to detect the robots in each frame, but every 3-7 for frames (every 90-210 milliseconds). The processing time and the number of false positives can be greatly reduced by adding the use of color-based methods and information about the location of the robot in the field (by reducing the search region area). The system can be further improved by performing a tracking of the robots.

References

1. Hjelmås, E., Low, B. K.: Face detection: A survey. Computer Vision and Image Understanding 83, 236–274 (2001)
2. Loncomilla, P., Ruiz-del-Solar, J.: Gaze Direction Determination of Opponents and Teammates in Robot Soccer. In: Bredenfeld, A., Jacoff, A., Noda, I., Takahashi, Y. (eds.) RoboCup 2005. LNCS (LNAI), vol. 4020, pp. 230–242. Springer, Heidelberg (2006)
3. Quinlan, M.J., et al.: The 2005 NUbots Team Report. In: Bredenfeld, A., Jacoff, A., Noda, I., Takahashi, Y. (eds.) RoboCup 2005. LNCS (LNAI), vol. 4020. Springer, Heidelberg (2006), http://www.robots.newcastle.edu.au/publications/NUbotFinalReport2005.pdf
4. Röfer, T., et al.: German Team 2005 Technical Report. RoboCup 2005, Four-legged league (February 2006), http://www.germanteam.org/GT2005.pdf
5. Ruiz-del-Solar, J., et al.: UChile Kiltros 2007 Team Description Paper. In: RoboCup 2007 Symposium, Atlanta, USA (CD Proceedings), July 9 – 10 (2007)
6. Schapire, R.E., Singer, Y.: Improved Boosting Algorithms using Confidence-rated Predictions. Machine Learning 37(3), 297–336 (1999)
7. Sung, K., Poggio, T.: Example-Based Learning for Viewed-Based Human Face Deteccion. IEEE Trans. Pattern Anal. Mach. Intell. 20(1), 39–51 (1998)
8. Verschae, R., Ruiz-del-Solar, J., Correa, M.: A Unified Learning Framework for object Detection and Classification using Nested Cascades of Boosted Classifiers. Machine Vision and Applications (in press, 2007)
9. Viola, P., Jones, M.: Fast and robust classification using asymmetric adaboost and a detector cascade. In: Advances in Neural Inform. Processing System 14, MIT Press (2002)
10. Wu, B., Ai, H., Huang, C., Lao, S.: Fast rotation invariant multi-view face detection based on real Adaboost. In: 6th Int. Conf. on Face and Gesture Recognition, pp. 79–84 (2004)

A Scalable Hybrid Multi-robot SLAM Method for Highly Detailed Maps

Max Pfingsthorn[1], Bayu Slamet[2], and Arnoud Visser[2]

[1] Jacobs University Bremen*, Campus Ring 1, 28759 Bremen, Germany
[2] Universiteit van Amsterdam, 1098 SJ Amsterdam, The Netherlands

Abstract. Recent successful SLAM methods employ hybrid map representations combining the strengths of topological maps and occupancy grids. Such representations often facilitate multi-agent mapping. In this paper, a successful SLAM method is presented, which is inspired by the *manifold* data structure by Howard et al. This method maintains a graph with sensor observations stored in vertices and pose differences including uncertainty information stored in edges. Through its graph structure, updates are local and can be efficiently communicated to peers. The graph links represent known traversable space, and facilitate tasks like path planning. We demonstrate that our SLAM method produces very detailed maps without sacrificing scalability. The presented method was used by the UvA Rescue Virtual Robots team, which won the Best Mapping Award in the RoboCup Rescue Virtual Robots competition in 2006.

1 Introduction

Simultaneous Localization and Mapping (SLAM) is a vital technology for autonomous mobile robots. Using a SLAM algorithm, the robot can keep track of its location by maintaining a map of the physical environment and an estimate of its position on that map. This provides a spatial context for the interpretation of current and past observations and enables higher level reasoning, control, and coordination. The map delivered may also provide an intuitive representation with which the robot can convey its findings to humans. For this purpose the visualization of the map may be augmented with any kind of extra information that the robot is able to infer.

As described by Thrun [1], SLAM algorithms can be roughly classified according to the map representation and the employed estimation technique. A very popular map representation is the occupancy grid [2]. Grid-based approaches typically require a large amount of memory, however they are able to represent the environment at arbitrary resolution and thereby have the potential to be highly detailed.

Graph-based representations on the other hand primarily map the topology of the environment. This results in a much more compact description and makes graph-based maps an attractive alternative when scalability to multiple robots is

---

* Formerly International University Bremen.

U. Visser et al. (Eds.): RoboCup 2007, LNAI 5001, pp. 457–464, 2008.

a concern. The low memory requirement of graph-based maps allows for low-cost information sharing between multiple robots, e.g. via a wireless network. Also, as topological graphs provide a direct description of the free-space regions and their interconnectedness, they significantly facilitate path planning algorithms. A disadvantage of such maps is that localization is limited to the nearest node due to the lack of more detailed information. Additionally, this absence of low-level geometric information precludes the rendering of detailed visualizations.

Given the individual strengths and shortcomings, researchers have tried to combine multiple representations in hybrid approaches [3,4,5,6,7]. These are potentially as scalable as topological approaches, and at the same time provide the same geometric detail as grid maps.

In this paper we present a hybrid SLAM method that combines grid-based and topological representations. The underlying data structure exhibits both desired properties: Our method produces highly detailed maps without sacrificing scalability. As part of our research presented in this paper, we participated in the Rescue Virtual Robots competition in the RoboCup World Championships of 2006 [8] where we demonstrated the presented approach. Our system supported teams of up to 8 robots that searched the computer-simulated emergency site [9,10] and jointly constructed a map of the environment. Our maps ranked highest after being evaluated on a number of aspects as described in [8] and earned the Best Mapping Award.

2   The Data Structure

The data structure the presented method is built on is inspired by the *manifold* concept conceived by Howard et al. [7]. It is a layered data structure with a topological organization at the global level and small detailed metric maps at the local level. Sensor observations are not integrated into small grid maps as done in other current approaches [5], but kept as raw data in the nodes for later processing. Edges represent known connections between poses where these sensor observations were made and contain a transformation between these two poses. Additionally, uncertainty information about these transformations are stored in the edges.

Such a data structure is a very natural observation-centric formulation of a map. It also explicitly includes the important factorization that was the key insight used in FastSLAM [11]: Sensor observations are conditionally independent given the robot path. The observations are kept explicitly connected to the easily correctable path encoded in the graph.

Effectively, the sensor data stored in a node is a self-contained piece of information. This means that node poses can be updated without much computational overhead, which is in direct contrast to grid-based maps. Also, the usual slight update of pose differences does not invalidate associated sensor readings or the general map integrity. The map stays usable for many algorithms, like path planners.

· Formally, the map m consists of a set of local sensor observations $\{\pi\}$, and a set of links $\{\phi\}$, with

$$\phi = \{i, j, \delta, \Sigma\}$$

where i and j are the indices of the two observations which are linked, δ is the estimated pose difference, and Σ is that estimation's covariance matrix.

For the multi-robot scenario, similar benefits apply. The displacement information present in the links is independent of any global coordinate frame, so collaborating robots can easily exchange parts of the graph without further processing.

3 Single-Robot SLAM

Given the above data structure, the main challenge is to gather the information necessary to construct the links ϕ.

Like in most current SLAM methods, a laser range scanner is used as the main source of mapping information. Odometry measurements are not used, the method exclusively relies on scan matching to estimate a robot's displacement. Thus, no explicit motion model must be developed. The corresponding covariance matrix can be computed either by sampling as described in [12] or directly from the scan matching result [13,14].

As long as the uncertainty is low enough, the information in the map is used to get an estimate of the current location. When new sensor data arrives, the scan matching algorithm is used to compare the current range scan to the scan stored at the current node. This results in a new location estimate. At the moment the uncertainty for the scan matching operation increases, a new node is created to store the current scan and a new link is created containing the latest displacement estimation.

We chose the Weighted Scan Matcher (WSM) by Pfister et al. [14], which belongs to the Iterative Closest Point (ICP) family. The IDC scan matcher by Lu and Milios [15] belongs to the same family and has been the most popular ICP scan matcher. However, WSM is known to outperform IDC significantly in both accuracy and speed, given that dense range scans are available [14]. In [16], we produced similar results in an extensive set of experiments specific to the simulated environment at the RoboCup Rescue Virtual Robots competition. In addition, computing the covariance matrix of the displacement given by WSM is easy.

Given the data structure and a means to acquire the needed displacement estimate, a naïve implementation would result in incremental SLAM. Without any further processing, a new laser range scan could be matched against the previous one (or a set of previous ones) and added to the graph. However, this often results in accumulated error, as shown in Figure 1a.

The main manifestations of this error are inconsistencies in overlapping map regions where the robot travelled in a loop. Overlaps in our map representation can be detected via feature extraction and comparison on a node by node basis, by mutual observations of robots in a multi-agent setting, or by re-observing artificial or natural landmarks. Due to error accumulation, the two position estimates of the overlapping regions are not likely to project to the same global coordinates.

In the USARSim [9,10] simulator used in the RoboCup Rescue Virtual Robots competition, victims and robot-placable RFIDs were uniquely identifiable. This made them perfect unique landmarks, and thus a great and reliable shortcut to loop detection.

Loop-Closing is the main way to reduce the accumulated error in the map. A new link is inserted into the graph by computing the displacement between the two overlapping nodes. The inconsistency introduced by the new link has to be resolved by incorporating this new information into the graph as a whole.

Re-matching each node to its respective neighbors incrementally, starting from the new link, produces good results and has been used in the literature before [7]. Figure 1b shows the corrected map using this approach.

It is also possible to close loops by optimizing the estimated global positions of graph nodes such that the estimated displacements stored in the links apply best to all nodes. Here, the probability distribution formed by the initial pose estimate δ and its covariance matrix Σ can be used to evaluate how good a certain global positioning fits a given link, as shown by Olson et al. [17].

4 Multi-robot SLAM

A graph-based map can be very easily shared with and communicated to collaborating robots. Only new nodes and corresponding links have to be transferred over a network connection to communicate map updates completely. New data only needs to be appended to the graph and connected to the right nodes. Scan matching does not have to be performed again.

Keeping multiple disconnected maps in memory is problematic in other successful SLAM methods. In the context of the graph however, this is trivial. We allow the graph to contain multiple disconnected components, one for each robot. Similarly, it is possible to start a new disconnected component when a robot looses track of its location, for example after falling down stairs. This is also an attractive solution to the "kidnapped robot problem".

The main challenge is merging the disconnected partial maps in a meaningful way. Incidentally, this process is very similar to loop closing. The same techniques can be used to detect overlaps in two disconnected maps. Map merging is done by computing a transformation between the two overlapping regions once by scan matching. Subsequently, one of the two maps is transformed as a rigid body and moved so that the two overlapping parts fit. Optionally, a loop closing operation may be run to refit the two maps for improved accuracy. An example is shown in the bottom of Figure 1.

Closing loops and merging partial maps can be delayed without impacting the continuity of the map. Only some collaboration has to be sacrificed for this significant reduction in immediate computational requirement. Some robots may explore the same area twice or may take less efficient routes through the

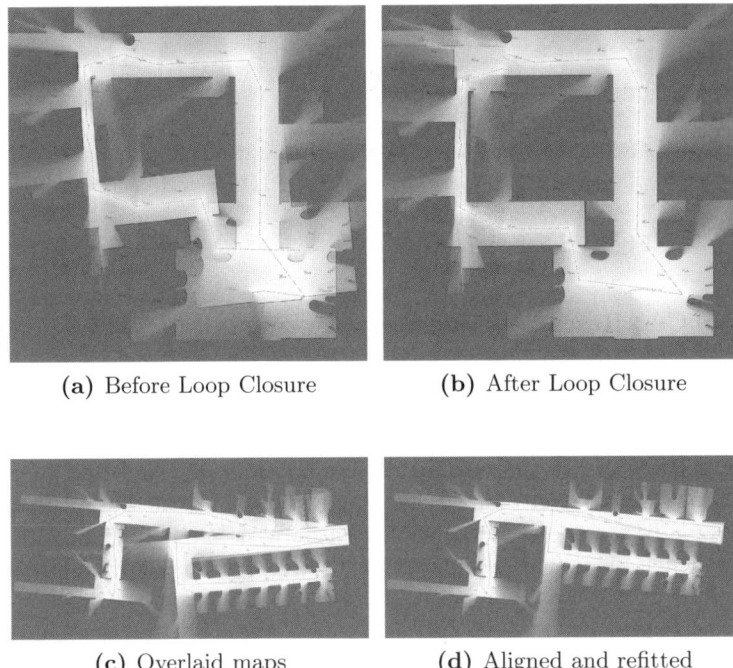

(a) Before Loop Closure   (b) After Loop Closure

(c) Overlaid maps   (d) Aligned and refitted

Fig. 1. Loop Closing and Map Merging

environment if maps are not merged immediately. However, map integrity is never at risk. This is interesting for Urban Search and Rescue as deferring these expensive operations saves time that can be used to further explore the emergency site. It is an open research question to decide when, and if, such operations should take place.

During the RoboCup Rescue Virtual Robots 2006 competition, and preliminary trials in the lab, our method was able to map a simulated area with up to 8 concurrently running robots. This high scalability in the multi-robot setting is mainly due to the above mentioned trivial map updates and deferrability of costly mapping operations.

5   Results

In this section, we show sample maps generated by our presented method for both simulated and real data. Simulated data was taken from the USARSim simulator used in the RoboCup Rescue Virtual Robots competition in 2006 [8,9,10]. The real robot data comes from a publicly available data set from the IROS 2006 Workshop "From sensors to human spacial concepts" [18].

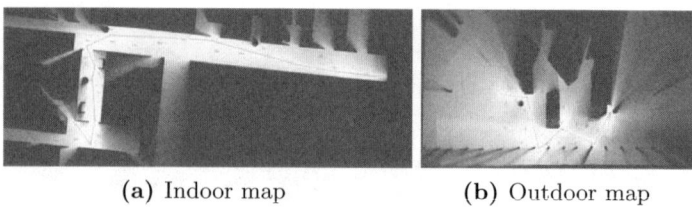

(a) Indoor map   (b) Outdoor map

Fig. 2. Some maps produced during the competition

5.1   RoboCup Rescue Virtual Robots Competition 2006

During the competition the maps provided by the participating teams were judged on their quality. The quality score had two components, a metric quality which was scored automatically and a topological quality which was assessed manually. The basis for the topological quality was how well the map corresponded to the actual environment, supplemented with bonuses for the utility of the map for a first responder. Utility increased with the ability for a person to determine which areas had been searched, where hazards may be located, and where victims were found. The more additional information was depicted on the map, the better the utility.

During the three qualification rounds (denoted by Q1-Q3) we managed to qualify for the semi-finals (denoted with Semi1 and Semi2). In these last two rounds our team of robots produced the maps displayed in Figure 2. A jury assessed the topological quality of our maps. In the competition, we achieved the maximum quality score by displaying very accurate and highly detailed maps.

The metric map quality was based on how well some artificial landmarks were localized. Our method achieved a root mean square (RMS) localization error of $0.2m$ and $0.02m$.

Subsequently, our team received the Best Mapping Award from the Virtual Robots competition as a special recognition of the high quality of our maps.

5.2   Results on Real World Data

The Cogniron data set [18] has been published on the Radish website [19]. The data-set has been acquired using a Nomad Scout robot in a home environment. The maps produced by our algorithm for this data-set are based exclusively on the laser range data, which were recorded using a SICK LMS-200 laser scanner.

Two maps generated from this data-set are shown in Figure 3. The odometry information was not used when constructing the maps. It should also be noted that our loop closure algorithm only works with landmark observations that could be acquired in the USARSim simulator. We have not implemented any detection algorithm for other landmarks or loops in general, so the maps presented were generated without explicitly closing any loops.

The Cogniron data set also includes the sensor logs for a run in which the robot traversed three loops. Despite these loops, and still without using odometry

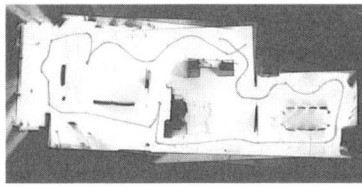
(a) Single Loop

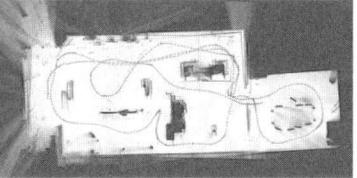
(b) Three Loops

Fig. 3. Corrected maps of the Cogniron data-set. The maps show very high detail by preserving small obstacles such as chairs and vases.

data or explicit closing loops, a highly consistent map is produced which is of comparable quality to the one of the former run.

6 Conclusion and Future Work

We presented a hybrid SLAM approach that combines grid-based and topological maps. Our approach thereby exhibits the best of both: Highly detailed maps are learned without limiting scalability to many robots.

Our system was used in the computer-simulated emergency sites of the Rescue Virtual Robot competition at the RoboCup World Championships of 2006, where it supported teams of up to 8 robots. The maps produced by our system earned the Best Mapping Award at this competition. Our experiments with real data show that these results carry over to real-world situations.

The presented work can be intuitively extended into 3D in the future. The graph structure, as well as all other algorithmic details presented, would remain unchanged, only a 3D scan matcher is needed.

Further research should investigate how current exploration algorithms can be adapted to work well with graph-based maps. This includes efficient ways to compute important information like frontiers [20]. Explicitly including meeting points in the plans to facilitate the active detection of overlapping regions and loops would significantly improve the resulting combined maps. Existing algorithms for grid maps might still apply with slight modifications.

References

1. Thrun, S.: Robotic Mapping: A Survey. In: Lakemeyer, G., Nebel, B. (eds.) Exploring Artificial Intelligence in the New Millenium, Morgan Kaufmann (2002)
2. Moravec, H.: Sensor fusion in certainty grids for mobile robots. AI Magazine 9, 61–74 (1988)
3. Thrun, S.: Learning metric-topological maps for indoor mobile robot navigation. Artificial Intelligence 99(1), 21–71 (1998)
4. Tomatis, N., Nourbakhsh, I., Siegwart, R.: Hybrid simultaneous localization and map building: Closing the loop with multi-hypotheses tracking. In: Proceedings of the IEEE International Conference on Robotics and Automation, Washington DC, USA, May 11 - 15 (2002)

5. Bosse, M., Newman, P., Leonard, J., Teller, S.: An Atlas Framework for Scalable Mapping. In: Proceedings of the IEEE International Conference on Robotics and Automation (ICRA) (2003)
6. Lisien, B., Morales, D., Silver, D., Kantor, G., Rekleitis, I., Choset, H.: The hierarchical atlas. IEEE Transactions on Robotics and Automation 21, 473–481 (2005)
7. Howard, A., Sukhatme, G.S., Matarić, M.J.: Multi-robot mapping using manifold representations. In: Proceedings of the IEEE - Special Issue on Multi-robot Systems (2006)
8. Balakirsky, S., Scrapper, C., Carpin, S., Lewis, M.: Usarsim: providing a framework for multi-robot performance evaluation. In: Proceedings of PerMIS 2006 (2006)
9. Carpin, S., Wang, J., Lewis, M., Birk, A., Jacoff, A.: High fidelity tools for rescue robotics: Results and perspectives. In: Proceedings of the 2005 RoboCup Symposium (2005)
10. Carpin, S., Lewis, M., Wang, J., Balakirsky, S., Scrapper, C.: Bridging the gap between simulation and reality in urban search and rescue. In: Lakemeyer, G., Sklar, E., Sorrenti, D.G., Takahashi, T. (eds.) RoboCup 2006: Robot Soccer World Cup X. LNCS (LNAI), vol. 4434, Springer, Heidelberg (2007)
11. Montemerlo, M., Thrun, S., Koller, D., Wegbreit, B.: Fastslam: A factored solution to the simultaneous localization and mapping problem. In: Proceedings of the AAAI National Conference on Artificial Intelligence, Edmonton, Canada, AAAI (2002)
12. Grisetti, G., Stachniss, C., Burgard, W.: Improved Techniques for Grid Mapping With Rao-Blackwellized Particle Filters. IEEE Transaction on Robotics 23(1), 34–46 (2007)
13. Lu, F., Milios, E.: Globally Consistent Range Scan Alignment for Environment Mapping. Autonomous Robots 4, 333–349 (1997)
14. Pfister, S.T., Kriechbaum, K.L., Roumeliotis, S.I., Burdick, J.W.: A Weighted Range Sensor Matching Algorithm for Mobile Robot Displacement Estimation. In: Proceedings of the IEEE International Conference on Robotics and Automation (ICRA) (2002)
15. Lu, F., Milios, E.: Robot Pose Estimation in Unknown Environments by Matching 2D Range Scans. Journal of Intelligent and Robotic Systems 18, 249–275 (1997)
16. Slamet, B., Pfingsthorn, M.: ManifoldSLAM: a Multi-Agent Simultaneous Localization and Mapping System for the RoboCup Rescue Virtual Robots Competition. Master's thesis, Universiteit van Amsterdam (2006)
17. Olson, E., Leonard, J., Teller, S.: Fast Iterative Optimization of Pose Graphs with Poor Initial Estimates. In: Proceedings of the IEEE International Conference on Robotics and Automation, pp. 2262–2269 (2006)
18. Zivkovic, Z.: IEEE/RSJ IROS 2006 Workshop: From sensors to human spatial concepts (2006)
19. Radish: The robotics data set repository (-)
20. Yamauchi, B.: A frontier based approach for autonomous exploration. In: Proceedings of IEEE International Symposium on Computational Intelligence in Robotics and Automation, Monterey, July 10-11, 1997 (1997)

A Force Sensor Made by Diaphragm Pattern Mounted on a Deformable Circular Plate

Alberto Tarizzo and Giuseppe Rella

Department of Automatica e Informatica
Politecnico di Torino, C.so Duca degli Abruzzi, 24 – 10129 Torino, Italy
tarizzo@isaacrobot.it, giuseppe.rella@gmail.com

Abstract. Measurement of the real distribution of pressure under the foot is an important challenge to obtain an efficacious stability control and a real dynamic walk. This paper is the result of the work of two students in mechanical and mechatronic engineering who have built a force sensor using a diaphragm pattern mounted on a deformable circular plate. The result of this study is a cheaper but accurate force sensor that will be mounted on the humanoid robot "I-2" of Politecnico di Torino. The design of the sensor in centred to analyse the deformation of a circular plate loaded at the center considering various edge conditions. It begin from the structural analyse of the plates considering different loads and edge conditions to obtain a deformation model as near as possible to reality. Final goal will be to obtain an output voltage proportional to the deformation of the plate.

Keywords: Circular plate, Loading condition, Edge condition, Strain gage.

1 Introduction

During development of new humanoid robot prototype "I-2", one of principal goal is to make a foot able to integrate four force sensors, one on each corner.

In order to obtain an efficient dynamic walk, the sensor must be accurate and with low hysteretic losses. Actual load-cells are big or heavy for this kind of application, and miniaturized sensors are too delicate for this use. After several tests a small deformable circular plate has been selected as part of this sensor. A small strain gage has been applied on the surface of the plate, in order to measure the deformation.

The design of the sensor is focused to analyse the deformation of a circular plate loaded at the center considering various edge conditions. It begin from the structural analyse of the plates considering different loads and edge conditions to obtain a deformation model as near as possible to reality.

Strain gage output is amplified by a small circuit mounted on the surface of the foot. In the end the complete sensor will provide a reasonable output voltage.

2 Symmetrical Bending of Circular Plates

Considering an element of the plate delimitated by the two meridian plane rotated of $d\vartheta$ and two cylinder of radius r e $r+dr$ respectively and imaging that plate is charged

U. Visser et al. (Eds.): RoboCup 2007, LNAI 5001, pp. 465–471, 2008.
© Springer-Verlag Berlin Heidelberg 2008

with a dispensed load q, the element will be affected by flexing moments and cut strains on its board. Radius of plate curvature in rz plane is R_r, radius of plate curvature in ϑz plane is R_ϑ. Unitary percentage deformations can be obtained: furthermore, for both planes, correspondent stress values are shown in (1).

$$\sigma_r = \frac{E}{1-v^2} z\left[\frac{1}{R_r} + v\frac{1}{R_\vartheta}\right] \quad \sigma_\vartheta = \frac{E}{1-v^2} z\left[\frac{1}{R_\vartheta} + v\frac{1}{R_r}\right] \tag{1}$$

Hence:

$$M_r = D\left[\frac{1}{R_r} + v\frac{1}{R_\vartheta}\right] \quad M_\vartheta = D\left[\frac{1}{R_\vartheta} + v\frac{1}{R_r}\right] \tag{2}$$

$$D = \frac{E}{1-v^2}\frac{h^3}{12} \tag{3}$$

In (3) flectional stiffness module for a plate with thickness h is expressed. The curvature radius R_r and R_θ if plate and load q present axial symmetry (q is orthogonally distributed respect considered plate) are tied trough inclination angle φ. Inclination angle is expressed in (4).

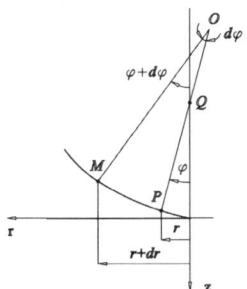

Fig. 1.

$$\varphi = -\frac{dw}{dr}. \tag{4}$$

Hence:

$$\frac{1}{r_r} = -\frac{d^2w}{dr^2} = \frac{d\varphi}{dr}. \tag{5}$$

r_r is curvature radius of surface on rz plane. The curvature radius θ_ζ in the point P correspond to the curvature radius in θ_z plane which coincides with the principal curvature in the plane that contains the perpendicular to surface in point P and is perpendicular to the rz plane. Principal curvature ρ_θ in θ_z plane for Meusnier's theorem is the distance PQ and correspond to $r_\vartheta \sin\varphi = PQ\sin\varphi = r$.

$$r_\vartheta = \frac{r}{\sin\varphi} \cong \frac{r}{\varphi} \tag{6}$$

Now unitary deformations can be written as in (7).

$$\varepsilon_r = \frac{z}{r_r} = z\frac{d\varphi}{dr} \qquad \varepsilon_\vartheta = \frac{z}{r_\vartheta} = z\frac{\varphi}{r}. \tag{7}$$

The point P is interested by a variation of r equal to $b=z\varphi$, so the circumference that was $2\pi r$ after deformation is $2\pi(r+b)$. ε_ϑ and ε_r can be extimated as in (8).

$$\varepsilon_\vartheta = \frac{2\pi(r+b)-2\pi r}{2\pi r} = \frac{b}{r} = z\frac{\varphi}{r} \qquad \varepsilon_r = \frac{dl'-dl}{dl} = \frac{zd\varphi}{dr} = z\frac{d\varphi}{dr} \tag{8}$$

During deformation, lateral walls of the place remain inside the space delimited by meridian planes that contain z axis of the plate and form between them $d\vartheta$ angle. If the rotation φ of the plate's element should be rigid, there will be the same $\overset{\!\!\vee}{\phi}$ rotation in both of element's faces. In deformation, faces remain inside meridian section: they will be affected only by the normal component of $\overset{\!\!\vee}{\phi}$ rotation indicated with φ_ϑ. This component has to be nullified, for this reason there is a relative rotation between the two lateral faces. After calculation of the curvature of elements with radius r_ϑ, equations shown in (2) can be changed as follows.

$$M_r = D\left[\frac{d\varphi}{dr} + v\frac{\varphi}{r}\right] = -D\left[\frac{d^2w}{dr^2} + v\frac{1}{r}\frac{dw}{dr}\right] \tag{9}$$

$$M_\vartheta = D\left[\frac{\varphi}{r} + v\frac{d\varphi}{dr}\right] = -D\left[\frac{1}{r}\frac{dw}{dr} + v\frac{d^2w}{dr^2}\right] \tag{10}$$

From rotation equilibrium around tangential direction, expression (16) is valid. Also, replacing M_r e M_ϑ in (11), (12) is obtained.

$$\frac{dM_r}{dr} + \frac{M_r - M_\vartheta}{r} + Q_r = 0 \tag{11}$$

$$\frac{d^3w}{dr^3} + \frac{1}{r}\frac{d^2w}{dr^2} - \frac{1}{r^2}\frac{dw}{dr} - \frac{Q_r}{D} = 0 \tag{12}$$

3   Circular Plate Loaded at the Center

When Q_r is represented by a function of r, equation (12) can be integrated without any difficulty in each particular case. For our application we consider the case of bending of a plate by shearing forces Q_0 uniformly distributed along the inner edge. From the value of shearing force per length-unit of a circumference of radius r and from (12) deflection is found after an integration. Constants of integration can be calculated from boundary conditions.

$$C_1 = \frac{P}{4\pi D}\left(\frac{1-\upsilon}{1+\upsilon} - \frac{2b^2}{a^2-b^2}\log\frac{b}{a}\right) \tag{13}$$

$$C_2 = -\frac{(1+\upsilon)P}{(1-\upsilon)4\pi D}\frac{a^2b^2}{a^2-b^2}\log\frac{b}{a} \tag{14}$$

$$C_3 = \frac{Pa^2}{8\pi D}\left(1+\frac{1}{2}\frac{1-\upsilon}{1+\upsilon} - \frac{b^2}{a^2-b^2}\log\frac{b}{a}\right) \tag{15}$$

Slope and deflection at any point of the plate are now calculable. In the limiting case where b is infinitely small, $b^2 \log(b/a)$ approaches zero and the constants of integration change. In this particular case, a value for w is obtained in (16).

$$w = \frac{P\cdot r^2}{8\pi D}\left[\frac{3+\upsilon}{2(1+\upsilon)}\left(a^2-r^2\right)+r^2\log\frac{r}{a}\right] \tag{16}$$

This coincides with the deflection of a plate without a hole and loaded at the center.

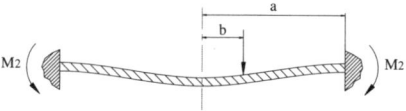

Fig. 2.

Applying one of the boundary condition it is possible to obtain the value of the deflection at any point of the plate shown in fig. 2.

$$w = \frac{P\cdot r^2}{8\pi D}\log\frac{r}{a} + \frac{P\cdot r^2}{16\pi D}\left(a^2-r^2\right) \tag{17}$$

4 Mechanical Details

The mechanical structure of the sensor is made directly on the foot and needs only a steel-made dowel stuck with Henkel-Loctite 638 into an aluminium alloy cylinder to prevent buckling effect.

Fig. 3. Sensor unit

5   Electronic Parts

5.1   The Strain Gage

The sensor used to measure the deformation is a strain gage with a full-bridge diaphragm pattern, an N2A-06-S102H-350 by VISHAY. Its internal circuit is presented in fig. 4.

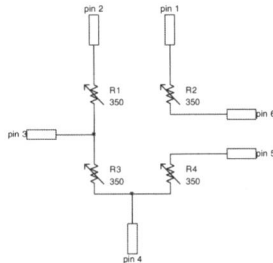

Fig. 4.

5.2   Pre-conditioning Circuit

In order to use this sensor with a reasonable output range a ±5V DC dual power supply is provided between pin 1 and pin 4. Also, a short circuit is needed between pin 1 and 2 and a trimmer has been connected between pin 5 and 6. This trimmer allows balancing temperature-dependent offset and other offsets of the system, setting a 0V output when the force applied is 0N. The value of this trimmer has to be comparable with the values of internal resistances of the sensor. For test a 10Ω multiturn trimmer has been used. Output is a differential voltage, proportional to the deformation of the plate and so to the linear force applied; amplification needs to be projected, because the mean value of the voltage, assuming mean input force to be 70N, is about 1mV – 1.5mV.

5.3   Amplification

A differential OPAMP, device INA118, has been used for a first treatment of sensor's output signal. Besides that, a linear amplification with a non-inverting TL081 circuit has been made. INA118 is used with a gain resistor of 560Ω. A 2.7µF capacitor has been put between output and reference pins, to obtain a pole around 100Hz. Furthermore, a TL081 circuit has been used in non-inverting configuration, giving static relation in (18).

$$\frac{V_{OUT}}{V_S} = \left(1 + \frac{50k\Omega}{560\Omega}\right) \cdot \left(1 + \frac{22k\Omega}{2.2k\Omega}\right) \tag{18}$$

A capacitor put in parallel with the 22kΩ resistor forces the system to have a low-frequency pole at about 10Hz; that is to avoid most of the noises and the force spikes that are not useful for measures. High-frequency poles of the system are neglected. Complete electronic part scheme is presented in fig. 5.

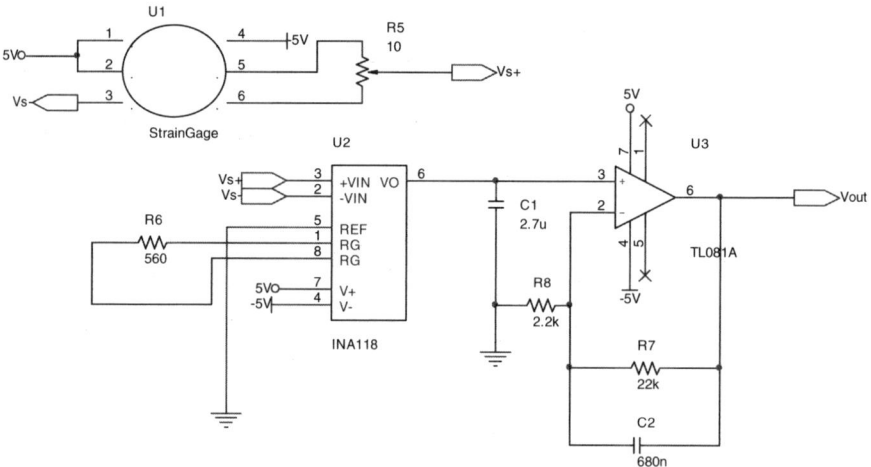

Fig. 5.

6 Results

A characteristic curve of the system has been plotted from test measures. An Ergal (Al7075) rectangular plate (weight 0.2Kg, 64mm x 80mm x 15mm) has been used to put some iron weights on the measure sting of the sensor.

The INL (integral non linearity error) is the measures we took is 20.4mV, and we have the maximum non linearity around 90N input. Making a very small saturation of the first values around the origin, we can consider (19) as final characteristic of the sensor. The (19) is the equation which should be implemented in a sensor reading system.

$$V_{OUT} = k \cdot F + m \quad k = 18.1 \frac{mV}{N} \quad m = -1.98mV \tag{19}$$

Main source of error of the system is the trimmer; its screw could be very sensible to shocks and variations of ambient temperature.

That is why a small value trimmer was used, however it could be valuable to calibrate the sensor every time you use it.

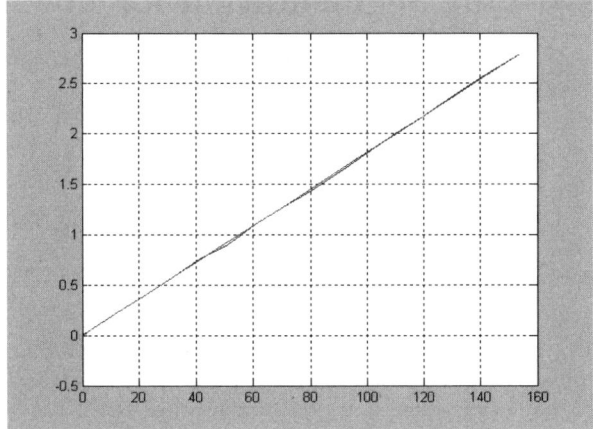

Fig. 6. Real (blue) and linear (red) response. X-axis is input force on dowel [N], Y-axis is output voltage [V].

The recommended range of temperature is 0 C° - 50 °C: inside this range sensor and amplifiers works with a linear behaviour. We saw also some hysteresis problems with the sensor, but this is considered not significant in low and high conditions of load, and has got a peak of 20mV in a medium range of load. This means that if we consider a sensor resolution of 1N our system is strong against hysteresis and other offsets and errors.

7 Conclusions

For a dynamic analysis we remand to future developments of this system. Considering the good output range of the system, the large input force range (good for an humanoid robot but also for other small-medium environments) and the good linear characteristic, we find out that this could be a sensor and electronic conditioning part used in many fields.

References

1. Curti, G.: Appunti delle lezioni del corso di tecnica delle costruzioni meccaniche, Politecnico di Torino (2003)
2. Curti, G.: Appunti delle lezioni del corso di costruzione di macchine, Politecnico di Torino (2003)
3. Timoshenko, S., Woinowsky-Krieger, S.: Theory of plates and shells, pp. 51–68. McGraw-Hill (1986)
4. Vishay Micro Measurements, Precision Strain gages. Vishay Micro-Measurements press (2003)
5. Vishay Micro-Measurements, Design consideration for diaphragm pressure transducers. Vishay Micro-Measurements press (2003)
6. http://www.vishaymg.com

Crossed-Line Segmentation for Low-Level Vision*

John Atkinson and Claudio Castro

Department of Computer Sciences
Universidad de Concepcion, Concepcion, Chile
atkinson@inf.udec.cl

Abstract. This work describes a new segmentation method for robotic soccer applications. The approach called crossed-line segmentation is based on the combination of region classification and a border detector which meet homogeneity criteria of medians. Experiments suggest that the method outperforms traditional procedure in terms of smoothing and segmentation accuracy. Furthermore, existing noise in the images is also observed to be reduced without missing the objects' borders.

Keywords: Color-based Segmentation, Robotic Vision, BLOBs.

1 Motivation

In the context of the RoboCup 4-legged competition, one of the challenging issues has been the dependence of the image segmentation methods on variations of the lighting conditions on the playfield [2]. Despite the fact that official competitions are held in highly controlled lighting, moving people, shadows, etc significantly affect the accuracy of the segmentation methods.

Most of the segmentation strategies [3,1,7] separate the color space by defining areas in which a single color can be identified. However, previous research [2] suggests that a huge amount of color sharing areas in the space depends on lighting conditions.

In this work, a new strategy for color-based segmentation which is a combination of the methods described above is proposed. This uses crossed-line filtering, with a special focus on efficiency issues. Thus we are able to provide effective segmentation at real-time rates. The approach shows several advantages whenever environment conditions do not allow using LUT segmentation properly including its better tolerance to lighting conditions and noise.

This paper is organized as follows: in section 2 the main issues and methods for color-based segmentation in robotic soccer vision system are discussed. Section 3 outlines our new approach to crossed-line segmentation. In section 4, we describe the performance of a system using the method and comparative assessments

---

* This research is partially sponsored by the National Council for Scientific and Technological Research (FONDECYT, Chile) under grant number 1070714 *"An Interactive Natural-Language Dialogue Model for Intelligent Filtering based on Patterns Discovered from Text Documents"*.

U. Visser et al. (Eds.): RoboCup 2007, LNAI 5001, pp. 472–479, 2008.

with other representative segmentation methods. Finally, some conclusions and further issues are drawn in section 5.

2   Related Work

Most of the color image segmentation methods require a significant amount of processing which is out of reach on the robots due to demanding speed requirements. For this, these methods use a *Look Up Table* (LUT) that makes reference to a single color value based on the value of one pixel of the target image [4,7]. Accordingly, the number of referenced values becomes less than the total number of colors. One problem with using LUT is that in order to make the processing faster, a big amount of additional memory is required.

Color-based segmentation has been tackled by using decision trees based classification techniques [7]. In particular the $C4.5$ algorithm is applied to separate a multi-dimensional space into different color classes in the YUV space. This uses previously classified pixels so as to isolate and generalize each generated cluster.

Because of these issues, adaptive segmentation techniques have shown to be promising for RoboCup. In particular, *Support Vector Machines* (SVM) have been used to clasify each color class separately. A single class is used to avoid classification errors as each sample to be segmented represents only a part of the space. SVM-based segmentation showed better results than LUT filtering by reducing errors in almost 18%. However, its tolerance to lighting changes has not been proved yet.

Overall, most of these segmentation methods require a set of samples consisting of preclassified colors extracted from representative images on the playfield's conditions. This process does not only take hours but also generates few samples of the candidate pixels obtained from the camera during a game. LUTs used by segmentation methods show huge gaps between points of each color class hence segmentation tries to generalize (i.e., spreed the points' influence) most of the sample set without missing the shape or the relation between each class. The idea here is based on the fact that for points belonging to a color class, the probability that their surrounding points are in the same class is high.

3   A New Crossed-Line Segmentation Method

A new color-based segmentation method is proposed for the 4-legged RoboCup competition. In order to assess its effectiveness, the procedure was compared with two traditional color-based segmentation strategies based on the results obtained from previous RoboCup 4-legged competitions: boundary (limit) segmentation and spatial influence segmentation [4].

Our proposed method, called *Crossed-line Segmentation* is based on region segmentation but uses a median-based homogeneity criterion and a color-difference border detector. The method works under the assumption that average points tend to be more stable than separated points providing that images are very noisy in spaces where regions of the same color can be identified.

In order to reduce noise, the method scans an image at one direction from one side to the other by assigning the average value of the same color to each point. The procedure stops whenever significant changes on the line are found. The crossed-line segmentation is so described as follows:

ALGORITHM Crossed-Line Segmentation
INPUT: image with features width (n), margin (m), median (X_c)
       points of one line of the image $\{x_1, x_2, ..., x_{n-1}, x_n\}$
       current position (i), starting point of the sequence (i_0)

   Segmentation starts at position $i = 1$
     (1) $X_c = x_i,\ \sum c = x_i,\ i_0 = i$
     (2) IF $i \leq n$ THEN
          $i = i + 1$
          GOTO (3)
        ELSE GOTO (4)
     (3) IF $|X_c - x_i| > m$ or (i is a border point) THEN
          $\forall x \in \{x_{i_0}, ..., x_i\}, x = X_c$
          GOTO (1)
        ELSE $\sum c = \sum x_{i_0}, ..., x_i,\ X_c = \sum c : (i - i_0)$
          GOTO (2)
     (4) FOR ALL x $\in \{x_{i_0}, ..., x_i\}, x = X_c$
            Advance one line
            GOTO (1)
END

The margin (m) is experimentally defined and the border (step (3)) is detected by using the border detector operator. Although the procedure works for one channel, this can also be applied to three YUV channels by computing the sums and averages for each channel. Whenever the absolute value of the difference between the median and the point exceeds the margin for the three channels, the median value is assigned to the predecessors of each channel. Once the image is segmented, compression is carried out by using RLE.

Applying this method has a smoothing effect on every point not meeting the separation rule. The procedure also complies with the region segmentation conditions [6] in which every point belongs to a surrounding region only if $|X_c - x_i| > m$ and this is not at the image's border.

The problem with using one direction (either horizontal or vertical) is that the average for one direction may be different from that for the orthogonal direction, though the color-based segmented with this average is likely to be the same. To deal with this issue, the strategy is applied twice at different directions and the result of each scan is stored into a temporary image containing the averages (figure 1). Furthermore, the algorithm's behavior is the same for the three cases. Results of the segmented image can be seen in figure 1. In order to define the margin m, different testings were performed by filtering three representative images. These represent the ball at one of the corners, a landmark and the ball

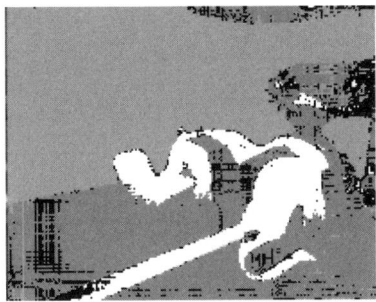

Fig. 1. Different applications of *Crossed-line Segmentation*: left to right, and downwards

in front of one arch. For every image, the margin's value is increased and the number, width and height of the BLOBs (*Binary Large Object*) representing the amount of noise is recorded.

Results of these experimental settings suggest that the method gets stabilized between margin values 20 and 70. For values above 70, width and height values get unstable. The margin should be kept big enough to reduce the noise and small enough to avoid removing small areas of interest, hence the margin used for our competition has been set to 25.

The influence of the border filter on the crossed-line segmentation was also assessed. For this, the previous three images were used again and the parameters used to set the margin were kept.

The method with and without border detection shows similar stable behaviors for the same margin settings (20-70). All the graphics show that width and height values are almost the same for the range 20 to 70. Both criteria become unstable whenever the margin exceeds 70. However, using borders shows better and stable behavior for values above 70, whenever the detection quit providing information on the differences of medians. For the remaining experiments, the method using the border detector was evaluated as this satisfies the region segmentation rules [6]. It is important to highlight that time efficiency was not an issue as the time spent by this operator is not significant for the current testings.

4   Comparative Experiments

In order to evaluate our crossed-line segmentation method and compare it with other segmentation methods, several experiments were carried out under different lighting conditions: fixed and variable. The aim was to assess metrics such as segmentation accuracy, noise and light tolerance, etc. Each method analized 41 images, scenarios and locations in which each SONY AIBO Robot gets involved during playing time. In addition, segmentation for 17 images using variable lighting, different from the previous set of images, was also considered.

Each segmented image was characterized based on numeric information obtained from the BLOBs representing color areas of interest. The target data included the number of BLOBs for every image, the number of BLOBs of a specific color, average width and height of the generated BLOBs, and the width and height of the biggest BLOB in the image. To assess the existing methods under ideal conditions, a list of manually defined BLOBs was defined. Next, crossed-line segmentation was compared with the median-filter segmentation by applying a 5x5 filter on the points of the image. This was then segmented using the LUT technique. Because of the method's image smoothing by using average values of a set of pixels, the results were roughly similar to the median filter.

In order to compare the proposed strategy with traditional median filters, a configuration setting was used by applying a 5x5 media filter on the image's points and then had it segmented via LUT. For the experiments, a set of images extracted from several localizations on the playfield was obtained [7] with different features.

Results of applying different segmentation methods for under fixed lighting can be seen in figure 2. Both LUT and boundary segmentation generate big amounts of noise, and a huge number of average BLOBs above 200 for each image with a average size of 5x5. Median filtering and LUT segmentation produced almost no noise because the image has been slightly smoothed above the median filter's value.

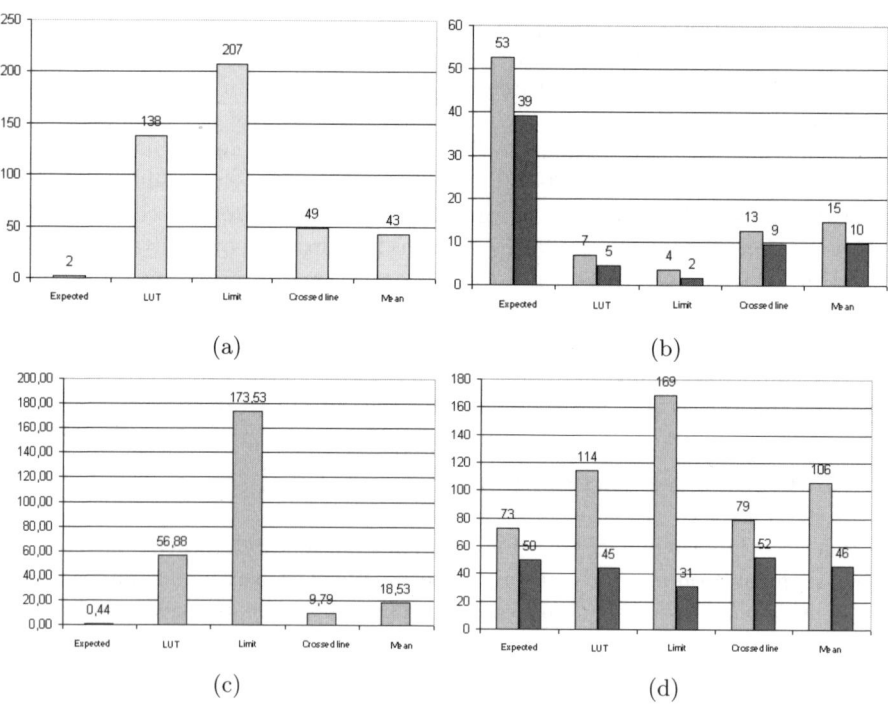

Fig. 2. Segmentation using fixed lighting: Average number of BLOBs per image (a), Average Width and Height of the BLOBs in pixels (b), Average Radius of BLOBs (c) and Average height of the Biggest BLOB (d)

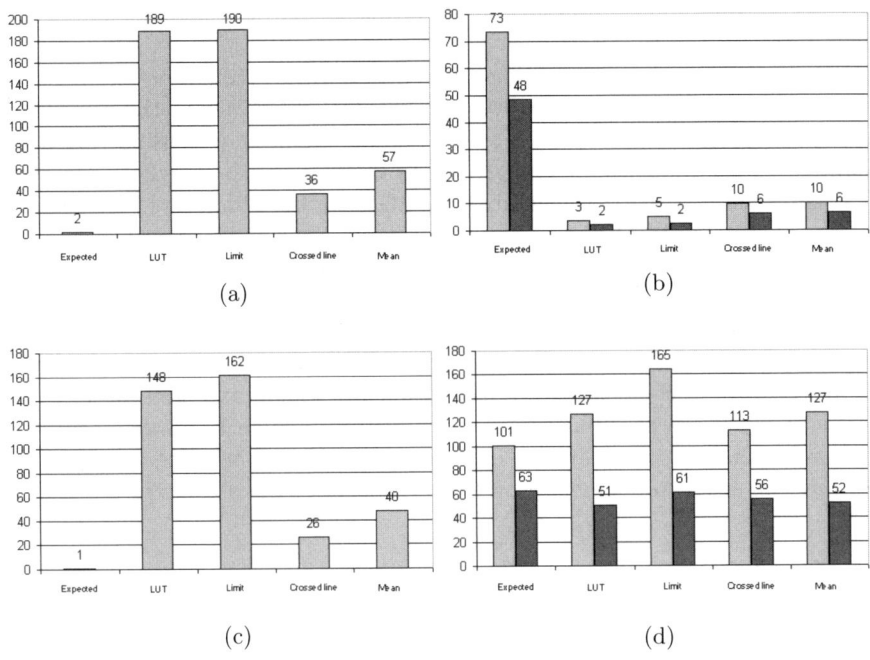

Fig. 3. Segmentation using variable lighting: Average number of BLOBs per image (a), Average Width and Height of the BLOBs in pixels (b), Average Radius of BLOBs (c) and Average height of the Biggest BLOB (d)

Crossed-line segmentation shows a similar behavior on noise reduction having 1/3 less of BLOBs than LUT segmentation alone. Both methods have the same average size of BLOBs, bigger than LUT and limit segmentation, which is due to the existing noise reduction: median filtering tends to concentrate big amounts of noise and transform this into *solid* areas by joining noise points which were previously separated.

Colorimetric distortion affected all the tested methods being the limit segmentation that having the worse performance. Figure 2(c) shows a median close to 174 blue BLOBs per image even when the ideal amount is less than one. This effect can also be observed for generalized LUT segmentation. Graphics in figure 2(c) also shows that settings for the median filter and the crossed-line segmentation have a smaller media value for blue BLOBs as small noise areas are concentrated into bigger regions and so low-intensity noise is removed by the filter. In addition, bigger decreases in line segmentation may be due to the fact that most of the classified points (i.e., blue for LUT segmentation) have been classified using colors not included in the table (i.e., black areas).

As for the height and width of the biggest BLOBs, line segmentation proved to get values closer to the ideal, whereas for limit segmentation, because of bigger areas of blue, the method produced width values that are far from the ideal sizes.

Variation in size can also be explained in terms of colorimetric distortion of blue which increases the BLOBs size.

Sampling using variable lighting showed a behavior similar to the previous case despite the fact that the quality of segmentation for each method decreased significantly. One explanation for this is the big changes of lighting conditions produced by the influence of sunlight on all the areas of the playfield.

A significant increase in the number of BLOBs for both limit and LUT segmentation (almost 200 BLOBs per image) can be seen at figure 3(a). Colorimetric distortion had an impact on the increase of the quality of BLOBs as seen at figure 3(c). However, some areas of the playfield are identified with incorrect colors caused by sunlight which, in turn, allows the green playfield area to be classified using white, blue and yellow colors. It is also important to stress that the average size of the biggest BLOBs for linear segmentation gets closer to the ideal size as for the previous samples. Unlike the median filter which smooths the border making their definition difficult by modifying their size, our approach generates an accurate segmentation for the border of the objects in the image.

5   Conclusions

A new method for image segmentation using a crossed-line filtering strategy was proposed. The approach shows several advantages whenever environment conditions do not allow using LUT segmentation properly. Under noisy images segmented via LUT, the method smooths most of the noise, specially in bigger areas of the same color. Other promising feature is its ability to remove noise from images with no loss of quality of the objects' borders.

This advantage is specially useful for RoboCup because for images very distant from arches and landmarks, a difference of 2 or 4 pixels may produce several centimeters of error in localizing objects. Unlike traditional smoothing techniques, our method does not only avoid border blurring problems but this stresses them. This is a key issue in RoboCup as for images very distant from arches and landmarks, a difference between 2 and 4 pixels may imply several centimeters of error to localize the objects. One of the drawbacks is that the strategy tends to erroneously classify smaller color areas or those with slight changes from one color to other as points show differences between colors that are lower than the selected margin. The procedure to analyze a full image runs on linear time and is very similar to spatial methods such as median filters. In addition, although all the segmentation methods are affected by colorimetric distortion, the crossed-line proved to be less subject to these variations.

References

1. Bruce, J., Balch, T., Veloso, M.: Fast and inexpensive color image segmentation for interactive robots. In: Proceedings of IROS-2000 (2000)
2. Chalup, S., Middleton, R., King, R.: The nubots' team description for 2005. In: RoboCup 2005: Robot Soccer World Cup VII (2005)

3. Chalup, S., Creek, N., Freeston, L., Lovell, N., Williams, M.-A.: When the nubots attack! - the nubots 2002 team report. Technical report, Newcastle Robotics Laboratory, The University of Newcastle, Australia (2002)
4. Michaela, F., Chalup Stephan, K., Oliver, C., Coleman Oliver, J., et al.: The nubots' team description for 2003. In: RoboCup 2003: Robot Soccer World Cup V (2003)
5. Hong, W., Georgescu, B., Zhou, X.S., Krishnan, S., Ma, Y., Comaniciu, D.: Database-guided simultaneous multi-slice 3d segmentation for volumetric data. In: Leonardis, A., Bischof, H., Pinz, A. (eds.) ECCV 2006. LNCS, vol. 3954, Springer, Heidelberg (2006)
6. Horowitz, S., Pavlidis, T.: Picture segmentation by a direct split-and-merge procedure. In: Int. Joint Conf on Pattern Recognition, pp. 424–433 (1974)
7. Quinlan, M.J., Chalup, S.K., Middleton, R.H.: runsiwft teamp description paper 2003. In: RoboCup-2003: Robot Soccer World Cup V (2003)
8. Ross, M.: Statistical motion segmentation and object tracking without a-priori models. In: 11th International Fall Workshop on Vision, Modeling, and Visualization (VMV 2006), Aachen, Germany (2006)

A Neural Network-Based Approach to Robot Motion Control

Uli Grasemann, Daniel Stronger, and Peter Stone

Department of Computer Sciences
University of Texas at Austin
Austin, TX 78712, USA
{uli,stronger,pstone}@cs.utexas.edu

Abstract. The joint controllers used in robots like the Sony Aibo are designed for the task of moving the joints of the robot to a given position. However, they are not well suited to the problem of making a robot move through a desired trajectory at speeds close to the physical capabilities of the robot, and in many cases, they cannot be bypassed easily. In this paper, we propose an approach that models both the robot's joints and its built-in controllers as a single system that is in turn controlled by a neural network. The neural network controls the entire trajectory of a robot instead of just its static position. We implement and evaluate our approach on a Sony Aibo ERS-7.

1 Introduction

Commercially available robots like the Sony Aibo usually come with built-in controllers that are designed to allow precise control over the robot's joint positions. In many applications, however, the goal is not simply to make the robot move to a given position, but rather to make it execute a given motion, i.e. to control the robot's position at all points in time. Furthermore, tasks like robot soccer, in which speed is an important factor, require the robot to execute motions like walks or kicks both precisely and at speeds close to the physical limits of the robot's effectors. The built-in controllers that come with robots like the Aibo are not designed for this task.

Most approaches that would allow precise control over a robot's motion require exact knowledge of all properties of the joints and motors involved, as well as the ability to bypass the built-in controllers and access the robot's effectors directly. Inexpensive, commercially available robots like the Sony Aibo usually do not meet these conditions.

In this paper, we explore an alternative approach to the problem, which uses neural networks that learn to predict the commands that are necessary in order to make the robot execute a predefined motion. The robot and its built-in joint controllers are both treated as part of system to be modeled. We implement and evaluate the proposed approach on a Sony Aibo ERS-7[1].

---

[1] http://www.aibo.com

U. Visser et al. (Eds.): RoboCup 2007, LNAI 5001, pp. 480–487, 2008.
© Springer-Verlag Berlin Heidelberg 2008

2   Background

Standard control theory [1] focuses on the task of finding a controller H that, given an observation of the state x of a system, provides an appropriate action at every time step such that the system eventually reaches a given target state. This paper considers the goal of enabling a robot to accurately execute a desired movement: a trajectory through its state space over time. The classical approach to robotic trajectory planning involves controlling the forces or torques exerted by the joints directly [2]. Applying this technique requires the parameters and specifications of the robot to be known, as well as low-level access to the robot's effectors. Inexpensive, commercially available robots like the Aibo usually meet neither of these conditions.

Fig. 1. A Sony Aibo ERS-7. The arrows point to the shoulder (1), abductor (2), and knee joint (3) of the Aibo's right front leg.

On the Sony Aibo robot, all control of the joints goes through the robot's API, which at the lowest level uses PID control [3]. However, previous work [4] suggests that each joint (at least on the Aibo ERS-210) can not be completely understood based on the theory of PID control. Nevertheless, because it is only possible to issue commands to the robot's joints through the PID controller, this paper considers that controller as part of the dynamical system, and therefore part of the problem. If the target is close to the actual position, for example, it will not move at maximum speed, even if maximum speed is required at the present part of the trajectory.

Several alternative approaches are commonly used to make a robot execute a given movement more reliably. For example, one way to increase precision is to slow down the movement of the robot so that the angle speeds involved are well below the maximum angle speeds possible. Another possibility is to search for a set of parameters used to create a sequence of angle requests, where an end goal like overall robot speed is used as a reward function [5,6].

Ideally, we would like some kind of an equivalent of classical controllers for motion control. As in standard control theory, there are two basic options. First, the equivalent of an open-loop controller would be a functional H_{open} that maps a desired trajectory T through the system's state space onto a sequence of appropriate actions U for any given time: $H_{open} : T \mapsto U$. This sequence can then be used on the robot in order to achieve the desired movement. Second, the equivalent of a feedback-controller would take as its input the present time t, the target trajectory T, and the current observation y of the robot's state x. Its output would be an appropriate action u that keeps the robot's motion sufficiently close to the target trajectory at any time:

$$H_{feedack} : (T, t, y) \mapsto u$$

In the real world, defining and finding such functionals is simplified by the fact that both time and the state space of a robot are effectively discrete, and the dependencies between the motor commands and the robot's trajectory are highly local. This allows us to define H_{open} in terms of a function h_{open} that maps a finite neighborhood $\pm l$ of T around a time t onto a single action u at the same time t: $h_{open} : (T(t-l), \ldots, T(t+l)) \mapsto u$. H_{open} is then defined by computing h_{open} for every discrete time step of T. For example, Stronger and Stone [4] construct such a function h_{open} for piecewise-linear trajectories by first constructing an empirical joint model of a Sony Aibo ERS-210, and then inverting that model to obtain h_{open}. By contrast, the approach described in this paper uses a neural network to obtain such functions directly using data acquired from a robot.

Finally, a wide range of previous work has also used neural networks to control robotic motion. Lewis et al. [7] show how neural networks can be used to approximate nonlinearities in the robot's dynamics. This method can be used for trajectory planning, but doing so requires direct control of the robot's motors, which is not available on many commercial robots. On a Sony AIBO, Billard and Ijspeert [8] use a neural network to generate qualitative variations on a type of motion, such as different gaits for walking. Angulo et al. [9] apply neuroevolution to a Central Pattern Generator (CPG) to demonstrate the emergence of a walking behavior. To the best of the authors' knowledge, these approaches have not been applied to the task addressed in this paper: performing accurate motion along an arbitrary trajectory.

3   A Neural Network-Based Approach to Motion Control

The basic idea behind our approach to robot motion control is simple: We use a neural network to predict which motor commands will cause the robot to execute a given movement. The robot's joints and their controllers together form the system we are trying to control.

Figure 2 illustrates the structure of our approach: A neural network maps a finite neighborhood of the

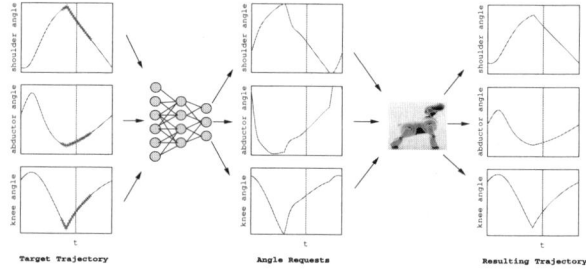

Fig. 2. The basic structure of the proposed motion controller at time t. A neural network maps a neighborhood of the target trajectory around t onto appropriate motor commands at time t. The motor commands are used by the Aibo and result in a motion close to the target trajectory.

target trajectory around the present time step onto a set of motor commands that is supposed to keep the robot on that trajectory. Computing the output of the neural network for each time step of a target trajectory gives a sequence of angle requests that can then be used by the robot to execute the desired motion. Note that the neural network plays the role of the function h_{open} defined in the previous section, implicitly defining an open-loop motion controller H_{open}.

The neural network, which represents the inverse of the system in question, can be learned directly from raw data: All that is needed is a sequence of angle requests U and the resulting movement T. For each time t, the neighborhood $T(t-l)\ldots T(t+l)$, together with the angle request $U(t)$, forms a training pattern for the network.

4   Experiments

The experiments reported in this paper were conducted using a Sony Aibo ERS-7, a commercially available four-legged robot with 17 degrees of freedom. All joints are equipped with PID controllers that cannot be bypassed to control the Aibo's movements on a lower level, and the precise specifications of both the Aibo's effectors and of the PID controllers are not documented. The ERS-7 has sensors on each joint that allow precise recording of the actual effects of any motion command. The motion commands are given in the form of one angle requests for each joint every 8ms. We used this maximum frequency in all experiments.

For the reported results, we focused on the task of controlling the Aibo's right front leg while the robot was not touching the ground. Figure 1 shows an ERS-7 and points out the joints involved in the reported experiments.

The first set of experiments served two separate purposes. First, it aimed to establish that the approach described in the last section leads to a significant improvement over just using the raw trajectory as motion commands. The experiment's second purpose was to find out whether a single neural network model of all joints involved performs better than having separate neural networks for each joint.

The first step was to create the neural network models of the Aibo's joints. As mentioned before we focused on controlling the Aibo's right front leg, which has three degrees of freedom: The shoulder, the abductor, and the knee (see Figure 1.) The Aibo was held in the air such that the leg never touched the ground.

4.1   Experiment I

We acquired training data for the neural networks by first creating a random continuous sequence of angle requests, then running those requests through an Aibo and recording the resulting movements using its sensors. Comparing the original angle requests and the resulting target trajectory in Figure 3 should give an impression of the kind of data used, although the data shown there were not part of the training set. Note how the actual trajectory lags behind the angle requests used to create it. Using about 80 seconds of training data, we then trained two different motion controllers for the Aibo's front leg: The first controller was intended to model all three joints at the same time using one neural network; the second controller used a separate network for each joint.

The input and output of the neural networks were exactly as described in section 3. Based on the estimated time it takes the Aibo to react to motor commands [4], we chose ±10 time steps as the size of the neighborhood on the target trajectory. This meant that the single-network model had 60 inputs (three joints × 20), and three output nodes (one for each joint.) The networks that modeled a single joint had 20 input nodes and one output node.

All networks were fully connected feedforward-networks with one hidden layer. The size of the hidden layer was the same as the input layer. The networks were trained for 2000 epochs using standard backpropagation with a small momentum term (0.2). The training rate was 0.1 for the first 1000 epochs, and 0.05 for the rest of the time. We used SNNS (the Stuttgart Neural Networks Simulator [10]) to create and train the networks.

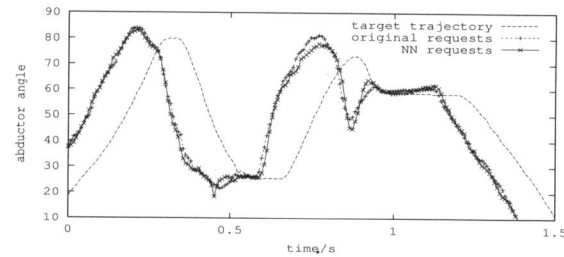

Fig. 3. Part of the test trajectory used in the first experiment (dotted line). Compare the original angle requests used to create the trajectory with the angle requests output by a neural network controller.

We then created a fresh sequence of random angle requests, and recorded the resulting movements of the Aibo's leg. Using these movements as a target trajectory, we used both neural network controllers independently to try and replicate the target trajectory on the Aibo.

In order to establish a reasonable baseline with which to compare our results, we also used the target trajectory as angle requests, after shifting it back by 12 time steps to allow for the lag. This is the equivalent of a controller that models only the time lag of the Aibo's joints. Additionally, we ran the original sequence of angle requests through the Aibo again, to find a practical upper performance limit due to motor and sensor precision.

Figure 3 shows part of the target trajectory for the Aibo's abductor joint, together with the original angle requests used to create it, and the angle requests that the single-network model thinks will reproduce the target trajectory. It seems like the requests created by the neural network stay reasonably close to the original.

Figure 4 compares the trajectories controlled by the two neural network controllers to the baseline trajectory. Both neural network controllers perform visibly better than the baseline, especially at sharp turns in the trajectory. Figure 5 compares the two neural network controllers. The height of each bar is the average Euclidean distance of the Aibo's foot from the target trajectory. The leftmost bar is the upper performance limit established by using the original set of angle requests again, and comparing the result to the target trajectory. The two bars in the center belong to the two neural network controllers. The bar on the right is the baseline error. The error bars denote the 95% confidence intervals for the distance averaged over 12 seconds. All differences are statistically significant ($p < .05$).

Figure 5 confirms the earlier impression that both neural network controllers perform well above the baseline. In fact, they both reduce the average error by more than half. Also, for the single network controller, the average distance from the target trajectory is

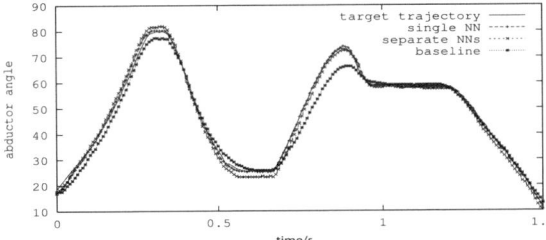

Fig. 4. Comparing trajectories controlled by neural networks with the baseline trajectory. Both neural network controllers perform visibly better than the baseline.

less than twice the upper performance limit defined by the Aibo's motor and sensor precision. Overall, our first experiment showed that both neural network controllers perform significantly above the baseline level, and the single network-controller is able to exploit the additional information it receives to outperform the controller using separate networks.

4.2 Experiment II

The second experiment also had two objectives. The first was to find out if the open-loop architecture chosen for the present implementation is able to handle trajectories outside the physical limits of the robot. Such trajectories are usually created to fool the built-in controllers into moving the joints faster than they would ordinarily, and would therefore be unnecessary given a working motion controller. However, it would still be useful to have a motion controller that, given a trajectory outside the physical constraints of the Aibo, creates the closest possible trajectory within the constraints.

The second objective was originally to make a quantitative comparison between our model and the ana-

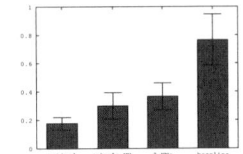

Fig. 5. The average Euclidean distance from the target trajectory (in cm) achieved by the neural network controllers are shown in the middle two bars

lytical model used by Stronger and Stone for the same task [4]. Since the results reported there were obtained using an earlier model of the Aibo, we attempted to implement the model on the new Aibo in order to make a quantitative comparison possible. However, early experiments revealed that the joint dynamics of the Aibo ERS-7 are sufficiently different from the earlier model as to make a direct implementation impossible. When the requested trajectory for a leg joint was set to a step function, the different joints exhibited qualitatively different and surprisingly erratic behaviors. The angle speeds changed unpredictably over time, and the joints' behavior did not appear to fall within the parameters of

Stronger and Stone's model. We believe that this in itself is a strong argument for an adaptive and more flexible approach to joint modeling.

Figure 6 shows the target trajectory used in this experiment. It is a half-ellipse with a period 65 timesteps, and could be realistically used for a fast walk on the Aibo ERS-7.

We used the single-network controller trained in the last experiment to try and reproduce the target trajectory on the Aibo. Figure 7 shows the resulting angle trajecto-

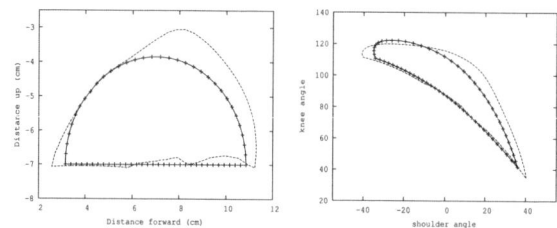

Fig. 6. The test trajectory used for the second experiment, in Cartesian coordinates (left), and in the Aibo's joint angle coordinates (right). The dotted lines are angle requests created by a neural network controller for this trajectory.

ries for the three joints involved. Like in the last experiment, the baseline curve was obtained using a controller that only compensates for the time lag between an angle request and the resulting motion. The trajectories of the shoulder and especially the abductor clearly show improvement over the baseline curve, while the trajectory for the knee is more or less the same as the baseline.

It would be reasonable to expect a corresponding improvement of the trajectory of the Aibo's foot in Euclidean space. However, no such improvement was observed. The average distance of the Aibo's foot from the target trajectory is about 8.5mm both for the baseline trajectory and for the one controlled by the neural network. This discrepancy can be understood as follows. When the neural network achieves an improvement over the baseline, the large knee angles moved the Aibo's foot close to the rotational axis of the abductor joint, which made the improvement in the abductor angle irrelevant in Euclidean space.

Notably, the neural network-based controller degraded gracefully when presented with a physically impossible trajectory, since the controller used to create the baseline curve still performs much better than using the raw target trajectory as angle requests.

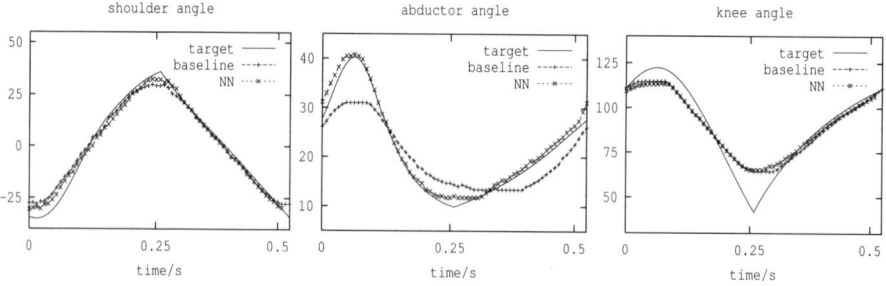

Fig. 7. The resulting angle trajectories for the shoulder, abductor, and knee joint. The shoulder and especially the abductor show improvement over the baseline trajectory.

5   Conclusion

This paper introduced a neural network-based approach to robot motion control. Using data recorded on a physical robot, we trained neural networks to predict which angle commands are necessary to make a robot execute a given movement. The built-in controllers for the robot's joints were treated as part of the system to be modeled and controlled.

We conducted two experiments, using a popular commercially available robot, the Sony Aibo ERS-7, as our experimental platform. The first experiment showed that the proposed approach is indeed able to bring a robot's motions significantly closer to the desired trajectory. In the second experiment, the neural network-controller failed to produce equally good results, but was nevertheless shown to degrade gracefully when presented with a target trajectory outside the robot's physical constraints.

Acknowledgements

This research is supported in part by NSF CAREER award IIS-0237699 and ONR YIP award N00014-04-1-0545. The authors thank Peggy Fidelman and Nate Kohl for useful discussions.

References

1. Bubnicki, Z.: Modern Control Theory. Springer (2005)
2. Sciavicco, L., Siciliano, B.: Modeling and Control of Robot Manipulators. McGraw-Hill Companies, Inc. (1996)
3. Johnson, M., Moradi, M. (eds.): PID Control: New Identification and Design Methods. Springer (2005)
4. Stronger, D., Stone, P.: A model-based approach to robot joint control. In: Nardi, D., Riedmiller, M., Sammut, C., Santos-Victor, J. (eds.) RoboCup 2004. LNCS (LNAI), vol. 3276, pp. 297–309. Springer, Heidelberg (2005)
5. Kim, M.S., Uther, W.: Automatic gait optimisation for quadruped robots. In: Australasian Conference on Robotics and Automation, Brisbane (2003)
6. Kohl, N., Stone, P.: Policy gradient reinforcement learning for fast quadrupedal locomotion. In: Proceedings of the IEEE International Conference on Robotics and Automation (2004)
7. Lewis, F., Jagannathan, S., Yesildirek, A.: Neural Network Control of Robot Manipulators and Nonlinear Systems. Taylor & Francis (1999)
8. Billard, A., Ijspeert, A.J.: Biologically inspired neural controllers for motor control in a quadruped robot. In: International Joint Conference on Neural Networks, vol. 6 (2000)
9. Angulo, C., Tellez, R., Pardo, D.: Emergent walking behaviour in an aibo robot. In: The European Research Consortium for Informatics and Mathematics (2006)
10. Zell, A., Mache, N., Hübner, R., Mamier, G., Vogt, M., Schmalzl, M., Herrmann, K.U.: Snns (stuttgart neural network simulator). In: Skrzypek, J. (ed.) Neural Network Simulation Environments, Kluwer Publishers (1993)

Dynamic Positioning Method Based on Dominant Region Diagram to Realize Successful Cooperative Play

Ryota Nakanishi, Kazuhito Murakami, and Tadashi Naruse

Graduate School of Information Science and Technology,
Aichi Prefectural University, Nagakute-cho, Aichi, 480-1198 Japan
im071016@cis.aichi-pu.ac.jp

Abstract. In this paper, we propose a new technique to compute, in real time, the positions of robots in a cooperative play such as the pass-and-shoot play. To evaluate the positioning of the robot, we use the *Dominant Region* (DR) diagram, which is a kind of a Voronoi diagram. In the DR diagram, the soccer field is divided into regions, each of which shows an area that a robot can reach faster than the other robots. This division is based on the time of arrival while the division by the Voronoi diagram is based on the distance of arrival. Though the DR diagram plays a primary role in the positioning of the robots, it has a serious problem of taking much computation time. To overcome this problem, we show an approximate calculation procedure to obtain the DR diagram, which realizes the real time computation, i.e. a computation within a frame time. Applying the approximate dominant diagram to the positioning of the robots for the pass play, we show, by the simulation study,

- the DR diagram can be calculated in real time,
- an appropriate position for the pass play can be obtained.

1 Introduction

The skills in robotic soccer are growing higher and higher by the year. In recent years, especially in Small Size League, the technique for achieving highly cooperative plays among robots are discussed[1,2]. In a cooperative play such as a pass play, the robot receiving the ball must get to where the pass comes through faster than any other robot. A systematic way to find such point is considering the robot's dominant region, i.e. the set of all points the robot can reach first or the territory of the robot.

Under the idealized assumption of infinite acceleration the time to reach a point is proportional to the distance to that point. In this case the Voronoi diagram [3] provides the dominant region, and the positioning of robots based on the Voronoi diagram is discussed in [4]. In the limited acceleration case, we must compute the arrival time of each robot for each point and select the fastest one[5]. In [5], the resulting figure is called the dominant region (DR) diagram. The computational problem of the DR diagram is to take much computation time. Consequently, real time computation of the DR diagram is required.

U. Visser et al. (Eds.): RoboCup 2007, LNAI 5001, pp. 488–495, 2008.

In this paper, we address the more realistic and much more challenging case of omnidirectional motion with limited velocity and acceleration. We propose an approximation algorithm that compute the dominant region diagram in submillisecond enabling real time operation. Simulation study shows that the positioning based on the approximate DR diagram works well to carry out a successful cooperative play.

2   Making an Approximate Dominant Region Diagram

Though the DR diagram gives the accurate territory of each robot, it takes much computation time. To employ the DR diagram in the real game of RoboCup, real time computation of the DR diagram is necessary. One of the solutions is to make an approximate DR diagram.

2.1   Calculation of an Approximate DR Diagram

To reduce the computation time, we calculate the arrival time for each of the selected points on the field. For each point other than the selected ones, we get the arrival time by interpolation. Using these arrival times, we can make the approximate DR diagram as follows[1].

Algorithm 1. Computing the approximate DR diagram

Let the maximal acceleration, the current velocity and the current position of the robot be $\mathbf{a}$, $\mathbf{v}$ and $\mathbf{P_x}$, respectively. (Each is a vector.) Since the robot can move in any direction, we consider the n acceleration vectors as shown in figure 1(a) to compute the possible future position. Let them be $\mathbf{a_1}$ ... $\mathbf{a_n}$. Assume that the field is divided into the grids.

Step 1. For each robot, compute the position of the robot for every Δt seconds from 0 to t_{max} according to the following equation

$$\mathbf{x_k} = \frac{1}{2}\mathbf{a_k}t^2 + \mathbf{v}t + \mathbf{P_x}, \quad k = 1, ... n, \qquad (1)$$

where, $t = l \cdot \Delta t, \quad l = 0, 1, ..., t_{max}/\Delta t$.

Step 2. For the points calculated in Step 1, make a triangulation as shown in figure 1(b). The arrival time of any grid point of the field in a triangle is calculated by the interpolation from the arrival times of its vertex points. (The triangulation will be done for each robot.)

Step 3. For each grid point, find the robot that has the minimal arrival time and put the grid point to be a territory of the robot.

We compare the difference between the Voronoi diagram and the approximate DR diagram by examples. Figure 2 (a) shows a polygon divided by the distance from a robot put at the right half of the field and figure 2 (b) shows a polygon

---

[1] The approximate DR diagram is discussed in [5], however the detailed algorithm is not shown there.

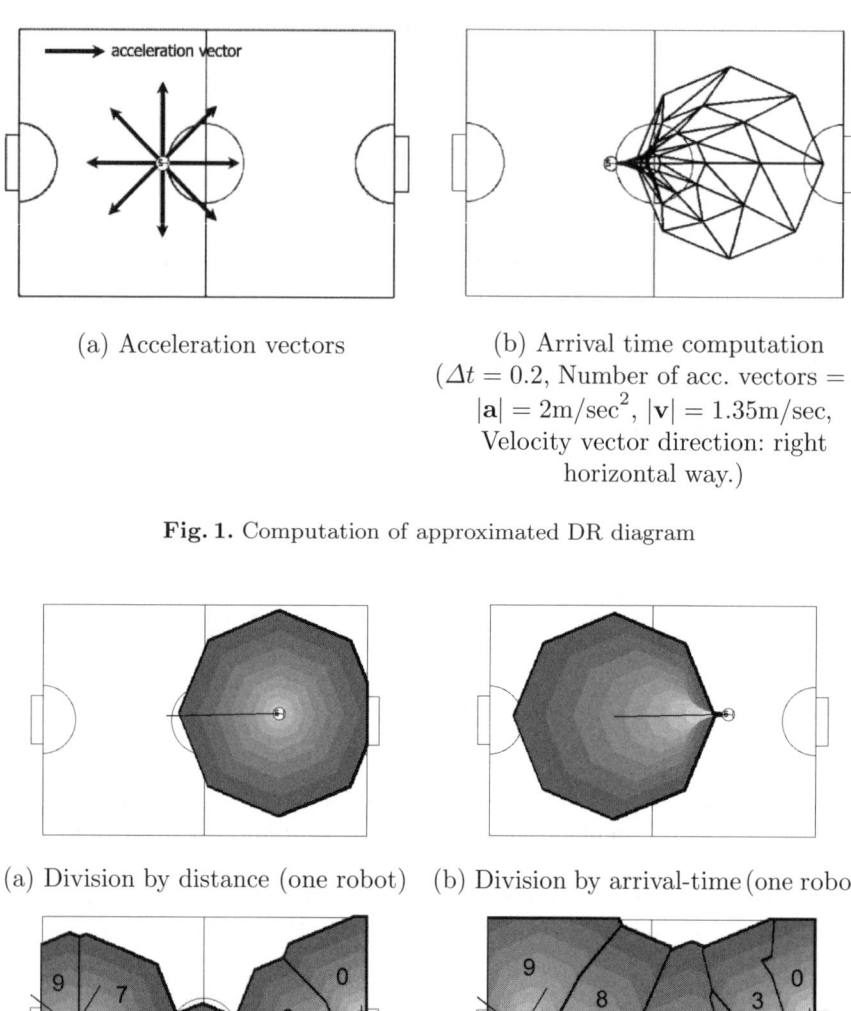

(a) Acceleration vectors

(b) Arrival time computation
($\Delta t = 0.2$, Number of acc. vectors $= 8$,
$|\mathbf{a}| = 2\text{m/sec}^2$, $|\mathbf{v}| = 1.35\text{m/sec}$,
Velocity vector direction: right
horizontal way.)

Fig. 1. Computation of approximated DR diagram

(a) Division by distance (one robot)   (b) Division by arrival-time (one robot)

(c) Voronoi diagram(ten robots)   (d) DR diagram(ten robots)

Fig. 2. Comparison between Voronoi diagram and DR diagram

divided by the arrival time of a robot put at the same point as in Fig. 2 (a).
Each polygon is shaded from light gray to deep gray according to the distance
or the arrival time from near to far. We call the polygon in Fig. 2 (b) the
dominant polygon. The dominant polygon is obtained by computing Step 1 of
Algorithm 1. Figure 2 (c) and (d) are the examples of the Voronoi diagram and

the approximate DR diagram when ten robots exist. The number in each region corresponds to the robot that dominates the region. Throughout this paper, the robots from 0 to 4 are opponents and the ones from 5 to 9 are teammates. The thin segment beginning from each robot shows a velocity vector. So is the ball. The end point of the thick segment beginning from the ball in Fig. 2 (d) is a point that robot 8 meets the ball since the DR diagram shows the arrival time.

3   Real Time Calculation of an Approximate DR Diagram

If we can calculate the DR diagram in real time, it will be a very useful tool for planning the robot action. In this section, we discuss the real time calculation of the approximate DR diagram.

According to the algorithm shown in the previous section, we can compute the approximate DR diagram. We show some computing results. Figure 3 shows the examples of the approximate DR diagram. We computed them under the environment of Debian Linux operating system and the Athlon 64 3500+ CPU with 512 MB main memory. The parameters in Fig. 3(a) are 32-way acceleration vectors, $\Delta t = 0.1$ seconds, time range from 0 to 1.5 seconds, and the 490 by 340 grid points. We get a fine diagram, however, it takes 143 seconds to compute. The parameters in Fig. 3(b) are 6-way acceleration vectors, $\Delta t = 0.2$ seconds, time range of 0 to 1 second, and the 98 by 64 grid points. We get a rough diagram but taking only 0.512 seconds. Unfortunately, even the latter case doesn't satisfy the computation within the frame time. Further reduction of the computation time is required, however, making the parameters rougher is not acceptable.

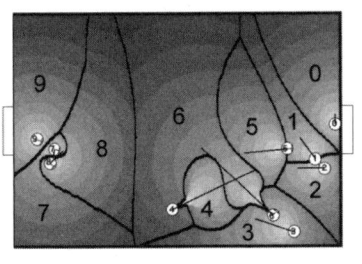
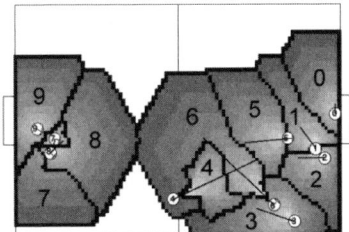

(a) Detailed calculation   (b) Rough calculation

Fig. 3. Two examples of DR diagram

In the soccer system, it is sufficient to know which robot will be able to get to the ball first. Therefore, one solution for the real time computation is an incremental computation of the DR diagram. In other words, we compute the *partial DR diagram* around the ball in the time range from 0 to t, and the ball locus. If the ball is in a dominant region of some robot, the computation finishes, otherwise the computation continues with increasing the time by Δt. An algorithm is shown below[2].

Algorithm 2. Incremental computation of the partial DR diagram
Assume that t_{min}, t_{max} and Δt are given constants.
Step 1. Set $t \leftarrow t_{min}$.
Step 2. Following to Step 1 of Algorithm 1, compute the position of each robot and make a dominant polygon. Also compute the ball position at time t.
Step 3. If the ball is in the dominant polygon of one robot, then the robot gets to the ball first. Otherwise, if it is in the dominant polygons of some robots, then compute the dominant polygon that has the minimal arrival time according to Step 2 of Algorithm 1. The robot corresponding to the dominant polygon gets to the ball first. Otherwise, go to the next step.
Step 4. $t \leftarrow t + \Delta t$. If $t > t_{max}$, then no robot can get to the ball within t_{max}, otherwise, go to Step 2.

In Step 2 in the algorithm above, the computed partial DR diagram should be memorized to avoid the superfluous computation in the next iteration.

We measured the computation time of Algorithm 2 under the following conditions.

1. $t_{min} = 0$, $t_{max} = 1.5$, $\Delta t = 0.1$ (Unit: second).
2. $|\mathbf{a}| = 2\text{m/sec}^2$, $|\mathbf{v}|$: any value, a number of acceleration vectors $= 8$, grid points $= 490 \times 340$.
3. The number of the robot is ten.

Table 1 shows the average computation time. It only takes submilliseconds. In the real games, since almost all the robots can get to the ball within 1 second, this result shows that it is sufficient to use Algorithm 2 for the real time computation of the partial DR diagram.

Table 1. Average computation time of Algorithm 2

Maximal time t_{max} (sec)	0.1	0.2	0.5	0.7	1.4	unlimited
Computation time (msec)	0.068	0.107	0.218	0.296	0.560	0.584

4   Some Positioning Algorithm Based on Partial DR Diagram

In this section, we show a positioning algorithm using the partial DR diagram. We consider the situation where teammate robot A would like to pass the ball to teammate robot B, however Robot B is being marked by an opponent, robot

---
[2] In case where the robot is moving, it is quite hard work to compute when the robot arrives at the ball under the omni-directional movement conditions.

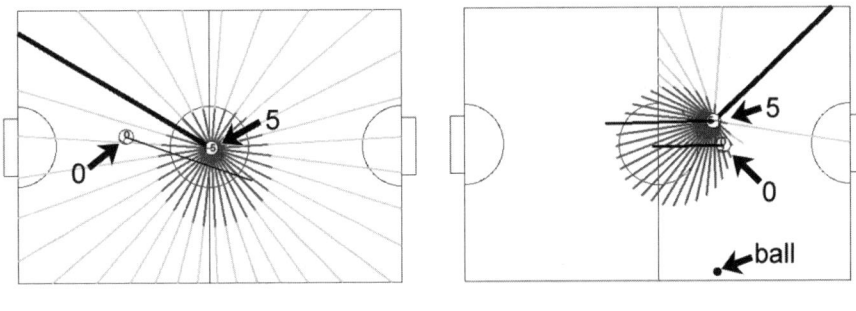

(a) Computation in Algorithm 3   (b) Computation in real game

Fig. 4. Candidate direction vectors to flee from marking of the opponent robot

X. Robot B should flee from the marking of robot X. Where should the robot go? The next algorithm gives the destination of robot B.

The basic idea is the following: if robot B is in the dominant polygon of robot X, select a direction which makes the distance between robot B and the boundary of the polygon shorter and the distance between robot B and the boundary of the field longer. The algorithm is given below.

Algorithm 3. Getting the destination to flee from the marking of the opponent robot

Let $TargetPos$ and $CurrentPos$ be the destination and the current position of robot B, respectively.

Step 1. For given time t, compute the dominant polygons of the opponent robots. If robot B is in the dominant polygons of some of them, let the robot that has minimal arrival time be robot X, i.e. robot X is the one that marks robot B. (Robot X is 0 and robot B is 5 in figure 4(a).) If there is no robot marking robot B, $CurrentPos$ is a destination, i.e. $TargetPos$.

Step 2. Compute the vector $C(\theta)$ from robot B to each vertex of the dominant polygon of robot X. (The dark thin segments in Fig. 4(a).)

Step 3. Extend each vector toward the boundary of the field. (The light thin segments in Fig. 4(a).) Let it be $E(\theta)$.

Step 4. Compute the following equation.

$$\theta_{max} = \arg\max_{\theta} \left(|E(\theta)|/|C(\theta)| \right)$$

and

$$TargetPos = CurrentPos + \frac{1}{2}(E(\theta_{max}) + C(\theta_{max}))$$

$E(\theta_{max})$ is shown in Fig. 4(a) as a thick segment, and the $TargetPos$ is a destination.

To make the $TargetPos$ stable, this computation is done once every N frame times, where N is decided by the experiment. (To get an optimal N is a future problem).

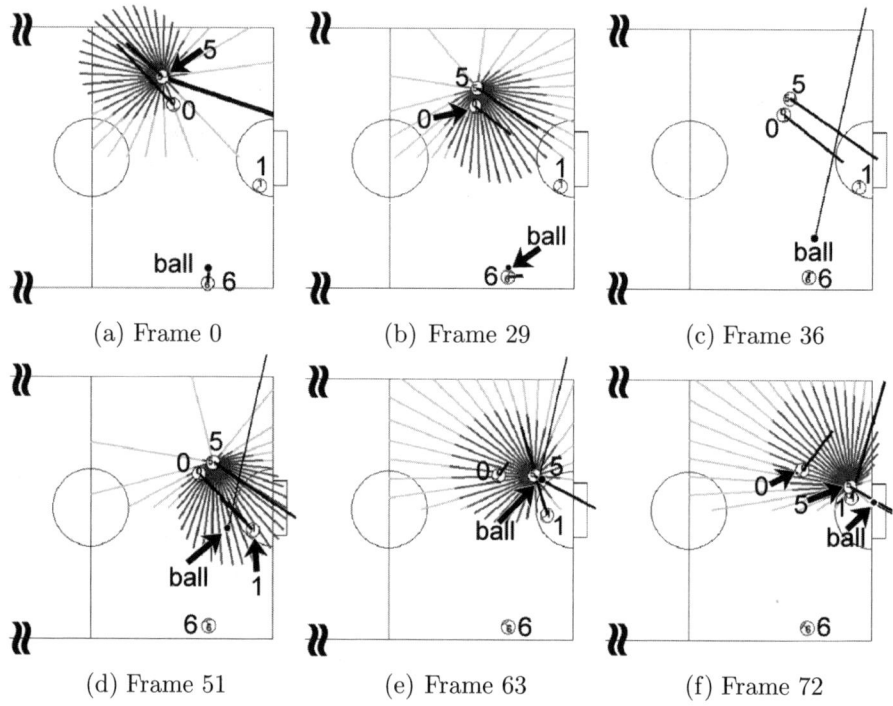

Fig. 5. An example of simulated game

In the real game, we compute the extended vectors within an area of attention. Figure 4(b) shows an example. In the figure, the extended vectors are only calculated within the right-upward area of the field, since the teammate robot would receive the ball at a position far from the ball and shoot it towards the goal.

We show an actual application example of Algorithm 3 in the game. Figure 5 shows the six time slices of a simulated game. Frame numbers are shown in the captions below figures. Each figure has the dominant polygon except the frame 36. This means that robot 5 is marked by opponent robot 0, and in the frame 36, there is no marking robot. A series of figures show the process of passing from robot 6 to robot 5 and then shooting of robot 5 to the goal by the direct play which we proposed in [2].

In figure 5(a), robot 5 was moving toward the left direction to escape from robot 0. (Black segment in front of each robot shows the velocity vector.) Robot 0 is faster than robot 5. Run-away direction computation gets robot 5 to move in the thick segment direction. In other words, it shows that robot 5 can run away from the marking of robot 0 in less time. Figure 5(b) shows the state just before robot 6 kicks the ball. In this time slice, the robots 5 and 0 are moving toward the down-right direction, since the run-away direction computed in Frame 0 is available. It is shown that the velocity of robot 5 is larger than robot 0, since robot 0 has moved toward upper-left direction in Frame 0. Figure 5(c) shows the state just after robot 6 kicked the ball and figure 5(d) shows the state the ball is moving halfway. Since the speed of robot 5 is faster than robot 0, robot 5 can get to the ball first. Figure 5(e) shows the state just after robot 5 gets to the ball and figure 5(f) shows the state the ball is kicked by direct play.

5   Concluding Remarks

In this paper, we proposed an algorithm that computes the partial dominant region diagram in real time and an algorithm that finds a run-away position of a teammate robot that is marked by an opponent robot. The latter algorithm efficiently uses the partial DR diagram to find the run-away destination. Simulation studies shows that the algorithm works well.

As further study, the need of improvement in the algorithm such as

- the optimization of the parameters in Algorithm 2,
- how to decide the stabilizer parameter N in Algorithm 3,

remain. The position evaluation methodology is an important issue that should be discussed in this field.

References

1. Murakami, K., Hibino, S., Kodama, Y., Iida, T., Kato, K., Naruse, T.: Cooperative Soccer Play by Real Small-Size Robot. In: Polani, D., Browning, B., Bonarini, A., Yoshida, K. (eds.) RoboCup 2003. LNCS (LNAI), vol. 3020, pp. 410–421. Springer, Heidelberg (2004)
2. Nakanishi, R., Bruce, J., Murakami, K., Naruse, T., Veloso, M.: Cooperative 3-robot passing and shooting in the RoboCup Small Size League. In: RoboCup international symposium 2006, CDROM (June 2006)
3. Preparata, F.P., Shamos, M.I.: Computational Geometry, An Introduction. Springer (1985)
4. Dashti, H.T., et al.: Dynamic Positioning based on Voronoi Cells. In: Bredenfeld, A., Jacoff, A., Noda, I., Takahashi, Y. (eds.) RoboCup 2005. LNCS (LNAI), vol. 4020, pp. 219–229. Springer, Heidelberg (2006)
5. Taki, T., Hasegawa, J.: Dominant Region: A Basic Feature for Group Motion Analysis and Its Application to Teamwork Evaluation in Soccer Games. In: RSFDGrC 2005, vol. 3641, pp. 48–57 (January 1999)

Introducing Physical Visualization Sub-league

Rodrigo da Silva Guerra[1], Joschka Boedecker[1], Norbert Mayer[1,2],
Shinzo Yanagimachi[3], Yasuji Hirosawa[3], Kazuhiko Yoshikawa[3],
Masaaki Namekawa[3], and Minoru Asada[1,2]

[1] Dept. of Adaptive Machine Systems,
[2] HANDAI Frontier Research Center,
Graduate School of Engineering, Osaka University, Osaka, Japan
[3] CITIZEN Co., Tokyo, Japan
{rodrigo.guerra,joschka.boedecker,norbert,asada}@ams.eng.osaka-u.ac.jp,
{yanagimachi,hirosawa,yoshikawaka,namekawam}@citizen.co.jp

Abstract. This work introduces the new sub-league of the RoboCup
Soccer Simulation League, called Physical Visualization. We show how
the fundamental collaborative concepts of this new sub-league shift essential research issues from the playing agents themselves to the development of a new versatile research and educational platform. Additionally,
we discuss benefits of this new platform in terms of standardization, flexibility and reasonable price. We also try to characterize and discuss the
place of this new sub-league within the RoboCup community. Finally,
competition formats and roadmaps are presented and discussed.

1 Introduction

Physical Visualization (PV for short) is candidate to be a new RoboCup Soccer Simulation sub-league. The sub-league is intended for fostering education,
research and development together with the RoboCup community. The PV is
based on a miniature multi-robot system which mixes reality and simulation
through an Augmented Reality (AR) environment. The project has a two-folded
focus: research and education. The main goals of the PV are:

- to gradually improve the platform so that it becomes a powerful and versatile
  standard for multi-agent research and education.
- to explore educational possibilities and real world applications based either
  on the system as a whole or on some parts of it (e.g. the robots alone).

We focused on versatility and affordability, taking advantage of well established industry technologies to allow the development of an inexpensive platform. In order to do that we used the know-how of the cutting-edge and low cost
watch technology as a basis for building an affordable miniature multi-robot system mixing reality and simulation. Three dominant characteristics of the project
are: (a) affordability, (b) standardization and (c) open architecture.

The rest of this paper is organized as follows: Section 2 gives more detailed
technical information on the current implementation of the system, section 3

U. Visser et al. (Eds.): RoboCup 2007, LNAI 5001, pp. 496–503, 2008.

introduces the sub-league's collaborative nature and discuss three different competitions using the system, and finally, section 4 discusses, from a wider perspective the place of this sub-league within RoboCup and gives some final remarks from the authors.

2   Technical Aspects

Robots obey commands sent by a central server through an IR beam, while their actual position and orientation is fedback to the server by a camera located on the top. Meanwhile a number of visual features are projected onto the field by using a flat display. This system merges characteristics and concepts from two of the most mature RoboCup leagues, Simulation and Small-Size [4], and adds a new key-feature: augmented reality.

All the robots are centrally controlled from one CPU but their decision making algorithms run on networked clients, making the robots behave autonomously virtually isolated from each other just like in simulation league. Position feedback is based on colored markers placed on top of the robots which are detected through a vision system in the same way used in small-size league. Robot control is based on strings of commands sent by modulated infrared signals (in this sense resembling U-league to some extent [1]).

2.1   The Position Feedback

The position of the robots (and eventually other objects, such as ball) is detected from the processing of high-resolution camera images. The computer vision system currently implemented can be divided into three main subsystems: (a) undistortion, (b) blob detection, and (c) identification & orientation. Each one is described in the following paragraphs.

Undistortion: Despite the fact of the PV robots being real three-dimensional objects occupying volume in space, the domain of possible locations for their bodies over the plane of the flat screen is known to be confined into a two-dimensional space. Because of that the calibration problem can be reduced, without loss of generality, to a plane-to-plane linear transformation from the plane of the captured image to the plane of field itself. This transformation is a single linear 3×3 matrix operator which defines a homography in the two-dimensional projective space (see figure 1).

Blob detection: After undistorted, the image is segmented into blobs of certain colors of interest. These colors are defined by a mask in the three-dimensional $Y \times U \times V$ space. Adjacent pixels, in a 8-neighborhood, belonging to the same color mask configure a single blob. The area (total amount of pixels) and center of mass (average (x, y) coordinates) of the blobs are extracted. Blobs whose mass values are not within a tolerance range from the expected are discarded. This procedure is used for finding the center of the colored marking patterns on the top of each robot – the red shape seen on figure 2-b.

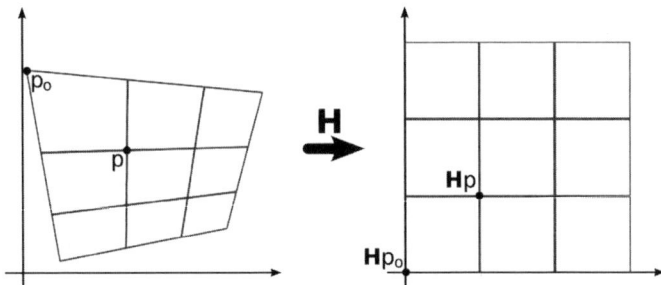

Fig. 1. Plane-to-plane projective undistortion based on homography transformation, where **H** is a 3×3 matrix operator and p and p_o are 3-dimensional vectors representing points in the two-dimensional projective space

Identification and orientation: The process here described is inspired on [5]. Once a potential blob is found, a radial pattern of colors is sampled within a predefined radius of its center. In figure 2-b these sampling locations are artificially illustrated by a closed path of little green dots. This pattern is cross correlated with a database of stored patterns, each of which uniquely defining a robot's identity. Let's denote $x(i)$ to be the color in the pattern x at the angle i. The cross-correlation r_{xy} is calculated accordingly to the equation 1 for each pattern y the database, and for each $\Delta\alpha$ in the interval $[0°, 360°)$. If, for a pattern x, the minimum value of $r_{xy}(\Delta\alpha)$, for any y and $\Delta\alpha \in [0°, 360°)$, exceeds a minimum threshold, then the corresponding y gives the identity of a robot, and $\Delta\alpha$ gives its orientation.

$$r_{xy}(\Delta\alpha) = \frac{\sum_{i=0°}^{360°} [(x(i) - \bar{x}) \cdot (y(i - \Delta\alpha) - \bar{y})]}{\sqrt{\sum_{i=0°}^{360°} (x(i) - \bar{x})^2} \cdot \sqrt{\sum_{i=0°}^{360°} (y(i - \Delta\alpha) - \bar{y})^2}} \tag{1}$$

2.2 Augmented Reality

The idea about the augmented reality setup is an extension of a previously published similar concept where robot ants would leave visually coloured trails of "pheromones" by the use of a multimedia projector on the ceiling of a dark room in a swarm intelligence study [6]. Huge improvements in versatility, flexibility, and standardization can be introduced by applying that concept into a more customizable system. The figure 2-a shows an illustrative drawing and figure 2-b shows an actual picture of our system in action. Given the reduced size and weight of the PV sub-league robots the application of a conventional flat display as the field becomes feasible – depending on the application, displays as small as 20-inches are more than enough. This adds much versatility to the system without adding much costs and without complicating the required setup.

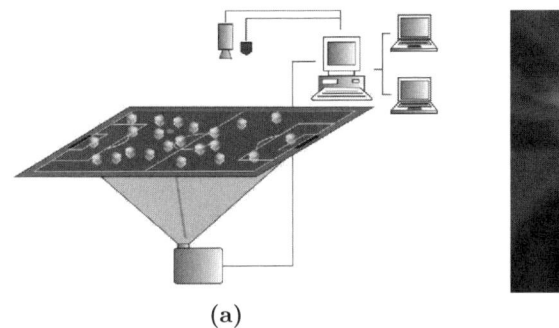
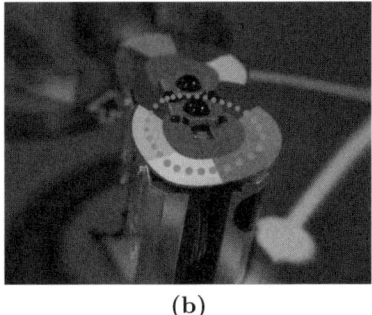

(a)   (b)

Fig. 2. On the left an illustration of the overall system including the feedback control loop (infrared transmitter, camera, server) and the augmented reality screen. On the right an actual close-up picture of two robots playing using such setup.

2.3   The Miniature Robot

Until now, a few developments have been made on very small sized robots, being ALICE one of the most prominent names (see [2] for a survey). The first versions of the miniature robot here used were originaly developed by CITIZEN as merchandize devices for demonstrating their new solar powered watch technologies [7]. Since March of 2006 three new prototype versions were already developed for matching the requirements of the sub-league. The most current version of the robot has dimensions of $18 \times 18 \times 22mm$, no sensors, an infrared receiver and is driven by two differential wheels. This first robot was purposely designed to have rather simplistic hardware configuration as a starting point, a seed, to be followed by numerous upgrades in the long term. The main robot components are (numbers in accordance to figure 3-b):

1. Motor – Customized from wristwatch motor unit. See further details in the dedicated sub-section 2.4.
2. Battery – Miniature one-cell rechargeable $3.7V$ lithium ion polymer battery with capacity of $65mAh$.
3. Control board – Currently based on the Microchip $8bit$ PIC18 family of microcontrollers, each robot comes equipped with a PIC18LF1220 which features 4kb of re-programmable flash memory.
4. IR sensor – An IR sensor is used in order to listen for commands from the PC. The sensor operates at the $40kHz$ bandwidth modulation (same of most home-appliance remote controls).
5. Body – The resistant durable body of the robot is micro-machined in aluminum using CITIZEN's high precision CNC machines.

2.4   The Micro Step Motor

Simply of-the-shelf wristwatch motors would not be able to bear with the torque requirements for moving the heavy body of the robot. For couping with that

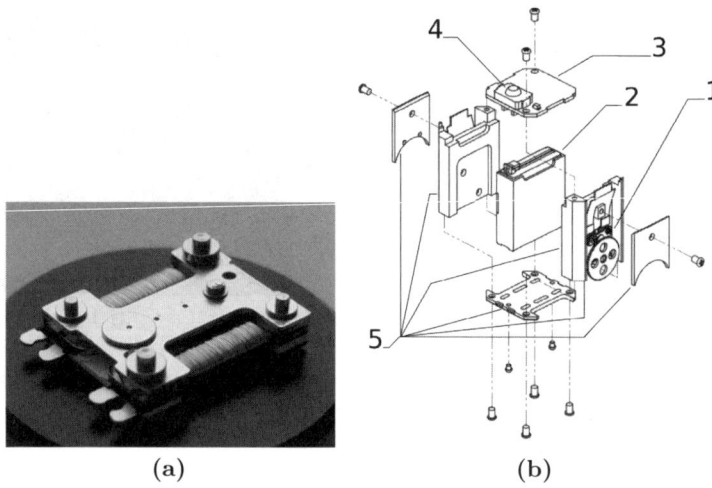

(a)   (b)

Fig. 3. On the left a close-up picture of the step motor, on the right an exploded view of revealing the robot parts

Table 1. Technical specifications of the step motors used in the miniature robots

Feature	Value
Dimensions (mm)	$7.0 \times 8.5 \times 1.9$
Configuration	2 coils $\times$ 1 rotor
Gear ratio	$1 : 240$
Torque $(gf \cdot cm$ at $2.8V)$	between 2.0 and 4.0
Power consumption at $200rps$ (mA)	between 4 and 12
Nominal rotation (rpm)	12.000
Direction	both standard and reverse

CITIZEN developed a new special class of step motors combining high-speed rotation and nano-scaled geared reduction.

2.5   Robot's Firmware and Control Protocol

The current control protocol was programmed in C and compiled using the proprietary MPLAB C18 compiler. All robots share the same firmware but dynamic IDs are be assigned so that commands to an individual robot can be discriminated. Each of the two wheels can be controlled to run at two different speeds, in both directions or stopped (total of 5 possible values). Commands have to be sent by the server to one robot at a time, in an ordered fashion. This implies that bigger number of robots result in longer control lags. Therefore the protocol format was designed so that the command could be sent in a very short time. The current command protocol has a length of $12bits$: ID ($5bits$), left command

($3bits$), right command ($3bits$), and bit parity check ($1bit$). Less frequently used instructions are multiplexed from a sequence of two or more commands.

3   Competitions with Cooperation

The original and dominant point of the proposed sub-league is its concept of collaboration towards the development of a central platform for the benefit of all. While in other leagues essential research issues are traditionally faced in the playing agents themselves (AI, biped walking, vision, etc.), in the PV the research issues are in the improvement of the system – in the development of the platform and its robots.

3.1   Electronics and Firmware Competition

Goal: Allow the evolution of the robot's technology and improve all non-software related aspects of the system.

Summary: Teams have the opportunity to contribute with new ideas for the electronic aspects of the system as well as robot's firmware. Those with background in fields more closely related to the hardware would be able to include in their projects the improvement of certain aspects of the system either for didactic purposes (e.g. class on microcontrollers) or for research. Meanwhile, teams with background in fields more related to computer science would be able to acquire valuable experience by accompanying or even contributing to these projects.

New electronic entries developments could be made on any of the current components of the system, including the robot, or by introducing a new electronic element to the system. All source code, CAD drawings, circuit schematics and documentation should be made available to other teams so that they can use and improve at their own.

In the control circuit of the robot several restrictions will be imposed regarding position of mounting holes, size and shape of the board, max bounding volume, limit weight and place of certain components would be applied in order to ensure compatibility with current micro-mechatronic architecture. For instance, connections to the motors would have pre-defined place and electrical characteristics that should remain unchanged. Within those constraints completely new architectures can be proposed.

Entries would be ranked according to a qualified review process preceding RoboCup and based on slide presentations realized during the event. The contributions from the winner of this competition would not necessarily become the new standard for the following years. Nevertheless, contributions published from winner and non-winner teams might be considered for incorporation depending of various criteria to be evaluated (e.g. practicability, price).

3.2   Educational AI Games Competition

Goal: Create a pool of interesting didactic software applications in the form of games using the system for educational purposes.

Summary: Entrants would come up with different game ideas using the system in which they teach concepts related to common subjects ranging from basic computer programming to very specialized topics related to multi-agent systems and artificial intelligence.

The entries would consist of the proposed games along with their source code, supporting tools or API (if any), documentation and accompanying teaching materials. In order to ensure that other teams could easily profit from these contributions the entries would need to be necessarily based on the current official system only. While the eventual introduction of accessories such as balls maze walls or colored objects would, in general, be permitted, no external specialized electronic devices would be allowed. Live demonstrations and poster presentations would be performed during the RoboCup event, and together with prior qualified reviewing would rank the entrant.

Again, just like in the competition described in sub-section 3.1, winner applications would not necessarily be incorporated as league games for the following years. On the other hand, contributions could be considered for incorporation regardless of the competition results, depending on their quality, topics covered and other criteria.

3.3   Rapid (Soccer) Team Development Competition

Goal: Allow undergraduate students to develop complete RoboCup teams of their own within the typically limited time window of their courses.

Summary: The teams would be based on a simplified didactic game framework allowing easy development requiring only a very limited amount of knowledge. All contestants would have an equally limited amount of time for the development of their teams, thus giving similar advantages to teams with limited time to spare. Game rules and supporting software would be officially released just a predefined amount of months before the RoboCup games.

This comes to fill the gap between RoboCup Junior and the other RoboCup Senior leagues. Refer to [1,3] for a previous attempt to include more undergraduate students, where a new league directed exclusively towards them was proposed (the U-league).

In the Rapid Soccer Competition institutions and laboratories equipped with the PV system would be able to let their alumni experiment their ideas into a RoboCup environment regardless of their time constraints (i.e. having more time to spare would post no advantage). Ideas previously introduced in the Educational AI Games competition could be entirely or partially incorporated into the system. Competitions would take the form of a tournament which would spam over the duration of the RoboCup event. At the end tutors would be invited to share the experiences they had when using the system within their courses and discuss improvements for the subsequent years.

4   Discussion

This paper introduced the main technical and conceptual characteristics of the PV sub-league. In particular, it was emphasized in the beginning of section 3, that the central innovative aspect of the proposed idea was the shift of focus from the playing agents to the shared system. Furthermore, the three proposed competitions showed in a more clear way how this collaboration shall be fostered toward the constant development of a versatile system for education and research.

References

1. Anderson, J., Baltes, J., Livingston, D., Sklar, E., Tower, J.: Toward an undergraduate league for robocup. In: Proceedings of RoboCup 2003: Robot Soccer World Cup VII. Lecture Notes In Artificial Intelligence, Springer (2003)
2. Caprari, G.: Autonomous Micro-Robots: Applications and Limitations. PhD thesis, Federal Institute of Technology Lausanne (2003)
3. Heintz, F.: Robosoc, a system for developing robocup agents for educational use. Master's thesis, Dept. of Computer and Information Science, Linkopings Univ. (2000)
4. Igarashi, H., Kosue, S., Kurose, Y., Tanaka, K.: A robot system for robocup small size league: js/s-ii project. In: Proceeding of RoboCup Workshop (5th Rim International Conference on Artificial Intelligence), pp. 29–38 (1998)
5. Shimizu, S., Nagahashi, T., Fujiyoshi, H.: Robust and accurate detection of object orientation and id without color segmentation. In: Bredenfeld, A., Jacoff, A., Noda, I., Takahashi, Y. (eds.) RoboCup 2005. LNCS (LNAI), vol. 4020, pp. 408–419. Springer, Heidelberg (2006)
6. Kazama, S.T., Watanabe, T.: Traffic-like movement on a trail of interacting robots with virtual pheromone. In: Proceedings of the 3rd International Symposium on Autonomous Minirobots for Research and Edutainment (AMiRE 2005), pp. 383–388 (2005)
7. Yoshikawa, K.: Eco-be!: A robot which materializes a watch's life. Nature Interface 8, 56–57 (2002)

A Real Time Vision System for Autonomous Systems: Characterization during a Middle Size Match

H. Silva, J.M. Almeida, L. Lima, A. Martins, and E.P. Silva

LSA - Autonomous System Laboratory
Institute of Engineering of Porto
Rua Dr Antonio Bernardino de Almeida 431 4200-072 Porto, Portugal
+351 22 834 0500 (ext. 1409)
{hsilva,jma,llima,amartins,eaps}@lsa.isep.ipp.pt
http://www.lsa.isep.ipp.pt

Abstract. This paper propose a real-time vision framework for mobile robotics and describes the current implementation. The pipeline structure further reduces latency and allows a paralleled hardware implementation. A dedicated hardware vision sensor was developed in order to take advantage of the proposed architecture. The real-time characteristics and hardware partial implementation, coupled with low energy consumption address typical autonomous systems applications. A characterization of the implemented system in the Robocup scenario, during competition matches, is presented.

1 Introduction

Artificial vision systems are primordial elements in robotics navigation, localization and perception. This is due, to the their great sensing capabilities and low cost. Furthermore embedded solutions with hardware parallel implementation are starting to surface allowing to solve one of the key problems in artificial vision, the high computational resources required.

Considering robotic vision to be a real-time problem, there are certain amount of functionalities, that the vision system must possess in order to face the highly dynamic environmental changes and be able to track external moving objects. Namely, all robotic vision frameworks have to deal with real time aspects and restrictions imposed on the robot modules. System designers must balance between using time and computational consuming methods, against more simple methods that allow low energy consumption and still have a good degree of robustness. Finally all environmental properties must be perceived, in order to control the vision system and allow the autonomous system to work.

As a consequence several vision software tend to be application oriented. In order to prevent this, a conceptual architecture is required. Most state-of-art frameworks for real time applications are mainly concern with the image processing modules[1][2]. There are other applications like active vision systems mounted in pan-tilt heads [3][4] or real-time human tracking methods for autonomous mobile robots[5]. Therefore, an overall real-time framework was developed to cope not only with the need to solve the image processing modules problems, but also through the use of a paralleled hardware

U. Visser et al. (Eds.): RoboCup 2007, LNAI 5001, pp. 504–511, 2008.

implementation[6] to be able to handle problems regarding image acquisition, power consumption and latency between frames.

One of the advantages of an modular framework, is that the different modules can be replaced or adapted to different applications and to different environmental conditions (indoor or outdoor, structured or not).

The modules are integrated in a pipeline structure architecture having two streams of data. One, the main data which concerns only the main modules, and the control data which connects the main modules with the auxiliary ones and provides the system with other type of self-adjusting and logging capabilities.

The framework here described was developed using the Robocup Middle Size League team ISePorto has a benchmark scenario, and has already been integrated in multiples autonomous vehicles [7][8] in different applications scenarios operating both indoors and outdoors.

Fig. 1. ISePorto Robot

This paper is structured in the following way. In the ensuing section, we present the overall architecture of the system followed by a shortened explanation of the architecture modules, there inputs and outputs. In section 3 the hardware embedded solution is presented and its modules detailed, followed by results of temporal analysis and quality of measures obtained. These results where obtained using the framework in a Robocup scenario, some during a real Robocup match. Finally in section 5 a conclusion regarding the overall system functionalities and future research will be presented.

2   Vision System Architecture

The proposed architecture, see Figure 2, follows an image processing pipeline approach. Where some of the layers can be hardware or software implemented. The pipeline starts with the acquisition layer (three left blocks) that takes care of all hardware related and image acquisition configurations, camera settings and so on.

The system is programmed to acquire frames from different types of devices (embedded, USB). It detects which type of camera is plugged in, and automatically starts to acquire frames using that device. This layer also contains the color interpolation module, whose function is to assigned color information to image pixels. The system can also integrate, logarithmic scales, IR and monochromatic image types.

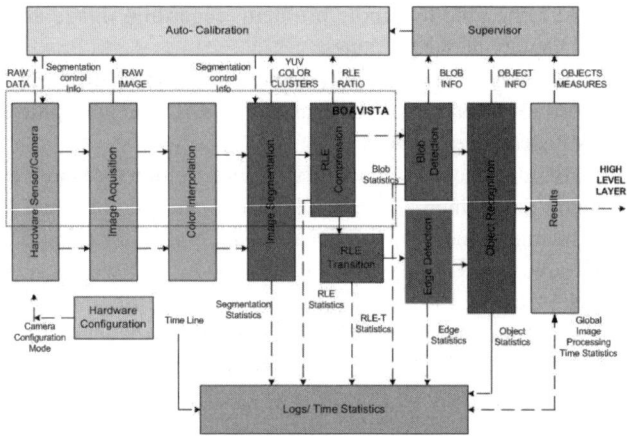

Fig. 2. Vision System Architecture

The image processing layer is responsible for the refinement and abstraction of the image data. After all image pixels have been given texture info, they will be segmented using the selected method. Currently, they are segmented into previously defined color clusters following a method proposed by[9]. Any other type of segmentation method, based on regions, histograms, clusters, neural networks, etc, can be used due to system modularity, by changing the segmentation module. We can for example use an hybrid approach like[10], that combines image segmentation with extraction of edge topological information. Afterwards, the pixels with color information will be compressed using run length encoding (RLE).

This compression method is used due to is effectiveness in conserving the information and reducing data size. Afterwards a run is conducted to look for previously sanctioned color transitions. When one of this transitions occurs, a color transition is created and stored. It will have similar information as a normal RLE: image position, color and number of pixels, but also deals with color transactions uncertainty. These may occur due to interpolation issues, occlusion or illumination problem.

One of the key points of the method is that the number of RLE transitions are not directly attached to the number of image pixels, but are attached to the number of image RLE. In our system is about 1/8 of the total computational cost of the RLE module and only stores 1/15 of the RLE data. After all types of RLE have been processed, the connected RLE will be grouped into similar color regions (BLOBS). Once all the RLE are grouped into BLOBS further processing is done in the top architecture layer, the high level data layer (Object Recognition block). The high level data layer is constituted by the modules that detect features for the robot localization and navigation sub-systems, thus closely related with the application. Lower level layers are relatively application independent and can be used in multiple autonomous scenarios. Image information at this stage already contains edge and blob identification, allowing particular object search. Besides the data processing modules, the system also has some auto-calibrations tools, that are used to help the vision system dealing with environmental changes. In our system a white balance calibration is done to allow perception of the illumination changes, some of the segmentation and color interpolation parameters are sent to a calibration module, that detects color clusters shifting. Statistical analysis on mean value for color pixels provides information for controlling camera parameters.

The architecture high level modules are being continually improved. Currently high level stereo is under development, this module will not work at pixel level but will merge high level objects information provided by the acquisition devices, leading to a change in some of the higher architecture modules. This method is less time consuming than merging camera information at pixel level.

3 Hardware Vision Sensor

The hardware embedded sensors are an emergent solution in robotics and autonomous systems applications. This is due to their hardware reconfiguration capabilities, low cost implementation, low energy consumption and low hardware concentration.

The vision systems built with reconfigurable and embedded hardware have some advantages over a standard vision system namely: cost, size, energy, computational resources and latency. It also possess some disadvantages like: processing capabilities, development cost and fixed point arithmetic.

These devices are a solution when dealing with hardware costs issues, several solutions exists using CCD and CMOS technologies. This one has the advantage of allowing more sensors addiction and with the use of field programmable gate arrays (FPGA) achieve low level processing with a low energy cost, taking advantages of the inner parallel processing architecture of this devices.

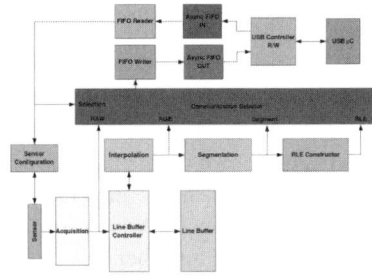

Fig. 3. Hardware Image Sensor Architecture

So taking this issues in consideration, the embedded vision sensor BOAVISTA was developed to free system resources from processing the most heaviest data. In order to do so, a FPGA platform is used to process the image on the fly from the CMOS sensor.

This sensor allows the implementation of lower architecture levels at a fraction of power required in standard CPUs by taking advantage of the inherent parallel nature of image information and architecture pipeline structure. The substantial power reduction constitutes a fundamental advantage to the use in autonomous systems namely in Robocup fields technology.

It is thus possible to implement advanced sensing capabilities in low power systems and widen the range and scope of applications. This image processing layer within the FPGA is divided into different modules, allowing access of the overall system to different kinds of data. As a result, the vision sensor can provide four types of data: raw data, RGB mode data, segmented data and RLE data.

In figure 4, an information processing pipeline time diagram description is presented. A maximum latency of 500 μs is achieved from the initial pixel acquisition in the CMOS sensor to the processed data reception at the user level application. This maximum latency includes processing and communication delays (USB bus). The interpolation, segmentation and RLE modules are similar to the software ones, for more information see[6].

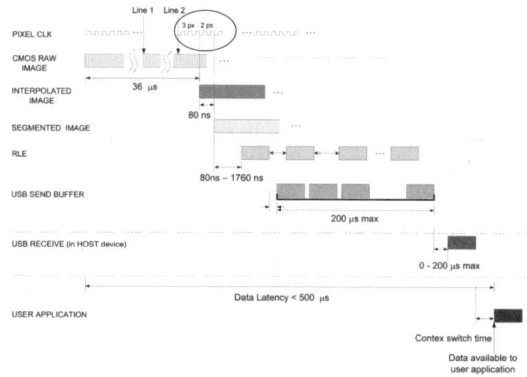

Fig. 4. Time diagram of the Embedded Hardware Vision Sensor

4   Results

The results presented in this paper where taken during a set of live Robocup Matches, and allows to establish and understand the type of data and resources that are needed to cope with a Robocup Vision System. This enables the use of this scenario has a benchmark scenario for other robotics applications.

The first experiment was to monitor the amount of data that each module of the vision system pipeline would generate during a match and how much time each processing modules consume. In order to do so we monitored the image processing modules during a Robocup Match. We can see the time analysis of each module by frame.

Table 1. Image Processing Modules Time Statistics (ms) during a Robocup Match

Image R.(320x240)	Seg+RLE Color T	Edges	Blobs
Mean	6.045	0.345	0.425	0.708
St.deviation	2.032	0.375	0.696	0.979

These experiments were made during a live Robocup Match, with a robot with two cameras Head and Kick (2.8mm, F1.4,maxvision MVL2810M) both at 320x240 resolution, placed at a vertical height of 60 cm and 50 cm accordingly.

The most time consuming module is the segmentation/RLE module. Which is never under 5 ms, due to the minimum computational time required to perform segmentation and run length encoding operations, to an image with a 320x240 resolution in a PIII 1.2Ghz Tualatin.

The other modules, have residual time consuming performances and do not influence significatively the overall performance of the robot vision system.

The second experiment is intended to characterized the amount of data generated in each of the processing modules. This tests are important, in order to understand which parts of the system should or should not be integrated in embedded hardware and what advantages could come from that migration. This experiment shows how the information flows through the pipeline. We can see that the most heaviest data is processed in the early stages of the pipeline. Which enforce our choice of migrating these modules to an hardware embedded solution, thus achieving the same results at a fraction of the power consumption cost.

Table 2. Image Processing Modules Dimensions Statistics during a Robocup Match

Image R.(320x240)	RLE	Color T	Edges	Blobs
Mean	1988	468.3	96.2	140.2
St.deviation	758.9	196.7	41.2	78.2

Furthermore, this statistical results obtained during a live Robocup Match, can also be used to help the vision system to obtain self-adjusting capabilities. When the RLE statistical numbers increase dramatically to an unexpected number during a large quantity of frames, that would indicate that the segmentation parameters were off-balanced and that the robot should re-adjust his camera settings or color configurations.

In a Robocup scenario there are some statical objects that can be used by the robot has landmarks in order to achieve self-localization. One of them are the field goals. In the third experiment bearing measures to a field goal were taken during a match.

In figure 5 bearing measures observed are shown for a 5 minute period of match. We can see that there are only seven false positive occurrences during that period. These false positive are ignored by the vision system due to the fact, that all bearing measures have a reliability factor that in the case of the field goal, is the bearing variation of the vertical post of the goal. This variation (see figure 5) is much higher in the false positive cases, more than 2 degrees that the normal standard deviation of a goal post inferior to 1 degree.

Other landmarks present in a Robocup field are the corner posts and the field lines. In figure 6a we can see a histogram, with the different range distances detection to a corner post. There is a great deal occurrences in the 2 meters area probably due to the robot movement.

In figure 6b a histogram showing field lines t-junction distance range of detections is displayed.

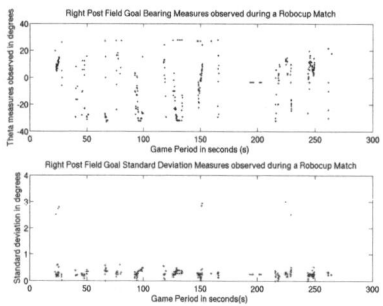
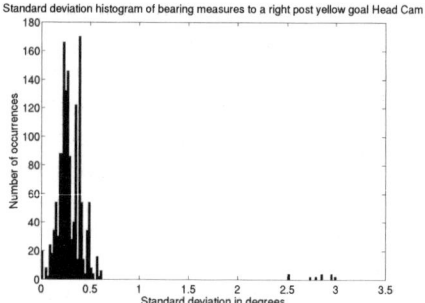
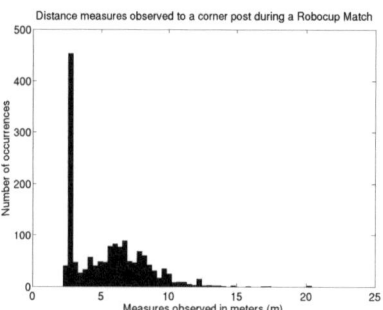
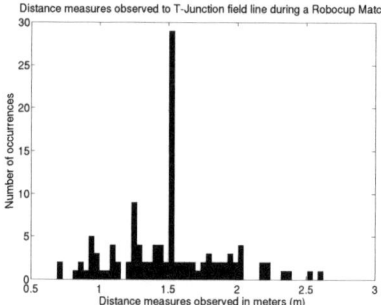

Fig. 5. Bearing measures to a goal post and standard deviation

Fig. 6. Max. Distance Range to a corner post in a Robocup field and distance measures observed to a t-junction

5   Conclusion

In this work we presented a real-time vision system for mobile robotics and autonomous systems applications. The presented framework architecture allows latency reduction in sensor data reception. Very low power consumption solutions can be integrated. Our proposed organization allows a hardware and software transparent implementation.

A dedicated hardware vision sensor was developed to implement the more time consuming processing steps, taking advantage of image information parallelism. A high performance programmable logic device (FPGA) was used to process data from a CMOS sensor capable of VGA resolutions at 60 fps. This vision sensor can use different image sensors with a higher frame-rate, resolutions and High Dynamic Range Image capabilities for used in outdoors applications.

The presented results, taken from a Robocup Match, allowed a overall evaluation of performance of the system, as well as, the characterization of the output data from each one of the vision processing modules.

Power consumption reduction was significant. It is now possible to segment and compress image for less than 1W. Information coherence is maintained through different levels of abstraction in the architecture with ploughable module integration. The vision architecture provided clear advantages to mobile robot navigation and advanced image perception systems, having also been applied to fire detection with Unmanned Aerial Vehicles.

References

1. CMVision, http://www.cs.cmu.edu/~jbruce/cmvision
2. Hager, G., Toyama, K.: X-Vision: A Portable Substrate for Real-Time Vision Applications. In: Computer Vision and Image Understanding, January, vol. 69(1), p. 2337 (1998)
3. Hai, Z., Kui, Y., Jindong, J.: A Fast and Robust Vision System for Autonomous Mobile Robots. In: Proceedings of IEEE Intelligent Conference on Robotics, Intelligent Systems and Image Processing 2003, China (2003)
4. Peiig, J., Skrikaew, A., Wilkes, M., Kawamura, K., Peters, A.: An active vision system for Mobile Robots. In: Proceedings of IEEE Intelligent Conference on Robotics, Intelligent Systems and Image Processing 2000, Takamatsu, Japan (2000)
5. Doi, M., Nakakita, M., Aoki, Y., Hashimoto, S.: Real Time Vision System for autonomous mobile robotics. In: IEEE International Workshop on Robot and Human Interaction Communication (2001)
6. Lima, L., Almeida, J.M., Martins, A., Silva, E.P.: Development of a dedicated hardware vision system for mobile robot navigation. In: Robotica 2004 International Conference (2004)
7. Martins, A., Almeida, J.M., Silva, E.P., Pereira, F.L.: Vision-based Autonomous Surface Vehicle Doccking Manoeuvre. In: MCMC 2006 7th IFAC Conference on Manoeuvring and Control of Marine Craft, Lisbon, Portugal (September 2006)
8. Martins, A., Almeida, J.M., Silva, E.P., Santos, F., Bento., D.: Forest Fire Detection with a Small Fixed Wing Autonomous Aerial Vehicle. IAV (submitted, 2006)
9. Bruce, J., Balch, T., Veloso, M.: Fast and Inexpensive Color Image Segmentation for Interactive Robots. In: IEEE/RSJ International Conf. On Intelligent Robots and Systems, vol. 3, pp. 2061–2066 (2000)
10. Pavlidis, T., Liow, L.: Integrating Region Growing and Edge Detection. IEEE Transactions on Pattern Analysis and Machine Intelligence 12(3) (March 1990)

ViRbot: A System for the Operation of Mobile Robots

Jesus Savage, Adalberto LLarena, Gerardo Carrera, Sergio Cuellar,
David Esparza, Yukihiro Minami, and Ulises Peñuelas

Bio-Robotics Laboratory, Department of Electrical Engineering
Universidad Nacional Autónoma de México, Unam
savage@servidor.unam.mx

Abstract. This paper describes a robotics architecture, the ViRbot,
used to control the operation of service mobile robots. It accomplish the
required commands using AI actions planning and reactive behaviors
with a description of the working environment. In the ViRbot architecture the actions planner module uses Conceptual Dependency (CD)
primitives as the base for representing the problem domain. After a command is spoken to the mobile robot a CD representation of it is generated,
a rule based system takes this CD representation, and using the state
of the environment generates other subtasks represented by CDs to accomplish the command. By using a good representation of the problem
domain through CDs and a rule based system as an inference engine,
the operation of the robot becomes a more tractable problem and easier
to implement. The ViRbot system was tested in the Robocup@Home [1]
category in the Robocup competition at Bremen, Germany in 2006 and
in Atlanta in 2007, where our robot TPR8, obtained the third place in
this category.

1 Introduction

In this paper is presented a mobile robot architecture, the ViRbot [2] system,
whose goal is to operate autonomous robots that can carry out daily service jobs
in houses, offices and factories. The ViRbot system has been tested with our
robot TPR8, see figure 1. This system divides the operation of a mobile robot
in several subsystems, see figure 2. Each subsystem has a specific function that
contributes to the final operation of the robot. Some of the layers of figure 2 will
be described in the following sections.

2 Virtual Environment

The virtual environment is visualized by a 3D system called ROC2. The simulation environment can be changed easily; it has multiple view ports of the
simulation; it has local or remote interaction (Internet); it can execute an user's
subroutines written in C/C++; simulation of the robot's movements and sensor's
readings can be provided also by the user using C/C++.

U. Visser et al. (Eds.): RoboCup 2007, LNAI 5001, pp. 512–519, 2008.
© Springer-Verlag Berlin Heidelberg 2008

Fig. 1. Robot TPR8, this robot has, for sensing the environment, a ring of sonars, a laser measurement system, a microphone and a vision system

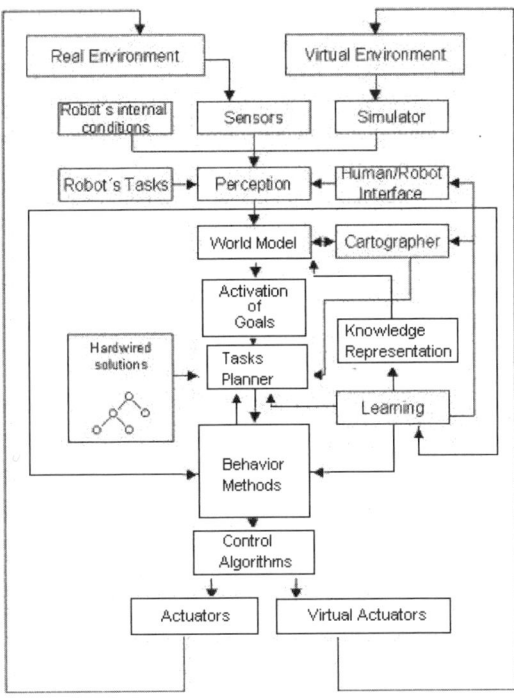

Fig. 2. The ViRbot System consists of several subsystems that control the operation of a mobile robot

3   Internal Sensors

The robots used in this research have the following internal sensors that reflect its internal state: wheel encoders and battery level sensor. Each of the sensors' values can be read any time during the operation of the robot.

4   External Sensors

The robots have the following external sensors that sense the surrounding environment: contact, reflective, infrared, video-cameras, microphones. Digital signal processing techniques are applied to the signals obtained from the video-cameras and microphones to be used in pattern recognition algorithms.

5   Simulator

The ViRbot system contains a simulator that provides the values of the internal and external sensors that the robot has, each simulated sensor has a mathematical model. Also the simulation of new sensors can be incorporated easily.

6   Robot's Tasks

Set of tasks that the robot needs to accomplish according to the time when it was programmed.

7   Human/Robot Interface

The Human/Robot Interface subsystem in the ViRbot architecture has tree modules: Natural Language Understanding, Speech Generation and Robot's Facial Expressions.

7.1   Natural Language Understanding

The natural language understanding module finds a symbolic representation of spoken commands given to a robot. It consists of a speech recognition system coupled with Conceptual Dependency [3] techniques.

Speech Recognition. For the speech recognition system it was used the Microsoft Speech SDK engine. One of the advantage of this speech recognition system is that it accepts continuous speech without training, also it is freely available and with C++ examples that can be customized as it is required. It allows the use of grammars, that are specified using XML notation, which constrains the sentences that can be uttered and with that feature the number of recognition errors is reduced.

Conceptual Dependency. Conceptual Dependency is a theory developed by Schank for representing meaning. This technique finds the structure and meaning of a sentence in a single step. CDs are especially useful when there is not a strict sentence grammar. One of the main advantages of CDs is that they allow rule based systems to make inferences from a natural language system in the same way humans beings do. CDs facilitate the use of inference rules because many inferences are already contained in the representation itself. The CD representation uses conceptual primitives and not the actual words contained in the sentence. These primitives represent thoughts, actions, and the relationships between them. Some of the more commonly used CD primitives are, as defined by Schank:

ATRANS: Transfer of ownership, possession, or control of an object (e.g. give.)
PTRANS: Transfer of the physical location of an object (e.g. go.)
ATTEND: Focus a sense organ (e.g. point.)
MOVE: Movement of a body part by its owner (e.g. kick.)
GRASP: Grasping of an object by an actor (e.g. take.)
PROPEL: The application of a physical force to an object (e.g. push.)
SPEAK: Production of sounds (e.g. say.)

Each action primitive represents several verbs which have similar meaning. For instance give, buy, and take have the same representation, i.e., the transference of an object from one entity to another. For example, in the sentence "Robot, go to the kitchen", when the verb "go" is found, a PTRANS structure is issued. PTRANS encodes the transfer of the physical location of an object, and it has the following representation:

(PTRANS (ACTOR NIL) (OBJECT NIL) (FROM NIL) (TO NIL))

The empty (NIL) slots are filled by finding relevant elements in the sentence. So the actor is the robot, the object is the robot (meaning that the robot is moving itself), and the robot will go from the living room to the kitchen (assuming the robot was initially in the living room). The final PTRANS representation is:

(PTRANS (ACTOR Robot) (OBJECT Robot) (FROM living-room) (TO kitchen))

CD structures facilitate the inference process, by reducing the large number of possible inputs into a small number of actions. The final CDs encode the users commands to the robot.

7.2 Speech Generation and Robot's Facial Expressions

The text to speech generation system used is called Festival [4] that is freely available. In human to human communication, facial expression plays an important role, so we consider that the same thing applies to human to robot communication. Thus, one of our robots contains a mechatronic head that shows simple expressions through movements of it, the opening of its mouth, the movement of its eyes and by modifying its eyebrows. The eyebrows are created using an array of LEDs that are turned on and off that creates different face expressions, see figure 3.

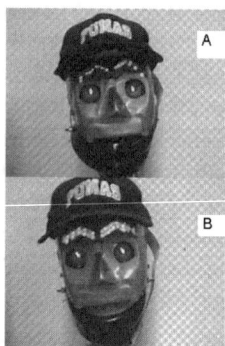

Fig. 3. In A, the robot's face shows that the robot did not understand the given command; in B, the robot's face shows that it understood the spoken command and that it is ready to execute it

8 Perception

The perception module obtains a symbolic representation of the data coming from the users and the robots sensors. The symbolic representation is generated by applying digital processing algorithms on the data coming from the sensors. With this symbolic representation a belief is generated.

8.1 Vision Subsystem

The vision subsystem consists of a robust implementation of an object tracker using a vision system that takes in consideration partial occlusions, rotation and scale for a variety of different objects. The objects are represented by feature points which are described in a multi-resolution framework, that gives a representation of the points in different scales. The interest points are detected using the Harris detector, and the description was based in an approximation coined SURF (Speeded-Up Robust Features) [5]. The ifound object is tracked by an Unscented Kalman Filter (UKF) [6].

9 Cartographer

This module has different types of maps for the representation of the environment: **Raw maps** are obtained by detecting the position of the obstacles using the robot's sonar and laser sensors. **Symbolic maps** contain each of the known obstacles defined as polygons, that consists of a clockwise ordered list of its vertexes. The Cartographer subsystem contains also topological and probabilistic (Hidden Markov Model) maps of the environment [7].

10   Knowledge Representation

A rule based system is used to represent the robot's knowledge, that is represented by rules, each one contains the encoded knowledge of an expert. In the ViRbot the rule based system CLIPS is used, an expert system shell, freely available and developed by NASA [8].

11   World Model and Activations of Goals

The belief generated by the perception module is validated by the cartographer and the knowledge representation modules, thus a situation recognition is created. Given a situation recognition, a set of goals are activated in order to solve it.

12   Hardwired Solutions

Set of hardwired procedures that solve, partially, specific problems. Procedures for movement, transference of objects, pick up, etc.

13   Task Planner

In the ViRbot planner subsystem there are two planning layers, the upper is the actions planner, based on a rule based system, and the lower layer the movements planner, based on the Dijkstra algorithm.

13.1   Actions Planner

The Robot is able to perform operations like grasping an object, moving itself from on place to another, etc. Then the objective of action planning is to find a sequence of physical operations to achieve the desired goal. These operations can be represented by a state-space graph. Thus, action planning requires searching in a state-space of configurations to find a set of the operations that will solve a specific problem. Internally, the actions Planner is built using CLIPS, that has an inference engine and this uses forward state-space search, that finds a sequence of steps that leads to a solution given a particular problem. Actions planning works well when there is a detailed representation of the problem domain. In the ViRbot architecture the actions planning module uses conceptual dependency as the base for representing the problem domain. After a command is spoken, a CD representation of it is generated. The rule based system takes the CD representation, and using the state of the environment it will generate other subtasks represented by CDs and micro-instructions to accomplish the command. The micro-instructions are primitive operations acting directly on the environment, such as operations for moving objects.

For example when the user says **"Robot, go to the kitchen"**, the following CD is generated:

(PTRANS (ACTOR Robot) (OBJECT Robot) (FROM Robot's-place) (TO Kitchen))

It is important to notice that the user could say more words in the sentence, like **"Please Robot, go to the kitchen now, as fast as you can"** and the CD representation would be the same. That is, there is a transformation of several possible sentences to a one representation that is more suitable to be used by an actions planner. All the information required for the actions planner to perform its operation is contained in the CD.

13.2   Movements Planner

If the command asks the robot to go from one room to another the movements planner finds the best sequence of movements between rooms until it reaches the final destination. Inside of each room the movements planner finds also the best movement path considering the known obstacles, that represent some of the objects in the room. Thus, the movements planner uses this information to find the best path avoiding the obstacles that interfere with the goal. The best path is found among several paths, according to some optimization criteria and using the Dijkstra algorithm. For the previous example where the robot was asked to go to the kitchen, the movements planner just needs to find the best global path between the Robot's place and the Kitchen, thus the action planner issues the following command to the movements planner:

(MOVEMENTS-PLANNER get-best-global-path Robot's-place to Kitchen)

And the answer of the movements planner is the following:

(best-global-path $Robot's - place\ place_1\ place_2...place_n\ Kitchen$)

Now a new set of PTRANS are generated asking the robot to move to each of the places issued by the planner:

(PTRANS (ACTOR Robot) (OBJECT Robot) (FROM $place_i$) (TO $place_j$))

For each of these PTRANS it is asked to the movements planner to find the best local path from $place_i$ to $place_j$:

(MOVEMENTS-PLANNER get-best-local-path $place_i$ to $place_j$)

And the answer of the movement planner is the following:

(best-local-path $node_1\ node_2...node_m$)

in which each $node_i$ is part of the topological map whose coordinates are x_i and y_i that the robot needs to reach.

14   Behavior Methods

After the movements planner finds the nodes where the robot needs to go, the Behavior subsystem tries to reach each of them, if it finds unexpected obstacles during this process it avoids them. The Behavior subsystem consists of behaviors based on potential fields methods and state machines, all these controls the final movement of the robot [9].

15 Control Algorithms and Real and Virtual Actuators

Control algorithms, like PID, are used to control the operation of the virtual and real motors. The virtual or the real robot receives the commands and executes them by interacting with the virtual or real environment and with the user.

16 Conclusions and Discussion

The ViRbot system was tested in the Robocup@Home category in the Robocup competition at Bremen, Germany in 2006 and in Atlanta in 2007, where our robot TPR8, obtained the third place in this category. Some videos showing the operation of the ViRbot system can be seen in *http://biorobotics.fi-p.unam.mx.*

References

1. Robocup@Home (2006), `http://www.ai.rug.nl/robocupathome/`
2. Savage, J., Billinhurst, M., Holden, A.: The ViRbot: a virtual reality robot driven with multimodal commands. Expert Systems with Applications 15, 413–419 (1998)
3. Roger, C.: Schank. Conceptual Information Processing. North-Holland Publishing Company (1975)
4. `http://www.cstr.ed.ac.uk/projects/festival/`
5. Bay, H., Tuytelaars, T., Van Gool, L.: SURF: speed up robust deatures. In: Proceedings of the ninth European Confernce on Computer Vision (May 2006)
6. Julier, S.J., Uhlmann, J.K.: Uncented filtering and nonlinear estimation. In: Proceedings of the IEEE 2004, vol. 93, pp. 401–422 (2004)
7. Thrun, Sebastian, Bucken, A.: Integrating Grid-Based and Topological Maps for Mobile Robot Navigation. In: Proceedings of the Thirteenth National Conference on Artificial Intelligence. AAAI Press (1996)
8. CLIPS Reference Manual Version 6.0. Technical Report Number JSC-25012. Software Technology Branch, Lyndon B. Johnson Space Center, Houston, TX (1994)
9. Muller, J.P., Pischeli, M., Thiel, M.: Modeling reactive behavior in vertically layered agent architectures. In: Wooldridge, M.J., Jennings, N.R. (eds.) ECAI 1994 and ATAL 1994. LNCS, vol. 890. Springer, Heidelberg (1995)

Grounded Representation Driven Robot Motion Design

Michael Trieu and Mary-Anne Williams

Innovation and Technology Research Laboratory
University of Technology, Sydney, NSW 2007 Australia
Michael.Trieu@gmail.com, Mary-Anne@it.uts.edu.au

Abstract. Grounding robot representations is an important problem in Artificial Intelligence. In this paper we show how a new grounding framework guided the development of an improved locomotion engine [3] for the AIBO. The improvements stemmed from higher quality representations that were grounded better than those in the previous system [1]. Since the AIBO is more grounded under the new locomotion engine it makes better decisions and achieves its design goals more efficiently. Furthermore, a well grounded robot offers significant software engineering benefits since its behaviours can be developed, debugged and tested more effectively.

Keywords: Robotics, Perception, Knowledge Representation.

1 Introduction

In order for a robot to achieve its objectives it must *ground* its representations: a grounded representation is one where the entities in the representation correspond *meaningfully* to the entities they represent [2, 7, 10]. In this paper we use a new grounding framework [12] to drive the design of a robot locomotion system and describe the value and benefits derived from that design. The main idea is that the grounding framework can not only be used at a theoretical level to analyse, evaluate and compare grounding capabilities in robots, but it also offers a practical guide to assist the design and construction of more reliable and adaptable robots. A major aim of modern science and engineering is to deploy dependable and flexible systems that are easy and cost effective to manage over their lifetime as their requirements and surrounding environment evolves and changes. Achieving this aim for systems like robots operating in complex and dynamic environments has proved to be extremely challenging. A poor understanding of *grounding* has been identified as major research bottleneck [2, 6, 7, 9, 10, 11]. In essence, *the grounding problem* is the challenge of designing and managing internal representations so that they meaningfully reflect the entities they are supposed to be representing. For example, how do we design representations of a robot's body so that it that can be appropriately managed by an intelligent control and behaviour system. Grounding involves building and maintaining coherent representations that correspond *meaningfully* to the entities they represent, whether the entities are physical, abstract, sensed, perceived, postulated, or simulated. A major challenge in addressing the grounding problem is not only in designing high quality representations as such but designing representations that are adaptable and conducive to change. In section 2 we describe the

U. Visser et al. (Eds.): RoboCup 2007, LNAI 5001, pp. 520–527, 2008.

grounding problem and the new grounding framework [12]. An AIBO robot system [1, 3] is described in section 3, in particular AIBO body, sensors and actuators, and the new Locomotion Engine. Section 4 describes the grounded representation guided design, and demonstrates how a grounding approach can drive design and development towards more resilient and reliable robotic systems.

2 The Grounding Problem

According to Brooks [3] "the world is its own best representation", and so if a robot only had to deal with the current state of the world then there is no necessity for representations. Since future world states in general are not a feature of the current world state, robots that need to plan and anticipate future world states in order to achieve their design goals require representations. Furthermore, the better a robot's representations are grounded, the more effectively it will achieve its goals, the more appropriate its behaviours and the higher the quality of its decisions. As a result robot representation design is a key area of interest in Artificial Intelligence. A grounded representation does not require that every entity in the representation be *linked* to a corresponding physical manifestation, but that a meaningful relationship exists between the entities in the representation and the entities being represented [12]. Maintaining a correspondence between representations of physical objects and the objects themselves is important but so too are the representations of object functionalities, relationships between objects, and as well as the descriptions of ways to interact with specific objects, etc. For the purpose of understanding grounding in robots, it is insightful to classify representations using the hierarchy of Gärdenfors [5] which describes the crucial relationships between three key representational entities: sensations, perceptions, and simulations. *Sensations* are immediate sensorimotor impressions, *perceptions* are interpreted/processed sensorimotor impressions, and *simulations* are detached representations, i.e. they are not tied to perceptions of the current state of the world. Sensations provide systems with an awareness of the external world and their internal world. They exist in the present, are localised in the body/system, and are modality specific, e.g. visual, auditory, not both. Perceptions encapsulate more information than raw sensorimotor information [2, 5]. Representations can be derived from information that has been gathered from a wide range of sources e.g. internal and external sensors, internal and external effectors, external instruments, external systems, etc. In this paper we focus on grounding representations derived from cued internal sensations and perceptions generated from a robot's body. The grounding framework [12] is motivated by the need to understand and build sophisticated systems such as robots that do (some of) the grounding themselves rather than systems that are completely grounded with the assistance of human grounding capabilities. It comprises five essential elements which can be as detailed as required for the purpose of the analysis:

1. *System Objectives* – description of the system objective, goals, tasks & activities
2. *Architecture of Grounding Capability* - a description of the underlying system architecture that supports or implements the grounding capability.
3. *Scope of the Analysis* - a detailed description of the scope of the analysis.
4. *Nature of the Grounding Capability* – relative to the grounding architecture.
5. *Groundedness Qualities* - includes a description of the pertinent groundedness qualities relative to each architectural component of the grounding capability.

All five components of the framework are related, e.g. the objectives and the scope will often determine how the qualities of groundedness are selected and assessed. The groundedness qualities identified as crucial to improving the Locomotion Engine are: faithfulness, correctness, transparency, accuracy, self-awareness, flexibility, adaptability and robustness.

3   AIBO Robot System and Locomotion Engine

The AIBO is a four legged robot developed by Sony. The main AIBO sensors and actuators are illustrated in Figure 1 below. Internal motors are used to move the AIBO body parts. The mouth has one degree of freedom; each leg has three, each ear one, and the tail two degrees of freedom.

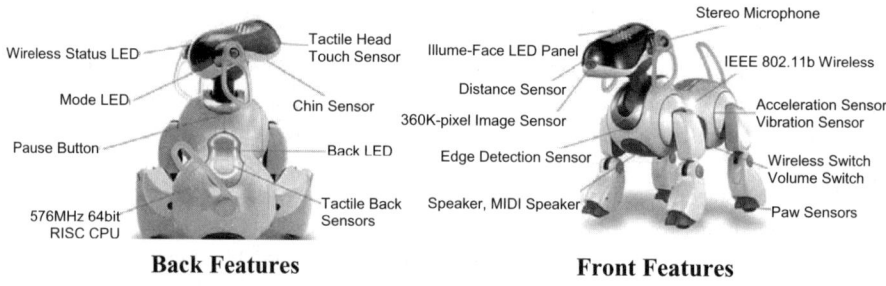

Back Features   **Front Features**

Fig. 1. AIBO sensors and actuators [Source: www.sony.com]

Our robotic system architecture [1, 4] is designed for soccer play on an AIBO platform. In the remainder of the paper we focus on the grounding of a robot's representations for improving the design of locomotion and behaviour. The Locomotion Engine [4] calculates and controls all locomotive movement of the robots. The robots walk with the bent fore-elbow stance adopted by most teams in the RoboCup Four-legged-league. The engine uses a static gait for all walking motions on the field. This involves the synchronous movement of diagonally paired legs; that is, the front left paw moves in synchronization with the back right paw while the other pair moves out of phase by half a step, as illustrated in Figure 2.

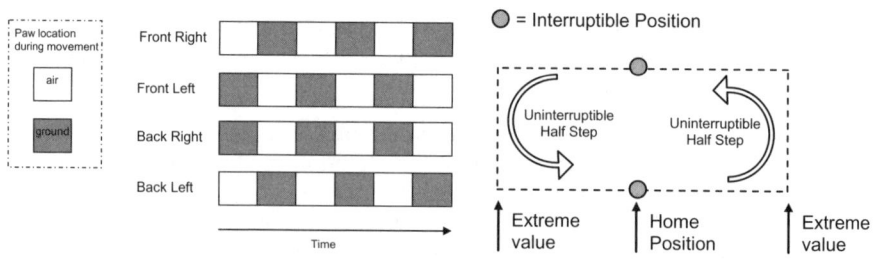

Fig. 2. Static gait   **Fig. 3.** Rectangular locus in detail

Each paw follows a particular locus and is interruptible at two specific points only as shown in Figure 3. A full step is when the paw moves around the locus and returns to its initial position and a half step is when the paw moves from one interruptible position to the other. This simple yet effective gait allows for desirable speed and stability as there is always two legs in contact with the ground at any given time. It is sometimes required that the robots move to a stable position before performing an action such as a kick. This stable position occurs when all four paws are situated at the home positions of their corresponding loci. The home position is the interruptible position located on the ground, which lies between the two extreme locus values as shown in Figure 3. The path each foot takes when in the air depends on the particular locus used. The locus path guides the robots paw through the air in a particular pattern and along the ground. Several successfully implemented loci including rectangle, ellipse, and raised rectangle each give rise to different walk types in terms of speed and stability. During the process of a step, the engine calculates the next point to move the robots paw on the chosen locus. It then uses this point to calculate the individual actuator joints by means of inverse kinematics and the parameters of the given walk. Controlling the walk stance, direction and speed requires five input parameters. These inputs are determined by an external module that commands the robot to complete an action. The input parameters are: type of walk/stance, forwards movement, strafe (sideways movement), turning movement, and speed. These input parameters are passed to the behaviour control engine when the robot decides to move. Each command is delivered as input parameters, which the engine uses to calculate the next uninterruptible half step to perform. The half steps are uninterruptible in that once they commence, they must complete regardless of any new commands. This allows the robot to take on a stable stance before commencing further action such as another half step or a kick. Each of these calculated steps incorporates a combinational movement of the Forward, Strafe and Turn directions. Each walk type corresponds to a particular set of unique parameters. They are fine-tuned for use in different situations by adjusting the many parameters associated with each leg. Having parameters associated with each leg allows total independent leg control and unique locomotive actions. The parameters for each of the four legs are shown in Figure 6: *Bounce Height* – the amount of bounce of each leg modeled over a sinusoid; *Shoulder Height* – the distance between the shoulder and the ground; *Step Height* – the maximum height at which the paw is lifted off the ground; *Step Position X* – the side distance between the paws home position & body; *Step Position Y* – the forward distance between paws home position & body.

4   Grounded Representation Driven Design

The use of a grounded representation design in the AIBO Locomotion Engine [1] led to major improvements to the robot soccer team [4]. The groundedness qualities we identified as being crucial to develop in the new design are (listed in order of priority): faithfulness, correctness, transparency, accuracy, self-awareness, flexibility, adaptability and robustness. Several new features were developed as a result of our grounded representation driven approach such as the following major enhancements to previous designs: compensation for asymmetric weight distribution, interpolated actions, action interrupt abilities, independent loci and variable locus points, independent binary file, and Delta factor for the ERS-7 model of the AIBO.

Compensation for asymmetric weight distribution: The robot movements are slightly biased to one side as a due to the uneven weight distribution of the robot which results from the off-centre placement of the robot's lithium battery. This has a significant impact on the walking direction of the robot if all four legs perform the same walking motions without calibration. An advantage of having independent representations for legs means that calibration of straight-line movement can be achieved by use of *factors* that allow each leg to have weighted movements, that is, particular legs are permitted to move more then others in a single step. The movements are weighted in units of percentage. There is a factor for every direction and leg combination, which gives rise to a large number of factor parameters (six factors × four legs), however this representation gives the AIBO the potential of achieving maximal control of its limbs. Each walk type has its own set of factors which are applied to each leg: Forward Factor, Backward Factor, Turn Right Factor, Turn Left Factor, Strafe Right Factor, and Strafe Left Factor. Without calibration, experiments show that the robot moves off to the right when trying to walk straight ahead. By adjusting the front Forward Factors so that the front left paw moves less than the front right paw, a slight pivot point is placed upon the front left paw. The pivot forces the robot to correct the biased movement, which results in straight-line movement as initially expected.

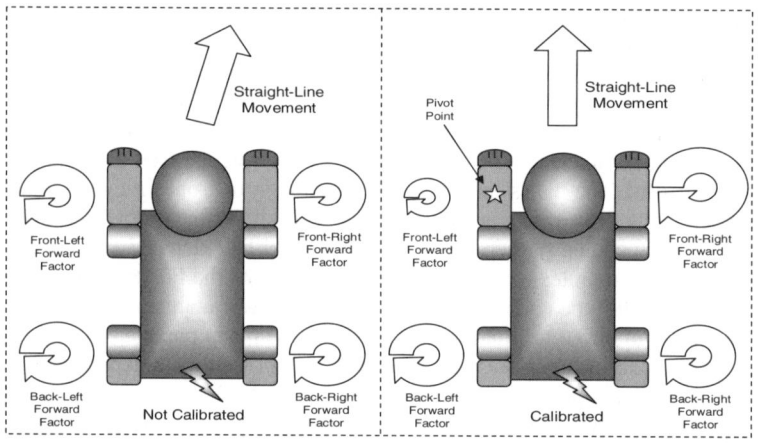

Fig. 4. Calibration using Independent Factors

These factors allow the AIBO to perform unique walks and kicks and other movements, which can be used for specific situations such as quick ball-handling maneuvers. The grounded control that the factors provide gives an unlimited freedom of configurable motions for game play, and ensures that the robots representations of motion are the well grounded in accordance with design criteria.

Interpolated Actions: All robot actions which include kicks, special movements and get up routines are made up of a number of position frames. Each position frame consists of a string of positions for every actuator on the robot. Actions on our previous design [1] were based on the *hold count* parameter on every frame. Sequencing through an action meant that the robot would snap to each position frame and hold it there according to the *hold count* in seconds. The new more grounded Locomotion Engine incorporates an extra parameter called the *interpolation count*. This parameter determines how fast (in seconds) each position frame is interpolated as demonstrated in the example below where knee angles are given in degrees. In the previous design FrontLeftKnee [1] had its *hold count* set to 0.8 seconds, and in the new design [4] it has its *interpolation count* at 0.5 seconds and *hold count* set to 0.3 seconds. Interpolated actions allow for smoother robot motion and hence improve the robots stability and grounding. Not only does this prevent slipping but also improves vision quality when the robot is in motion. From a grounding perspective the interpolation version is more faithful to the motion of the robot than the un-interpolated version. Furthermore, not only did this allow smooth actions but also a less dependency on the behaviour module to move to points between two positions. This provided the locomotion engine more control over the robots movements and hence increased the AIBO's self-awareness.

Action Interruptibilities: The new more grounded Locomotion Engine allows each action to fall in one of the following four categories: (i) Not Interruptible, (ii) Head Interruptible, (iii) Legs Interruptible, and (iv) Action Interruptible. The interruptibility of an action allows actions and walks the ability to override other actions when necessary. This allows complex sequences of actions to be performed during runtime. *Not Interruptible* means that the action cannot be interrupted by anything once it is started with the exception of the get up routine. *Head interruptible* allows only the head to be controlled by another action, even if the overriding action contains other movements besides the head, only the head movements will be executed with everything else continuing with the previous action. *Legs Interruptible* allows only the legs to be controlled by another action, e.g. if the robot wished to complete a head kick and walk forward at the same time, the head kick must be set to *Legs Interruptible*. *Action Interruptible* allows the entire robot to be overridden by another action or walk. Action interpolation and interruption capabilities together endow the AIBO with an important highly grounded stability awareness. In our previous version; actions, kicks and walks were managed by a behavior module that included knowledge of when to change from one action to another. If the behaviour module was not well grounded then it did not do this correctly or in an orderly fashion, and as a result the robot often became unstable. In the new grounded version actions, kicks, and walks each had new interruptibility parameters which prevented an action to occur during another at crucial moments. This ensures that the robot remains stable.

Independent Loci and Variable Locus Points: The previous locomotion system was limited in that there were only three loci shapes to choose for the walk engine as explained earlier, and all legs were required to use the same locus. In our new Locomotion Engine the front and back legs have independent loci with each locus containing a number of user defined points. This flexibility ensures that the Locomotion Engine is able to find a fast walk using reinforcement learning algorithms. Each point of the locus has three parameters, *X, Y and Time*. *X* and *Y* determine the location of the point and *Time* determines the percentage of time between specific points. The number of locus points can also be chosen to create many different shapes, thereby increasing the flexibility of the AIBO's movements.

Delta Factors: The new grounded Locomotion Engine is implemented on the ERS-7 and its predecessor AIBO 210. The ERS-7's body is in a polished plastic casing and the leg surface it much smoother than the AIBO 210s, which lead to a number of problems when ERS-7s were turning. In order to keep the ERS-7 robots grounded an additional factor was introduced into the walking parameters, namely the *Delta factor*. The Delta factor essentially determines how much skid steering the robot should complete depending on the amount of turn required as illustrated in Figure 5. By using skid steering, the robot is essentially moving like a tank when turning. To turn

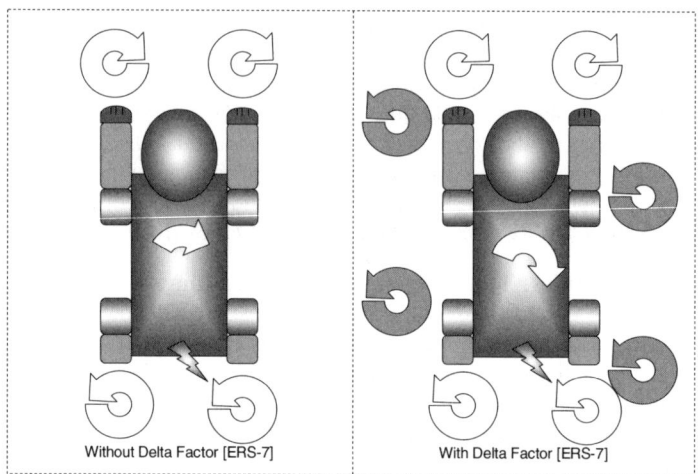

Without Delta Factor [ERS-7]   With Delta Factor [ERS-7]

Fig. 5. No Delta Factor and Delta Factor – Turning Right Example

right as shown in Figure 8, the legs on the right must move as if the robot is moving backwards whereas the legs on the left must move as if the robot was going forwards. The Delta factor determines how much of this behaviour occurs whilst turning.

The introduction of the Delta factor lead to a heightened awareness of walk calibration so that when the AIBO's control module instructed it to move forward then it did so in a straight line. In our new grounded locomotion engine each limb has calibration factors that are customized to each different walk. The introduction of factors removed the previous need for the behaviour module to compensate for miscalibrations and allowed it to specify motion without having to take weight distribution into account since the new more grounded Localization Engine that directly since it had a heightened awareness of the relationship between body movements and higher level goals in terms of movement. The Delta factor determines how much of this behaviour occurs whilst turning.

Reconfigurable Walks, Actions and Kicks: The new grounded Locomotion Engine is capable of adding and updating all walks, actions and kicks on-the-fly. The configuration is completed via text files which can be updated via wireless communication while the robot is in play. In this way the robots movements are detached from specific client applications. This flexible configuration allows any application that can alter text files the ability to configure and create any new walks, actions or kicks. In terms of groundedness this additional feature not only increases the AIBO's flexibility but also its transparency therefore enhancing its grounding capability.

Locomotion Trainer: In order to meet the grounding requirements for transparency we built the so called Locomotion Trainer V2 which is a Microsoft Windows based tool that is used to calibrate and remotely control the robots. It allows the creation of actions and walks. All parameters can be changed on the fly and the results can be observed immediately. Successful sets of parameters can be saved and stored into *.txt files for use on the robots. The Trainer can also monitor the odometry readings. Calibration of walk factors can be completed using this program.

Learning to Walk: A major advantage of the new grounded Locomotion Engine is that is offers significantly more scope for learning than the previous version since the locomotion parameters are more grounded within the AIBO's body and also more transparently integrated into the robots control system. The new grounded design supports a wide range of algorithms including genetic algorithms, reinforcement learning, and human assisted learning. Unsurprisingly, we found that a combination of approaches led to the best results. Speed was not the only important factor in determining the machine learning techniques. Other factors included smoothness so that the robot's camera did not bounce around causing difficulties with vision and stability so that the robot would not fall over, particularly when coming to a halt.

5 Conclusion

In this paper we have advocated a grounded representation driven approach to robot design based on a new grounding framework [12], and we illustrated the following benefits using an AIBO Locomotion Engine: (i) heightened awareness of stability and calibration, (ii) improved ability to respond to falling over, (iii) improved ability to turn with skid steering, and (iv) improved ability to *learn* new actions. The grounding driven design exhibits a range of desirable properties such as faithfulness, correctness, transparency, accuracy, self-awareness, flexibility, adaptability and robustness. The AIBO representations were more faithful to the robot body state, i.e. the actual body part locations against measured locations of actuators; they are also more correct and accurate. The design offered new flexibilities and as a result supported a high degree of adaptability in locomotion and higher level behaviours. The AIBO attained a higher level of awareness in the new design and due to the grounded motion, and behaviour was more robust and resilient.

References

1. Agnew, N., Brownlow, P., Dissanayake, G., Hartanto, Y., Heinitz, S., Karol, A., Stanton, C., Trieu, M., Williams, M.-A., Zeman, A.: Robot Soccer World Cup 2003: The Power of UTS Unleashed (2003), http://www.unleashed.it.uts.edu.au
2. Barsalou, L.: Perceptual symbol systems. Behavioural & Brain Sciences 22, 577–660 (1999)
3. Brooks, R.: The engineering of physical grounding. In: Erlbaum, L. (ed.) Proc of the Annual Meeting of the Cognitive Science Society, Hillsdale, pp. 153–154 (1993)
4. Chang, M., Dissanayake, G., Gurram, D., Hadzic, F., Healy, G., Karol, A., Stanton, C., Trieu, M., Williams, M.-A., Zeman, A.: Robot World Cup Soccer 2004: The Magic of UTS Unleashed (2004), http://www.unleashed.it.uts.edu.au/
5. Gärdenfors, P.: How Homo became Sapiens. Oxford University Press (2003)
6. Gärdenfors, P., Williams, M.-A.: Building rich and grounded robot world models from sensors and knowledge resources. In: Proc. of International Symposium on Autonomous Minirobots for Research and Edutainment (2003)
7. Harnard, S.: The symbol grounding problem. Physica D 42, 335–346 (1990)
8. Karol, A., Williams, M.-A.: Distributed sensor fusion for object tracking. In: Proceedings of the International RoboCup Symposium, Springer (2005)
9. Newell, A.: Physical symbol systems. Cognitive Science 4, 135–183 (1980)
10. Searle, J.: Minds, brains and programs. Behavioral & Brain Sciences 3, 417–457 (1980)
11. Sharkey, Jackson.: Grounding computational engines. AI Review 10, 65–82 (1996)
12. Williams, M.-A., Gärdenfors, P., McCarthy, J., Karol, A., Stanton, C.: A Framework for Grounding Representations. In: IJCAI Workshop on Agents in Real-Time and Dynamic Environments (2005)

Design of Design Methodology for Autonomous Robots

Eli Kolberg[1], Yoram Reich[2], and Ilya Levin[3]

[1] Bar-Ilan University, School of Engineering, Ramat-Gan, 52900, Israel
kolbere@eng.biu.ac.il
[2] Tel-Aviv University, Faculty of Engineering, Tel-Aviv, Israel
yoram@eng.tau.ac.il
[3] Tel-Aviv University, School of Education, Tel-Aviv, Israel
ilia1@post.tau.ac.il

Abstract. We present a methodology for deriving design methodology for autonomous robots. We designed this methodology in the context of a robotics course in high schools. The motivation for designing this new methodology was improving the robots' robustness and reliability and preparing students for becoming better designers. The new methodology proved to be highly successful in designing top quality robots. In the methodology design, we explored and adapted design methods to the specific designers, the nature of the product, the environment, the product needs, and the design context goals. At the end of this comprehensive design, we selected a synergetic integration of six methods to compose the methodology for this product context: conceptual design, fault tolerant design, atomic requirements, using fuzzy logic for the control of robotics systems, creative thinking method, and microprogramming design.

1 Introduction

In this paper, we deal with the design of robotics systems. Within this research, we developed a new design methodology for robotics systems [1]. In order to conduct this research we needed teams that actually designed. In order to test several design methodologies in a large-scale comparative study to get reliable and valid results, we had to choose an environment that provides such scale. Consequently, the research could not be implemented in industry because it is not possible to interrupt the ongoing work of many engineers in industry. An alternative environment, where a learning process takes place and has a more structured setting than industrial product development, is the education system. We decided to conduct the research among senior students majoring in science from four high schools, who within a robotics course [2], build autonomous mobile robots for participation in an international robotics contest. We discovered after several years of conducting this course with conventional design methodology that consistent problems were manifested [3].

The primary goal of the course was to teach the subject of robotics to high school students. The following were the course overall objectives: 1. Acquiring technical knowledge; 2. Acquiring a system thinking approach; 3. Improving skills of problem solving, decision making, and learning; 4. Developing critical and creative thinking abilities; 5. Experiencing development of a product, with time and budget restrictions; 6. Developing teamwork skills; 7. Improving students' design skills; and 8. Improving students' perception of technology.

U. Visser et al. (Eds.): RoboCup 2007, LNAI 5001, pp. 528–539, 2008.
© Springer-Verlag Berlin Heidelberg 2008

Until now, we did not take into account the subject of the design course, namely robotics. Fortunately, robotics products are classic examples of contemporary designs; therefore, the subject – robotics – does not change our analysis. It merely fixes the task of acquiring technical knowledge to deal with robotics related subjects.

The main goal of our research was to develop a new context dependent integrative design methodology for robotics systems, and to measure its success in an existing high school robotics course context.

We used the following design methods as a selection tool: Function-means trees, FMEA (Failure Mode and Effect Analysis), failure analysis, QFD (Quality Function Deployment), Pugh's concept selection, and AHP (Analytic Hierarchy Process) [4]. As an outcome we decided to integrate six design methods in the new design methodology [1]: 1. Conceptual design; 2. Fault tolerant design; 3. ASIT creative thinking; 4. ATR design; 5. Microprogramming design; and 6. Fuzzy logic control design. We introduced the new design methodology for robotics systems into the course. We tested the performance of the methodology in the years 2003-2005 [5].

Figure 1 describes the roadmap of this study. It is composed of theory development and course design followed by course implementation. The results of the course feed back into the theory development and the course was redesigned. The theory underlying the course design is a synthesis of ideas, drawn from different disciplines: engineering design, robotics, learning paradigms, engineering education, project base learning, contest oriented design, and learning by design. These disciplines provide the guidance in the course design, its implementation, and testing. The course design starts from requirements that are translated into course goals to be addressed by a design of the design methodology to be integrated into a detailed curriculum, which is implemented and tested. The results lead to reflection that helps improve our understanding and course design.

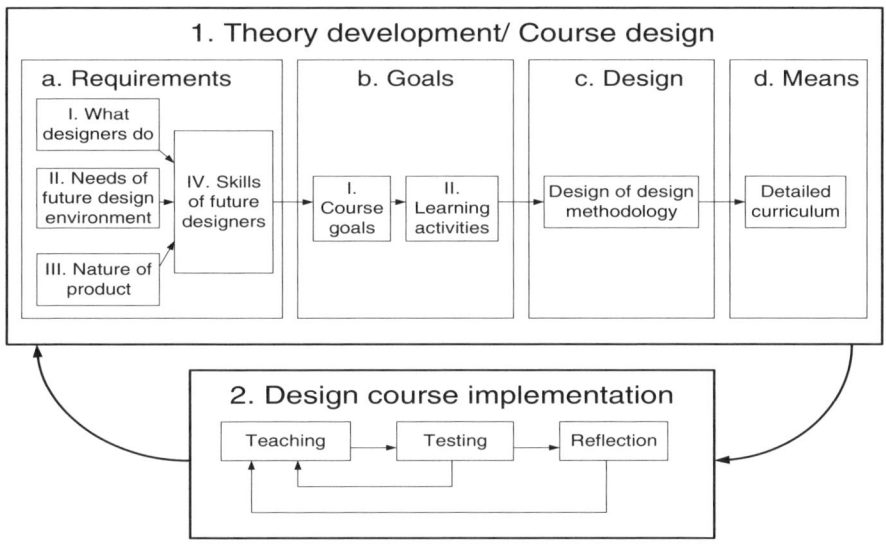

Fig. 1. Roadmap for designing designers

While there have been many studies on the design of curriculum in education and other fields (e.g., [6], [7], [8]), there is no single large-scale study on teaching design that was tested in a controlled experiment and produced conclusive results as the present study.

2 Method

2.1 Design of the Design Methodology (1c in Figure 1)

Since the students had no background knowledge in design, and since they had to complete the course with a quality design in order to compete in the competition, we decided to teach them enough design methods that would allow them to design and build excellent robots. We have also used these methods to teach other general concepts, such that imprecise information could lead to very precise behavior, as in fuzzy logic.

Design environment should allow for a meaningful design experience. The design of a complex product as a mobile robot allows for such experience. The contest supplies both time limits for project completion and environment for testing the results.

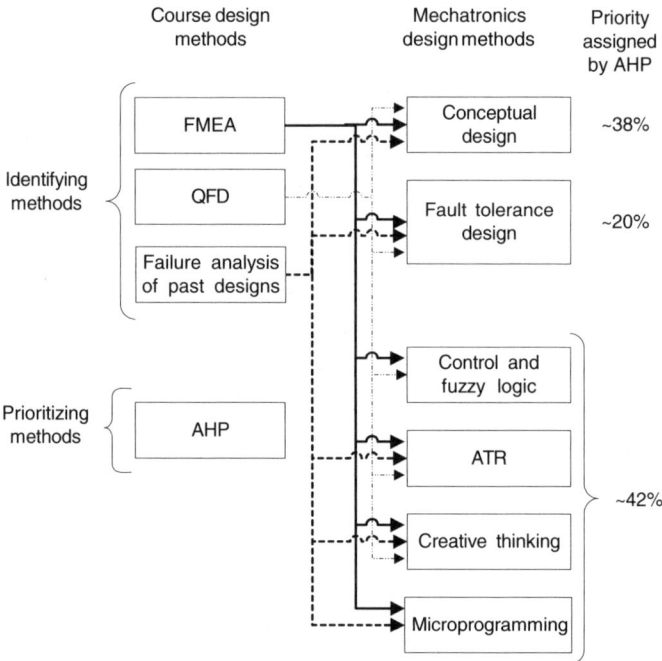

Fig. 2. Course design methods leading to robotics design methodology

Design methods are seldom taught in high schools. Moreover, in spite of their importance, it is even uncommon to teach them in universities. By and large, universities focus on analytical rather than synthesis skills. Our reasons for teaching design methods stem from the course's 2nd to 7th objectives. Moreover, design methods serve as guidelines that help students (as well as designers in real design projects) focus on the critical features when developing any engineering product. In addition, design methods glue technology to science. They help students realize the relations between the different science subjects learned at school and between science and the engineering work in robotics. These are central to understanding robotics as a discipline.

Further support for systematic teaching of design methods arises from feedback obtained from previous courses on the subject. One observation was that ignorance of design methods prevented effective use of expensive equipment purchased to support new technology and science related courses. Another observation was that lack of knowledge about design methods led to numerous occasions in which teams designed robots that violated simple engineering practice, resulting in quick robot failures. While product success is not mandatory for course success, these easily avoidable failures led to students and teacher disappointments, which are undesired. The logic for designing the design methodology and selecting the design methods is presented in figure 2.

2.2 First FMEA Analysis

We did a FMEA analysis for of the common design methods in general ([9], [10]) and specific to the context. We will discuss these methods and present in summary the considerations for disqualifying or accepting each method for further analysis. The design methods were:

1.   Selection design – It involves choosing one item from a list of similar items. There is a need to evaluate the potential solutions versus our specific requirements to make the right choice. This is done much more seriously with conceptual design methods. Consequently, we disqualified this design method.

2.   Product architecture design deals with the arrangement of the physical elements of the product to carry out its required functions. This is important for any complex system, so this method is kept for further analysis.

3.   Configuration design deals with how to assemble all the designed components into the complete product. As the robot has components that have to be assembled, we decided to move this method to next stage.

4.   Parametric design identifies the attributes of parts in the design configuration, which become the design variables for parametric design. The objective is setting values for the design variables that will produce the best possible design considering both performance and manufacturability. As the robot's subsystems have different attributes and optimal performances are needed, we decided to move this method to further analysis.

5.   Original design – Any time the design requires the development of a process, component, or assembly not previously in existence, it calls for original design. This sort of design being original does not supply tools for doing specific design and thus it is not relevant here. We disqualified it from further analysis.

6.   Conceptual design (CD) is one of the two most critical steps in product development. It is the basic design method; it places things into order. It allows to realize the big picture, and to see the important factor out of the large amount of data. It also can be used to divide the work between team members quite effectively and complete the project on time. Hence, we decided to move it to the next stage for further analysis.

7.   Concurrent design deals with cross-functional design team, where skills from the functional areas are embedded in the team. This allows for parallel design. This mainly

refers to heavy designed products where for example the manufacturing process development group starts its work as soon as the shape and materials for the product are established, and the tooling development group starts its work once the manufacturing process has been selected. In our case, we talk about small teams where there are no design skills differences among team members, there are no manufacturing or tooling teams, and there is no much meaning to parallel development. That is why we disqualified this method for further analysis.

8.   Atomic requirements (ATRs) design – Atomic (which cannot be further divided to two or more requirements) Requirements design is a tool that helps to understand the functionality and debugging requirements; it allows to divide the requirements into very fundamental, thus simple to understand requirements. It also helps to clearly identify unnecessary, overlapping, or conflicting requirements, isolate bug areas, and make clear what is to be done to implement the requirements. In the debugging mode, and problem solving, each requirement can be tested easily and separately. It is an efficient communication tool between people of different backgrounds. It is suitable for making more modular and more convenient to debug and fix robot systems. Hence, we decided to move it to the next stage of further analysis.

9.   Ergonomic design deals with interaction between people and the product. As some interaction occurs between the team members and the robot in the testing and operating the robot, we decided to move this method to the next stage of further analysis.

10.   Microprogramming (uP) design is common with products that include a microprocessor or a microcontroller. It allows for designing the robot control by moving between two different representations that make it easy for designing, debugging, and coding, simultaneously [11]. Microprogramming design presents the duality between two representations of control schemes and that even though it is more "natural" to use one to describe the robot operation, it is better to use another in order to be more robust and efficient. It also shows a way for being more effective when for example it is possible to combine two or more control schemes and save resources. Generally, it shows duality in two representations and understand that different representations are suited to different needs – a powerful problem-solving principle. The robot's control was based on a microcontroller, so we moved this method to next stage.

11.   Industrial design is concerned with the visual appearance of the product and the way it interfaces with the customer. These two are irrelevant to our robot, and thus will not be considered further.

12.   Fault tolerance (FT) design is crucial for creating robust products and it is inseparable method of every good design. It brings insight of the difference between products that are designed according to requirements, and robust products that can sustain faults up to a certain degree. It also introduces the possible faults during the design phase which improves the ability to identify and overcome problems. This influences on being more careful when design the robot parts, for example, the sensors array. It also demonstrates that in unstructured environments, no design could survive without making it robust to faults because it is usually impossible to foresee all potential situations. We moved it to the next stage.

13.   ASIT (Advanced Systematic Inventive Thinking) creative thinking design is a systematic method for creative thinking, which is designed especially for problem solving. It is important when a solution to a non-trivial problem is needed. By using

this method, it is possible to solve complicated problems. It seems fundamental in all design stages. We moved it to further analysis.

14.   Design for serviceability is concerned with the ease with which maintenance can be performed on a product. Products often have parts that are subject to wear and that are expected to be replaced at periodic intervals. That calls for a maintenance service. The robots built by the students are not a product that is intended for an extended use. There is no need for periodic service like oil replacement in cars. The part of design for easy access for parts replacement is covered in conceptual design. Hence, we decided to disqualify this method for further analysis.

15.   Fuzzy logic (FL) helps in simplifying things related to motors control. It is more straightforward and can be checked in an easy way, compared to other control methods. It is more intuitive to the students and is faster in implementation than other control methods. Fuzzy logic control design is used successfully in industry and we thought it would be adequate to move it to the next stage.

16.   Design for the environment is concerned with issues such as recycling, environmentally friendly materials, product waste minimization, packaging recovery, and noise reduction. Some of the robot parts are reused from previous years' materials, so we moved this method to next stage.

17.   Detail design is the way to realize the product. We will move this method for further analysis.

18.   Design for manufacturability – As the robots will not be manufactured beside for the project, we disqualified this method from moving to the next stage.

19.   Usability design – here the designer fits the product to user's physical attributes and knowledge, simplify user tasks, and make the user controls and their functions obvious. This is irrelevant for the course autonomous mobile robot, so we disqualified this method from further analysis.

20.   Design for reliability is quite similar to fault tolerance design (clause 12), which makes it redundant. That is why we disqualified it from further analysis.

After this session, 11 methods remained as candidates: conceptual design, ergonomic design, product architecture design, atomic requirements design, microprogramming design, fault tolerance design, parametric design, configuration design, ASIT creative thinking design, fuzzy logic control design, and detailed design.

2.3   QFD Analysis

We performed QFD analysis for selecting the design methods according to the criteria presented in table 1. The criteria were treated as the requirements and the design methods as the engineering characteristics.

Table 1. Robot's performance evaluation criteria

	Criteria		Criteria
1	Success in the contest	8	Fast navigation to all rooms
2	Driving well in corridor	9	Overcoming uneven floor
3	Making 90 and 180 degrees turns	10	Obstacle avoidance
4	Driving well in reverse mode	11	Non tethered robot operation
5	Finding a white line on a black background	12	Sound activation of the robot
6	Finding a lit candle in a room	13	Navigation from each room back to starting point
7	Fast extinguishing of a lit candle		

Table 2 presents the QFD analysis [10] for choosing the appropriate design methods. Based on the criteria, the "whats" are listed in room 1. Room 4 lists the various design methods that should be checked against the criteria. Next, we turn to room 2. The *criteria importance* was established by interviewing teachers and mentors, and allocating the views along a 1-5 scale, where 5 is the highest. The previous years' robots were ranked according to the way in which they satisfied requirements, on a 1-5 scale, and subsequently, the planned robots were rated against the requirements. In room 3, the ratio between the planned to previous robots is called the *improvement ratio*. The product of *criteria importance* x *improvement ratio* gives the *total improvement ratio*. The *relative weight* is each value of *total improvement ratio* weight divided by the sum of all values of importance weight. The relationship matrix, room 6, tells us how each design methods help attain the criteria list. Here a strong impact is worth 9, a medium high impact 5, a medium low impact 3, and a weak impact 1. The importance of the design methods in room 7 is determined by multiplying each of the cells in the matrix by its *relative weight* and summing each column to give the *absolute importance*. The *relative importance* in room 8 is obtained by dividing the *absolute importance* by the sum of all absolute importance values. Six methods rank highest and almost twice as high as the next in line: conceptual design, fault tolerance design, atomic requirements design, ASIT creative thinking method, use of fuzzy logic in robot control, and microprogramming design.

Table 2. QFD analysis of design methods

		Conceptual design	Ergonomic design	Product architecture design	Atomic requirements design	Microprogramming design	Fault tolerance design	Parametric design	Configuration design	ASIT creative thinking method	Use of fuzzy logic in robot control	Detail design	Criteria importance	Previous years robots	Planned robot	Improvement ratio over previous robots	Total Improvement ratio	Relative weight
1.	Performance	9	1	9	9	9	9	9	9	9	9	9	5	4	5	1.3	6.5	0.058
2.	Real time hardware failure resistance	9	1	1	5	5	9	1	3	5	9	1	5	2	5	2.5	12.5	0.111
3.	System simplicity	9	1	3	9	9	5	1	1	9	9	1	4	3	4	1.3	5.2	0.046
4.	Flexibility	9	1	9	5	9	9	5	1	9	9	1	4	2	5	2.5	10.0	0.089
5.	Robot reliability	9	1	3	9	5	9	5	5	5	9	3	5	3	5	1.7	8.5	0.076
6.	Software modularity	5	1	1	9	9	9	5	1	5	3	1	3	2	5	2.5	7.5	0.067
7.	Robot testing ability	9	1	9	9	9	9	1	1	9	9	3	4	2	5	2.5	10.0	0.089
8.	Fast hardware fixing	9	3	9	9	1	5	1	3	9	1	1	4	2	4	2.0	8.0	0.072
9.	Ability of upgrading	9	3	9	1	9	9	9	1	9	9	9	2	2	5	2.5	5.0	0.045
10.	Cost saving	9	1	1	9	9	9	1	1	9	9	3	3	1	5	5.0	15.0	0.134
11.	Ease of transferring the subject matter	9	1	1	9	5	5	1	3	9	9	1	5	3	5	1.7	8.5	0.076
12.	Short learning time	5	3	3	9	5	5	1	1	9	9	1	5	5	5	1.0	5.0	0.045
13.	Ease of use	9	1	5	9	9	9	1	1	9	9	3	5	5	5	1.0	5.0	0.045
14.	Can be modified to high school students	9	1	3	9	9	9	3	3	9	9	3	5	5	5	1.0	5.0	0.045
	Absolute importance	8.53	1.32	4.16	7.82	7.17	8.03	2.84	2.37	7.97	8.00	2.6	60.8				111.7	0.998
	Relative importance	0.14	0.02	0.07	0.13	0.12	0.13	0.05	0.04	0.13	0.13	0.04						

2.4 Failure Analysis and Maim Problems Encountered with Previous Robots and Possible Solutions

Another method used for selecting the design methods was failure analysis where poor design practice was analyzed. We reviewed many of the previous robots available description and data, including interviews with teams, reading project reports, observing failures of robots from previous competitions, and getting the robots performance in local and international competitions. Upon organizing and sorting the data, we found the following as the main problem issues.

1. Need for several hardware and software changes and modifications. We found that it was common among many teams to totally redesign their robot more than once. The most appropriate solution to this kind of problem would be implementing conceptual design methods.

2. Malfunction equipment. Sometimes robots are not qualified in their trial runs due to malfunctioning equipment. A solution to this could be to introduce checkers that identify sensor failure and upon identifying the above failure, change the position control.

3. We observed that high school students in general had difficulties in designing reliable robot speed and position control. The students had difficulties to calculate or experimentally find the proper gains of the PID control loop and were not aware what was happening with the robot control. They knew the formality needed for implementing the control but they knew neither the essence of it nor how to decide on proper gains. In some cases, the improper gain values caused the robot to be too slow or too fast, and as a consequence the robot hit the wall. The use of fuzzy logic control could remedy these difficulties.

4. When students reached the design stage, they stated the robot requirements among their team members in an ambiguous way. There was also lack of ability to test and debug the robot, because of contradicting or unclear requirement definitions. The ATR method would address these problems.

5. Occasionally, the teams did not overcome the problems properly. Solving these problems was possible if the students would apply a creative thinking method like ASIT.

6. The last noticeable group of problems was the difficulty to follow and debug an ASM algorithm; in many occasions the students did not cover all possible situations. Using an FSM algorithm might inherently prevent these situations. In FSM, all states and transitions must be declared and taken care off, and it is easy to find an uncovered situation. Hence, translating the ASM to FSM is important, and was done by microprogramming techniques, which also allow for integrating several sub algorithms and saving code.

2.5 Another FMEA Analysis of Adapting the Design Methods for High School Students

In order to reduce the chances of failing with this methodology in the particular context in which it was implemented, we exercised FMEA and tried to think of the issues that could make it fail and to produce some counter measures. As the students were inexperienced, we had to adapt the methodology to their lack of engineering mathematics skills and experience. This led to preparing an appropriate training course for these students and designing the curriculum to enable logical teaching to the students. Another critical issue that we faced was the modification of industry methods to suit the teaching environment of a high school where students lack prerequisite knowledge. Next, we will describe the modifications made to each of the design methods for their inclusion in the course material.

Conceptual design: The teaching of conceptual design requires no prerequisite knowledge; however, the time constraint forced a short version that was modified to suit the needs of the students. The stages of problem definition, and identifying customer needs with subjects, such as how to interview the customer, making focus groups, preparing customer surveys and handling customer complaints, were not taught because contest rules can be regarded as stating the problem and covering the customer needs. Only a small part of benchmarking was taught, as there was no identical commercial product to test against. There were robots from the year before, which were analyzed by the teams in comparison to their robots.

Creative/inventive thinking: ASIT was taught completely as it requires no special background and could easily be taught to the students in a short time. Another assisting factor in using ASIT was that we had an accessible simple training material that could be distributed to students for home practice.

Fuzzy logic: As the designers were high school students, no intensive mathematics background was introduced. The fuzzy logic (FL) control subject was introduced to the students as a technical straightforward procedure. The students learned to create the variable membership functions, adapted to the capabilities of the microcontroller they used; derive the fuzzy rules; and receive the output variable for further processing.

Robot control: Robot control was taught using an innovative teaching method built upon the use of dual representations. The method was taught without the intensive mathematical manipulations. It is further explained through the microprogramming subject.

Atomic Requirements: This method was taught completely. It requires no special background and could easily be taught to the students in a short time.

Microprogramming: Microprogramming is an approach to teaching a number of subjects related to computer hardware. We adapted microprogramming for designing robotics systems. The main idea of this adaptation is based on considering a robotics system to be a composition of two units: *a control unit* and an *operational unit* [12]. The operational unit of the system includes such building blocks as motors, sensors, lamps, manipulators, etc. The control unit receives information from the operational unit and produces a sequence of control signals that results in executing desired operations by the operational unit. Usually microprogramming is a subject that is studied at the undergraduate level. It is built on a number of strong prerequisites including introductory logic design and programming. For introducing the subject into the high school robotics course, we have developed a specific "microprogramming curriculum" including a number of well formulated formal notations and definitions. The curriculum skips some technical details connected to specific computer architecture. Further, the presented microprogramming concept includes only a Finite

State Machine (FSM) based microprogrammed controller, and not the classical Wilks architecture. It allows presenting the concept of microprogramming in a simpler manner and makes it practically productive for the process of robotics design.

Fault Tolerant Design: All fault tolerant (FT) components are useful in robotics systems design. Particularly, the robotics design described in this study includes one necessary component for fault tolerant design, which is the self-checking design. Within high school curriculum, the self-checking design was based on the development of specific redundant units, so-called checkers. The main goal of the checker is to prevent entering incorrect data to the control and operation parts of the robotics system. Students are able to construct checkers for robotics systems by using a number of standard solutions for the checkers design. These solutions are based on fundamental principles of self-checking design: fault secure and self-testing property. Students had to develop an appropriate checker and also prove it's self-checking.

Within the current course design, the mathematics involved with FL, microprogramming and FT was too complicated. Yet, even by eliminating the mathematical details, there was sufficient benefit to teach these methods and use them. We considered teaching neural networks control but found it too complicated and of little importance. We also considered teaching 3D modeling and schematic software but the teaching overhead and the software cost would not justify their inclusion.

2.6 The Chosen Methods for the Design Methodology and AHP Analysis of Design Methods Selection

To conclude, besides the general confidence about introducing design methods into the classroom, we used three guidelines to design the design methodology to teach: (1) addressing poor design practice by previous years' teams; (2) introducing methods that had high impact on attaining course's goals; and (3) avoiding complex methods. The six methods selected are complementary and cover the complete design process; they include: Conceptual design, ASIT creative thinking method, ATR design, Fault tolerance design, Microprogramming design, and Fuzzy logic control design.

Within the scope of the possible robotics design methods, these have an important role or influence over the product quality and its performance in field conditions. Moreover, these methods allow appreciating issues beyond the original goals. For example, fuzzy logic allows appreciating that mathematics is not always about precise numbers. In fact, a great deal of engineering reasoning is qualitative and imprecise [13]. Fuzzy control demonstrates that imprecise concepts lead to very robust behavior that is relatively easy to attain.

Subsequent to identifying the design methods, two experts used AHP [4] to prioritize the methods in order to allocate them the necessary teaching resources. It was agreed that conceptual design is the most important method (importance 42% out of 100% for one expert and 34% for the second). The method that was secondly important was fault tolerance and testability (19% and 22%, respectively). The expert agreed on the following four methods but differed in the order of importance that they assigned to each method. Nevertheless, the expert assessment and our own judgment were quite consistent. After the relative importance evaluation, and given the stringent teaching hours limit, we decided to teach subsets of these design methods that deemed critical to the robot design or that would contribute significantly to other course goals. The findings and the experience from the first year of conducting this research approve the effectiveness of these methods.

3   Implementation, Evaluation and Validation of the Design Methodology

The resulted methodology was taught fully or partially in three schools and traditional design study was taught in a fourth school. The teaching in all of the four schools was conducted in parallel. The parallel tracks ended with measurement, evaluation, and comparison between the four schools.

In order to evaluate and validate the methodology we followed the principles presented by Nevo [14], and Marshall [15]. The following points support the design methodology validation.

1. Nevo's structure for evaluation recommendation items like evaluation background, the conceptual frame, the questions which the evaluation tried to answer on, and methods, and outcomes, are covered and described in this paper.
2. All the design methods included in the methodology are known and proved to be efficient and effective.
3. The criteria for testing the design methodology are clear and can easily be tested.
4. The methods within the design methodology are complete and orthogonal.
5. The testing and validation process was done in large scale within three years: pilot study in the year 2003, full implementation in the year 2004, and transfer implementation in the year 2005. In each phase, a careful research work was done. The implementation in four schools among 127 students and 7 teachers further supports the validation of the methodology.
6. All results are measurable.
7. The methods in the methodology were chosen from a larger list. Moreover, as we state that the methodology is context dependent; it might and probably will change in the case of different contexts.

4   Discussion

Through careful design, implementation, and testing, we developed a design methodology for robotics systems. We described the process of deriving the design methodology and the importance of context dependent design. In this case, we made adaptations related to the context of the design, namely: high school students, high school environment, the product, and the contest. Each of the six methods had its own special contribution to the design, to the product of the design – the mobile robot- and to the students. The students were aware of the design methods they learned and we observed that the students developed abilities to apply the proper design methods to specific problems they encountered in the project.

We argue that design is context dependent. We showed that taking into account the specific design conditions leads to a tailored design methodology. We believe that with the same design methodology approach, technology courses can be taught in universities and industry, yielding even more profound benefits to designers.

References

1. Reich, Y., Kolberg, E., Levin, I.: Designing designers. In: Proceedings of International conference on engineering design, ICED 2005, Melborne (2005)
2. Verner, I., Waks, S., Kolberg, E.: Upgrading Technology Towards the Status of High School Matriculation Subject: A Case Study. Journal of Technology Education 9(1), 64–75 (1997)
3. Kolberg, E., Reich, Y., Levin, I.: Project-Based High School Robotics Course. International journal of Engineering Education 19(4), 557–562 (2003)
4. Saaty, T.L.: The Analytic Hierarchy Process. McGraw-Hill, New York (1980)
5. Reich, Y., Kolberg, E., Levin, I.: Designing contexts for learning design. International Journal of Engineering Education 22(3), 489–495 (2006)
6. Barrows, H.: How to Design a Problem Based Curriculum for the Pre-Clinical Years. Springer, N.Y (1985)
7. Clark Jr., E.T.: Designing and Implementing an Integrated Curriculum: A Student-Centered Approach. Holistic Education Press, Brandon (1997)
8. Diamond, R.M.: Designing and Assessing Courses and Curricula: A Practical Guide, Jossey-Bass, San-Francisco (1998)
9. Ullman, D.G.: The Mechanical Design Process. McGraw-Hill, N.Y (1992)
10. Dieter, G.E.: Engineering Design: A Materials and Processing Approach. McGraw-Hill, Singapore (2000)
11. Levin, I., Kolberg, E., Reich, Y.: Robot Control Teaching with a State Machine Based Design Method. International Journal of Engineering Education 20(2), 202–212 (2004)
12. Baranov, S.: Logic Synthesis for Control Automata. Kluwer Academic Publishers, Netherlands (1994)
13. Subrahmanian, E., Konda, S.L., Levy, S.N., Reich, Y., Westerberg, A.W., Monarch, I.: Equations aren't enough: Informal modeling in design. AI EDAM 7(4), 257–274 (1993)
14. Nevo, D.: Useful Evaluation: Evaluating Educational and Social Projects, Massada, Israel (Hebrew) (1989)
15. Marshall, C.: Goodness criteria: are they objective or judgment calls? In: Guba, E.G. (ed.) The paradigm dialog, Goodness Criteria, Sage publications, California (1990)

Opponent Provocation and Behavior Classification: A Machine Learning Approach

Ramin Fathzadeh[1], Vahid Mokhtari[1], and Mohammad Reza Kangavari[2]

[1] Mechatronics Research Laboratory
Department of Computer Engineering, Islamic Azad University,
Qazvin Branch, Qazvin, Iran
{fathzadeh,mokhtari}@qazviniau.ac.ir
http://www.mrl.ir/
[2] Department of Computer Engineering, Iran University of
Science and Technology, Tehran, Iran
kangavari@iust.ac.ir

Abstract. Opponent Modeling is one of the most attractive and practical arenas in Multi Agent System (MAS) for predicting and identifying the future behaviors of opponent. This paper introduces a novel approach using rule based expert system towards opponent modeling in RoboCup Soccer Coach Simulation. In this scene, an autonomous coach agent is able to identify the patterns of the opponent by analyzing the opponent's past games and advising own players. For this purpose, the main goal of our research comprises two complementary parts: (a) developing a 3-tier learning architecture for classifying opponent behaviors. To achieve this objective, sequential events of the game are identified using environmental data. Then the patterns of the opponent are predicted using statistical calculations. Eventually, by comparing the opponent patterns with the rest of team's behavior, a model of the opponent is constructed. (b) designing a rule based expert system containing provocation strategies to expedite detection of opponent patterns. These items mentioned are used by coach, to model the opponent and generate an appropriate strategy to play against the opponent. This structure is tested in RoboCup Soccer Coach Simulation and MRLCoach was the champion at RoboCup 2006 in Germany.

1 Introduction

Multi Agent System is one of the sub-disciplines of artificial intelligence which was introduced for the purpose of defining the rules and principles for developing complex systems and provides a mechanism for cooperating the agents [1], [2]. In real-time environments, multi agent systems need agents that are able to act automatically as members of a team. Modeling opponent in our multi agent system environment predicts the future behaviors of opponent and proposes an appropriate counteraction [3]. RoboCup is an MAS environment and opponent modeling plays a crucial role in this context. In this domain, every team is defined as a group of autonomous agents which are connected to a server and play a simulated soccer [4]. Coach agent of the team, receives the complete and noiseless information from the field and in order to enhance the performance sends messages in format of the standard coach language, called CLang, to its players [5], [6].

U. Visser et al. (Eds.): RoboCup 2007, LNAI 5001, pp. 540–547, 2008.

Recently emphasizing on opponent modeling, coach competition has regulation changed been, so that coach becomes in charge of identifying the weaknesses and strengths of the opponent *(patterns)*, from other behaviors of the opponent *(base strategy)*. The 2006 coach competition rule defines pattern and base strategy as:

Pattern: A *simple behavior* that a team performs which is *predictable* and *exploitable* for the coaches.

Base strategy: The general strategy of the test team regardless of the pattern in it.

To exemplify this, a pattern may be a sequence of consecutive passes between some particular players, clearing the ball to the outside of penalty area by defenders, or any different formation of players between pattern and base strategy

Our work is focused on opponent modeling and online pattern identification. For this purpose, MRLCoach receives the previous plays of the opponent as two log files of pattern and base strategy, and by analyzing them identifies the events occurred such as pass, shoot, dribble, etc. Then for pattern recognition, *chi-square test* [7], [8], issued to analyze the possible relation between an event and a sequence of previously occurred events. The eventual model of the opponent could be a collection of multiple identified patterns. Now, using a *radix tree* [9], we compare constructed models from the pattern and base strategy log files and store the differences as the final model of the opponent in the *model repository*. Finally, coach makes models for each of the pattern and base strategy log files. *Opponent Provocation* could be considered as a new problem in MAS environment. In RoboCup coach domain, this goal means for each identified behavior of opponent, a strategy is constructed to activate this behavior in online game. Hence *Rule Based Expert System* is recruited to expose the appropriate strategies for opponent provocation. Thereupon, in online mode, by observing the opponent behavior, coach looks for an appropriate strategy for it in *Knowledge Base*. Once an instance is found, it is sent to the players. In online mode, observing the live game, coach exposes an online model of the opponent and compares it with the stored models in repository. When a matching online model is identified, coach reports it as the current opponent model to the server. The remainder of this paper is organized as follows: At first, we introduce the soccer server environment, second, a 3-tier learning architecture for predicting and exposing the opponent behavior is presented. Afterwards, we explain how rule based expert system proposes a proper strategy against the opponent team and how the process of learning is accomplished in online game. Continuing on, section 4 presents the results of our experiments in details. The final section of this paper is devoted to the future works.

2 The Environment

The RoboCup simulation league uses the Soccer Server System [4] to simulate the field and the players. Each player has to be a unique process that connects via a standard network protocol to the server. The players receive video and audio information every 150 msec over the network and can issue primitive actions like kick, dash, turn, turn-neck and say every 100 msec. The server processes the actions of the players and generates new visual information. The rest of information consists of the distances and angles to other players, the ball and landmarks. The players can only perceive objects that are in their field of vision and both the visual information and the execution of the actions are noisy. Additionally, the accuracy and amount of sensory information decreases with the distance of objects. Communication between the clients is only allowed if it passes via the server, considering the fact that bandwidth and hear range are limited.

An extra client on each team can connect as a coach, who can see the whole field and send strategic information to clients when the play is stopped, for example for a free-kick. In the soccer server, a coach agent has three main advantages over a standard player. First, a coach has given a noise-free omniscient view of the field at all times. Second, the coach is not required to execute actions in every simulator cycle and can, therefore, allocate more resources to high-level considerations. Third, in competition, the coach has access to log-files of past games played by the opponent, which can provide to important strategic insights.

3 Coach Framework

Coach agent behavior comprises of two phases: In *Opponent Behavior Acquisition* phase, raw data is received from the environment, and events of the game are identified using statistical calculation. Then the opponent behaviors are classified as patterns. In *Opponent Provocation* phase, pattern recognition process is expedited. Here a rule based expert system is being used to opponent provocation. Figure 1 shows the general architecture of coach.

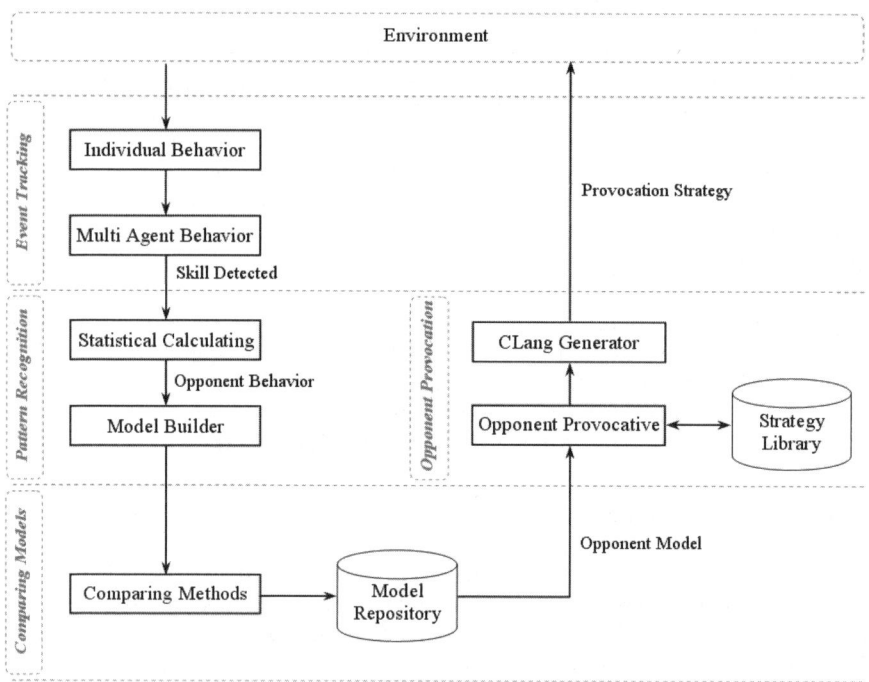

Fig. 1. MRLCoach architecture: event tracking, pattern recognition, comparing models and opponent provocation

3.1 Opponent Behavior Acquisition

Before starting the game, coach is provided with a set of prepared patterns and base strategies of the opponent's past games. By analyzing this information, coach exposes the occurred events. Then, using the chi-square test, expected patterns in a log file are identified. Constructed model of the opponent is a set of these patterns. This model is compared with the model created from base strategy. Their difference is stored as a final model of the opponent for online use. During the game, online coach receives the match's information and analyzes play of the opponent with similar methods used in offline phase and compares the created model with those already existing in the repository and reports a matching one as the current opponent model to the server.

The main goal of coach is to mine the opponent behavior. For this purpose, we classify the possible behaviors in a simulated match to different classes such as formation, pass, shoot, dribble, hold, etc., some of which are divided into subclasses. For instance, the pass class has the following 3 subclasses: *direct pass*, *pass graph* and *closed pass graph* [10], [11].

The opponent modeling process is comprised by *event tracking*, *pattern recognition* and *comparing models*.

Event Tracking. The first step in modeling the opponent is to detect the events occurred in a game. Event tracking consists of breaking problem down into two individual and multi agent behaviors [12]. For tracking these behaviors, raw data including play mode, positions and velocities of both the players and the ball are gathered from the field. Then by identifying the ball owner in every cycle, the individual behaviors, namely pass, shoot, dribble, hold and intercept are exploited. After identification of the individual behaviors, in a next higher level, multi agent behaviors such as formation, defending system, offending system and pass graphs are recognized.

Pattern Recognition. A pattern is a sequence of events appeared in a game sufficient number of times, which is predictable and exploitable. In order to treat the sequence of events identified at the previous section as patterns, we have recruited the chi-square test:

$$\chi^2 = \sum_{i=1}^{k} \frac{(O_i - E_i)^2}{E_i} \tag{1}$$

Which O_i is the observed frequency of an event, E_i is the expected frequency of an event and k is the number of random variables.

As an example for recognition of pass pattern, 29 observed passes for the player 2 are shown in the frequency table 1.

Table 1. Use of chi-square test in recognition of pass pattern

	Player 3	Player 4	Player 5	Total
Observed Pass (O)	4	9	16	29
Expected Pass (E)	2.9	5.8	20.3	29
(O-E)²/ E	0.417	1.765	0.910	3.092

The columns of this table contain players receiving the pass. The first and the second rows have respectively the numbers of observed and expected passes. Based on our experience in recognition of pass pattern, we consider the expected frequency to be at least 70% of the total of observed passes to the player with maximum number of receives. The total in the third row is the calculated chi-square value. Now, we should compare the chi-square value we calculated, $\chi^2=3.092$, with the χ^2 value read from the *table of* χ^2 [13], with n-1 *degrees of freedom* (where n is the number of *categories* which is the number of pass receiver players, 3 in our case). So we have only 2 degrees of freedom. From the χ^2 table, we find the critical value of 5.99 with *probability*=0.05.

Because the calculated value of 3.092 is less than 5.99, our assumption is though acceptable. This means in 95% of the cases the calculation of pass pattern is significance.

Comparing Models. A model of the opponent is a set of detected patterns. We store this model using *radix tree ADT*. For each of the pattern or base strategy log files, a radix tree is created. Afterwards, it is necessary to compare these radix trees to suggest the final opponent model. The difference between *pattern radix tree* and *base radix tree* determines the final model of the opponent that is stored in a third radix tree in model repository. Actually each node of this tree is a node existing in pattern radix tree but not in base radix tree.

After all the possible models of the opponent are exposed from the log files in the offline section, in the online mode, we are to identify the current model of the opponent in real-time. To accomplish this, in each game, by receiving the information from the field, events and patterns of the opponent are identified with similar methods used in offline; then an online model of the opponent is created. Unlike the offline mode where model of the opponent is identified by comparing the log files of pattern and base strategy, in online mode, a current model is compared to a collection of previous models of the opponent. Hence, some similarities or conflicts between patterns are possible. To preclude erroneous reports, our policy is to store the similar or conflicting patterns in a specific table. To deal with these similarities or conflicts, we carry on computations until they are distinguished. In this case, if these conflicts are not settled until the end of the game, reporting is not allowed.

3.2 Opponent Provocation

One of the other policies applied in online section is the selection of a suitable strategy to motivate the opponent players to disclose the expected patterns. For this purpose, we have used rule based expert system architecture to provide a provoking strategy for opponent players. Rule-based systems are computer systems that use rules to provide recommendations or diagnoses, or to determine a course of action in a particular situation or to solve a particular problem [14], [15]. To design such rule-based architecture, the patterns identified in offline mode are considered as antecedents and the consequents are strategies for opponent provocation which are built with the assistance of a human expert. We store these ordered pairs of patterns and strategies as rules in our *Strategy Library*. To present an appropriate strategy, the *Forward Chaining* method is used. In a way that by receiving observations from the environment, we search in strategy library for a rule whose condition part is identical to these observations. In this case, this rule is triggered and its action part is sent to the players as provoking strategy. Therein one of the outstanding problems of this system is *Conflict Resolution*. This means that it is probable that more than one rule are qualified to be fired. There are several conflict resolution strategies, such as choosing the most recent activated, the least frequently triggered, etc. The most suitable conflict resolution strategy is priority-based: assigning a priority value to every rule and selecte the fired rule with the highest priority. In the case that several rules have the highest priority value, a random selection is performed. To deal with this problem we have benefited from prority-baseed conflict resolution.

With this method, coach can select the best possible strategy to provoke opponent players. This increases the accuracy of reporting pattern and speeds up pattern recognition. Figure 2 illustrates some used provocation strategies.

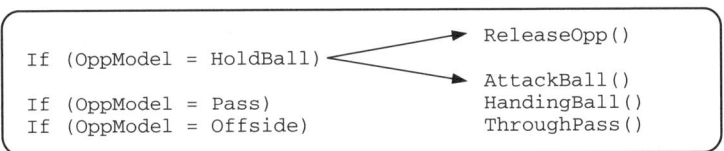

Fig. 2. Example of provocation strategies

What we have mentioned here about opponent provocation could be considered as a novel approach in Opponent Modeling.

Two examples are given here to clarify the idea.

– Let's assume that the opponent pattern is offside trap. In this case, we should put our players in offside situation. Therefore the candidate strategies which have the properties and can be used to activate opponent behavior could be "move forward" or "through pass". Meanwhile the game, by observing the first occurrence of offside trap, if the offside fact is found in strategy library, "move forward" or "through pass" strategy is activated.
– Let the opponent behavior be a simple direct pass e.g. a pass from player 9 to player 10. For this case, our strategy is "handing ball to player 9 of opponent".

Eventually these strategies are advised to players in the format of standard coach language. This structure is completely implemented and tested at the RoboCup competitions. In the following section, our experiments are explained in full detail.

4 Experimental Result

The MRL team acquired the 1st place among 10 participated teams in RoboCup 2006 competition. This competition had 3 rounds, each consisting of 4 iterations. In every round, nearly 15 to 20 patterns are fed to the log analyzer. The participants are responsible for creating these patterns. According to coach regulation, teams should provide at least 3 patterns for each round and active patterns for the iterations are also selected by them. Log analyzer has an average of 5 minutes to process a pattern.

Meanwhile the game, online coach should identify and report the activated patterns within 10 minutes. The score of a team depends on both the number of correct detected patterns and the time of report. In the first round, our team placed second. After making some slight modification to the parameters in the algorithms used to reduce the noise, we attained the first place in the second round and could pass to the final round without any wrong reports. Although most of the patterns in the final round were chosen from the patterns which other teams had prepared, our team took the greatest score. MRL detected 10 correct patterns from 18 activated patterns that equal the sum of all identified patterns of other teams in this round. The final round results and the ranks of teams are depicted in table 2.

Table 2. Total scores in the final round of the competition for the top 4 finishers

Team	Iteration 1 Score	Rank1	Iteration 2 Score	Rank2	Iteration 3 Score	Rank3	Iteration 4 Score	Rank4	Final Rank
MRL	67583.59	1	40857.6	1	51311.25	1	91578.75	1	4
UT Austin	44649.99	2	-4739.40	4	9548.5	3	9543.75	2	11
Caspian	11200.0	3	-4000.0	3	15000.0	2	-8000.0	4	12
Pasargad	-9282.0	4	2475.199	2	-9282.0	4	5418.0	3	13

The results of RoboCup 2006 competitions showed that MRL had well-deserved victory for being champion. And despite of lots of similarities and conflicts between the patterns, we had the least wrong reports number among all the teams, in the way that in the final round from a total of 26 wrong reports just one of them was ours.

5   Conclusion and Future Work

In this paper, we presented a novel architecture for modeling the opponent in coach competition. MRLCoach is an agent that is fully implemented and has been successfully tested in RoboCup competition. Providing this learning structure, MRL team took the 1st place in RoboCup 2006 coach competition. Our unparalleled performance in the competition has convinced us that the recipe for our success had been our capability in the handling of the noises and conflicts. Pattern categorization and noise handling are of prominent factors in our success in the competition. The trick in advising our players to motivate the opponent players to demonstrate the patterns had also assisted us in identifying the opponent behaviors simpler and sooner.

Opponent provocation could be considered as a novel approach in MAS for achieving a specific goal. This method can be applied to arenas such as military applications and other adversarial domains in order to give strategies to provoke and trap the enemy. In the future, our study will be focused on optimization of opponent provocation system to expedite opponent modeling and make this process more accurate.

References

1. Russell, S., Norvig, P.: Artificial Intelligence: A Modern Approach, 2nd edn. Prentice-Hall, Inc. (1995)
2. Weiss, G.: Multiagent Systems a Modern Approach to Distributed Artificial Intelligence. MIT Press (1999)
3. Riley, P.: Coaching: Learning and Using Environment and Agent Models for Advice., Ph.D. dissertation, School of Computer Science, Carnegie Mellon University, Pittsburgh, PA 15213-3891 (2005)
4. Murray, J., Noda, I., Obst, O., Riley, P., Stiffens, T., Wang, Y., Yin, X.: RoboCup Soccer Server User Manual for Soccer Server 7.07 and later (2002)
5. Riley, P., Veloso, M.: An Overview of Coaching with Limitations. In: Proceedings of the Second Autonomous Agents and Multi-Agent Systems Conference, pp. 1110–1111 (2003)
6. Riley, P., Veloso, M.: Coaching Advice and Adaptation. In: Polani, D., Bonarini, A., Browning, B., Yoshida, K. (eds.) RoboCup 2003: The Sixth RoboCup Competitions and Conferences. Springer, Berlin (2004)
7. Rohlf, F.J., Sokal, R.R.: Statistical Tables, 3rd edn. W. H. Freeman and Company, New York (1995)
8. Banks, J., Carson, J.S.: Discrete-Event System Simulation. Prentice-Hall Inc. (1984)
9. Sedgewick, R.: Algorithms. Addison-Wesley (1983)
10. Fathzadeh, R., Mokhtari, V., Shahri, A.M.: Coaching With Expert System towards RoboCup Soccer Coach Simulation. In: Bredenfeld, A., Jacoff, A., Noda, I., Takahashi, Y. (eds.) RoboCup 2005. LNCS (LNAI), vol. 4020. Springer, Heidelberg (2006)
11. Kuhlmann, G., Stone, P., Lallinger, J.: The Champion UT Austin Villa 2003 Simulator Online Coach Team. In: Polani, D., Browning, B., Bonarini, A., Yoshida, K. (eds.) RoboCup 2003: Robot Soccer World Cup VII. Springer, Berlin (2004)
12. Stone, P.: Layered Learning in Multi-Agent Systems., Ph.D. dissertation, School of Computer Science, Carnegie Mellon University, Pittsburgh, PA 15213-3891 (1998)
13. Shannon, R.E.: Systems Simulation: The art and science, p. 372. Prentice Hall, Englewood Cliffs (1975)
14. Buchanan, B.G., Shortliffe, E.H.: Rule-Based Expert Systems: The MYCIN Experiments of the Stanford Heuristic Programming Project. Addison Wesley (1984)
15. Biondo, S.J.: Fundamentals of Expert System Technology: Principles and Concepts, Intellect (1990)

Robust Color Classification Using Fuzzy Reasoning and Genetic Algorithms in RoboCup Soccer Leagues

Alireza Kashanipour[1], Amir Reza Kashanipour[2], Nargess Shamshiri Milani[2], Peyman Akhlaghi[2], and Kaveh Khezri Boukani[2]

[1] Mechatronic Research Labratory
Dept. of Electrical and Computer Engineering
Azad University, Qazvin, Iran
alireza.kashanipour@gmail.com
[2] Dept. of Electrical Engineering
Sahand University of Technolegy,Tabriz, Iran
{kashanipour,kaveh.khezri}@Gmail.com,
{Pakhlaghi,N.Milani}@ieee.org

Abstract. Color segmentation is typically the first step of vision processing for a robot operating in a color coded environment, like RoboCup soccer, and many object recognition modules rely on that, in this paper we present a method for color segmentation that is based on fuzzy logic. Fuzzy sets are defined on the H, S and L components of the HSL color space and provide a fuzzy logic model that aims to follow the human intuition of color classification. The membership functions used for the fuzzy inference are optimized by genetic algorithms. The method requires the setting of only a few parameters and has been proved to be very robust to noise and light variations, allowing for setting parameters only once. The approach has been implemented on MRL middle size robots, and successfully experimented in the numbers of the friendly matches of the Middle size in the 2006's games.

Keywords: Color classification, Image segmentation, fuzzy classification, Genetic Algorithms, Soccer robot.

1 Introduction

Many color vision systems require the first step of classifying pixels in a given image into a discrete set of color classes. This early vision step plays an important role in soccer robot because RoboCup soccer is a color-coded environment, where colors are used to define principal objects introducing specific concepts for robots. Recognition and positioning of colored beacons and goals in the field are used for self-localization and reactive behaviors, while the recognition of the orange ball feeds behaviors and coordination tasks.

Good color segmentation allows for easy implementation of object recognition and localization, most of the robot vision systems are based on fast and accurate implementation of such process. Most of the teams recognize and locate objects from a rough segmentation (e.g. [1]), applying more sophisticated recognition techniques (e.g., region growing) at a later stage. However, this second approach may be less

U. Visser et al. (Eds.): RoboCup 2007, LNAI 5001, pp. 548–555, 2008.
© Springer-Verlag Berlin Heidelberg 2008

reliable or require more computational resources. According to [2], fuzzy approaches for image segmentation can be categorized into four classes: segmentation via thresholding, segmentation via clustering, supervised segmentation, and rule-base segmentation. Among these categories, rule-base approaches are able to take advantage of application dependent heuristic knowledge, and model them in the form of fuzzy rules. In [3], a set of fuzzy rules are established based on fuzzy variables, which are associated with the membership values of pixels obtained by the fuzzy c-mean clustering approach (FCM) [4] and the possibilistic c-mean clustering approach (PCM) [5], to construct a correction matrix for modifying the fuzzy partition matrix. In [6], a fuzzy reasoning method in conjunction with genetic algorithms is employed for color image segmentation through region merging.

In this paper we present an approach to color segmentation that has been used for RoboCup soccer in the Middle size league (also applicable for other usages). The aim of our work is to retrieve images according to their dominant(s) color(s) expressed through linguistic expressions and implementing in our vision system. In this paper a Genetic Algorithms for an automatic production of the optimized fuzzy rules has been employed. The results of performing this method on experimental data have demonstrated the efficiency of the method.

2 Color Spaces

MRL robots have many limitations that must be considered in the development of their vision system: low computational power and low quality of the color image. Therefore, many of the state-of-the-art segmentation techniques in the literature cannot be implemented in real-time application and do not have optimal results.

HSL space (Hue, Lightness and Saturation) is a space that characterizes the color directly thanks to its hue. In this space saturation corresponds to the quantity of "white" in the color and lightness corresponds to the light intensity of the color. Thus, the identification of color is made in two steps: first H, then L and S. the HSL space can be represented through a cylinder or a bi-cone (figure 1). The hue H is an angle, it means that its definition interval loops (0 and 255 are the same points). The "pure" red ((255,0,0) in RGB space) corresponds to an angle equal to 0 for h, a saturation s equal to 255 and a lightness l equal to 128.

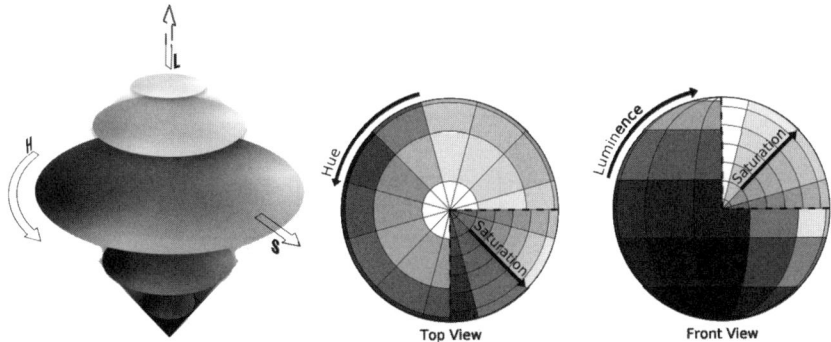

Fig. 1. The HSL space (wikipedia.org)

In our work, we limit ourselves to the nine fundamental colors defined by the set T representing a good sample of colors (dimension H):

$T = (red, orange, yellow, green, cyan, blue, purple, magenta, pink)$

T corresponds to the seven colors of Newton [7] to which we have added color pink and color cyan, that are included in the rainbow color set.

3 Color Representation

Another important point about spaces is the problem of uniformity of the scale. HSL space is quite convenient for our problem but it is a non UCS (uniform color scale) space [8]. Indeed our eyes don't perceive small variations of hue when color is green ($h = \pm 85$) or blue ($h = \pm 170$) while they perceive it very well with orange ($h = \pm 21$) for example. Thus to model the fact that the distribution of colors is not uniform on the circle of hues, Truck *et al.* propose to represent them with trapezoidal or triangular fuzzy subsets [9]. Several other works have been done in the field of none uniformly distributed scales: for example, Herrera and Martinez use fuzzy linguistic hierarchies with more or less labels, depending on the desired granularity [10].

Similarly, [9] associate colors with fuzzy sets. Indeed, for each color of T they built a membership function varying from 0 to 1 (f_t with $t \in T$). If this function is equal to 1, the corresponding color is a "true color" (cf. figure 2).

For each fundamental color, the associated interval is defined according to linguistic names of colors. For some colors, we obtain a wide interval. It is the case for the colors "green" and "blue" which are represented by trapezoidal fuzzy subsets.

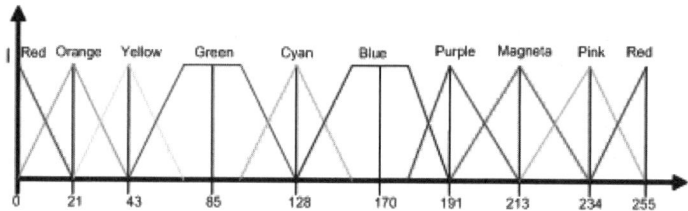

Fig. 2. The dimension H

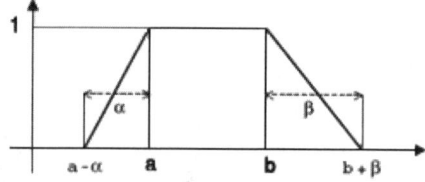

Fig. 3. Trapezoidal fuzzy subset

A trapezoidal fuzzy subset is usually denoted (*a; b;* α; β) (cf. figure 3) and when the kernel is reduced to one point, it is a triangular subset denoted by (*a;* α; β) since a= *b* [11].

To complete the model, it is necessary to take into account the two other dimensions (L and S). Each colorimetric qualifier is associated to one or both dimension(s). To facilitate the process, each dimension interval is divided into three sub-intervals: *low value*, *average value* and *strong value*. Thus, we obtain six "one dimension-dependent" qualifiers and nine "two dimension-dependent" qualifiers [12] denoted by *Q*.

Q = {somber, dark, deep, gray, medium, bright, pale, light, luminous}.

Figure 4 shows the nine "two dimension-dependent" qualifiers.

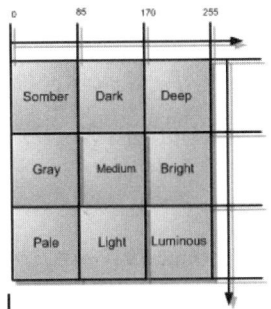

Fig. 4. Fundamental color qualifiers

Each qualifier of *Q* is associated to a membership function varying between 0 and 1 ($\widetilde{f}_q$ with $q \in 2\ Q$). Every function is represented through the 3 dimension-set (*a; b; c; d;* α; β; γ; δ) (cf. figure 5) As for the hues, the intersection point value of these functions is also supposed equal to 1/2 (cf. figure 6).

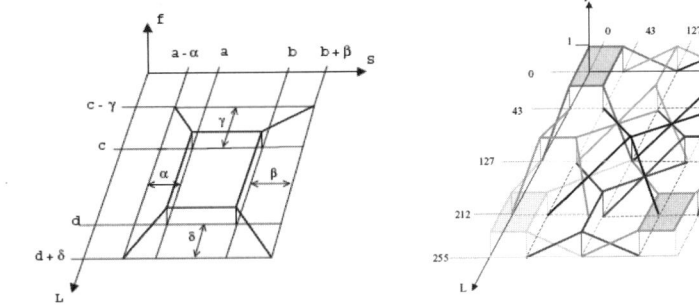

Fig. 5. Trapezoidal 3-D fuzzy subset   **Fig. 6.** Dimensions L and S

The membership function of any qualifier q is defined below:

$$\forall q \in Q,\ \tilde{f}_q(l,s) = \begin{cases} 1 & if \quad a \le s \le b \quad and \quad c \le l \le d \\[2mm] 0 & if \quad s \le a - \alpha \quad or \quad s \ge b + \beta \\ & or \quad l \le c - \lambda \quad or \quad l \ge d + \delta \\[2mm] \dfrac{l - (c - \lambda)}{\lambda} & if \quad c - \gamma < l < c \quad and \quad \alpha l - \gamma s \le \alpha c - \gamma a \\ & and \quad \beta l + \gamma s \le \beta c + \gamma b \\[2mm] \dfrac{(d + \delta) - l}{\delta} & if \quad d < l < d + \delta \ and \quad \beta l - \delta s > \beta d - \delta b \\ & and \quad \alpha l + \delta s > \alpha d + \delta a \\[2mm] \dfrac{s - (a - \alpha)}{\alpha} & if \quad a - \alpha < s < a \quad and \quad \alpha l - \gamma s > \alpha c - \gamma a \\ & and \quad \alpha l + \delta s \le \alpha d + \delta a \\[2mm] \dfrac{(b + \beta) - s}{\beta} & if \quad b < s < b + \beta \quad and \quad \beta l + \gamma s > \beta c + \gamma b \\ & and \quad \beta l - \delta s \le \beta d - \delta b \end{cases} \tag{1}$$

4   Optimization of Membership Functions by Genetic Algorithms

In this section, we explain the method of optimizing the membership functions by genetic algorithms. Since the parameters of fuzzy membership function has a great effect on performance of our classification thus by using genetic algorithms we can optimize fuzzy parameters to minimize the error in fuzzy classification.

Genetic algorithms are search algorithms based on the mechanics of natural selection and natural genetics [13]. Genetic algorithms have three operations: a) crossover operation, b) mutation operation, c) selection operation.

Each membership function of H, S and L forms a trapezoid in this method. One of the membership functions is encoded to the three real numbers defined by the trapezoid. We regard these real numbers as a gene. Figure 7 shows the encoding of the membership functions.Because three membership functions are required by each reference vector in HSL space, the number of genes needed to express the classification rule is three times the number of vector in HSL space. The classification rule is encoded as a string using these genes. First, the string shaving random genotypes are taken as the initial population. In the crossover operation, some pairs of strings are selected as parents. One of the parents is selected according to its classification abilities, and another is selected at random. A crossover position is randomly selected on each pair of strings. Two new offspring strings are made by swapping real numbers after that position. In the mutation operation, each gene of the offspring is changed for a random real number with an occurrence probability. Before the selection operation, the strings are evaluated with the training samples. Evaluation is done using an objective function called

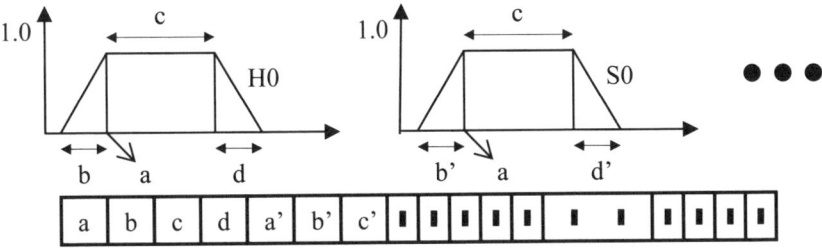

Fig. 7. Encoding of the membership function

The fitness function:

$$f(k) = W_0 E_{0k} + W_1 E_{1k} + W_2 E_{2k} + W_3 E_{3k} \tag{2}$$

Where F(k) is fitness value of the k-th string: E_{0k} is the classification rate, E_{1k} is average of the difference from μ for the correct color to the other μ; E_{2k} is the average of the difference from the membership value for the correct color of each parameter to the same value for the other colors; E_{3k} is the average width of the membership function; W_0, W_1, W_2 and W_3 are the eight for each parameter.

In the selection operation, the strings are selected according to their fitness values. Some strings whose fitness value is low are removed from the population. This operation keeps the size of the population constant. This method requires a search of a wide solution space because a real number is used as a gene. Therefore, we use the random search operation with the three operations; each string can search the narrow space around itself before the selection operation. The number of searching trials and the size of the Space searched are kept constant when a string finds a string more suitable than itself; the genes of the string are changed to these of the more suitable string. This operation makes it possible to search a wide space in detail without a large size of the population. Simulations of classification were performed using data measured by a prototype color sensor-whose measuring geometries were 0 illuminated and 45 received. Optimization was performed for classification of the dealt colors. Classification rate: 100%, where output fluctuation: ±0.2 the same simulations were performed with the conventional method using permissible ranges. As result classification rate: 88.4% from the comparison of these results, we confirmed the effectiveness of this method.

5 Experimental Result

The Training Data have been captured from colorful images of middle size robots in 9 color sets (Red, Orange, Yellow, Green, Cyan, Blue, Purple, Magenta and Pink), under various lights intensities in our lab. The goal was to make this segmentation resistant in different lights and make sure that this segmentation is still performed correctly in variation of light.

Fuzzy system had three inputs each representing one of the dimensions of HSL. For the dimension H, 10 membership functions, for L and S, 3 and for the output 9 permanent membership functions were considered. Experimental results of some of test images are reported. In order to show the effectiveness of this method to object classification in field of MSL matches, we compared the classification ability of this method with that of the conventional one. The colors of object to be classified were relatively analogous. When objects contain different substances, they must be separated. In the case of highly analogously colored object, the classification is difficult because of the color and light non-uniformities.

The causes of the non-uniformity of the GA's parameters were as follows: population size was 100; number of offspring created by crossover operation was 50 for each generation; and occurrence probability of mutation was 5%. The classification rule that had a maximum fitness value at the 100th generation was evaluated for its classification ability. In the simulations 100 datum for each color were classified.

The same simulations were performed by the conventional method. In this method, the permissible ranges were obtained from a minimum and a maximum value of some classification trials. Therefore, the permissible ranges were yielded by the training samples used for this method. The simulation results were compared with those of this method. The results show that the proposed method is more robust against color non-uniformity and light illuminating non-uniformity. When color and light fluctuations were 5 % and 15 %, respectively, the classification rates in this method were 100% and 96. 3 %, respectively; the rates in the conventional method were 91.4 % and 78.3%, respectively.

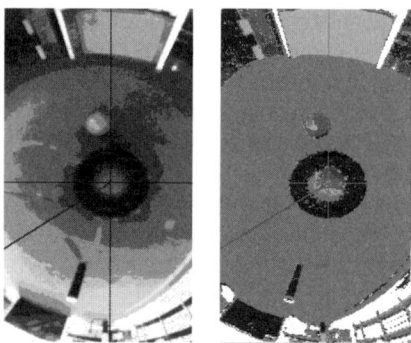

Fig. 8. Sample of color classification by this method in MRL

6 Conclusion

We have developed a robust vision for MRL robots with use of fuzzy sets and GA. This improved vision would help our Behavior Management System the possibility to work with hierarchies of informed behaviors, improved coordination mechanism, improved self-localization and improved color classification to solve the problem of shadow blades on the field.

Effectiveness of this method compared to the conventional method is clear. When color and light fluctuations were 5 % and 15%, the classification rates in this method were 100% and 96. 3 %, respectively; the rates in the conventional method were 91.4% and 78.3%, respectively. Thus, changing the color or light of robots surrounding environment can be a possible threat for its rival. Color calibration can be undertaken more quickly, as the calibration method encourages the human trainer to identify all possible pixel values for each color of interest, rather than avoiding those that may cause misclassification (e.g. those that occur in shadow or on the borders of different objects within the image). Recently, object recognition has also improved, not only due to image segmentation performance, but because of object recognition routines which can result in recognition of different levels of color uncertainty indicated by core colors, maybe colors, and unknown colors.

Acknowledgement

The authors would like to thank Dr. Mohammad Ebrahim Shiri, Faculty of Amirkabir Univ. of Technology for his great support and leading during this work.

References

1. Lovell, N.: Illumination independent object recognition. In: Bredenfeld, A., Jacoff, A., Noda, I., Takahashi, Y. (eds.) RoboCup 2005. LNCS (LNAI), vol. 4020, Springer, Heidelberg (2006)
2. Bezdek, J.C., Keller, J., Raghu, K., Pal, N.H.: Fuzzy models and algorithms fo. Pattern recognition and image processing. Kluwer Academic Pubkishers, Boston (1999)
3. Tolias, Y.A., Panas, S.M.: On applying spatial constraints in fuzzy image clustering using a fuzzy rulebased system. IEEE Signal Processing Letters 5, 245–247 (1998)
4. Bezdek, J.C.: Patten recognition with fuzzy objective function algorithms, pp. 65–85. Plenum Press, New York (1981)
5. Krishnapuram, R., Keller, J.M.: A possibilistic approach to clustering. IEEE Pans, Fuzzy Systems 1, 98–110 (1993)
6. Makrogiannis, S., Eeanomou, G., Fatopoulos, S.: A fuzzy dissimilarity function for region based segmentation of color images. int. j. pattern recogn. artif. Intell. 15, 255–267 (2001)
7. Roire, J.: Les noms des couleurs, Pour la science, Hors sfierie, no. 27 (2000)
8. Truck, I.: Approaches symbolique et floue des modificateurs linguistiques et leur lien avec ’aggregation, Ph.D. Thesis, Universit_e de Reims Champagne-Ardenne, France (December 2002)
9. Truck, I., Akdag, H., Borgi, A.: A Symbolic Approach for Colorimetric Alterations. In: Proceedings of EUSFLAT 2001, Leicester, England, pp. 105–108 (September 2001)
10. Herrera, F., Martinez, L.: A model based on linguistic two-tuples for dealing with multi-granularity hierarchical linguistic contexts in multiexpert decision-making. IEEE transactions on Systems, Man and Cybernetics, Part B 31(2), 227–234 (2001)
11. Bouchon-Meunier, B.: La Logique Floue et ses Applications. Addison-Wesley (1995)
12. Truck, I., Akdag, H., Borgi, A.: Using Fuzzy modifiers in Colorimetry. In: Proceedings of the 5th World Multi conference on Systemic, Cybernetics and Informatics, SCI 2001, Orlando, Florida, USA, pp. 472–477 (2001)
13. Sakurai, M., Kurihara, Y.: Color Classification Using Fuzzy Inference and Genetic Algorithm. In: Proc. IEEE int. conf. on Fuzzy systems, vol. 3, pp. 1975–1978 (1994)

Compliance Control for Biped Walking on Rough Terrain

Masaki Ogino[1,2], Hiroyuki Toyama[2], Sawa Fuke[1,2], Norbert Michael Mayer[1,2], Ayako Watanabe[2], and Minoru Asada[1,2]

[1] JST ERATO Asada Project, Yamada-oka 2-1, Suita, Osaka 565-0871, Japan
[2] Osaka University, Yamada-oka 2-1, Suita, Osaka 565-0871, Japan

Abstract. In this paper, we propose a control system that changes the compliance based on the walking speed to stabilize biped walking on rough terrain. The proposed system changes walking modes depends on its walking speed. In the downhill terrain, when the walking speed increases, the stiffness of the ankle in the support phase is controlled so as to brake the increased speed. In the uphill terrain, when the walking speed decreases, the stiffness of the waist joint is controlled and the desired trajectory for the supported leg is shifted so as not to falls down backward. To validate the efficiency of the proposed system, the stability of walking with the proposed system is examined in the two dimensional dynamics simulation. It is shown that the robot with the proposed system can walk in the more variable rough terrain and with the broader walking speed than without changing the stiffness of the joints.

1 Introduction

Biped walking algorithms are divided roughly into two categories; the model-based walking and dynamics based walking. In the former algorithm, the precise parameters of a robot and the environment are needed to calculate the control parameters such as zero moment point. However, this strategy needs a robot to sense the surface of the floor in advance precisely. Other groups realizes the walking on rough terrain with special mechanisms in the foots. Yamaguchi et al. [1] developed the foot mechanism with which a robot can sense the relative position and absolute inclined angle of the ground. With that mechanism, they realized the walking on the terrain with different levels in real time control. Hashimoto et al. [2] developed a parallel-linked biped walker with the semi-active adaptive ground mechanism that realizes the stable support area on the ground with small different levels. Both strategies extended in model-based approach realized the walking rough terrain only in the very limited way; statical walking with special mechanisms in foots.

However, humans seem to realize rough terrain walking in very different ways from these approaches. Human walking seems to utilize the dynamics of the body efficiently without precise sensing of the ground state [3]. The approach to realize a biped walking by using the dynamics of the body is called dynamics-based walking approach. Owaki et al. investigates the effect of the non-linear springs

U. Visser et al. (Eds.): RoboCup 2007, LNAI 5001, pp. 556–563, 2008.

on robustness of passive running [4]. Taga et al. [5] proposed a CPG (Central Pattern Generator) for biped walking, and showed that it can realize the robust walking on the flat floor and uniform slopes thanks to global entrainment of the body, control and environment dynamics. Miyakoshi [6] proposed the memory based control with which a robot can walk on known slope and the rolling slope. However, in these studies, the setting of the rough terrain is very limited and there have been few studies to investigate the possibility of dynamics based walking on rough terrain.

The rough terrain we treated in this paper has the random different levels and gradients that are relatively small to the robot body. On this terrain, the proposed controller enables a robot to walk stably without sensing the ground state by utilizing the compliance control.

In the following, first, the basic idea of the controller is introduced. Then, the effectiveness of the proposed controller is shown in the simulation experiments. Finally, the discussion and conclusion is given.

2 Walking on Rough Terrain

In daily life, there are various types of surfaces in the ground. Depending on the difference of the levels or the tilting angle of the ground, human changes its walking pattern. Here, we categorizes the rough terrain into two categories. The first category of rough terrain is the small rough terrain, in which the difference of the levels is relatively small compared with the length of the body. To stabilize the walking on this type of the floor, the controller should have the feedback property to go back to the normal walking automaticaly. The second category is the large rough terrain like the staircase or the steep slope. For walking on this type of terrain, human should know the state of the floor in advance by the visual information. In this paper, we treat the two dimensional walking for the first type of floor, in which a robot can automatically recover its walking against the small disturbance.

One of the reasons for falling down during walking on rough terrain is the excess decrease or increase of the kinetic energy caused by the difference levels. The proposed controller can compensate this energy disturbance so that it prevents a robot from falling down in forward or backward way. In the following, first, the basic controller for walking in the flat floor is introduced. Then, the rough terrain whose difference of levels are small relative to the body are classified into three groups;

rough downslope: The angle of the inclination of the ground changes randomly, but always negative.

rough upslope: The angle of the inclination of the ground changes rondomly, but always positive.

rough terrain: The angle of the inclination changes rondomly, positively and negatively.

and the stabilizing control modes for these groups are introduced.

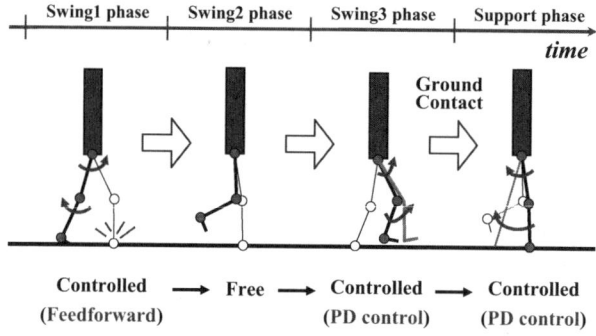

Fig. 1. The phases of walking controller

2.1 Base Control

The basic walking controller changes the control method depending on the walking phase (Fig. 1). In the first phase of the swing phase, the constant torque is applied to the waist and the knee joints of the swing leg. In the second phase, no torques are applied and the swing leg moves only by the inertial force. In the third phase, the proportional-derivative (PD) control is applied to the waist and knee joints to realize the landing posture that is determined in advance. In the stance phase, PD control is used to bring the stance leg backward. This control method can realize the torque pattern similar to the human walking [7].

The torque τ applied to the joint is given by the following equation,

$$\tau = -K_p(\theta - \theta_d) - K_v(\dot{\theta} - \dot{\theta}_d) + \tau_d, \tag{1}$$

where θ and $\dot{\theta}$ are the current position and speed of the joint angle, θ and $\dot{\theta}$ are the desired position and speed of the joint angle, and K_p and K_v are the gains for PD control, respectively. The desired angle in PD control θ_d is calculated by the following simple cosine function,

$$\theta_d = \begin{cases} \frac{\theta_f - \theta_0}{2}(1 - \cos\frac{\pi t}{T}) + \theta_0 & (t < T) \\ \theta_f & (t \geq T) \end{cases} \tag{2}$$

where θ_0 and θ_f are the initial and end angle of the joint in the phase, respectively, and t and T are the current time and the transition time to the next phase. This controller realizes a stable walking in the flat floor. However, on rough terrain, a robot easily falls down. In order to realize the stable walking on rough terrain, the compliance property is added to the control in the stance phase, as explained in the following section. The complicance (stiffness) of the joint angles are realized by changing the gains of PD control, K_p and K_v.

2.2 Control for Rough Downslope

In the rough downslope, in which the inclination angle of the ground changes rondomly in negative value, the typical cause of falling down is the excess increase of the kinetic energy. To suppress the increase of walking speed, when the body of the speed exceeds certain threshold, the stiffness of the foot joint in the first stance phase is made high. The stiffness of the joint angle is determined by the gain of PD control, thus the control is simply described by the following equation,

$$K_p = \begin{cases} K_{3p}^{down} & (V > V^{down}) \\ K_{3p} & (V \leq V^{down}) \end{cases}, \tag{3}$$

where K_p is the proportional gain of the foot joint in the stance leg, V is the walking speed of the body, V^{down} is the threshold of the walking speed, and K_{3p}, K_{3p}^{down} are the high and low constant values.

2.3   Control for Rough Upslope

In the rough upslope, in which the inclination angle of the ground changes rondomly in positive value, a robot falls down backward because of the excess decrease of the kinetic energy. To prevent the decrease of the kinetic energy, when the walking speed is lower than the certain threshold, the desired joint angle of the conroller in the stance leg phase is changed. The stance leg phase is divided into the first half and the last half phases.

Control for the first half of the stance phase. When a robot is described by a simple inverted pendulum, walking is modeled as ascending the potential energy. When the initial speed is low and the kinetic energy is lower than the potential energy, the robot falls down backward. However, if the stance leg is made shorter and the hight of the center of mass is made lower, the potential energy to get over becomes lower. Thus, to make the height of the center of mass lower, when the walking speed becomes lower, the controller bends the knee joint more with high gain of PD control. However, the gain of the waist and foot joints are made lower. This lower stiffness enables a robot to keep the trunk upright and makes foot fit to the ground without the detail information about the rough terrain. When the vertical position of the body x_{body} proceeds the supporting point x_{heel}, the PD gains of waist and foot are got back to high gain.

Control of the last half of the stance phase. When the walking speed is still low in the last half of the stance phase, further control is applied. Another solution to prevent the falling down backward is to increase the walking energy by extending the ankle joint in the last half of the stance phase. It is necessary to set the PD gain of the ankle joint not so high so that the weight shifts smoothly. Moreover, the desired posture at the end of the stance phase is also changed as the stance leg does not go so backward (to make the relative position of the body to the supporting point higher than usual.) Just after this control phase, the feed forward torques of the first swing phase that are applied to waist and knee joints are augmented than usual so that the swing leg contacts with the ground. The gain control in the rough upslope can be sumarized as Fig. 2.

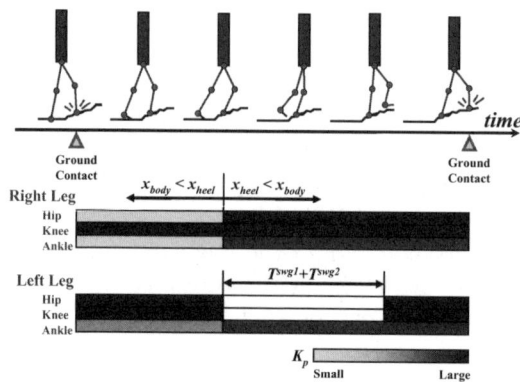

Fig. 2. Change of K_p in upslope phase

2.4   Rough Terrain

The walking system for general rough terrain can be constructed by integrating the controlers for rough upslope and downslope above mentioned. The integrating controller for one leg is described in Fig. 3 (a). In this figure, "upslope" and "upslope'" are the walking modes for the first and last half of the stance leg in upslope, respectively. Whether the controller enters the control phase "upslope'" depends on the walking speed, V'_h, which is the walking speed when the opposite leg contacts with the ground. Thus, the controllers for right and left legs interacts each other as shown in Fig. 3 (b). In the following section, the effectiveness of the proposed controller is examined in the simulation experiments.

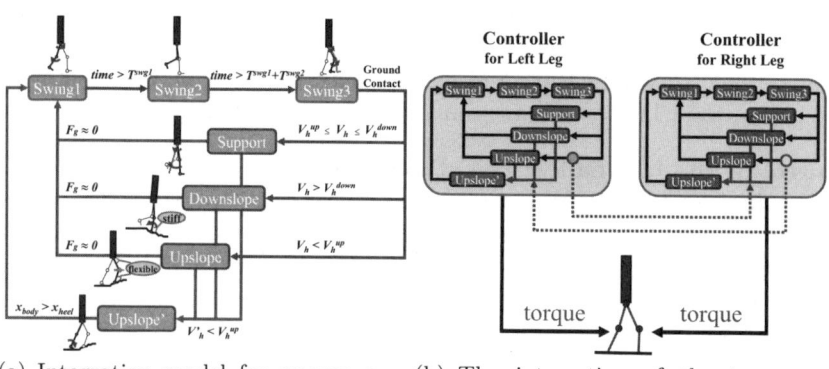

(a) Integration model for uneven surface for one leg

(b) The interaction of the two controllers

Fig. 3. Walking control system for rough terrain

3   Experimental Results

3.1   Simulation Setting

Controller. Fig. 4 shows the robot model used in the simulation experiments. The robot consits of 7 links; one upper body, two thighs, two shanks and two soles.

Here, "Downslope1" and "Downslope2" are the control phases "Downslope phase" in case of $V > V^{down}$ and $V < V^{down}$, respectively. "Upslope1" and "Upslope2" are the control phases "Upslope phase" in case of $x_{body} < x_{heel}$ and $x_{body} > x_{heel}$, respectively. "Upslope'1" is the first swing phase "Swing1" just after "Upslope phase'". "Upslop'3" is the third swing phase "Swing3" just after "Upslope phase". The joint angles are set to 0 [deg] when the robot stands upright, and anticlockwise is set as the positive direction. The desired posutre in each phase $\theta_f = 25$ [deg] and $\dot{\theta}_f = 0$ [deg/sec] are given.

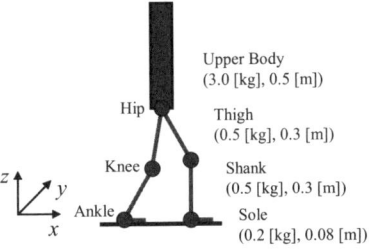

Fig. 4. Robot model

The thresholds for phase transition in respective control phases are $V^{down} = 0.6$ [m/sec], $V_h^{down} = 0.85$ [m/sec], and $V_h^{up} = 0.6$ [m/sec]. The PD gains to keep the trunk upright are $K_w p = 5000$ [Nm/rad] and $K_w v = 10$ [Nm sec/rad]. The limitation of the torques in each joint is set as 10 [Nm].

Rough terrain. The rough terrains used in the simulation experiments are constructed by the polygonal lines. The i-th ground position $(X(i), Z(i))$ is defined by the following equations,

$$X(i) = X(i-1) + X_0 R \tag{4}$$
$$Z(i) = Z(i-1) + Z_0(R - 0.5) \tag{5}$$

where R is the random number from 0 to 1. X_0 and Z_0 are the constant values that determines the degree of the roughness of the ground.

Fig. 5(a) shows the averaged walking steps with and without the proposed controller in relation to the roughness degree, Z_0 and $X_0 = 0.05[m]$. The number of maximam steps is 20 and 10 trials are examined in each roughness. The graph shows the robustness improves compared with the walking without the proposed controller. Fig. 5(b) is the time sequences of the walking with and without the proposed controller when the roughness of the ground is set as $X_0 = 0.05$ and $Z_0 = 0.023$ [m].

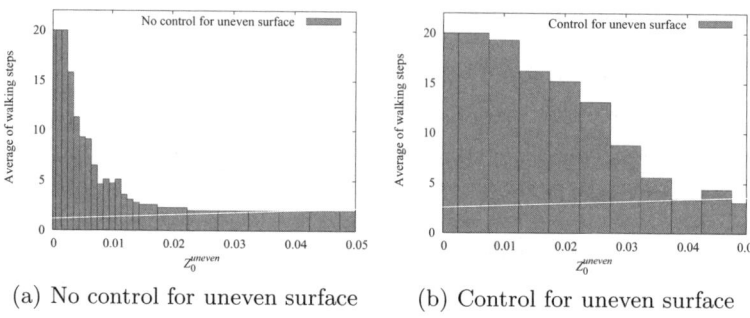

(a) No control for uneven surface   (b) Control for uneven surface

Fig. 5. Average of walking steps for uneven surface

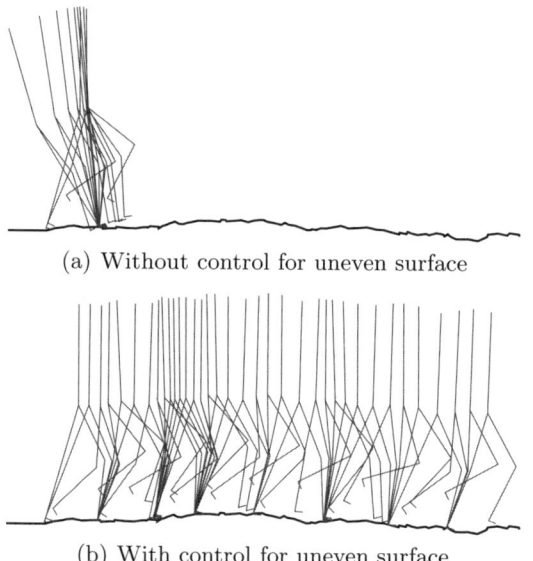

(a) Without control for uneven surface

(b) With control for uneven surface

Fig. 6. Walking on uneven rough surface

4   Conclusion

This paper proposes a walking controller that enables a robot to walk on the rough terrain by changing the compliance of joints without sensing the state of the surface of the ground.

To apply the proposed controller to a real robot, there are many problems to be left. First, the current control model is restricted to two dimensional. The motions on the frontal plane affects to that on the sagittal plane. It is the next challenge to develop the controller for the stabilization of the motions on the sagittal plane. The second problem is the slips on walking. The current simulator does not consider the slips between the feets and the ground. In the real situation, the steeper the slope is, the more the robot slips. Third, the most difficult problem is the actuators. In this paper, we modeled the simple PD controller and the various types of the stiffness are realized by changing the gains of PD controller. However, the real DC motors that are usually used in humanoid robots are difficult to realize the low stiffness. Moreover, it is difficult to design the robot that utilizes the dynamics of the body with DC motors because power/weight ratio of DC motors are not good. The pneumatic actuators may be the possible candidates for dynamics based humanoid robots [8].

References

1. Yamaguchi, J., Kinoshita, N., Takanishi, A., Kato, I.: Development of a dynamic biped walking system for humanoid -development of a biped walking robot adapting to the humans living floor. In: Proceedings of the 1996 IEEE International Conference on Robotics and Automation, pp. 232–239 (1996)
2. Hashimoto, K., Lim, H.O., Takanishi, A.: Develoopment of foot system of biped walking robot capable of maintaining four-point contact. In: Proceedings of the 2005 IEEE/RSJ International Conference on Intelligent Robots and Systems, pp. 1464–1469 (2005)
3. Mochon, S., McMahon, T.A.: Ballistic walking. Journal of Biomechanics 1(46), 9–14 (2002)
4. Owaki, D., Ishiguro, A.: Enhancing stability of a passive dynamic runnning biped by exploiting a nonlinear spring. In: Proceedings of the IEEE/RSJ International Conference on Intelligent Robots and Systems, pp. 4923–4928 (2006)
5. Taga, G.: A model of the neuro-musculo-skeletal system for human locomotion i emergence of basic gait. Biological Cybernetics 73 (1995)
6. Miyakoshi, S.: Memory-based bipedal walking control on the slope and uneven surface. In: Proceedings of the Annual Conference of the Robotics Society of Japan (in Japanese), CDROM (2006)
7. Ogino, M., Hosoda, K., Asada, M.: Learning energey-efficient walking with ballistic walking. In: Kimura, H., Tsuchiya, K., Ishiguro, A., Witte, H. (eds.) Adaptive Motion of Animals and Machines, pp. 155–164. Springer, Tokyo (2006)
8. Takuma, T., Hosoda, K., Ogino, M., Asada, M.: Stabilization of quasi-passive pneumatic muscle walker. In: Proceedings of the Fourth International Symposium on Human and Artificial Intelligence Systems, pp. 370–375 (2004)

Author Index

Printing: Mercedes-Druck, Berlin
Binding: Stein + Lehmann, Berlin

Lecture Notes in Artificial Intelligence (LNAI)